# PHYSICAL PROPERTIES OF HIGH TEMPERATURE SUPERCONDUCTORS I

T0182172

# PHYSICAL PROPERTIES OF HIGH TEMPERATURE SUPERCONDUCTORS I

**Editor**
## Donald M. Ginsberg
*Professor of Physics*
*University of Illinois at Urbana-Champaign*

 **World Scientific**
*Singapore • New Jersey • London • Hong Kong*

*Published by*

**World Scientific Publishing Co. Pte. Ltd.**
P O Box 128, Farrer Road, Singapore 9128

*USA office*: World Scientific Publishing Co., Inc.
687 Hartwell Street, Teaneck, NJ 07666, USA

*UK office*: World Scientific Publishing Co. Pte. Ltd.
73 Lynton Mead, Totteridge, London N20 8DH, England

PHYSICAL PROPERTIES OF HIGH TEMPERATURE SUPERCONDUCTORS I

ISBN    9971-50-683-1
        9971-50-894-X (pbk)

Printed in Singapore by Utopia Press.

*This book is dedicated to the many scientists and students who have risen to the high $T_c$ challenge.*

# PREFACE

The discovery in 1986 of high temperature superconductivity has produced a flood of experimental and theoretical publications, approximately 10,000 of them. This book maps out some of the main currents in an ocean of papers. We concentrate on the experimentally determined physical properties of these materials, and compare these properties with the predictions of theoretical models.

It was not easy to choose the topics for this book, which is necessarily limited in size. Chapters which would have been desirable include nuclear magnetic and quadrupole resonance, magnetic ordering, microstructure, electron tunneling, photoemission, and thermoelectric properties in the mixed state. We expect this book to be the first in a series of volumes to be published by World Scientific on the physical properties of high temperature superconductors. Later volumes will incorporate some of the missing subjects, as well as further developments in some of the topics which are covered here.

The authors of this book have made a significant sacrifice, taking time from their own research during a period of very great activity. They have done a job which is extremely difficult because of the deluge of preprints and articles. There is no way the authors could be aware of all the worthy manuscripts which have been written, and certainly it has been impossible for them to include references to all the literature in the field. Indeed, only by leaving out a great deal could they give each chapter a form which can be readily grasped. We ask you, the reader, to be understanding of these problems if you discover that some of your own work has gone unmentioned.

Donald M. Ginsberg
*Urbana, Illinois*
*Nov. 1, 1988*

# CONTENTS

# 1
# INTRODUCTION, HISTORY, AND OVERVIEW
# OF HIGH TEMPERATURE SUPERCONDUCTIVITY

Donald M. Ginsberg

*Physics Department and Materials Research Lab.*
*University of Illinois at Urbana-Champaign*
*Urbana, IL 61801, USA*

In this introductory chapter we relate the historical background of high temperature superconductivity and indicate the main new phenomena which have been observed and some of the most important fundamental questions raised by them. The later chapters of this book describe some of these phenomena and questions in more detail, and give more references.

The history of high temperature superconductivity is briefly reviewed in Section I. The crystal structures are described in Section II. In Sections III and IV, we call attention to related compounds and background references, respectively. Section V presents some fundamental experimental results, and an attempt is made to determine whether the high temperature superconductors are fundamentally different from the classic superconductors (sometimes called the conventional superconductors). Section VI is a discussion of three of the most unusual properties of the high temperature superconductors: high superconducting transition temperature $T_c$, short coherence lengths, and large anisotropy. In Section VII, we consider some important fundamental questions to be investigated and list some of the experimental methods that may help answer those questions. Finally, Section VIII contains a summary, along with a few words of caution on the need for good sample quality and thorough sample characterization.

## I. A BRIEF HISTORY OF THE FIELD

Materials have recently been discovered that exhibit superconductivity up to much higher temperatures than anyone had dared to hope. Since 1911, when Kamerlingh Onnes discovered superconductivity in mercury at 4.2K, the highest observed values of $T_c$ gradually moved upward, thanks to the work of people such as B. T. Matthias, J. K. Hulm, J. E. Kunzler, and T. H. Geballe.[1,2] Finally, in 1973, J. R. Gavaler observed that sputtered films of $Nb_3Ge$ began to become superconducting at 22.3K,[3] and this was soon pushed up to 23.2K by L. R. Testardi et al. by altering the sputtering conditions slightly.[4] In spite of great efforts to increase this limit further, it stood as the record until 1986. In that year, J. G. Bednorz and K. A. Müller[5] observed that a lanthanum barium copper oxide began its superconducting transition as it was cooled below 35K. For this discovery, which opened the way for all of the subsequent work on high temperature superconductors, Bednorz and Müller received the Nobel Prize in Physics in 1987.[6]

The work of Bednorz and Müller was at first greeted with some skepticism. People were aware that observations on other compounds of possible superconductivity at elevated temperatures had not been fruitful. For example, large diamagnetic anomalies had been reported[7] for CuCl at temperatures as high as 250K; this excited much interest, although the investigators who discovered the effect felt that a claim of superconductivity would have been "speculative". In fact, perfect diamagnetism and zero resistance were never obtained in CuCl, so its superconductivity has never been decisively proven. Another source of skepticism concerning the significance of Bednorz and Müller's data was the failure of the electrical resistance of their La-Ba-Cu-O compound to reach zero until the temperature had been reduced to approximately 11K.

Their results were, nevertheless, confirmed late in 1986 at the University of Tokyo[8] and at the University of Houston.[9] Early in 1987, groups at the University of Tokyo,[10] Bell Communications Research (Bellcore)[11] and AT&T Bell Laboratories[12] found that substitution of Sr for Ba in the La-Ba-Cu-O compound raises $T_c$ to approximately 40K. Subsequent efforts to raise $T_c$, carried out mainly in China, Japan, and the United States, were successful, culminating in the announcement by M.-K. Wu and his group at the University of Alabama at Huntsville and C. W. Chu at the University of Houston,[13] of the first material capable of becoming superconducting in liquid nitrogen; it turned out to be $YBa_2Cu_3O_{7-\delta}$, with $T_c$ a few degrees above 90K. The discovery was soon verified at Bellcore.[14]

The exact value of $T_c$ depends on the method of heat treatment and oxidation. It should be noted that some people report the "onset temperature", where the resistance begins to fall steeply, but others report the temperature where the resistance becomes immeasurably small. Previously, people had usually reported $T_{1/2}$, where the resistance fell to half of its value at the onset temperature. The values given here refer to the onset temperature, although it is not always possible to determine this parameter precisely. In well made samples, it does not differ very much from $T_{1/2}$.

Following the discovery of these extraordinarily high superconducting transition temperatures, two families of compounds were discovered with even higher values of $T_c$. First came the announcement, by H. Maeda et al.[15] at the Tsukuba Laboratories in Japan, of superconductivity in a Bi-Sr-Ca-Cu-O compound with an onset at 120K in the resistance transition and at 110K in the transition toward perfect diamagnetism[16] which is also characteristic of superconductors. This was soon followed by a similar announcement by A. M. Hermann and Z. Z. Sheng at the University of Arkansas of a Tl-Ba-Ca-Cu-O compound with an onset temperature near 140K in the resistive transition and of 118K in the diamagnetic transition.[17,18]

## II.  STRUCTURE

While the La, Bi, and Tl compounds contain planes of Cu and O atoms, the Y compounds[19,20] have both planes and chains of Cu and O. (See Figure 1.)  A great deal of work has focussed on the roles of these planes and chains in the yttrium compounds. It is now known that the planes play the major role in generating superconductivity, while the chains act as electron reservoirs which can be filled or emptied either by changing the oxygen stoichiometry or by other types of doping.[21-25] If the number of oxygen atoms per formula unit is reduced to 6.5 or 6.7 (the exact value is in question), the yttrium compound's $T_c$ falls to 55 or 60K.[26-28]  There is a tendency for the oxygen atom vacancies to occupy a single chain, and the $YBa_2Cu_3O_{7-\delta}$ compounds with $0 < \delta < 1$ tend to have ordered arrays of completely oxygen-depleted chains.[28-31] (When $\delta = 1$, there are no chains.)

It is possible to obtain a series of Bi or Tl compounds with varying stoichiometry, reflecting the possibility of inserting a varying number of Cu-O planes into the crystal structure's unit cell. For each new Cu-O plane, an adjacent Ca-O plane is also introduced. The general

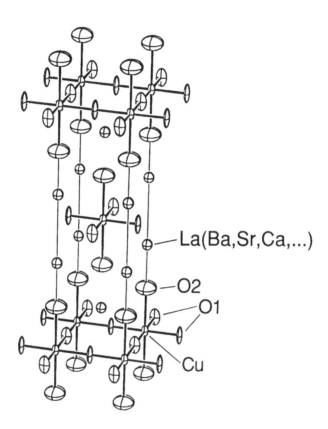

Figure 1a.   The structure of $La_{2-x}A_xCuO_4$, where A = Ba, Sr, Ca,...
Figure supplied by J. D. Jorgensen.

6

Figure 1b. The structure of $YBa_2Cu_3O_7$. Taken from Ref. 20b, by M. A. Beno, L. Soderholm, D. W. Capone II, D. G. Hinks, J. D. Jorgensen, J. D. Grace, and I. K. Schuller.

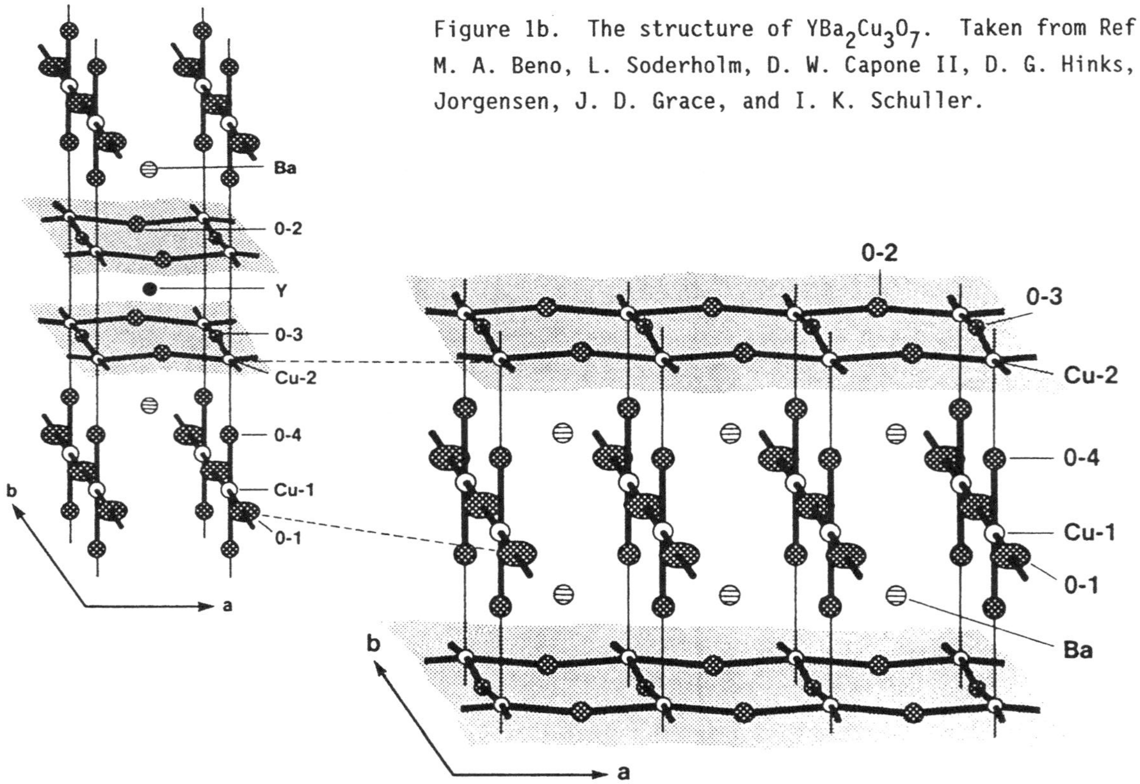

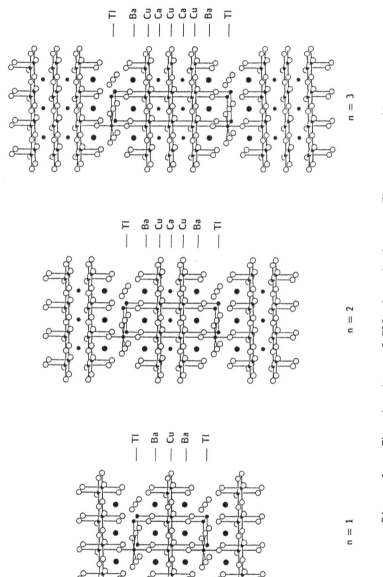

n = 1          n = 2          n = 3

Figure 1c.  The structure of $TlBa_2Ca_{n-1}Cu_nO_{2n+3}$.  The corresponding bismuth compound has the same structure, with Tl replaced by Bi. Figure supplied by A. W. Sleight and C. C. Torardi.

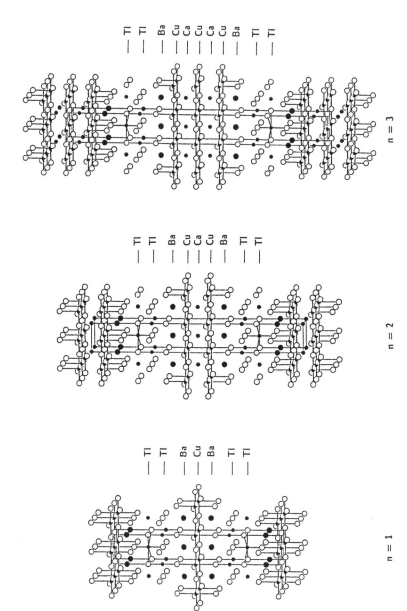

Figure 1d. The structure of $Tl_2Ba_2Ca_{n-1}Cu_nO_{2n+4}$. The corresponding bismuth compound has the same structure, with Tl replaced by Bi. Figure supplied by A. W. Sleight and C. C. Torardi.

chemical formula for the Bi and Tl compounds is $B_yA_2Ca_{n-1}Cu_nO_x$, where B is Bi or Tl, A is Sr for the Bi compound and Ba for the Tl compound, and n is the number of Cu-O layers per unit cell: n = 1, 2, 3, 4... [32] The subscript y is 1 or 2, denoting single or double (adjacent) layers of Tl-O; for the superconducting Bi compounds, y = 1. By increasing n, $T_c$ is raised at least to 120K for a Bi compound[33-35] and to 125K for a Tl compound ($Tl_2Ba_2Ca_2Cu_3O_x$ holds the high-$T_c$ record at this time[16]), and it might be possible to reach a $T_c$ of 180K this way.[32] (A report of $T_c$ = 162K in one of the Tl compounds is still unconfirmed, as are all other reports of $T_c$ values greater than 125K in various compounds.)

It is proving difficult to produce single-phase samples of the Bi and Tl compounds. A mixture of phases is usually seen, each having its own number of Cu-O planes and Ca-O planes per unit cell. This "syntactic intergrowth" has made it difficult to do good physics experiments.

The first observations of superconductivity in the new materials have usually been made on polyphase samples. The stoichiometry and even the crystal structure of the superconducting components of these samples were determined with great rapidity, frequently by several laboratories, within one to three days after the initial observations. In each case, neutron diffraction measurements[36] are required to determine the position of the oxygen atoms with certainty, since x-ray diffraction data are insensitive to the position of atoms containing a small number of electrons. The La compounds have a tetragonal structure. The Y compounds with approximately 7 oxygen atoms per formula unit are orthorhombic, and almost always exhibit twinning. It was speculated that the twin planes in these $YBa_2Cu_3O_{7-\delta}$ samples are a source of high temperature superconductivity, but this idea has seemed less convincing since the discovery of even higher $T_c$ values in the Bi and Tl compounds, which do not exhibit twin planes. Furthermore, Raman effect data on untwinned $YBa_2Cu_3O_{7-\delta}$ show the presence of superconductivity.[37]

On numerous experimental and theoretical grounds, it is believed that charge transport and superconductivity in the La, Y, Bi, and Tl compounds are dominated by holes on the oxygen sublattice in the Cu-O planes.[38-41]

## III. RELATED COMPOUNDS

Investigations of superconducting oxides have been carried on for some time. Before 1986, the highest $T_c$ for any oxide was announced by A. W. Sleight in 1975: $BaPb_{.73}Bi_{.27}O_3$ becomes superconducting at 13K.[42] It is probable that we should consider this compound as one of the high-temperature superconductors, since it shares with the other members of this group an abnormally high value of $T_c$, compared with the Sommerfeld constant $\gamma$, the coefficient of the linear term (the electronic part) of the normal-state specific heat.[43] Indeed, on a plot of $T_c$ vs. $\gamma$, the high-temperature superconductors fall on a single curve.[44] It should be noted, however, that $BaPb_{.73}Bi_{.27}O_3$ does not have isolated planes or chains of Cu and O atoms. The discovery that compounds with chemical formulas $Ba_{1-x}K_xO_{3-y}$ are superconducting,[45] and that the onset $T_c$ can be as high as 29.8K[46] has raised the question of whether they are similar to the other high-temperature superconductors, even though they have a cubic structure (at least, at room temperature), no Cu atoms, and display variable-range hopping (and therefore localized electronic states) in the normal state.[47] The work on various low-temperature superconducting oxides has recently been reviewed.[48]

Since the initial work by Bednorz and Müller, the search for new high-$T_c$ superconductors has proceeded mostly by following a simple strategy: the substitution of a chemical element by another with similar chemical properties, as indicated by the periodic table of the elements. In choosing substitute elements, care must be given to the ionic radii[49] if one wants to keep the same crystal structure. These considerations have been useful in the development of perovskites in general and the new superconductors in particular. See Figure 2, in which some ionic radii are plotted. The Figure makes it evident, for example, why one would try doping with Ca atoms in place of Cu atoms, or Bi in place of Pb. The effect of pressure on $T_c$ provides a suggestion about possible substitutions; if $T_c$ increases with pressure, this suggests making an atomic substitution to decrease the size of the

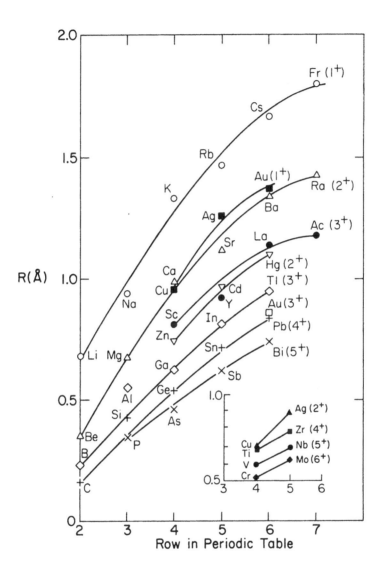

Figure 2.  Ionic radii of some of the chemical elements.  These data
are useful in selecting atomic substitutions to attempt.  Plotted from
data given in Ref. 49; see that reference for the radii of other ions
also.  Graph provided by T. R. Lemberger.

structure's unit cell.

By means of chemical substitutions, a large number of compounds have been made having properties similar to those of the parent compounds which we have listed.  For each possible combination of cations and their stoichiometry, however, a complete exploration of the possibility of high-temperature superconductivity requires a thorough testing of compounds with different amounts of oxygen, and with different methods of oxidation.  In addition, the sequence of heat treatments of the ingredients may be critical, since some of the necessary intermediate products may be metastable, or may attack the crucible or other intermediate products.  Because one can never try all possible heat treatments and oxidation procedures, one can never be sure that a cation stoichiometry already tried will not produce a useful superconductor some day.

It should be noted that many physicists, including the authors of this book, refer to a high temperature superconductor as a "ceramic" only if it is polycrystalline, whereas ceramists use the word "ceramic" to refer to any inorganic, nonmetallic solid, whether it is polycrystalline or a single crystal.  The high temperature superconductors are indeed nonmetallic, in that they have normal-state resistivities that are very high (but not infinite), and in that they are brittle.

There have been reports of superconductivity at temperatures higher than 125K.  Values up to room temperature and even beyond 500K have been claimed.  These reports, however, have not passed the usual tests:

1.  The material must show a decrease of the electrical resistance to a value too small to be measured with reasonably good equipment.
2.  The material must show magnetic flux expulsion when it is cooled in a constant magnetic field (the Meissner effect).
3.  The material must be stable.
4.  The production of the material and the effects observed must be reproducible at different times and in different laboratories.

## IV.  BACKGROUND REFERENCES

Discussing high-temperature superconductivity forces one to use the
ideas and vocabulary gained from investigations of the classic superconductors.  Fortunately, there are many good books available for
reference at the introductory,[50] more advanced,[51,52] or encyclopedic[53]
level.  Some major developments in recent years are treated in review
articles, for example on fluctuations,[54] the interactions between
magnetism and superconductivity,[55] and two classes of superconductors
possibly related to the new ones:  Chevrel compounds[56,57] and heavy-
Fermion compounds.[58]

## V.  SOME FUNDAMENTAL RESULTS AND THEIR INTERPRETATIONS

Theory helps us to select experiments to perform, to relate different types of properties to each other, to understand the significance of observations, and to predict new phenomena.  However, a
microscopic theory of high temperature superconductivity is proving to
be very difficult to develop.  Different schools of thought abound, but
not many specific predictions have been made which can be tested in the
laboratory.  Investigators interpreting the observations frequently
fall back on general theorems of thermodynamics[59] or statistical
mechanics, or use the assumption that our theoretical understanding of
the classic superconductors is also applicable.  It is frequently
assumed that the Bardeen-Cooper-Schrieffer (BCS) theory[60] (or a strong-
coupling variant of it) can be employed, since it successfully accounts
for the properties of the classic superconductors.  The binding of
electrons into Cooper pairs would probably be mediated by more than
just the electron-phonon interaction,[61] but the rest of the theoretical
treatment might survive intact.  Since Gor'kov showed long ago[62] that
the Ginzburg-Landau (GL) theory, phenomenological up to that time,
could be derived from the BCS theory near $T_c$, we frequently use the
theoretical tools arising from GL theory, such as the classic work by
Abrikosov on quantized magnetic flux vortices (fluxoids).[63]

In this chapter, I use "the GL theory" to denote the theory of superconductivity published by Ginzburg and Landau in 1950.[64] The term "GL theory" is frequently used by other authors, however, to denote more generally any phenomenological theory employing an expansion in successively higher powers of the order parameter(s) and the spatial and time derivatives thereof, and dropping the highest order terms. For example, theories of that type can be written down to describe pairing states other than singlet or s-wave. Similarly, I use "the BCS theory" to refer to the theory in the original paper by Bardeen, Cooper, and Schrieffer,[60] with the assumptions made there:  electron-phonon coupling as the sole interaction responsible for superconduc-tivity, pairs weakly bound (compared with the maximum phonon energy), and an isotropic electronic state.  Frequently "the BCS theory" is used by other authors to denote any theory of superconductivity involving singlet pairing of the electrons, regardless of the origin of the attractive interaction between the electrons, the strength of that interaction, or anisotropy.  I will refer to the former as "the BCS theory", and will refer to the latter, more general, use of the term as "some variant of the BCS theory."

The evidence that some variant of the BCS theory is applicable stems from a number of observations:

a.) Electron pairing is seen in Shapiro steps displayed by Joseph-son tunnel junctions biased with both ac and dc voltages.  These steps have the voltage spacing $h\nu/2e$ which is expected if the electrons are bound into Cooper pairs.[65-67]  Also, the flux quantum has been directly measured and is $hc/2e$, again indicating pairing.[68,69]  The quantized vortex lattice has been seen in $YBa_2Cu_3O_{7-\delta}$ by an imaging technique,[70] and it is found that each vortex carries one quantum of flux, $hc/2e$.

b.) By measurements of electron tunneling and electromagnetic absorption, an energy gap has been observed in the distribution of energy levels available to the system, and it is of the same order of magnitude at temperatures $T \ll T_c$ as that predicted by the BCS theory, $3.53\ k_BT_c$.  The observed values initially extended over a range from about 2 to 14 $k_BT_c$, but more recent determinations have tended to lie in the range 4 to 8 $k_B T_c$. Values somewhat larger than 3.53 $k_B T_c$ are known to be associated with strong electron-phonon coupling (or strong electron-other virtual excitation coupling);[71] in the BCS theory, the binding energy of a Cooper pair is assumed to be weak compared with the maximum phonon energy. By nuclear quadrupole resonance on the Cu nuclei, two different values of the energy gap have been observed in $YBa_2Cu_3O_{7-\delta}$.[72] It is now believed that the larger value is associated with the planes and the smaller with the chains.[73]

c.) Nuclear magnetic resonance data on some of the $^{17}O$ nuclei (their site assignment is in question) in $YBa_2Cu_3O_{7-\delta}$ show that the $1/T_1$ relaxation rate rises to a peak just below $T_c$ and then falls toward zero.[74] This is just the behavior observed in aluminum by Hebel and Slichter long ago,[75] the behavior predicted by the BCS theory.

d.) The temperature dependence of the electromagnetic penetration depth, measured by muon spin resonance,[76,77] has at least the qualitative features predicted by the BCS theory.

e.) High-resolution x-ray diffraction data show no evidence for a coupling of an orthorhombic distortion to the superconducting order parameter; there is no significant change in the crystal's structure at $T_c$.[19] This is at least consistent with a spherically symmetrical order parameter. (It could have only cylindrical symmetry.)

f.) Josephson tunneling has been observed to occur between a Pb-Sn alloy and $YBa_2Cu_3O_{7-\delta}$.[78] Such an effect is thought to be impossible between a singlet-paired and a triplet-paired superconductor.[79] Since a Pb-Sn alloy is singlet-paired, $YBa_2Cu_3O_{7-\delta}$ is also thought to be singlet-paired, as predicted by the BCS theory.

On the other hand, there are arguments to support the contention that the BCS theory (in its original form) cannot apply. These arguments are:

a.) Observed values of $T_c$ on the order of 90K or higher are probably too high to be explained by the BCS theory unless one assumes that some interaction much stronger than the electron-phonon interaction is at work.[61]

b.) A linear term in the specific heat vs. temperature curve has been observed in the La, Y, and Tl compounds at low temperatures,[80-82] (including the tetragonal phase of $YBa_2Cu_3O_{7-\delta}$[83]) as well as in $BaPb_{1-x}Bi_xO_3$,[84] whereas the BCS theory predicts an exponentially small electronic specific heat if $T \ll T_c$. (See Figure 3.) This prediction is a consequence of a theoretically predicted gap in the spectrum of energy levels available to the electronic system in the superconducting state, a gap with a nearly temperature independent width when $T \ll T_c$. (We speak of "the electronic system" for brevity, although the electrons could be interacting with non-electronic virtual excitations.) Of course, we cannot yet be sure that the observed linear term in the specific heat is electronic. It could be associated with an undetected impurity phase, or with some sort of "glassy" behavior, associated with the atomic positions. (The linear term is absent for nonsuperconducting $La_{2-x}Ba_xCuO_4$ with small x,[85] and possibly for the Bi-Sr-Ca-Cu-O compounds.[86])

c.) Data obtained from Raman scattering experiments indicate that $YBa_2Cu_3O_{7-\delta}$ may have an energy region which is partially depleted of states, but is not entirely empty.[87] The same conclusion follows from infrared absorption measurements; see the Chapter in this book by Timusk and Tanner. Thus, the energy gap might vanish along certain lines or at certain points in momentum space, i.e., for electrons with momentum along certain directions. An energy gap with nodal lines or points would remind us of the heavy-fermion superconductors, ($UBe_{13}$, $CeCu_2Si_2$, $U?t_3$, and related compounds).[58] Such a behavior of the energy gap could also explain the linear term in the specific heat that we have just described. It may also be related to the observation of an approximately linear dependence of the thermal conductivity of $YBa_2Cu_3O_{7-\delta}$ which has been seen near 0.1K.[88,89] Evidence against nodes in the energy gap, however, is provided by tunneling experiments from various groups, showing a reasonably well defined energy gap with very few, if any, states in it. A possible explanation for this contradiction might identify states showing up in a Raman effect experiment, but not in a tunneling experiment, since each of these involves transition probabilities (matrix elements) as well as the density of electronic

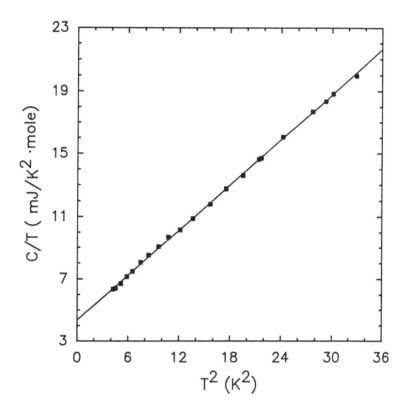

Figure 3. Specific heat, C, data for polycrystalline $YBa_2Cu_3O_{7-\delta}$. C/T is plotted against $T^2$. The absence of a rise in the data at low values of $T^2$ indicates an absence of magnetic impurities. The nonzero intercept indicates a linear term in the temperature dependence of the specific heat at low temperatures. Data of M. E. Reeves, S. E. Stupp, and D. M. Ginsberg.

states.

d.) High-resolution specific heat measurements near the superconducting transition temperature disclose that the superconducting order
parameter may have more than two real (one complex) components.[90,91]
This would violate "the GL theory" (using the term in the narrow sense
described above), which is a consequence of the BCS theory. This
observation needs to be verified, however, before it can be accepted.

## VI.  THE MOST UNUSUAL FUNDAMENTAL PROPERTIES OF THE NEW MATERIALS

The new superconductors have three most unusual fundamental properties:  large $T_c$, short coherence lengths, and large spatial anisotropy.

### A.  Large $T_c$

The large $T_c$ values open the way for many new applications which
were not previously attractive economically. From a fundamental
viewpoint, when $T_c$ is large, many types of excitations, such as
phonons, are present in the upper temperature ranges of the superconductivity. The presence of these excitations, if they break Cooper
pairs, is expected to affect some of the properties of these new
compounds, such as $T_c$ and the critical current density $J_c$, which is the
maximum current density which the material can carry without dissipation.[92-94]

Figure 4 shows the thermal conductivity of $YBa_2Cu_3O_{7-\delta}$ as a function
of temperature. Because $T_c$ is an appreciable fraction of the Debye
temperature (approximately 375K) in this high-temperature superconductor, there are many phonons present at $T_c$, and they contribute much
more to the thermal conductivity than do the conduction electrons. As
the temperature is reduced below $T_c$, the binding energy of each Cooper
pair of electrons, $2\Delta$, increases. It is impossible for a phonon with
energy less than this binding energy to break a pair. Thus, the phonon
mean free path rises as T falls, and the thermal conductivity rises, as
seen in Figure 4. (As T is lowered further, fewer and fewer phonons
are present, so the thermal conductivity begins to fall again.) Such a

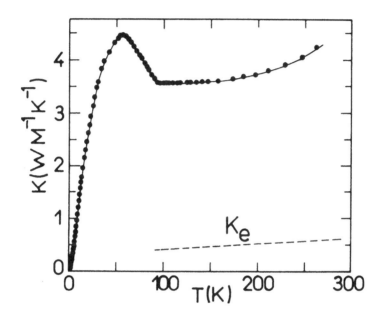

Figure 4. Superconducting transition of polycrystalline $YBa_2Cu_3O_{7-\delta}$, as seen by measurements of thermal conductivity. The dashed line indicates an upper limit for the part of the thermal conductivity contributed by the electrons; down to the neighborhood of $T_c$, the main part is contributed by the phonons. The rise in the thermal conductivity as T is reduced below $T_c$ is a sign of reduced scattering of phonons by electrons as the electrons become bound into Cooper pairs. Data of C. Uher and A. B. Kaiser, Ref. 129.

rise of the thermal conductivity as T falls below $T_c$ is seen for only a few of the classic superconductors, because most of them have such low superconducting transition temperatures that the number phonons which are present at $T_c$ is too small to dominate the thermal conductivity.

## B.  Short Coherence Lengths

The short coherence lengths observed in the high temperature superconductors lead to several unusual properties. We are talking here mainly about the Ginzburg-Landau coherence length $\xi_{GL}$. This parameter is the range of propagation of a disturbance in the __magnitude__ of the superconducting order parameter (the density of superconducting electron pairs). Another coherence length, the Pippard coherence length[95] $\xi_p$, would presumably also be short; it is the range of propagation of a variation in the __phase__ of the order parameter (and therefore in the supercurrent). Experimental determinations of $\xi_p$ are indirect and subtle, and have not yet been performed on high-temperature superconductors.

In these new materials, $\xi_{GL}(T)$ at $T \ll T_c$ is typically between 0.5 and 30 Å, depending on the crystallographic direction of the electron momentum, the substance, the type of experiment, and the assumptions made in interpreting the data. $\xi_{GL}(T)$ is expected to become larger as T is raised toward $T_c$. Near $T_c$, we expect that $\xi_{GL}(T)$ to have the temperature dependence $\xi_{GL}(0)/(1 - T/T_c)^{1/2}$. In the classic superconductors, $\xi_{GL}(0)$ is typically a few thousand Angstroms, unless it is decreased by alloying or defects.

Because $\xi_{GL}(T)$ is much shorter than the electromagnetic penetration depth in high-$T_c$ materials, they are all type II superconductors,[51,52,63] i.e., they form quantized magnetic vortices (fluxoids) when exposed to a large magnetic field, and they have  extremely high low-temperature values of the upper critical field $H_{c2}$, at which they are forced into the normal state. Also associated with the short coherence length is a weak pinning of the fluxoids,[96] compared with the pinning in the classic superconductors. This decreased flux pinning diminishes the size of $J_c$.

The reduced size of $\xi_{GL}(T)$ may be responsible for an observed melting of the lattice of quantized fluxoids:[96,97] as the sample is warmed up, the fluxoid lattice's shear modulus goes to zero at a temperature below $T_c$. This melting may account for an upward curvature (i.e., a positive second derivative) in the graph of the temperature dependence of the magnetic field at which electrical resistance reappears, although such a curvature is also seen for some of the classic superconductors with layered structures.[98] This melting of the fluxoid lattice has cast doubt on published values of $\xi_{GL}(T)$ calculated from presumed determinations of $H_{c2}$ from resistance measurements, but values of $\xi_{GL}(T)$ are available from magnetization measurements.[99] The shape of the resistance-vs.-temperature curve for one of the new superconductors in a magnetic field has also been interpreted in terms of reduced flux pinning.[100]

Although the flux pinning is weak at temperatures comparable with $T_c$, it is large enough at lower temperatures to make measurements of the lower critical field $H_{c1}$ very difficult, so values in the literature for $H_{c1}$ of $YBa_2Cu_3O_{7-\delta}$, for example, range from 1 to 5000 Oe. Flux pinning is responsible for the measured magnetic susceptibility being different when measured on a sample cooled in a constant magnetic field than on one cooled in zero field and exposed to the field after the lowest temperature is reached. (See Figure 5.) In the former case, the fluxoids move out of the sample as it is cooled, whereas in the latter, they move into it as it is heated; in each case, flux pinning impedes the motion. These two types of data are frequently referred to as "Meissner data" and "zero-field-cooled data," respectively. A plot of the behavior of the sample in true thermodynamic equilibrium would be located somewhere between those showing these two types of data. If the temperature were swept upward at an infinitesimal rate the true thermodynamic state would be explored.

Another effect of the shortness of the coherence length is that a sample held just above $T_c$ has small regions, of size about $\xi_{GL}(T)$, undergoing noticeable fluctuations into the superconducting state. Conversely, such a sample held just below $T_c$ has small regions fluc-

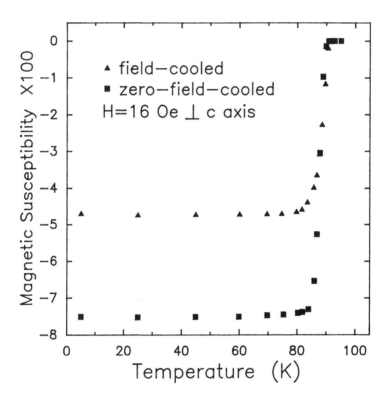

Figure 5.  Superconducting transition of a single crystal of YBa$_2$Cu$_3$O$_{7-\delta}$, as seen by measurements of magnetic susceptibility (plotted in Gaussian units).  The two sets of data differ because of flux pinning.  Data provided by J. P. Rice, E. D. Bukowski, and D. M. Ginsberg.

tuating into the normal state. The fluctuations are made likely by the small number of electrons in each of these regions. Thus, cooling one of these materials causes a transition to the superconducting state which appears to be gradual, even for homogeneous samples. (See Figure 6.) The effects can be seen in the specific heat,[89-91] conductivity,[101-105] and magnetic susceptibility.[106] The width of the transition is usually too small to be observed in the older (classic) superconductors, but it was first observed and interpreted long ago in one of them, amorphous bismuth.[107]

Yet another effect of the shortness of $\xi_{GL}$ in the high-temperature superconductors is that many experimental data are determined by a thin surface layer (of thickness $\xi_{GL}$) of the sample. Many groups have observed that the chemical composition and crystal structure of such a thin surface layer may be different from that inside the sample, making it difficult to learn about the bulk of the sample from observations such as electron tunneling or photoemission. The shortness of the coherence length also decreases the coupling between the superconducting fluctuations in different parts of the material at an atomic level, and may therefore reduce some of the effects of impurity atoms.

### C.  Large Anisotropy

The large spatial anisotropy exhibited by the new superconductors poses fundamental questions, and has important implications for the viability of many applications. The theory of superconductivity in a highly anisotropic lattice is in a primitive state because the classic superconductors are almost all nearly isotropic,[108] except for:

a.) layered films,

b.) the intercalated graphites[109] and other layered compounds, such as $AS_2$, $ASe_2$, and $ASSe$, where A is a cation,[98] and

c.) the one-dimensional compound $Tl_2Mo_6Se_6$.[110,111]

Anisotropy effects in the classic superconductors have been reviewed.[108] The role of the anisotropy of the electron effective mass is well understood[112,113], but the roles of the anisotropies of the

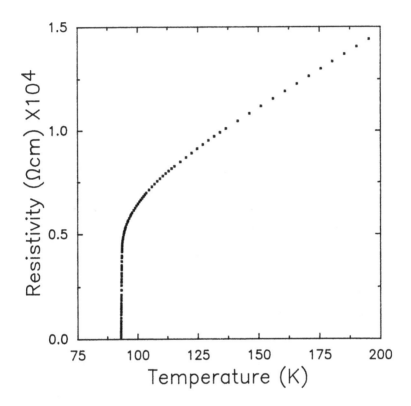

Figure 6.  Superconducting transition of a single crystal of
$YBa_2Cu_3O_{7-\delta}$, as seen by measurements of electrical resistance.
The rounding of the curve near $T_c$ shows thermodynamic fluctuations of
small regions into the superconducting state.  Data provided by
T. A. Friedmann, J. P. Rice, J. Giapintzakis, and D. M. Ginsberg.

various types of electron scattering have not been highly developed.

Apparently the Bi compounds are more anisotropic than the La or the Y compounds,[114,115] and the Tl compounds are likely to be the most anisotropic of all.[116]

## VII. SOME IMPORTANT FUNDAMENTAL QUESTIONS TO BE INVESTIGATED

Many of the experiments performed on the high-$T_c$ superconductors are designed to answer fundamental questions about their properties. Some of these questions are:

a.) What is the nature of the electron pairing and the symmetry of the superconducting order parameter? In a material with chains and planes, would the superconductivity be best described in terms of two order parameters, one for the chains and one for the planes? Are there more subtle causes for there being two order parameters?[117,118] If there is more than one order parameter, how are they coupled to each other? Among experimental methods helping to answer these questions are measurements of fluctuation effects in the specific heat, conductivity, and diamagnetism. Also useful in this regard are nuclear resonance data and observations of the effects of doping on the Cu-O planes or the chains. Attention needs to be given to the moment of the probability distribution in the energy gap parameter $\Delta$ (magnitude of the order parameter) which is measured by each experimental technique.

b.) Are there nodes in the superconducting energy gap for any directions of the electron momentum? To answer this question, many types of data are useful, such as the low-temperature specific heat, muon spin resonance, Raman scattering, infrared absorption, electron tunneling, and ultrasonic attenuation. One needs to work out theoretically what moments of the direction-dependent anisotropic energy gap are measured by each experimental method.

c.) Are the electron pairs bound by an interaction which is strong-coupling (compared with energy of the virtual excitations which provide the electron pairing)? The ratio of $\Delta C$, the jump[119,120] in the specific heat at $T_c$, to $\gamma T_c$; the temperature dependence of the penetration depth and the thermal conductivity; the size of the energy gap parameter $\Delta$ compared with $k_B T_c$ as observed in tunneling or transport experiments; and the change in the ultrasonic velocity at $T_c$ are all useful in determining the strength of the interaction binding each Cooper pair.

The ratio $\Delta C / \gamma T_c$ is approximately the BCS value 1.43 in weak-coupling superconductors, but is generally larger[121] (or perhaps sometimes smaller[71]) for strong-coupling superconductors. (Measured values for the classic superconductors range from about 1.4 to 2.7.[122]) For the high-temperature superconductor $YBa_2Cu_3O_{7-\delta}$, $\Delta C$ has been measured[123], but the specific heat constant $\gamma$ has not. (The lattice specific heat dominates the electronic specific heat near $T_c$, and measurements of $\gamma$ for $T \ll T_c$ cannot be made, since we do not have magnetic fields large enough to force these materials into the normal state at low temperatures.) If one estimates $\gamma$ from the normal-state magnetic susceptibility (ignoring Stoner enhancement of the suscep-tibility and effective mass enhancement of electronic specific heat) then one finds that the coupling is weak, and this leads one to conclude[124] (on the basis of theoretical work[71]) that the excitations which mediate the binding of electrons into Cooper pairs must be very large compared with $k_B T_c$. The mediating excitations therefore could not be (mainly) phonons, and would have to be electronic in nature.

The temperature dependence of $\Delta(T)/\Delta(0)$ would not tell us about the strength of the coupling if the new superconductors behave like the classic ones.[125,126]

d.)  Do the new superconductors have properties characteristic of 3-dimensional systems, or do they behave as 1- or 2-dimensional systems because of the chains or planes of Cu and O atoms? The effective dimensionality may depend on temperature.

e.)  What is the mechanism of superconductivity? Is it mediated primarily by spin or charge fluctuations (correlations, virtual excitations) rather than by phonons? Perhaps different materials are made superconducting by different mechanisms. For example, in the Cu-containing materials, a magnetic mechanism might be important, whereas in $Ba_{1-x}K_xBiO_{3-y}$, a magnetic mechanism is not expected, but a charge

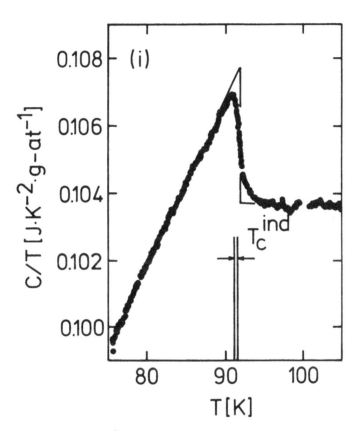

Fig. 7. Superconducting transition of polycrystalline $YBa_2Cu_3O_{7-\delta}$, as seen by measurements of the specific heat. The jump in the specific heat as T is reduced below $T_c$ is a sign of an energy gap which increases from zero. $T_c^{ind}$ shows the inductively measured transition temperature, 10% to 90% points. Data provided by A. Junod, A. Bezinge, and J. Muller (Ref. 120).

mechanism could be responsible. Measurements of neutron scattering, nuclear resonance, and electron tunneling are particularly powerful in helping to answer these questions.

f.) As a pure, insulating material is doped to increase its conductivity, do electron pairs form before superconductivity sets in at temperatures $\ll T_c$? Do quantum fluctuations of some type destroy superconductivity, even though the pairs are formed? Various thermodynamic and transport experiments will be useful in answering these questions.[127]

g.) Does the theory need to be modified if the geometric mean of the GL coherence lengths associated with the three crystallographic directions turns out to be less than $n^{-1/3}$, where n is the number density of conduction electrons in the normal state?

h.) Which atoms contribute electrons dominating the superconducting properties, and which atoms are mainly electron reservoirs? Various doping experiments help here.[21-25]

i.) The normal-state phases of the high-temperature superconductors exhibit a host of important phenomena, including antiferromagnetism, charge-density waves, strong correlations among the conduction electrons, and variable-range hopping. Some of these properties remind us of the organic superconductors[128] and the heavy-Fermion superconductors.[57] How can we explain the interesting normal-state properties of the new materials? Fundamental considerations are involved. For example, do the electrons form a Fermi liquid in the normal state, so that a Fermi surface would be observable? We learn much from the metal-insulator transition, which appears to be characteristic of the high-temperature superconducting materials as their stoichiometry is changed slightly from that which optimizes $T_c$. What are the effects of strong magnetic coupling seen by neutron scattering and by resonance experiments? Why are magnetic transitions frequently induced by slight doping? What roles do antiferromagnetism and magnetic frustration play? Are the magnetic properties and the electronic transport properties affected primarily by different types of excitations? What effects do twin boundaries or phase boundaries have on the electronic properties? Why do the high temperature superconductors have normal state resistivities which are both high and temperature dependent? Why do they have temperature dependent Hall coefficients? The normal state will have to be explained before the superconducting properties can be fully understood. j.) For the Bi and Tl compounds, why is $T_c$ raised by increasing the number of adjacent pairs of Cu-O and Ca-O planes in the unit cell? Is the effect caused mainly by an altered density of electronic states at the Fermi level in the normal state, or is it caused mainly by a change in the strengths of the interactions responsible for the superconductivity?

k.) How do magnetic vortices in these type II superconductors interact with each other, how are they pinned, and how do they move in response to the force exerted on them by the passage of a current through the material?

## VIII. SUMMARY AND CAUTIONS

The high temperature superconductors have many surprising properties. While efforts are being made to discover materials with even higher values of $T_c$, a large-scale effort is under way to understand those we already have. The theoretical problems are immense, since strong electron correlation effects and strong magnetic coupling among the atoms and the conduction electrons are likely to be vital parts of the story. Efforts are being made to determine the ways the BCS theory will have to be modified to account for the properties of these new materials. It is possible that this theory will survive essentially intact, but with a more comprehensive electron interaction than the electron-phonon interaction by itself. On the other hand, both the normal and superconducting properties of the high-$T_c$ superconductors may demand that our ways of thinking about solids be drastically modified.

These materials are among the most complex ever studied seriously by solid state physicists. Chemical reactions during synthesis frequently lead to multi-phase samples, with various impurities. Pure, homogeneous, single crystals are required to investigate properly the anisotropic properties, such as the normal-state resistivity, the

superconducting critical current density, and the critical fields. It
is clear that excellent characterization of samples is required if the
significance of experimental results is to be correctly evaluated. It
is also clear that newly observed effects need to be verified before
they can be taken seriously.

Our experience with the classic superconductors has speeded our
learning about the high temperature superconductors. We have embarked
on a fascinating voyage of discovery.

## ACKNOWLEDGMENTS

The author is deeply indebted to many colleagues, students, and
teachers, past and present; they are too numerous to list here. He is
also grateful to the National Science Foundation's Division of Materials Research for providing continuous research support, even when many
people believed that further work on superconductivity was unlikely to
prove interesting. The data plotted in Figs 3, 5, and 6 were acquired
in research supported by the National Science Foundation's Division of
Materials Research under Grants DMR-87-14555 and DMR-86-12860.

REFERENCES

1.  B. T. Matthias, American Scientist $\underline{58}$, 80 (1970).

2.  A. Khurana, Physics Today $\underline{40}$, No. 4, 17 (1987).

3.  J. R. Gavaler, Appl. Phys. Lett. $\underline{23}$, 480 (1973).

4.  L. R. Testardi, J. H. Wernick, and W. A. Royer, Solid State Commun. $\underline{15}$, 1 (1974).

5.  J. G. Bednorz and K. A. Müller, Z. Phys. B $\underline{64}$, 189 (1986).

6.  J. G. Bednorz and K. A. Müller, Revs. Mod. Phys. $\underline{60}$, 585 (1988).

7.  C. W. Chu, A. P. Rusakov, S. Huang, S. Early, T. H. Geballe, and C. Y. Huang, Phys. Rev. B $\underline{18}$, 2116 (1978).

8.  S. Uchida, H. Takagi, K. Kitazawa, and S. Tanaka, Jpn. J. Appl. Phys. $\underline{26}$, L1 (1987).

9.  C. W. Chu, P. H. Hor, R. L. Meng, L. Gao, Z. J. Huang, Y. Q. Wang, Phys. Rev. Lett. $\underline{58}$, 403 (1987).

10. K. Kishio, K. Kitizawa, S. Kanbe, I. Yasuda, N. Sugii, H. Takagi, S. Uchida, K. Fueki, and S. Tanaka, Chem. Lett. Japan (1987), p. 429.

11. J. M. Tarascon, L. H. Greene, W. R. McKinnon, G. W. Hull, and T. H. Geballe, Science $\underline{235}$, 1373 (1987).

12. R. B. VadDover, R. Cava, B. Batlogg, and E. Rietman, Phys. Rev. B $\underline{35}$, 5337 (1987).

13. M. K. Wu, J. R. Ashburn, C. J. Torng, P. H. Hor, R. L. Meng, L. Gao, Z. J. Huang, Y. Q. Wang, and C. W. Chu., Phys. Rev. Lett. $\underline{58}$, 908 (1987).

14. J. M. Tarascon, L. H. Greene, W. R. McKinnon, and G. W. Hull, Phys. Rev. B $\underline{35}$, 7115 (1987).

15. H. Maeda, Y. Tanaka, M. Fukutomi, and T. Asano, Japan. J. Appl. Phys. $\underline{27}$, L209 (1988).

16. S. S. P. Parkin, V. Y. Lee, E. M. Engler, A. I. Nazzal, T. C. Huang, G. Gorman, R. Savoy, and R. Beyers, Phys. Rev. Lett. $\underline{60}$, 2539 (1988).

17. Z. Z. Sheng and A. M. Hermann, Nature $\underline{332}$, 55 (1988).

18. A. M. Hermann, Z. Z. Sheng, D. C. Vier, S. Schultz, and S. B. Oseroff, Phys. Rev. B $\underline{37}$, 9742 (1988).

19. J. E. Greedan, A. H. O'Reilly, and C. V. Stager, Phys. Rev. B $\underline{35}$, 8770 (1987).

20. W. I. F. David, W. T. A. Harrison, J. M. F. Gunn, O. Moze, A. K. Soper, P. Day, J. D. Jorgensen, D. G. Hinks, M. A. Beno, L. Soderholm, D. W. Capone II, I. K. Schuller, C. U. Segre, K. Zhang, and J. D. Grace, Nature $\underline{327}$, 310 (1987); M. A. Beno, L. Soderholm, D. W. Capone II, D. G. Hinks, J. D. Jorgensen, J. D. Grace, and I. K. Schuller, Appl. Phys. Lett. $\underline{51}$, 57 (1987).

21. J. M. Tarascon, W. R. McKinnon, L. H. Greene, G. W. Hull, and E. M. Vogel, Phys. Rev. B $\underline{36}$, 226 (1987).

22. Y. Maeno, T. Tomita, M. Kyogoku, S. Awaji, Y. Aoki, K. Hoshino, A. Minami, and T. Fujita, Nature $\underline{328}$, 512 (1987).

23. J. M. Tarascon, P. Barboux, P. F. Miceli, L. H. Greene, G. W. Hull, M. Eibschutz, and S. A Shunshine, Phys. Rev. B $\underline{37}$, 7458 (1988).

24. G. Xiao, M. Z. Cieplak, D. Musser, A. Gavrin, F. H. Streitz, C. L. Chien, J. H. Rhyne, and J. A. Gotaas, Nature $\underline{332}$, 238 (1988).

25. Y. Tokura, J. B. Torrance, T. C. Huang, and A. I. Nazzal, Phys. Rev. B $\underline{38}$, 7156 (1988).

26. J. M. Tarascon, W. R. McKinnon, L. H. Greene, G. W. Hull, B. G. Bagley, E. M. Vogel, and Y. LePage, in High Temperature Superconductors, D. V. Gubser and M. Schluter (Materials Research Society, Pittsburgh, PA, 1987), p. 65.

27. R. J. Cava, B. Batlogg, C. H. Chen, E. A. Rietman, S. M. Zahurak, and D. Werder, Phys. Rev. B $\underline{36}$, 5719 (1987).

28. Y. Kubo, T. Ichihashi, T. Manako, K. Baba, J. Tabuchi, and H. Igarashi, Phys. Rev. B $\underline{37}$, 7858 (1988).

29. J.-L. Hodeau, P. Bordet, J.-J. Capponi, C. Chaillout, and M. Marezio Physica C $\underline{153-155}$, 582 (1988).

30. R. M. Fleming, L. F. Schneemeyer, P. K. Gallagher, B. Batlogg, L. W. Rupp, and J. V. Waszczak, Phys. Rev. B $\underline{37}$, 7920 (1988).

31. C. Chaillout, M. A. Alario-Franco, J. J. Capponi, J. Chenavas, P. Strobel, and M. Marezio, Solid State Commun. $\underline{65}$, 283 (1988).

32. P. Haldar, K. Chen, B. Maheswaran, A. Roig-Janicki, N. K. Jaggi, R. S. Markiewicz, and B. C. Giessen, Science <u>241</u>, 1198 (1988).

33. R. M. Hazen, C. T. Prewitt, R. J. Angel, N. L. Ross, L. W. Finger, C. G. Hadidiacos, D. R. Veblen, P. J. Heaney, P. H. Hor, R. L. Meng, Y. Y. Sun, Y. Q. Wang, Y. Y. Xue, Z. J. Huang, L. Gao, J. Bechtold, and C. W. Chu, Phys. Rev. Lett. <u>60</u>, 1174 (1988).

34. R. M. Hazen, C. T. Prewitt, R. J. Angel, N. L. Ross, L. W. Finger, C. G. Hadidiacos, D. R. Veblen, P. J. Heaney, P. H. Hor, R. L. Meng, Y. Y. Sun, Y. Q. Wang, Y. Y. Xue, Z. J. Huang, L. Gao, J. Bechtold, and C. W. Chu, Phys. Rev. Lett. <u>60</u>, 1174 (1988).

35. M. A. Subramanian, C. C. Torardi, J. C. Calabrese, J. Gopalakrishnan, K. J. Morrissey, T. R. Askew, R. B. Flippen, U. Chowdhry, and A. W. Sleight, Science <u>239</u>, 1015 (1988).

36. See, for example, J. D. Jorgensen, M. A. Beno, D. G. Hinks, L. Soderholm, K. J. Volin, R. L. Hitterman, J. D. Grace, and I. K. Schuller, Phys. Rev. B <u>36</u>, 3608 (1987).

37. F. Slakey, S. L. Cooper, M. V. Klein, J. P. Rice, and D. M. Ginsberg, Phys. Rev. B, to be published.

38. J. M. Tranquada, S. M. Heald, A. R. Moodenbaugh, and M. Suenaga, Phys. Rev. B <u>35</u>, 7187 (1987).

39. J. M. Tranquada, S. M. Heald, and A. R. Moodenbaugh, Phys. B <u>36</u>, 5263 (1987).

40. Z.-X. Shen, J. W. Allen, J. J. Yeh, J.-S. Kang, W. Ellis, W. Spicer, I. Lindau, M. B. Maple, Y. D. Dalichaouch, M. S. Torikachvili, J. Z. Sun, and T. H. Geballe, Rev. B <u>36</u>, 8414 (1987).

41. N. Nucker, J. Fink, J. C. Fuggle, P. J. Durham, and W. M. Temmerman, Phys. B <u>37</u>, 5158 (1988).

42. A. W. Sleight, J. L. Gillson, P. E. Bierstedt, Sol. State Commun. <u>17</u>, 27 (1975).

43. B. Batlogg, Physics Today <u>40</u>, No. 4, 23 (1987). In this article, the point for Y-Ba-Cu-O is plotted incorrectly; it should be shifted to the right; then it falls on the dashed line.

44. J. C. Phillips, Phys. Rev. B <u>36</u>, 861 (1987).

45. L. F. Mattheiss, E. M. Gyorgy, and D. W. Johnson, Jr., Phys. Rev. B $\underline{37}$, 3745 (1988).

46. R. J. Cava, B. Batlogg, J. J. Krajewski, R. Farrow, L. W. Rupp, Jr., A. E. White, K. Short, W. F. Peck, and T. Kometani, Nature $\underline{332}$, 814 (1988).

47. B. Dabrowski, D. G. Hinks, J. D. Jorgensen, R. K. Kalia, P. Vashishta, D. R. Richards, D. T. Marx, and A. W. Mitchell, Physica C $\underline{156}$, 24 (1988).

48. A. W. Sleight, in High-Temperature Superconducting Materials, edited by W. E. Hatfield and J. H. Miller, Jr. (Dekker, NY, 1988), chapter 1.

49. F. S. Galasso, Structure, Properties, and Preparation of Perovskite-Type Compounds (Pergamon, Oxford, 1969).

50. For example, A. C. Rose-Innes and E. H. Rhoderick, Introduction to Superconductivity (Pergamon, NY, 1978).

51. M. Tinkham, Introduction to Superconductivity, (McGraw-Hill, NY, 1975 and Krieger, NY, 1980).

52. P. G. de Gennes, Superconductivity of Metals and Alloys, (Benjamin, New York, 1966).

53. R. D. Parks, Superconductivity (Dekker, N.Y., 1969).

54. W. J. Skocpol and M. Tinkham, Rep. Prog. Phys. $\underline{38}$, 1049 (1975).

55. M. B. Maple, in Magnetism, Vol 5, edited by H. Suhl (Academic, NY, 1973), Chapter 10, p. 289.

56. O. Fischer, Appl Phys. $\underline{16}$, 1 (1978).

57. K. Yvon, in Current Topics in Materials Science, Vol. 3, edited by E. Kaldis (North-Holland, 1979), Chapter 2.

58. G. R. Stewart, Revs. Mod. Phys. $\underline{56}$, 755 (1984).

59. Interesting examples are found in M. M. Fang, J. E. Ostenson, D. K. Finnemore, D. E. Farrell, and N. P. Bansal, preprint, and in D. K. Finnemore, R. N. Shelton, J. R. Clem, R. W. McCallum, H. C. Ku, R. E. McCarley, S. C. Chen, P. Klavins, and V. Kogan, Phys. Rev. B $\underline{35}$, 5319 (1987).

60. J. Bardeen, L. N. Cooper, and J. R. Schrieffer, Phys. Rev. $\underline{108}$, 1175 (1957).

61. W. L. McMillan, Phys. Rev. $\underline{167}$, 331 (1968).

62. L. P. Gor'kov, Zh. Eksperim. i Teor. Fiz. 36, 1918 (1959) [Soviet Phys. - JETP 9, 1364 (1959)].

63. A. A. Abrikosov, Zh. Eksperim. i Teor. Fiz. 32, 1442 (1957) [Soviet Phys. - JETP 5, 1174 (1957)].

64. V. L. Ginzburg and L. D. Landau, Zh. Eksperim. i Teor. Fiz. 20, 1064 (1950).

65. T. Yamashita, A. Kawakami, T. Nishihara, Y. Hirotsu, and M. Takata, Japan. J. Appl. Phys. 26, L635 (1987), and T. Yamashita, A. Kawakami, T. Nishihara, M. Takata, and K. Kishio, ibid. 26, L671 (1987).

66. T. J. Witt, Phys. Rev. Lett. 61, 1423 (1988).

67. D. Esteve, J. M. Martinis, C. Urbina, M. H. Devoret, G. Collin, P. Monod, M. Ribault, and A. Revcolevschi, Europhys. Lett. 3, 1237 (1987).

68. R. H. Koch, C. P. Umbach, G. J. Clark. P. Chaudhari, and R. B. Laibowitz, Appl. Phys. Lett. 51, 200 (1987).

69. C. E. Gough, M. S. Colclough, E. M. Forgan, R. G. Jordan, M. Keene, C. M. Muirhead, A. I. M. Rae, N. Thomas, J. S. Abell, and S. Sutton, Nature 326, 855 (1987).

70. P. Gammel, D. J. Bishop, G. J. Dolan, J. R. Kwo, C. A. Murray, L. F. Schneemeyer, and J. V. Waszczak, Phys. Rev. Lett. 59, 2592 (1987).

71. F. Marsiglio, R. Akis, and J. P. Carbotte, Phys. Rev. B 36, 5245 (1987).

72. W. W. Warren, Jr., R. E. Walstedt, G. F. Brennert, G. P. Espinosa, and J. P. Remeika, Phys. Rev. Lett. 59, 1860 (1987).

73. C. H. Pennington, D. J. Durand, D. B. Zax, C. P. Slichter, J. P. Rice, and D. M. Ginsberg, Phys. Rev. B 37, 7944 (1988).

74. K. Ishida, Y. Kitaoka, K. Asayama, H. Katayama-Yoshida, Y. Okabe, and T. Takahashi, preprint.

75. L. C. Hebel and C. P. Slichter, Phys. 113, 1504 (1959); L. C. Hebel, Phys. Rev. 116, 79 (1959).

76. G. Aeppli, R. J. Cava, E. J. Ansaldo, J. H. Brewer, S. R. Kreitzman, G. M. Luke, D. R. Noakes, and R. F. Kiefl, Phys. Rev. B 35, 7129 (1987).

77.  D. R. Harshman, G. Aepli, E. J. Ansaldo, B. Batlogg, J. H. Brewer, J. F. Carolan, R. J. Cava, M. Celio, A. C. D. Chaklader, W. N. Hardy, S. R. Kreitzman, G. M. Luke, D. R. Noakes, and M. Senba, Phys. Rev. $\underline{36}$, 2386 (1987).

78.  J. Niemeyer, M. R. Dietrich, and C. Z. Politis, Z. Phys. B $\underline{67}$, 155 (1987).

79.  O. S. Akhtyamov, Zh. Eksperim. i Teor. Fiz. Pis'ma $\underline{3}$, 284 (1966). [Sov. Phys. JETP Lett. $\underline{3}$, 183 (1966).]

80.  M. E. Reeves, T. A. Friedmann, and D. M. Ginsberg, Phys. Rev. B $\underline{35}$, 7207 (1987).

81.  B. D. Dunlap, M. V. Nevitt, M. Slaski, T. E. Klippert, Z. Sungaila, A. G. McKale, D. W. Capone, R. B. Poeppel, and B. K. Flandermeyer, Phys. Rev. B $\underline{35}$, 7210 (1987).

82.  L. E. Wenger, J. T. Chen, G. W. Hunter, and E. M. Logothetis, Phys. Rev. B $\underline{35}$, 7213 (1987).

83.  S. von Molnar, A. Torressen, D. Kaiser, F. Holtzberg, and T. Penney, Phys. Rev. B $\underline{37}$, 3762 (1988).

84.  C. E. Methfessel, G. R. Stewart, B. T. Matthias, and C. K. N. Patel, Proc. Natl. Acad. Sci. U.S.A. $\underline{77}$, 6307 (1980).

85.  K. Kumagai, Y. Nakamichi, I. Watanabe, Y. Nakamura, H. Nakajima, N. Wada, and P. Lederer, Phys. Rev. Lett. $\underline{60}$, 724 (1988).

86.  R. A. Fisher, S. Kim, S. E. Lacy, N. E. Phillips, D. E. Morris, A. G. Markelz, J. Y. T. Wei, and D. S. Ginley, preprint.

87.  S. L. Cooper, M. V. Klein, B. G. Pazol, J. P. Rice, and D. M. Ginsberg, Phys. Rev. B $\underline{37}$, 5920 (1988).

88.  C. Uher and J. L. Cohn, preprint.

89.  U. Gottwick, R. Held, G. Sparn, F. Steglich, H. Rietschel, D. Ewert, B. Renker, W. Bauhofer, S. von Molnar, M. Wilhelm, and H. E. Hoenig, Europhys. Lett. $\underline{4}$, 1183 (1987).

90.  S. E. Inderhees, M. B. Salamon, N. Goldenfeld, J. P. Rice, B. G. Pazol, D. M. Ginsberg, J. Z. Liu, and G. W. Crabtree, Phys. Rev. Lett. $\underline{60}$, 1178 (1988).

91.  M. B. Salamon, S. E. Inderhees, J. P. Rice, B. G. Pazol, D. M. Ginsberg, and N. Goldenfeld, Phys. Rev. B $\underline{38}$, 885 (1988).

92.  J. Appel, Phys. Rev. Lett. $\underline{21}$, 1164 (1968).

93. P. A. Lee and N. Read, Phys. Rev. Lett. 58, 2691 (1987).

94. T. R. Lemberger and L. Coffey, Phys. Rev. B 38, 7058 (1988).

95. A. B. Pippard, Proc. Roy. Soc. (London) A216, 547 (1953).

96. Y. Yeshurun and A. P. Malozemoff, Phys. Rev. Lett. 60, 2202 (1988).

97. P. L. Gammel, L. F. Schneemeyer, J. V. Waszczak, and D. J. Bishop, Phys. Rev. Lett. 61, 1666 (1988).

98. For a review, see J. A. Woollam, R. B. Somoano, and P. O'Connor, Phys. Rev. Lett. 32, 712 (1974).

99. M. M. Fang, V. G. Kogan, D. K. Finnemore, J. R. Clem, L. S. Chumbley, and D. E. Farrell, Phys. Rev. B 37, 2334 (1988).

100. M. Tinkham, Phys. Rev. Lett. 61, 1658 (1988).

101. P. P. Freitas, C. C. Tsuei, and T. S. Plaskett, Phys. Rev. B 36, 833 (1987).

102. N. Goldenfeld, P. D. Olmsted, T. A. Friedmann, and D. M. Ginsberg, Solid State Commun. 65, 465 (1988).

103. B. Oh, K. Char, A. D. Kent, M. Naito, M. R. Beasley, T. H. Geballe, R. H. Hammond, A. Kapitulnik, and J. M. Graybeal, Phys. Rev. B 37, 7861 (1988).

104. S. J. Hagen, Z. Z. Wang, and N. P. Ong, Phys. Rev. B 38, 7137 (1988).

105. T. A. Friedmann, J. P. Rice, J. Giapintzakis, and D. M. Ginsberg, preprint.

106. W. C. Lee, R. A. Klemm, and D. C. Johnston, preprint.

107. J. S. Shier and D. M. Ginsberg, Phys. Rev. 147, 384 (1966).

108. Anisostropy Effects in Superconductors, edited by H. W. Weber (Plenum, Oxford, 1977).

109. R. Clarke and C. Uher, Adv. Phys. 33, 469 (1984).

110. T. Mori, Y. Yokogawa, A. Kobayashi, and Y. Sasaki, Solid State Commun. 49, 249 (1984).

111. R. Brusetti and P. Monceau, Solid State Commun. 66, 181 (1988).

112. V. L. Ginzburg, Zh. Eksperim. i Teor. Fiz. 23, 236 (1952).

113. K. Takanaka, Solid State Commun. 42, 123 (1982).

114. S. Martin, A. T. Fiory, R. M. Fleming, L. F. Schneemeyer, and J. V. Waszczak, Phys. Rev. Lett. 60, 2194 (1988).

115. O. Laborde, P. Monceau, M. Potel, P. Gougeon, J. Padiou, J. C. Levet, and H. Noel, Solid State Commun. <u>67</u>, 609 (1988).

116. J. H. Kang, K. E. Gray, R. T. Kampwirth, and D. W. Day, preprint.

117. A. E. Ruckenstein, P. J. Hirschfeld, and J. Appel, Phys. Rev. B <u>36</u>, 857 (1987).

118. G. Kotliar, Phys. Rev. B <u>37</u>, 3664 (1988).

119. M. Decroux, A. Junod, A. Bezinge, D. Cattani, J. Cors, J. L. Jorda, A. Stettler, M. Francois, K. Yvon, O. Fischer, and J. Muller, Europhys. Lett. <u>3</u>, 1035 (1987).

120. A. Junod, A. Bezinge, and J. Muller, Physica C <u>152</u>, 50 (1988).

121. D. J. Scalapino, in Ref. 51, Chapter 10, p. 449.

122. R. Meservey and B. B. Schwartz, Ref. 51, Chapter 3, p. 117.

123. S. E. Inderhees, M. B. Salamon, T. A. Friedmann, and D. M. Ginsberg, Phys. Rev. B <u>36</u>, 2401 (1987).

124. J. Bardeen, D. M. Ginsberg, and M. B. Salamon, in <u>Novel Superconductivity</u>, edited by S. A. Wolf and V. Z. Kresin (Plenum, NY, 1987), p. 333.

125. S. Bermon and D. M. Ginsberg, Phys. Rev. <u>135</u>, A306 (1964).

126. R. F. Gasparovic, B. N. Taylor, and R. E. Eck, Solid State Commun. <u>4</u>, 59 (1966).

127. Questions in this paragraph were suggested by Nigel Goldenfeld.

128. K. Bechgaard and D. Jerome, Scientific American <u>247</u>, No. 1, 52 (1982).

129. C. Uher and A. B. Kaiser, Phys. Rev. B <u>36</u>, 5680 (1987).

# 2
# THERMODYNAMIC PROPERTIES, FLUCTUATIONS, AND ANISOTROPY OF HIGH TEMPERATURE SUPERCONDUCTORS

M.B. Salamon

*Department of Physics and Materials Research Lab.*
*University of Illinois at Urbana-Champaign*
*Urbana, IL 61801, USA*

## I. INTRODUCTION

Since the remarkable discovery of high temperature superconductors[1] there has been an enormous effort to elucidate their fundamental properties. Speculations as to the underlying mechanisms have abounded, seemingly limited only by the imagination of the researcher. In this chapter, we critically explore recent results on thermodynamic properties in order to establish some boundary conditions on possible theories of the underlying mechanism. We focus on $YBa_2Cu_3O_{7-x}$ (123), about which the most is known, and for which reliable data are available on both twinned single crystals and aligned grains. Recent data[2] on the Tl-Ba-Ca-Cu-O materials are sufficiently similar to those on the 123 compound to justify the assertion that the 123 material is typical of high-temperature oxide superconductors.

It is not our goal here to summarize all of the thermodynamic data available on the high temperature superconducting oxides, nor even all on the 123 compound. The reader is referred to one of several reviews for that purpose.[3,4] Rather, we will attempt to extract from the body of data available a consensus of the experimental facts and to discuss them in both phenomenological and microscopic (i.e., BCS[5]) contexts. The main focus will be on extracting Ginzburg-Landau parameters and on discussing fluctuation effects. A final section will examine briefly the question of low temperature properties.

## II. THERMODYNAMIC PROPERTIES OF HIGH TEMPERATURE SUPERCONDUCTORS

Almost concurrent with the discovery of high temperature super-conductors, researchers sought to characterize their superconductive properties.[6] This occurred on two levels--phenomenological and microscopic--with considerable confusion in distinguishing between the two. The phenomenology of superconductivity is based on the well-known Ginzburg-Landau model[7] (a remarkable extension of the Weiss mean-field theory of magnetism), which is extremely useful in describing the macroscopic properties of all types of superconducting materials. The microscopic level is, of course, based on the Bardeen-Cooper-Schrieffer (BCS) theory,[5,8] a quantum mechanical description of the instability of a Fermi liquid to the formation of Cooper pairs. Strictly speaking, the BCS theory deals with the ideal case of a clean metal with weak electron-phonon coupling. Corrections for strong-coupling effects and finite electron mean-free path (dirty superconductor) have been incorporated into the BCS formalism. A connection between the phenomenological Ginzburg-Landau approach and the BCS theory was provided by Gor'kov,[9] who derived the G-L equations from BCS theory close to $T_c$.

There has been a tendency for researchers working on high temperature superconductivity to switch indiscriminately between phenomenological and microscopic formulae. This is rather dangerous. While we might expect the phenomenological approach to be applicable, and for modifications (for example, including more intricate ordering) to be relatively straightforward, it is far from clear at the outset how well the BCS theory will survive. It is the goal here to extract from the experimental data available at this writing the phenomenological parameters of the GL model and then to them compare with BCS expressions.

## A. Ginzburg-Landau Model

The Ginzburg-Landau approach to ordered phases introduces the idea of an order parameter $\psi(\mathbf{r})$ to describe the degree of superconductivity at each point and a free-energy functional $F(\psi)$, from which the actual free-energy of the system may be calculated. In the original argument, the true free energy is set equal to the minimum of $F(\psi)$ with respect to both $\mathrm{Re}(\psi)$ and $\mathrm{Im}(\psi)$. This is equivalent to evaluating the functional integral for the free-energy by the method of steepest-descents. The Ginzburg-Landau functional forms the basis for the modern theory of critical phenomena, but with the crucial difference that the steepest-descents method is replaced by the more appropriate renormalization group approach.[10]

The usual starting functional can be written in the form[11]

$$F = F_{no} + (1/V)\int ( \ H^2/8\pi + a|\psi|^2 + b|\psi|^4/2 \ +$$

$$+ \ \Sigma_j(1/2m_j{*})|(-i\hbar\nabla_j - e{*}A_j/c)\psi|^2)dV, \tag{1}$$

where $F_{no}$ is the normal state free energy, $\mathbf{A}$ is the vector potential, and the summation is over the principal axes of the condensate effective mass tensor $\mathbf{m}{*}$. The coefficient $a(T)$ changes sign at $T_c$, and may be approximated as $a(T) = \alpha\tau$, where $\tau \equiv T/T_c - 1$. The uniform solution in zero applied field is well known to be

$$|\psi|^2 = -a(T)/b, \tag{2}$$

and

$$F - F_{no} = H_c^2/8\pi = -a^2(T)/b. \tag{3}$$

$H_c(T)$ is the thermodynamic critical field. From Eq. (3) it follows directly that the slope of the critical field at $T_c$ is directly related to the step change $\Delta C$ of the heat capacity at $T_c$ through

$$(dH_c/dT)_{T_c} = (4\pi\Delta C/T_c)^{1/2} \tag{4}$$

It is customary to define $n_p \equiv \alpha|\tau|/b$, although this notation tacitly (and correctly for BCS) assumes that the order parameter in some sense measures the density of superconducting pairs in the BCS theory. The G-L functional Eq. (1) can be written in terms of the dimensionless variables,

$$\psi_o = \psi/\sqrt{n_p}; \quad h = H/H_c; \quad A_j' = A_j/(\sqrt{2}H_c\lambda_j), \text{ and } x'_j = x_j/\lambda_j, \tag{5}$$

as

$$F = F_{no} + (1/V')(H_c^2/8\pi)\int\{h^2 - 2|\psi_o|^2 + |\psi_o|^4$$
$$+ \Sigma_j|[(\nabla_j'/i\kappa_j) - A_j']\psi_o|^2\}dV'. \tag{6}$$

The Ginzburg-Landau penetration depths are defined as

$$\lambda_j = m_j{}^*c^2/4\pi n_p e^{*2}, \tag{7}$$

while the corresponding Ginzburg-Landau parameters are

$$\kappa_j = \sqrt{2}H_c e^*/\lambda_j{}^2\hbar c. \tag{8}$$

Note that, in the absence of a field, the variational minimum of Eq. (6) leads[8] to the differential equation

$$d^2\psi_o/dr_j{}^2 = (\kappa_j/\lambda_j)^2(\psi_o - \psi_o{}^3), \tag{9}$$

indicating that variations in the order parameter occur on a length scale defined as

$$\xi_j = \lambda_j/\kappa_j. \tag{10}$$

In this case, the subscript on the Ginzburg-Landau parameters refers to direction, not the more usual parameters,[12] $\kappa_1, \kappa_2$, and $\kappa_3$.

The Ginzburg-Landau approach is capable of treating inhomogeneous superconductivity in the presence of a magnetic field. It can be shown quite simply that the surface energy for forming a normal region within a superconducting material is positive for $\kappa < 1/\sqrt{2}$ (type I super- conductor) and negative (type II) otherwise. This causes type II superconductors to be unstable to the formation of vortices, each of which generates a quantum of magnetic flux

$$\Phi_0 = hc/e* = 2.07 \times 10^{-15} \text{ Wb};\qquad(11)$$

the numerical value assumes $e* = 2e$. The instability occurs at the "lower critical field" $H_{c1}$ at which the vortex lines would be separated by the penetration depth; the superconducting state is destroyed when the normal cores of the vortices overlap. The latter effect occurs at the "upper critical field" $H_{c2}$ for which the vortex separation is of order the Ginzburg-Landau coherence length.

In this chapter we are concerned primarily with the properties of $YBa_2Cu_3O_{7-x}$, samples of which are generally twinned. Consequently, the mass values in the a-b plane cannot be separated. Therefore, we assume that $m_x* = m_y* = m_\parallel$ and define $m_z* = m_\perp$, where the crystalline c axis coincides with z. Because the G-L model is only valid near $T_c$, we focus on the slopes of the critical fields:

$$dH_{c2\perp}/dT = \sqrt{2}\kappa_\parallel \, (dH_c/dT) = -\Phi_0/(2\pi T_c \xi_\parallel^2),\qquad(12)$$

$$dH_{c2\parallel}/dT = (2\kappa_\perp \kappa_\parallel)^{1/2}(dH_c/dT) = -\Phi_0/(2\pi T_c \xi_\perp \xi_\parallel),\qquad(13)$$

where the derivatives are evaluated at the transition temperature and the directions on the left-hand side refer to the axis along which the field is applied. The expressions for the lower critical field are

$$dH_{c1\perp}/dT = \ln(\kappa_\parallel)(dH_c/dT)/\sqrt{2}\kappa_\parallel = -\Phi_0 \ln(\kappa_\parallel)/4\pi T_c \lambda_\parallel^2,\qquad(14)$$

$$dH_{c1\parallel}/dT = \ln(\kappa_\parallel \kappa_\perp)(dH_c/dT)/2(2\kappa_\parallel \kappa_\perp)^{1/2}$$

$$= -\Phi_0 \ln(\kappa_\parallel \kappa_\perp)/8\pi T_c \lambda_\parallel \lambda_\perp.\qquad(15)$$

A number of authors have combined these results to extract pheno- menological parameters for $YBa_2Cu_3O_{7-x}$, the only high-temperature superconductor for which adequate crystalline data are available. A set of parameters that fits all the data has proven elusive. Recently, however, it has become clear that, due to flux-lattice melting effects,[13] the slopes of the upper critical field, Eqs. (12) and (13), are underestimated when the resistivity[14] or ac susceptibility[15] are used. At the same time, a large number of experiments, summarized by Fisher et al.,[3] have converged on the value $\Delta C = 3.6 \pm 0.4$ J/mol K for the heat capacity step at $T_c$. Larger values have been reported, but these have tended to extrapolate the lambda-like portion of the data, rather than include fluctuation effects (see Section III). Substituting this value into Eq. (4) and using $T_c = 92$ K, we find

$$dH_c/dT = -22 \pm 2 \text{ mT/K} \tag{16}$$

Two recent determinations of the lower critical field, by torque methods[16] and from static magnetization data,[17] agree that

$$H_{c1\perp}/H_{c1\parallel} = 4.3 \pm 0.2. \tag{17}$$

From Eqs. (12) and (13) it follows that

$$\epsilon = (dH_{c2\parallel}/dT)/(dH_{c2\perp}/dT) = (\kappa_\perp/\kappa_\parallel)^{1/2}. \tag{18}$$

Combining Eqs. (14)-(17), we find that

$$\epsilon \ln(\kappa_\parallel)/[\ln\epsilon + \ln(\kappa_\parallel)] = 4.3 \tag{19}$$

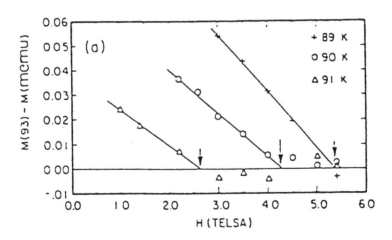

Fig. 1 The difference between normal state and superconducting state magnetization. The arrows mark the values of $H_{c2\perp}(T)$. (From Ref. 18.)

The upper critical field has been determined by Fang, et al.[18] from the onset of irreversibility in field-aligned powders of $YBa_2Cu_3O_{7-x}$. They interpret their results in terms of a $|\tau|^{1/2}$ curve, for which the slope at $T_c$ is infinite. However, their data are consistent with the value $dH_{c2\perp}/dT = -2.5\pm0.5$ T/K. (See Fig. 1.) Substitution of this value along with Eq. (16) into Eq. (12) fixes the basal plane Ginzburg-Landau parameter to be $\kappa_\| = 80\pm20$. Substituting this into Eq. (19), we find $\epsilon = 6.1\pm0.3$. With this result, plus one critical field, it is possible to calculate the remaining parameters. We use the measured value $H_{c1\|} = 23.5\pm0.5$ mT from references 16 and 17. The input values for the analysis are tabulated in Table I.

Table I. Input values for determination of Ginzburg-Landau parameters for crystalline $YBa_2Cu_3O_{7-x}$. The value $T_c = 92$ K is assumed. The symbols $\|$ and $\perp$ refer to directions parallel and perpendicular to the crystallographic ab plane.

| Quantity | Value | Source |
|---|---|---|
| $-(dH_c/dT)_{T_c}$ | $22 \pm 2$ mT/K | $\Delta C = 3.6 \pm 0.4$ J/mol K (Ref. 3) |
| $H_{c1\perp}/H_{c1\|}$ | $4.3 \pm 0.2$ | Ref. 16 and 17 |
| $H_{c1\|}$ | $23.5 \pm 0.5$ mT | Ref. 16 and 17 |
| $-(dH_{c2\perp}/dT)_{T_c}$ | $2.3 \pm 0.5$ T/K | Ref. 18 |

In Table II, we have calculated the various parameters of the Ginzburg-Landau model for $YBa_2Cu_3O_{7-x}$. Where single-crystal data are available, they have been included, as have some results which depend on the geometric mean of the anisotropic parameters. For comparison, we include results from a similar analysis[19] for a conventional layered superconductor, $NbSe_2$. $YBa_2Cu_3O_{7-x}$, in addition to having a higher transition temperature, is considerably more anisotropic than $NbSe_2$, and has much smaller coherence lengths. As we shall see in Section III, fluctuation properties depend on the inverse sixth power of the coherence lengths, making them much more pronounced in $YBa_2Cu_3O_{7-x}$ than in other layered materials, such as $NbSe_2$. It should be noted that both the extremely large mass ratio and the extremely small zero-temperature coherence length suggest that the two-dimensional aspects are extremely important. Recent calculations have considered two-dimensional Ginzburg-Landau theories with Josephson coupling between the layers,[20,21] and may be a more appropriate starting point for the phenomenological analysis than the one used here.

## B. BCS Analysis

Having determined a consistent set of Ginzburg-Landau parameters, we are now in a position to consider the extent to which they can be related to microscopic properties of the material through the BCS theory. The necessary connections were first demonstrated by

Table II. Values of the Ginzburg-Landau parameters for $YBa_2Cu_3O_{7-x}$ from the input values of Table I. Experimental values are given where known, along with comparable values for a conventional layered superconductor, $NbSe_2$. The notation is the same as in Table I.

| Quantity | $YBa_2Cu_3O_{7-x}$ | | $NbSe_2$ |
|---|---|---|---|
| | Calculated | Measured | |
| $\kappa_\parallel$ | $74 \pm 20$ | | 9 |
| $\kappa_\perp$ | $2750 \pm 900$ | | 100 |
| $m_\perp/m_\parallel = \kappa_\perp/\kappa_\parallel$ | $37 \pm 10$ | | 11 |
| $\xi_\parallel$ | $1.2 \pm 2$ nm | | 7.7 nm |
| $\xi_\perp$ | $0.2 \pm 0.04$ nm | $0.3^a$ nm | 2.3 nm |
| $\lambda_\parallel$ | $89 \pm 20$ nm | | 69 nm |
| $\lambda_\perp$ | $550 \pm 200$ nm | | 230 nm |
| $H_{c1\perp}$ | $90 \pm 20$ mT | $95^b$-$110^c$ mT | |
| $H_{c1\parallel}$ | $21 \pm 6$ mT | $24^{b,c}$ mT | |
| $\xi_{av} = (\xi_\parallel^2 \xi_\perp)^{1/3}$ | $0.7 \pm 0.2$ nm | $0.7^d$ nm | |
| $\lambda_{av} = (\lambda_\parallel^2 \lambda_\perp)^{1/3}$ | $160 \pm 60$ nm | $140^e$ nm | |
| $-(dH_{c2\parallel}/dT)_{T_c}$ | $14 \pm 5$ T/K | | |
| $-(dH_{c1\parallel}/dT)_{T_c}$ | $0.2 \pm 0.06$ mT/K | | 5 mT/K |

[a]Ref. 39  [b]Ref. 17  [c]Ref. 16  [d]Ref. 30
[e]D.R. Harshman, et al., Phys. Rev. B**36**, 2386 (1987); A.T. Fiory, A.F. Hebard, P.M. Mankiewich, and R.E. Howard, (to be published).

Gor'kov,and were subsequently corrected for the effects of finite
electronic mean free path. A fairly complete tabulation of the
relationships between BCS and Ginzburg-Landau parameters for general
values of the electronic mean-free path is found in a paper by Orlando,
et al.[22] These expressions are for an isotropic superconductor, and we
will use spatial averages of the values above, for comparison. This
updates a previous study[23] which, using data from ceramic samples,
concluded that $YBa_2Cu_3O_{7-x}$ is in the intermediate regime, where the
mean free path is comparable to the low-temperature BCS coherence
length. Therefore, this material cannot be treated in either the clean
or dirty limits.

The BCS theory assumes that electron-electron coupling is weak. In
the strong coupling limit, numerical corrections to the theory are
needed. These correction factors connect BCS and strong-coupling (SC)
expressions for any quantity Z (such as $Z = H_c$) through

$$Z^{SC} = \eta_Z(T)Z^{BCS}. \tag{20}$$

These factors, which depend on a single characteristic frequency $\omega_0$ in
an Einstein approximation, were discussed and tabulated by Orlando et
al.[22] The strong-coupling factor for the heat capacity jump was
discussed in the context of high temperature superconductors by
Marsiglio et al.[24] As noted by Orlando et al., the values of the
various $Z^{BCS}$ can be related to only four independent parameters: the
resistivity $\rho$, the electronic specific heat coefficient $\gamma$, the Fermi
surface area S, and $T_c$. Usually, $\gamma$ can be determined separately, for
example by low-temperature measurement at fields above $H_{c2}$. That is
not practical for $YBa_2Cu_3O_{7-x}$. Consequently, we select that combination of $\gamma$ and $\omega_0$ that comes closest to the spherically averaged
Ginzburg-Landau parameters. The results are given in Table III. As
may be seen from Table III, Gor'kov's transport coefficient $\lambda_{tr}$, which
is the ratio of the BCS coherence length to the normal-state mean free
path, is of order unity, confirming our earlier conclusion[23] that
$YBa_2Cu_3O_{7-x}$ is intermediate between clean and dirty limits. The
significant strong-coupling effects can be seen, for example, in the
enhancement of the ratio $\Delta C/\gamma T_c$ from its weak-coupling value of 1.43.

Following Quadar and Salamon,[23] we can estimate the exchange enhancement of the susceptibility from the data in Table III, and from it, calculate that the triplet-pairing interaction $g_t \approx .19$. The strong-coupling frequency provides a suitable cut-off frequency and permits an estimate of the critical temperature if the pairing were mediated by ferromagnetic spin fluctuations:

$$T_c \approx 1000K \exp(-1/g_t) \approx 5K \qquad (21)$$

The analysis leading to Table III shows that no reasonable choice of isotropic parameters for the BCS equations reproduces the geometrical means of the anisotropic Ginzburg-Landau parameters. This situation is particularly noticeable in the values of the upper-critical fields, as reflected in the very large $\kappa_{av}$ in the Ginzburg-Landau analysis. This

Table III. Comparison of phenomenological Ginzburg-Landau parameters with isotropic, strong-coupling BCS quantities calculated with corrections for finite mean-free path. The assumed values of the parameters are: resistivity at $T_c$ $\rho = 200$ $\mu\Omega$ cm; $T_c = 90$ K; electron density n = $6 \times 10^{21}$ cm$^{-3}$; and the electronic specific heat coefficient $\gamma = 18.2$ mJ/mol K$^2$.

| Quantity | BCS Calculation | Ginzburg-Landau Value |
|---|---|---|
| $\xi_{av}$ | 0.9 nm | 0.7 ± 0.2 nm |
| $\lambda_{av}$ | 134 nm | 160 ± 60 nm |
| $\kappa_{av}$ | 150 | 245 ± 60 |
| $-dH_c/dT$ | 19 mT/K | 22 ± 2 mT/K |
| $-(dH_{c2}/dT)_{av}$ | 4.6 T/K | 8 ± 3 T/K |
| $H_c(0)$ | 1.1 T | 1.0[a] T |
| $\Delta C/\gamma T_c$ | 2.23 | 2.21 |
| $2\Delta(0)/k_B T_c$ | 3.9 | |
| $\lambda_{tr}$ | 0.77 | |
| $\omega_o$ | 1000 K | |
| m*/m | 6.1 | |
| $\xi_{BCS}$ | 1.7 nm | |

[a]We use the parabolic law to match the critical field slope at $T_c$.

may reflect the strongly two-dimensional nature of the material which renders a three-dimensional BCS analysis inadequate. The remaining parameters are all within experimental uncertainty. The small values of the low temperature BCS and Ginzburg-Landau coherence lengths are troubling. The electron-electron distance at the assumed density is 0.66 nm, almost equal to $\xi_{av}$ and not much smaller than $\xi_{BCS}$. This has led Ranninger and coworkers[25] to consider models of real-space pairing to describe the new high temperature superconductors. Clearly, even if the usual BCS picture remains valid, such small values of the coherence lengths will make fluctuations much more important here than in conventional superconductors. This is the subject of the next section.

## III. FLUCTUATIONS

The superconducting transition belongs to the family of second order phase transitions for which the order parameter is continuous at the critical temperature. Fluctuation effects typically dominate such transitions; indeed, understanding these led to the development of the scaling theory of critical behavior and eventually, of renormalization-group methods.[10] Long before these theories were developed, Ginzburg[26] provided an elegant, but simple, criterion for the relative importance of critical fluctuations. In the case of a superconductor, a typical fluctuation into the normal state at $T < T_c$ occupies a volume spanned by the spatially averaged Ginzburg-Landau coherence length, $\approx \xi_{av}^3(T)$. Forming such a region costs condensation energy which is of order $H_c^2(T)$ per unit volume. Such fluctuations are unimportant so long as the energy cost is much greater than thermal energies; that is, if

$$H_c^2(T)\xi_{av}^3(T) \gg k_B T. \qquad (22)$$

Using the temperature dependence of the critical field and coherence length from Eqs. (3), (8) and (11), we obtain a form of the Ginzburg criterion

$$(1-T/T_c) \gg \tau_G \equiv (1 - T_G/T_c) = (1/2)[k_B T_c/H_{co}^2 \xi_{av}(0)^3]^2; \qquad (23)$$

a more proper derivation[27] results in the factor of 1/2 on the right hand side. At temperatures that satisfy Eq. (23), fluctuations do not dominate, and mean-field theory provides a reasonable approximation to the thermodynamic behavior of the superconductor. For a typical Type II superconductor (Nb, $T_c = 9.2K$; $H_{co} = 2$ kOe; $\xi_o = \xi_{av}(0) = 4 \times 10^{-6}$ cm), we find that $\tau_G = 2 \times 10^{-11}$, making the mean-field approximation valid at all realizable temperatures. By way of contrast, a typical magnetic transition has a condensation energy of order $k_B T_C$ per spin, in which case the Ginzburg reduced temperature $\tau_G$ is approximately the inverse of the number of spins encompassed by the zero-temperature coherence volume. Since this is typically of order unity in magnetic problems, mean-field theory is never a valid approximation.[28]

To see how the Ginzburg criterion changes with superconducting parameters, we use standard weak-coupling BCS results. The zero-temperature critical field can be written as

$$H_c^2(0) = 4\pi n(E_F)\Delta^2, \tag{24}$$

where $n(E_F)$ is the density of states per unit volume at the Fermi surface and $\Delta$ is the low temperature value of the energy gap. For a free electron gas we have $n(E_F) = (9/8\pi r_s^2 E_F)$, where $r_s$ is the radius of the volume occupied by one electron. Since the gap is $1.76 k_B T$ in BCS theory, Eq. (23) can be rewritten for a BCS superconductor as

$$\tau_G(BCS) = (.003)(E_F/k_B T_c)^2 (r_s/\xi_o)^6. \tag{25}$$

Again considering Nb, for which $E_F/k_B T_c \approx 6 \times 10^3$ and $r_s/\xi_o = 4 \times 10^{-3}$, we get $\tau_G(BCS) \approx 4 \times 10^{-10}$, somewhat larger than the value given above due to the use of BCS values. The oxide superconductors, as noted in Table III, are characterized by very small values of $\xi_o \approx 0.7$ nm , large $r_s \approx 0.35$ nm and, of course, large $T_c$. The analysis leading to Table III gives the ratio $E_F/k_B T_c \approx 25$, so that the approximate Ginzburg reduced temperature is $\tau_G(123) \approx 0.006$. Using estimates for $YBa_2Cu_3O_{7-x}$ that are less tied to BCS weak-coupling expressions ($H_{co} = 1$ T, $\xi_o = 1$ nm, $T_c = 92$ K), we find that $\tau_G(123) = 0.01$, which is clearly in the experimentally accessible temperature range. Another way to see this is to note that the closeness of $\xi_o$ and $r_s$ means that the Cooper pair size in the BCS theory is not large enough to average out the graininess of the electron gas.

At values of reduced temperature $|\tau|>>\tau_G$, the usual BCS and Ginz-burg-Landau results are approximately valid. Much closer to $T_c$, there will be a shift from the mean-field transition temperature and a cross over to critical exponents, as discussed by Lobb.[27] What should be expected in the intermediate temperature regime? Strictly speaking, a proper approach would involve a renormalization-group solution valid very close to $T_c$ plus appropriate "corrections to scaling." No such theory has been attempted. In its place, the so-called Gaussian fluctuation theory is used, which treats small fluctuations around the mean-field solutions.[29] The reader is cautioned not to confuse this approximation with the "Gaussian model," an unrealistic model[10] which allows fluctuations above $T_c$ to be arbitrarily large. While the two coincide above the transition temperature, it is understood that the Gaussian fluctuation model fails as the Ginzburg temperature is approached.

Fluctuations are, of course, most directly observed in the thermo-dynamic properties. The Gaussian fluctuation contribution to the specific heat is treated as a textbook example,[10] and can be shown to be of the form

$$C_{fl}{}^+(T) = C^+\tau^{-2+d/2},\tag{26}$$

and

$$C_{fl}{}^-(T) = C^-|\tau|^{-2+d/2},\tag{27}$$

for $\tau$ positive and negative respectively. Here, $C^+ = (nk_BT/16\pi\xi_{av}(0)^dD^{3-d})$, d is the dimensionality of the embedding space, n is the number of independent components of the order parameter, D is the physical size of a sample lower-dimensional (d < 3) sample, and

$$C^- = 2^{d/2}C^+/n.\tag{28}$$

(Note that this formula is wrong in Ref. 10.)[30] By independent compo-nents we mean here the number of terms in the free energy functional that are allowed in a model with O(n) symmetry. Above $T_c$ this func-tional may be viewed as a paraboloid in the n+1 dimensional space of order parameter values and amplitude. Fluctuations occur in all n directions from the minimum. Below $T_c$, the surface becomes an n+1 dimensional "sombrero." It costs energy to change the value of that combination of parameters that represents the order of the system ("longitudinal fluctuations"); the other n-1 modes are "transverse," or "gapless," and cost no energy to excite. Thus, the ratio $C^+/C^-$ can be used to determine the structure of the order parameter and therefore, the nature of the superconducting pairing mechanism.[30]

Evidence for fluctuation contributions to the specific heat of a twinned crystal of $YBa_2Cu_3O_{7-x}$ has been reported by Inderhees et al.[30] A definite upturn in the heat capacity was observed below $T_c$, a feature that cannot arise from a distribution of transition temperatures. The data were fitted to the sum of a BCS-like step and the fluctuation contributions Eqs. (26) and (27). The lattice background, which is much larger than the superconducting contribution, was approximated by a linear contribution near $T_c$. The fit is shown in Fig. 2. It is clear

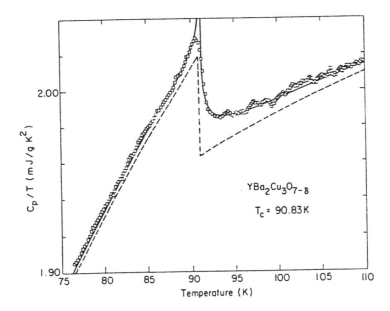

Fig 2. Heat capacity of a single crystal of $YBa_2Cu_3O_{7-x}$. The dashed line is the sum of lattice background and BCS contributions. The solid line includes Gaussian fluctuations. (From Ref. 57.)

from a superficial observation that the fluctuation contribution above $T_c$ exceeds that below, forcing n to be larger than 2 if d > 2. The effective number of order parameter components, defined as

$$n_{eff} = 2^{d/2}(C^+/C^-), \qquad (29)$$

was found to be 7±2 for d = 3. The Ginzburg-Landau coherence length at zero temperature (the geometric mean of the anisotropic coherence lengths) is 0.8 nm, which compares well with the Ginzburg-Landau value in Table III. Similar data have been reported recently by Fossheim et al.[31], (Fig. 3), and some evidence for peaking was obvious in the best polycrystalline samples studied by Junod, et al.[32] As yet, no similar data have been reported for other high temperature superconducting (HTSC) materials. It must be pointed out that the fits involve

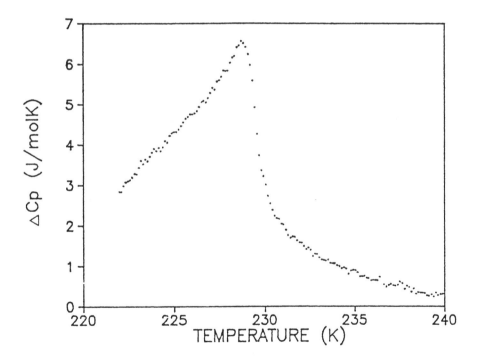

Fig. 3. Heat capacity of $YBa_2Cu_3O_{7-x}$ crystals after subtraction of an estimated lattice contribution. Compare with Fig. 2. (From Ref. 31.)

considerable uncertainty, since neither the lattice background nor the underlying BCS heat capacity are accurately known.

It has been pointed out by several authors[33] that a simple $O(n)$ model cannot be used to describe an anisotropic superconductor with a multicomponent order parameter. Unlike a true theory of critical phenomena, the Gaussian approximation does not give a universal value of the ratio $C^+/C^-$, but one that depends on the particular symmetry of the Ginzburg-Landau functional. This was considered in detail by Annett, et al.[34], who showed that a wide variety of $n_{eff}$ values are possible. For orthorhombic symmetry, they found that only triplet p-wave pairing is consistent with the large value found experimentally. If the orthorhombic distortion of the 123 structure can be ignored, a d-wave singlet state and three triplet states agree with measurements.

A number of other authors have considered fluctuation effects in high temperature superconductors. Bulaevskii and Dolgov[35] have considered the effect of strong coupling on the amplitude of the fluctuation contribution. They find that the amplitude is enhanced in the strong coupling limit by a factor $(3/2)(1+\Lambda)$, where $\Lambda$ is the strong-coupling coefficient. Inclusion of this factor would increase slightly the size of the coherence length deduced from heat capacity fluctuations. Sobyanin and Stratonnikov[36] have argued that fluctuations associated with 2D ordering along twin boundaries enhances the heat capacity above the bulk $T_c$. Some early measurements[30] did, indeed, show signs of a double transition but more recent results[31] on better samples give no evidence for a second transition. Nonetheless, the possibility of a significant contribution from twin boundary states in $YBa_2Cu_3O_{7-x}$ cannot be ruled out.

Fluctuations above $T_c$ manifest themselves in other thermodynamic and transport properties. Precursors of the diamagnetic susceptibility, often called "parasusceptibility" have long been known.[37] Their observation is more ambiguous than the heat capacity fluctuations since a small volume fraction of superconducting material with a transition temperature higher than the average can make a large diamagnetic contribution if it forms a loop normal to the probing magnetic field.

The diamagnetic contribution was calculated within the GL model by

Schmidt[38] and is written by Tinkham[8] as

$$\chi_{fl} = -(\pi^2 k_B T \xi_{av}{}^2/\Phi_o{}^2 V)(S/s)^2,$$ (30)

where $\Phi_o$ is the flux quantum; V, the volume of a typical fluctuation;
S, its cross-sectional area normal to the field, and s, the perimeter
of that area. In 3D, and for 2D systems with the field normal to
plane, the ratio S/s is proportional to the coherence length $\xi_{av}(T)$.
The fluctuation volume, however is proportional to $\xi_{av}(T)^d D^{3-d}$, where D
is a characteristic length in the 3-d constrained dimensions. [For the
layered high temperature superconductors, D may be the interplanar
spacing; more conventionally, it is the film thickness.] This gives an
overall dependence of $\xi_{av}(T)^{4-d} \sim \tau^{(d/2-2)}$. Note that, unlike heat
capacity fluctuations, susceptibility fluctuations are reduced in the
high-$T_c$ materials relative to conventional superconductors with longer
coherence lengths.

Extensive measurements of the fluctuation diamagnetism have been
made on many superconductors, and are summarized Tinkham's text.[8]
Recently, the fluctuation diamagnetism has been reported[39] for
$YBa_2Cu_3O_{7-x}$. For $\tau < 0.02$, a $\tau^{-1/2}$ divergence of the diamagnetic
susceptibility is observed, changing to a $\tau^{-1}$ behavior farther from $T_c$.
This cross-over from 2D to 3D behavior suggests that $\xi_\perp \approx 0.3$ nm as
given in Table II. Lee et al.[39] find the effective number of degrees
of freedom from the amplitude of the susceptibility to be $n_{eff} = 4.8$,
much as found from the heat capacity. However, they argue that the
presence of two sheets per unit cell doubles the number of fluctuating
components, so that $n_{eff}$ should be halved. This has been disputed by
Goldenfeld.[40] There has been no serious attempt to analyze the
deviations from perfect diamagnetism below $T_c$ in terms of fluctuations,
since the behavior is then dominated by flux pinning effects.

The effect of fluctuations on the electrical conductivity has
received by far the most theoretical and experimental attention. The
original predictions are due to Aslamazov and Larkin[41] (AL) and are
given by

$$\sigma_{3D}' = (e^2/32 h \xi_o) \tau^{-1/2};$$ (31)

$$\sigma_{2D}' = (2\xi_{av}(T)/D)\sigma_{3D}'. \tag{32}$$

Lawrence and Doniach[42] proposed an interpolation scheme for layered superconductors which compares the characteristic layer spacing D with the coherence length $\xi_\perp(T)$, in the direction normal to the layers:

$$\sigma_{LD}' = \sigma_{3D}'[1 + (D/2\xi_\perp(T))^2]^{-1/2}. \tag{33}$$

The AL conductivity is caused by the acceleration, during their lifetime, of superconducting pairs formed transiently above $T_c$. An additional contribution was noted by Maki[43] which involves the anomalous increase in the normal-state conductivity in the presence of superconducting fluctuations. With a correction for pair-breaking effects calculated by Thompson,[44] this so-called Maki-Thompson term makes a contribution in 2D given by

$$\sigma_{MT}' = (e^2/3\hbar D)(\tau-\delta)^{-1}\ln(\tau/\delta), \tag{34}$$

where $\delta$ is the fractional temperature shift caused by the pair-breaking effect. Extensive data are available on the fluctuation conductivity in conventional superconductors.[45]

A number of investigators have attempted to fit the temperature dependence of the electrical conductivity to various of the above expressions for the paraconductivity. The analysis is complicated by the possible effects of weak links, filamentary current paths, and locally large current densities. The best recent work has been done on crystals of $YBa_2Cu_3O_{7-x}$ which show an abrupt resistive and magnetic transition at $T_c$, and on textured thin films.[46] These crystals, as with the heat capacity and susceptibility samples, are multiply twinned. Hagen, et al.[47] analyzed their data in terms of the 2D AL model, achieving good fits for $0.06 \leq \tau \leq 1.9$, but with a wide variation in D. The effective transition temperature used in the definition of $\tau$ was 3K to 5K below the observed zero resistance point. Friedmann et al.[48] have recently analyzed their resistivity data using all possible models (3D-AL, 2D-AL, LD, and 2D-AL plus Maki-Thompson). They obtain the best fit with the Lawrence-Doniach model; the results are shown in Fig. 4 for various fitting ranges. While the fits are

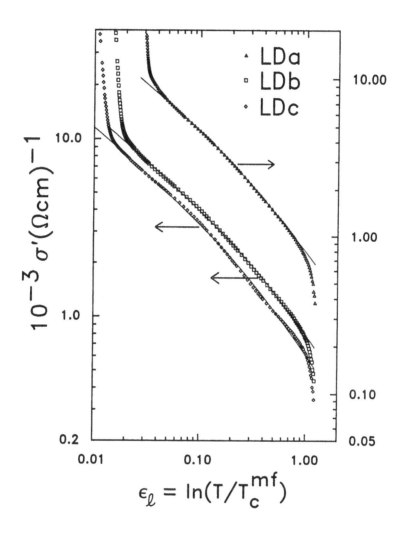

Fig. 4  Fit of the paraconductivity of a $YBa_2Cu_3O_{7-x}$ crystal to the Lawrence-Doniach form.  Different curves are the results of different ranges of data used in the fit.  (From Ref. 48.)

quite good, the parameters do not seem reasonable. The effective
interplanar spacing D = 0.2 nm is less than half of the crystallographic spacing. The zero temperature coherence length normal to the
planes is of order 0.05 nm, much smaller than estimates in Table II,
and too small to be a physically meaningful measure of any interaction
range. Both the cross-over temperature and the effective $T_c$ are below
the zero resistance point.

Recent results on the thermoelectric power of twinned crystals also
raise some doubt about the resistivity data. Using a novel, high
resolution ac technique, Howson et al.[49] found a sharp peak in the
thermopower just above $T_c$. In all samples, smaller peaks were found in
the range of temperatures over which the resistivity decreases. This
result strongly suggests a majority phase with a high transition
temperature connected in series with one or more minority phases, each
with a distinct but well-defined $T_c$. Anomalies in the thermopower were
predicted by Maki[50] to arise from fluctuation effects, but are too
small to explain the observed effects when conventional values of the
parameters are used. Howson et al. suggest, in analogy with the AL
calculation of the paraconductivity, that a singular contribution to
the thermopower might arise from the density gradient of transient
pairs accompanying a temperature gradient.

The unexpected parameters derived from the fluctuation conductivity
call into question the entire Gaussian fluctuation picture of the
transition. Fluctuation effects, when they dominate, tend to push the
actual critical temperature below, rather than move it above, the mean-
field value. A tendency to diverge at a temperature below the actual
transition is characteristic of a "weakly first order" transition, such
as found near a tricritical point. There is no evidence for such a
phase diagram for the high temperature superconductors. While it is
entirely possible that these effects reflect poor sample quality,
inhomogeneities, or the effect of numerous twin boundaries, we must be
alert to the possibility that the thermodynamics is significantly
different from that of a conventional BCS superconductor.

The effect of magnetic fields on physical properties provides
further insight into the superconductivity of high-$T_c$ materials. The

primary effect of a magnetic field on a type II superconductor is to shift the transition to a lower temperature along the line $H_{c2}(T)$, retaining its second-order character. Indeed, the observation of a shift in the heat capacity jump with an applied field[51] provided some of the earliest evidence for type II superconductivity. The effect of a magnetic field on fluctuations near $T_c$ was first considered by Lee and Shenoy[52] for dirty type II superconductors. The line $H_{c2}(T)$ is defined by the transition temperature of the lowest Landau level, with the other Landau levels having transition temperatures at successively lower temperatures spaced approximately by $\Delta T \approx -H/(dH_{c2}/dT)_{T_c}$. If the experimental temperature range is within $\Delta T$ of the transition, the fluctuations contributing to the heat capacity are essentially one-dimensional and are, consequently, enhanced over the higher dimensional expressions in Eqs.(26) and (27). In particular, it was shown that, in the limit $h/\tau_H \gg 1$,

$$C_H^+(3D) \rightarrow C_{H=0}^+(3D)h/|\tau_H|,  \tag{35}$$

where $h = -H/[T_{co}(dH_{c2}/dT)_{T_c}]$ and $\tau_H \approx T/T_c(H)-1$. The full expression for the fluctuation heat capacity can be cast in a scaling form valid for all fields (and the usual n = 2 BCS order parameter):

$$C_H^+(3D) = (k_B/8\pi\xi_o^3)h^{-1/2}f(x),  \tag{36}$$

where $x = \tau_H/h$. The function $f(x)$ has the limiting values $f(x) \rightarrow x^{-1/2}$ for $x \ll 1$ and $f(x) \rightarrow x^{-3/2}$ for $x \gg 1$, recovering Eqs. (26) and (35) respectively. Recently, Quadar and Abrahams[20] extended this analysis to consider Josephson coupled superconducting layers, and mapped the dimensional cross-over effects as functions of interlayer coupling and applied field. A more detailed treatment of the large field regime was reported by Thouless[53] and was confirmed experimentally for Nb by Farrant and Gough[54]

A second effect of the magnetic field is to expand the critical region. In the dirty limit and for d = 3, the Ginzburg criterion Eq. (23) takes the form

$$\tau_G(H=0) \approx r_s^4/\xi_o L^3  \tag{37}$$

60

given by Farrell;[55] here L is the mean free path. In the presence of a field, Lee and Shenoy argue that

$$\tau_G(H) = h^{2/3}[\tau_G(H=0)]^{1/3},\qquad(38)$$

which represents a significant broadening of the true critical regime in the presence of magnetic fields.

What regime is appropriate for $YBa_2Cu_3O_{7-x}$? Early estimates[15] gave $dH_{c2}/dT \approx -0.5$ T/K in $YBa_2Cu_3O_{7-x}$ for fields along the c-axis. In a field of 5 T, we would have h $\approx$ 0.1, making it possible to observe enhanced fluctuations within 10 K of $T_c$. Experiments on polycrystal-line samples by Fisher et al.[56] showed a <u>reduction</u> in the size of the heat capacity anomaly with little shift in position. Because $dH_{c2}/dT$ was reported to be much larger for fields in the basal plane, it was thought that the broadening reflected the powder-average response. However, single crystal data[57] with fields parallel to the c-axis are very similar to the ceramic data. These are shown in Fig 5 for a range of magnetic fields. Comparable data for a conventional superconductor would show a displacement of the anomaly with a slight enhancement in the peak.[54]

Equation (36) has the general form of a scaling law. Its form can

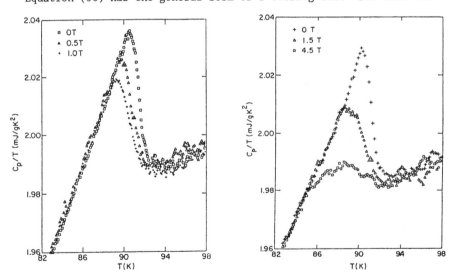

Fig. 5 Variation of the heat capacity with applied magnetic fields along the c-axis. Same sample as Fig. 4. (From ref. 57.)

be determined generally by considering the way in which the magnetic field, temperature, and heat capacity change when the length scale of the problem is changed. The field enters the Ginzburg-Landau functional Eq. (1) through the vector potential, which in turn scales in the same way as the gradient operator. Thus, the field will vary as the inverse square of the length scale. The temperature scale is determined by the temperature dependence of the coherence length. A general form of this dependence is $\xi_{av}(T) \propto \tau^{-\nu}$, where $\nu = 1/2$ in mean-field theory. It is straightforward to show that Eq. (36) now takes the form

$$C_H^{\pm}(d) = h^{d-2/\nu} g(y) \tag{39}$$

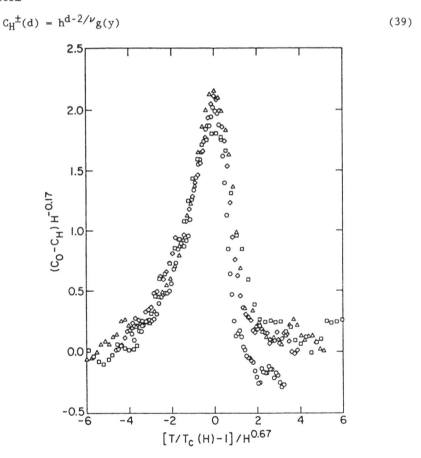

Fig. 6 Data of Fig. 5 scaled according to Eq. (39). These indicate that $\nu \approx 3/4$. (From Ref. 57.)

where now $y = \tau/h^{1/2\nu}$ and $g(y)$ includes the appropriate amplitudes for a d-dimensional material. This form can describe either the true critical regime or Gaussian fluctuations. So long as $\nu$ does not change, Eq. (39) remains valid as $H\to 0$, causing the difference $C_o^{\pm} - C_H^{\pm}$ to have the same scaling form as Eq. (16). The data of Fig. 5 have been plotted in scaling form in Fig. 6  The exponents are consistent with the choice $\nu = 3/4$, but with neither the mean-field values $\nu = 1/2$ nor the XY value $\nu = 2/3$. Very similar values of $C_o^{\pm} - C_H^{\pm}$ have been deduced from the magnetically determined free energy curves by Athreya et al.[58] and comparable data have been reported for $Tl_2Ba_2Ca_2Cu_3O_{10}$ by Finnemore et al.[2] That these data do not fit the mean-field predictions may suggest that the Gaussian fluctuation picture is inappropriate; in fact, the heat capacity data in zero field have also been fitted[57] with a power law consistent with this value of $\nu$. If the usual[10] $\epsilon$-expansion or $1/n$-expansion results were to remain valid for a multicomponent superconducting order parameter, this value of $\nu$ would suggest $4 \le n \le 6$.

Further evidence for the inapplicability of Gaussian fluctuations follows from the field dependence of the resistivity. Tinkham[59] demonstrates that the field dependence of the resistivity can be deduced by shifting the zero-field resistivity curve by an amount proportional to $H^{2/3}$, the same scaling observed in the specific heat. This procedure implicitly assumes that there is no downward shift of $T_c$ in fields up to 9 T, and further, that the zero field curve is already broadened by the same mechanism. This analysis puts the true $T_c$ at the point where the magnetoresistance vanishes.

A number of theoretical suggestions have been made to resolve the inconsistencies with conventional models. Tsuneto[21] suggested that vortex pairs or rings are the important fluctuations and treated them in the context of a Kosterlitz-Thouless transition. Moore[60] regarded the irreversibility line as a melting transition of the flux lattice, and argued that no truly ordered state exists in finite fields. A provocative Monte Carlo calculation by Ebner and Stroud[61] found that, for parameters relevant to $YBa_2Cu_3O_{7-x}$, the transition temperature is depressed by 30% due to fluctuations, and that the field-dependence of the heat capacity is qualitatively similar to experimental results. The near constancy of $T_c(H)$ for H along the c-axis of $YBa_2Cu_3O_{7-x}$, along with the very small values of the c-axis coherence length deduced from the resistivity data, require lower estimates of the spatially averaged zero-temperature coherence length than early results indicated. Because of the strong dependence of the Ginzburg temperature on the coherence length, it appears increasingly likely that Gaussian corrections to mean field theory are not adequate to describe the behavior of the high temperature superconductors close to $T_c$.

## IV. LOW TEMPERATURE HEAT CAPACITY

One of the hallmarks of the BCS theory is the existence of an energy gap in the electronic excitation spectrum and its manifestation in an exponentially small contribution to the low temperature heat capacity. More complicated pairing, such as the triplet pairing suggested for heavy Fermion materials, can leave points or lines of the Fermi surface ungapped. These lead to power-law contributions to the heat capacity. It is characteristic of both the $La_2CuO_4$-based (214) materials[62] and the 123 materials that a residual _linear_ contribution is observed. This was seized upon[63] as support for radically different physical properties of these materials. However, recent results[64] on the Bi-Ca-Sr-Cu system show no such linear terms.

The magnitude of the low-temperature linear contribution varies widely with sample preparation. Low temperature upturns in plots of $C/T$ vs $T^2$ are sometimes observed, as are various maxima at temperatures of order 10 K. Summaries of the large body of available data are found in the reviews, references 3 and 4. It has been pointed out that a number of nearby phases, such as $BaCuO_{2+x}$ in the 123 material, have large pseudolinear low temperature heat capacities. Magnetic impurities and, especially, tunneling states are other possible sources of such contributions. Typical values for the coefficient of the linear term are 3 mJ/mol $K^2$ in $YBa_2Cu_3O_{7-x}$ and the 214-materials. This should be compared with the estimated _total_ electronic specific heat coefficient used in Table III, $\gamma$ = 18 mJ/mol $K^2$. Clearly, a substantial fraction of the Fermi surface would have to remain ungapped to provide a contribution of the observed magnitude.

Linear contributions to the heat capacity are frequently observed in glassy materials as a result of a distribution of "tunneling states." However, it was pointed out[65] that on a per unit volume basis the observed value of $3 \times 10^{-5}$ $J/cm^3$ $K^2$ is an order of magnitude larger than is typical of either dielectric or metallic glasses. However, Golding, et al.[66] find a roughly logarithmic temperature dependence of the sound velocity of $YBa_2Cu_3O_7$ at low temperature--a signature of tunneling states. The size of the effect gives an independent estimate of the density of states of the two-level systems, and leads to a linear heat capacity contribution with a coefficient of 1.4 mJ/mol $K^2$, comparable to observed values.

While it is not possible to rule out a fundamental origin for the T-linear low temperature heat capacity, it does not appear likely. A recent systematic study of the 124 system[67] showed that the linear contribution remains constant over a composition range spanning the superconducting regime, but vanishes at stoichiometry. Similar results were reported[63,68] for single crystals of $YBa_2Cu_3O_{7-x}$. Finally, the absence of a linear contribution in the Tl- and Bi-Ca-Sr-Cu-based systems shows that the presence of a linear contribution to the heat capacity is not a necessary condition for high temperature superconductivity.

## V.  CONCLUSIONS

We have demonstrated that recent data on crystalline samples of $YBa_2Cu_3O_{7-x}$ bring all measured properties into concordance with an anisotropic Ginzburg-Landau model--at least within experimental error. The geometric mean of these anisotropic parameters, except for the upper critical field data, can be fitted with strong-coupling BCS expressions in the regime intermediate between clean and dirty limits. This analysis is imperfect, since the extreme anisotropy of the Ginzburg-Landau results point to a model based on Josephson coupled 2D layers.

Major problems occur in understanding the detailed thermodynamic behavior near $T_c$. While the Ginzburg criterion seems to suggest that corrections to mean-field theory are adequate to describe the data, a Gaussian approximation fails to predict the proper amplitudes of the fluctuation contributions, leads to unreasonably small values of the zero-temperature coherence lengths in the resistivity data, and fails to describe the dependence of fluctuation effects on applied magnetic field. A recurring theme is the existence of lines in the field-temperature diagram of the form $-\tau \propto H^{2/3}$. This led to suggestions[69] that the materials act as "superconductive glasses," but it now seems very likely that this behavior reflects a fundamental scaling relation between field and temperature.

Nearly a half century elapsed between the discovery of supercon-ductivity and its theoretical explanation. During that half century, a body of experimental data and phenomenological inferences were developed that framed the possible theoretical approaches. Working on a much compressed time scale, we now seem to be developing a comparable experimental and phenomenological context for an eventual theory of high temperature superconductivity. It seems from this experimental perspective that a successful theory will have to include extreme anisotropy, evident in the Ginzburg-Landau analysis, and will have to address the unusual power-law relationship between magnetic fields and temperature that are obvious in many properties.

## VI. ACKNOWLEDGEMENTS

Much of the author's work which is described in this chapter was supported in part by the National Science Foundation, through the Materials Research Laboratory Program Grant DMR-8612860. The author gratefully acknowledges the hospitality of the National Institute of Standards and Technology, where this work was completed.

# References

1.  J.G. Bednorz and K.A. Muller, Z. Phys. **64**, 189 (1986); M.K. Wu et al. Phys. Rev. Lett. **58**, 908 (1987).

2.  D.K. Finnemore, M.M. Fang, and D.E. Farrell (to be published).

3.  R.A. Fisher, J.E. Gordon, and N.E. Phillips, J. Supercond. (in press).

4.  H.E. Fischer, S.K. Watson and D.G. Cahill, Comments on Cond. Matt. Phys. **XIV** (to be published).

5.  J. Bardeen, L.N. Cooper and J.R. Schrieffer, Phys. Rev. <u>108</u>, 1175 (1957).

6.  R.J. Cava, B. Batlogg, R.B. van Dover, D.W. Murphy, S. Sunshine, T. Siegrist, J.P. Remeika, E.A. Rietman, S. Zahurak, and G.P. Espinosa, Phys. Rev. Lett. **58**, 1676 (1987).

7.  For an excellent summary, see N.R. Werthamer *in Superconductivity*, edited by R. Parks (Dekker, New York, 1969)pp. 321-370.

8.  M. Tinkham, *Introduction to Superconductivity*, (McGraw-Hill, New York, 1975)

9.  L.P. Gor'kov, Zh. Eks.i. Teor. Fiz. **36**, 1918 (1959) [Sov. Phys.-- JETP **9**, 1364 (1959)].

10. S-K Ma, *Modern Theory of Critical Phenomena*, (Benjamin, New York, 1976).

11. V.L. Ginzburg, Physica C **153-155**, 1617 (1988).

12. A.L. Fetter and P.C. Hohenberg in R. Parks, <u>op. cit.</u>, pp. 817-923.

13. Y. Yeshurun and A.P. Malozemoff, Phys. Rev. Lett. **60**, 2202 (1988); A.P. Malozemoff, T.K. Worthington, Y. Yeshurun, F. Holtzberg, and P.H. Kes (to be published).

14. Y. Iye, T. Tamegai, H. Takeya, and H. Takei, Jpn. J. Appl. Phys. Part 2 **26**, L1057 (1987); Y. Iye, et al. Physica C **153-155**, 26 (1988).

15. T.K. Worthington, W.J. Gallegher, and T.R. Dinger, Phys. Rev. Lett. **59**, 1160 (1987).

16. L. Fruchter, C. Giovanella, G. Collin, and I.A. Campbell, Physica C **156**, 69 (1988).

17. A.P. Malozemoff, this volume.

18. M.M. Fang, V.G. Kogan, D.K. Finnemore, J.R. Clem, L.S. Chumbley, and D.E. Ferrell, Phys. Rev. **B37**, 2334 (1988).

19. P. de Trey, S. Gygax, and J.-P. Jan, J. Low Temp. Phys. **11**, 421 (1973).

20. K.F. Quadar and E. Abrahams (to be published).

21. T. Tsuneto (to be published).

22. T.P. Orlando, E.J. McNiff, Jr., S. Foner, and M.R. Beasley, Phys. Rev. **B19**, 4545 (1979).

23. K.F. Quadar and M.B. Salamon, Solid State Commun. **66**, 975 (1988).

24. F. Marsiglio, R. Akis, and J.P. Carbotte, Phys. Rev. **B36**, 5245 (1987).

25. J. Ranninger, R. Micnas, and S. Robaszkiewicz, Ann. de Physique (Paris) (to be published).

26. V.L. Ginzburg, Fiz. Tverd. Tela. **2**, 2031 (1960) [English translation: Sov. Phys.--Solid State **2**, 1824 (1960)].

27. C.J. Lobb, Phys. Rev. **B36**, 3930 (1988).

28. L.P. Kadanoff, W. Gotze, D. Hamblen, R. Hecht, E.A.S. Lewis, V.V. Palciauskas, M. Rayl, J. Swift, D. Aspnes and J. Kane, Rev. Mod. Phys. **39**, 395 (1967).

29. V.Y. Skocpol and M. Tinkham, Rept. Prog. Phys. **38**, 1049 (1975).

30. S.E. Inderhees, M.B. Salamon, N. Goldenfeld, J.P. Rice, B.G. Pazol, D.M. Ginsberg, J.Z. Liu, and G.W. Crabtree, Phys. Rev. Lett. **60**, 1178(1988); erratum: ibid. **60**, 2445 (1988).

31. K. Fossheim, O.M. Nes, and T. Laegreid, Int. J. of Mod. Phys. B (to be published).

32. A. Junod, A. Bezinge, and J. Muller, Physica A **152**, 50 (1988).

33. P. Muzikar, Phys. Rev. Lett. **61**, 479 (1988).; H. Brand and M.M. Doria, Phys. Rev. Lett. **61**, 480 (1988).

34. J.F. Annett, M. Randeria, and S.R. Renn, Phys. Rev. (to be published).

35. L.N. Bulaevskii and O.V. Dolgov, Solid State Commun. **67**, 63 (1988).

36. A.A. Sobyanin and A.A. Stratonnikov, Physica C **153-155**, 1681 (1988).

37. J.P. Gollub, M.R. Beasley, R.S. Newbower, and M. Tinkham, Phys. Rev. Lett. **22**, 1288 (1969).

38. A. Schmidt, Phys. Rev. **180**, 527 (1969).

39. K. Kanoda, T. Kawagoe, M. Hasumi, T. Takahashi, S. Kagoshima, and T. Mizoguchi, J. Phys. Soc. Japan **57**, 1554 (1988); W.C. Lee, R.A. Klemm, and D.C. Johnston (to be published).

40. J. Annett, N. Goldenfeld, and S. Renn (private communication).

41. L.G. Aslamazov and A.I. Larkin, Fiz. Tverd. Tela **10**, 1104 (1968). [Sov. Phys.-Solid St. **10**, 875 (1968).]

42. W.E. Lawrence and S. Doniach, <u>Proc. 12th Int'l Conference on Low Temperature Physics</u>, edited by E. Kanda (Keigaku, Tokyo, 1970) p. 361.

43. K. Maki, Prog. Theor. Phys. **39**, 897 (1968).

44. R.S. Thompson, Phys. Rev. **B1**, 327 (1970).

45. R.A. Craven, G. A. Thomas and R.D. Parks, Phys. Rev. **B7**, 157 (1973).

46. B. Oh, K. Char, A.D. Kent, N. Naito, M.R. Beasley, T.H. Geballe, R.H. Hammond, A. Kapitulnik, and J.M. Graybeal, Phys. Rev. **B37**, 7861 (1988).

47. S.J. Hagen, Z.Z. Wang, and N.P. Ong (preprint, 1988).

48. T.A. Friedmann, J.P. Rice, and D.M. Ginsberg (preprint, 1988).

49. M.A. Howson, M.B. Salamon, T.A. Friedmann, S.E. Inderhees, J.P. Rice, D.M. Ginsberg, and K.M. Ghiron (to be published).

50. K. Maki, J. Low Temp. Phys. **14**, 419 (1974).

51. F.J. Morin, J.P. Maita, H.J. Williams, R.C. Sherwood, J.H. Wernick, and J.E. Kunzler, Phys. Rev. Lett. **8**, 275 (1962).

52. P.A. Lee and S.R. Shenoy, Phys. Rev. Lett. **16**, 1025 (1972).

53. D.J. Thouless, Phys. Rev. Lett. **34**, 946 (1975).

54. S.P. Farrant and C.E. Gough, Phys. Rev. Lett. **34**, 943 (1975).

55. R.A. Ferrell, J. Low Temp. Phys. **1**, 241 (1969).

56. R.A. Fisher, J.E. Gordon, S. Kim, N.E. Phillips, and A.M. Stacey, Physica C **153-155**, 1092 (1988).

57. M. B. Salamon, S.E. Inderhees, J.P. Rice, B.G. Pazol, D.M. Ginsberg, and Nigel Goldenfeld, Phys. Rev. **B38**, 885 (1988); M. B. Salamon, S.E. Inderhees, J.P. Rice, B.G. Pazol, and D.M. Ginsberg, J. de Physique (to be published).

58. K. Athreya, O.B. Hyun, J.E. Ostenson, J.R. Klemm, and D.K. Finnemore, (preprint, 1988).

59. M. Tinkham, Phys. Rev. Lett. **61**, 1658 (1988).

60. M. A. Moore (preprint, 1988).

61. C. Ebner and D. Stroud (preprint, 1988).

62. M.E. Reeves, T.A. Friedmann, and D.M. Ginsberg, Phys. Rev. B **35**, 7207 (1987); B.D. Dunlap, M.V. Nevitt, M. Slaski, T.E. Klippert, Z. Sungaila, A.G. McKale, D.W. Capone, R.B. Poeppel, and B. K. Flandermeyer, Phys. Rev. B **35**, 7210 (1987); L.E. Wenger, J.T. Chen, G.W. Hunter, and E.M. Logothetis, Phys. Rev. B **35**, 7213 (1987).

63. P.W. Anderson, Science **235**, 1196 (1987).

64. R.A. Fisher, S. Kim, S.E. Lacey, and N.E. Phillips, Phys. Rev. (to be published).

65. S. von Molnar, A. Torresson, D. Kaiser, F. Holtzberg, and T. Penney, Phys. Rev. **B37**, 3762 (1988).

66. B. Golding, N.O. Birge, W.H. Haemmerle, R.J. Cava, and E. Rietman, Phys. Rev. **B36**, 5606 (1987).

67. K. Kumagai, Y. Nakamichi, I. Watanabe, Y. Nakamura, H. Nakajima, N. Wada, and P. Lederer, Phys. Rev. Lett. **60**, 724 (1988).

68. H.B. Brom, J. Baak, A.A. Menovsky, and M.J.V. Menken (to be published).

69. K.A. Muller, M. Takashige, and J.G. Bednorz, Phys. Rev. Lett. **58**, 1143 (1987).

# 3

# MACROSCOPIC MAGNETIC PROPERTIES OF HIGH TEMPERATURE SUPERCONDUCTORS

A.P. Malozemoff

*IBM Thomas J. Watson Research Center*
*Yorktown Heights, NY 10598-0218, USA*

# I. INTRODUCTION

No aspect of the new high-temperature (high-$T_c$) oxide superconductors is more controversial or in a more rapid state of evolution than the study of their macroscopic magnetic properties. At the time of this writing, important new experimental data are appearing almost daily, and many novel interpretations are being proposed. A coherent interpretation does not yet exist. One might wonder about the utility or feasibility of a review of these magnetic properties at this time.

Nevertheless many characteristic and interesting features are now becoming apparent, particularly as results on a variety of new materials systems become available. These features include the unusual irreversibility line in the H-T plane, the possible melting of the vortex lattice above this line, the huge non-exponential time-relaxation of the magnetic properties and its ageing effect, the discrepancy of magnetic and transport critical current densities in ceramic material and the weak-link structure which this implies, the almost exponential falloff of the critical current density with temperature and field, the unusual field-dependence of the field-cooled susceptibility, the unexpected broadening of the resistive transition in a magnetic field, and so forth. These features will surely form the basis for future interpretations. Thus an overview, particularly of the diverse experimental results, should be of value, along with at least some brief introduction to the present theories.

These remarkable phenomena are of interest in themselves. In addition, understanding them is essential for identifying basic parameters like the coherence length and magnetic penetration depth, and their asymmetry and temperature dependence, which are vital for pinning down the underlying mechanism of high-temperature superconductivity.

These, then, are the questions which will be the focus of this review: First, to what extent are the experimentally observed macroscopic magnetic properties similar to - or different from - those of conventional superconductors? One thesis of this review is that in fact many properties **are** very different from those of conventional superconductors, implying interesting new physics as well as having an important impact on applications.

Second, assuming we can decipher these magnetic properties, what can we say about the fundamental bulk material parameters? It will be another thesis of this review that the upper critical fields $H_{c2}$ and the coherence lengths $\xi$ derived from them are poorly known, in spite of an enormous number of studies already in the literature; so this area will require considerable further work. There are also problems in determining the lower critical fields $H_{c1}$ and the related magnetic penetration depths $\lambda$, but recently it appears that the data are converging.

The two most fundamental characteristics of superconductivity are the disappearance of resistance and the expulsion of magnetic flux. Thus it is no wonder that along with resistance measurements, magnetic measurements are among the most common in the high $T_c$ literature, beginning with the historic first magnetic confirmation of superconductivity in ceramic LaBaCuO by Bednorz, Takashige and Müller[1]. Magnetic measurements are particularly convenient for characterizing the new materials because they are non-contact measurements and thus avoid problems in electrically contacting the materials through surface oxide layers or in handling delicate single crystals. Thus, with the convenience of magnetic measurements, the multitude of new phenomena observed and the impact on our fundamental understanding of the superconducting mechanism, it is no wonder that the magnetic high-$T_c$ literature which has accumulated in only two years is utterly gigantic.

To limit the scope of this review, the focus will be specifically on "macroscopic" magnetic measurements, by which is meant studies of the Meissner, screening and vortex effects, rather than on the "microscopic" aspects, such as high-temperature susceptibility, diamagnetic fluctuations, or NMR and neutron studies of local moments. Also excluded are the many studies of rare earth or transition metal doping, which have led to interesting new magnetic phases at low temperatures, or to suppression of superconductivity. Instead the focus here will be on the standard and best characterized copper oxide materials: $La_{2-x}Ba_xCuO_{4-\delta}$ and $La_{2-x}Sr_xCuO_{4-\delta}$, $YBa_2Cu_3O_{7-\delta}$ ($T_c \simeq 90K$), $Bi_2Sr_2Ca_1Cu_2O_\delta$ ($T_c \simeq 85K$) and $Tl_2Ba_2Ca_2Cu_3O_\delta$ ($T_c \simeq 125K$), which, for simplicity, will be denoted LaBaCuO, LaSrCuO, YBaCuO, BiSrCaCuO and TlBaCaCuO throughout the text. However both ceramic and bulk crystalline forms of these materials will be considered because of the great importance of granularity in copper oxide superconductivity.

Microwave studies are also closely related to the topic of this review. But they will not be described here, even though many phenomena found in dc or low frequency magnetic measurements have been confirmed in this way. Muon spin rotation data are discussed, but only as they bear on the penetration depth. The magnetic properties are also closely related to transport properties. For example, the hysteretic magnetic behavior of Type II superconductors is largely determined by the critical current density. While there will be no general review of the countless studies of critical current density, the resistive determination of the upper critical fields and some general features of the temperature and field dependence of the critical current density will be described.

Even with these limitations, the literature remains huge and fascinating. Some measure of the extraordinary volume of material is given by the following numbers of publications, on this topic alone, in key journals and conference proceedings during the past two years: Europhysics Letters,[1-13] starting with the first Bednorz-Takashige-Müller work on magnetic properties: 13 articles; Physical Review Letters[14-33]: 20 articles; Physical Review B[34-107]: 74 articles; Solid State Communications[108-140]: 32 articles; Physica C Proceedings of the Interlaken Conference[141-215]: 75 articles; Applied Physics Letters[216-241]: 26 articles; and many others.[242-330]

The review is structured as follows: In the next section (II), the experimental data are reviewed, with particular emphasis on phenomena which are novel in the context of conventional low temperature superconductivity. First, in IIA, the basic zero-field-cooled, field-cooled and remanent moment measurements are described, from which Müller et al.[16] first introduced the key novel concept of an irreversibility line. Interesting behaviors are also found in the field-cooled susceptibility and the remanent moment. In IIB, the closely related hysteresis loops are discussed. From these data, the classic Bean[331] critical-state model permits a determination of the critical current density, which turns out to be very different, at least in ceramic material, from that obtained in direct transport measurements. This discrepancy has led to the now well-established weak-link or granular model of the ceramic materials.

In IIC, the magnetic relaxation effects are presented: Surprisingly they are comparably large in both ceramics and bulk crystals. In IID, magnetic torque data, as well as other phenomena associated with sample motion and rotation, such as mechanical oscillator experiments and levitation, are described. In IIE, ac susceptibility experiments are shown to indicate that measurements originally interpreted in terms of the upper critical field most likely show some kind of dynamical vortex transition.

Another longstanding mystery is the broadening of the resistive transition with magnetic field, even in the best bulk crystals. These and other more recent data which undermine the previous determinations of $H_{c2}$ and point rather to flux flow are summarized in IIF . Near $T_c$, the magnetization becomes reversible, permitting an an alternative determination of $H_{c2}$, described in IIG .

Underlying all these macroscopic measurements is the Abrikosov vortex array. The magnetic decoration experiments, described in IIH, are thus of special importance in giving direct visual information about the array and its possible pinning mechanisms. Muon spin rotation is also sensitive to the vortices, and the resulting measurements of the magnetic penetration depth are given in II-I.

Section III turns to a comparison with available theory. The basic theme here is introduced in IIIA, namely the propensity of these small-coherence-length superconductors to form weak link structures. This is abundantly verified in the behavior of high-$T_c$ ceramics, but the nature of the bulk crystals or individual crystalline grains is less clear. In fact, a controversy which permeates this review is whether the bulk high-$T_c$ materials should be thought of as granular, or whether they are simply conventional bulk type II superconductors whose unusual properties arise from weak flux pinning and possible flux-lattice melting. The surprising result of the experimental summary in Section II is that many properties characteristic of the ceramic material are also observed in the best quality crystals!

The possible granularity of bulk crystals is a central question because, as mentioned earlier, magnetic measurements provide us some of the most fundamental superconducting parameters, and one must know whether the deduced parameters are characteristic of weak links or of the bulk material.

In this context, IIIB reviews the basic ideas of the Ebner – Stroud[332] superconducting glass model, which emerges from a picture of randomly located grains interconnected with Josephson weak links. This model explains many of the experimental phenomena. By contrast, IIIC describes a more conventional flux pinning picture[333,334] of these superconductors but introduces the notion of giant flux creep[28] and thermally activated flux flow[31,296] which can account for the irreversibility line. IIID gives a brief summary of yet another approach to understand the irreversibility line, as a flux-lattice melting[26,27,33] or vortex-glass freezing.[313]

Finally, Section IV proceeds from the hypothesis that bulk crystals are not granular but show the properties of true bulk Type II superconductivity. In IVA, the measurements of the upper critical field $H_{c2}$ are reviewed. The present state of knowledge is unsatisfactory because most of these measurements probably represent not $H_{c2}$, but rather the irreversibility line. By contrast, the determinations of the lower critical field $H_{c1}$ (IVB) and magnetic penetration depth (IVC) reveal YbaCuO to be a not-very-anisotropic superconductor having properties consistent with conventional singlet BCS pairing.

The reader is assumed to be familiar with the basic principles of Type II superconductivity[335] and with the anisotropic structure and materials characteristics of the high-$T_c$ superconductors, reviewed in the other chapters of this book. Field directions are always given with respect to the c-axis perpendicular to the copper oxide planes. Cgs units are generally used.

## II. EXPERIMENTAL RESULTS

IIA. Dc magnetic measurements, as a function of temperature

IIA-1. The irreversibility line

In the first study devoted specifically to the unique magnetic properties of the oxide superconductors, Müller, Takashige and Bednorz[16] measured the temperature dependence of the dc magnetization and immediately made the key observations which have remained the focus of attention to the present day. Using a standard SQUID (Superconducting Quantum Interference Device) magnetometer, they first cooled their LaBaCuO ceramic sample to low temperature in the small background field of the apparatus. They subsequently applied a field and measured the diamagnetic moment as a function of slowly increasing temperature, obtaining the "zero-field-cooled" curve labeled 1 in Fig. 1. They then slowly reduced the temperature back down through the superconducting transition, obtaining the "field-cooled" curve labeled 2; this gave a considerably smaller diamagnetic moment. Finally they turned off the field at low temperature and observed a positive moment which gradually decreased to zero with increasing temperature.

Müller et al. noted several key features of the data. First, while the zero-field-cooled magnetization appeared to be unstable and to relax slowly downward in magnitude with time, the field-cooled magnetization appeared to be in equilibrium within experimental accuracy. The remanent magnetization also relaxed downward. These relaxations were nonexponential in time (see Fig. 1 and IIC). Second, they observed that above a certain temperature, indicated by point D in the figure, the field-cooled and zero-field-cooled curves as a function of temperature merged into a common reversible behavior. Studying this merging-point as a function of the applied field H of the experiment, they found that the temperature shifted with field according to

$$1 - (T/T_c^{\star}) \propto H^q \ . \tag{1}$$

They found a $T_c^{\star}$ noticeably lower than the bulk superconducting transition temperature $T_c$ and a power q close to 2/3.

Such an effect is familiar in analogous measurements on magnetic spin glasses. Mean-field spin glass theory by de Almeida and Thouless[336] had predicted just such a law; so Müller et al.[16,253] dubbed this a "quasi-de-Almeida-Thouless" line. As pointed out by Morgenstern et al.[253] and Vinokur et al.,[306] the spin-glass derivation is probably not strictly relevant to the superconducting case. So we will henceforth use the more general terminology "irreversibility line" to describe the phenomenon, although a distinction should be made between this dc irreversibility line and the presumably related irreversibility line determined from ac susceptibility and resistivity measurements, described in IIE-2 and IIF-1 below.

Such effects have now been confirmed, with minor variations, in all the major families of ceramic high-$T_c$ superconductors.[35,45,81,118,131,133,169,202,276] The power q can vary from as low as 1/2 to as high as 3/4, and in YBaCuO, $T_c^{\star}$ in Eq. 1 appears to approach $T_c$, the bulk zero-field superconducting transition temperature. However the exact position of the line has considerable ambiguity because the zero-field-cooled and field-cooled curves approach each other asymptotically as a

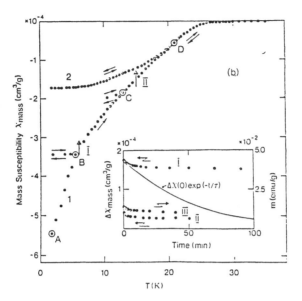

Fig. 1.    Zero-field-cooled (1, A-B-C-D) and field-cooled (2) susceptibility of a
LaBaCuO ceramic as a function of temperature.    Point D marks the
irreversibility temperature for 300 Oe applied field.    The inset shows the
non-exponential time decay of the susceptibility.    (After Müller et al., Ref.
16.)

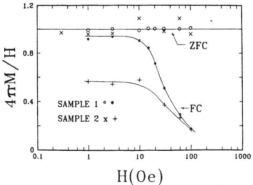

Fig. 2.    Low temperature zero-field-cooled (ZFC) and field-cooled (FC) susceptibilities of two YBaCuO crystals as a function of applied field perpendicular to the c-axis.    The different low-field saturations of the field-cooled
Meissner fraction indicate different levels of oxygenation of the 40-50 μm
thick crystals.    (After D. C. Cronemeyer and F. Holtzberg, private communication.)

function of temperature; so the determination depends on the experimental resolution. There is also some evidence that remanent moments can persist even above the dc irreversibility line in a certain field range (IIA-4). Irreversibility lines determined from resistivity measurements (IIF-1) are also ambiguous because they depend on the voltage criterion. The most precise measurements come from the ac susceptibility (IIE-2), and these will be used for comparison to theory in III below.

It might have been thought that these features arose from the grain-boundary weak-links of the ceramics and were not characteristic of bulk material. Nevertheless, right from the start, Müller et al.[16] suggested on the basis of field-scaling arguments (see IIIB) that the bulk crystal grains were responsible. Indeed, Yeshurun and Malozemoff[28] recently found an irreversibility line with an $H^{2/3}$ dependence and substantial magnetic relaxation (see IIC) in an YBaCuO crystal. Yeshurun et al.[276] also found an irreversibility line shifted to much lower relative temperatures in a BiSrCaCuO crystal, consistent with many other indications of the greater reversibility in this system.[228]

In summary, the irreversibility line, whatever its origin, is perhaps the most important of the novel experimental phenomena observed in the high-$T_c$ superconductors.

IIA-2. Zero-field-cooled moment: the screening effect

Let us next take a closer look at the zero-field-cooled, field-cooled and remanent moments. It is important to recognize the basic difference between the zero-field-cooled and field-cooled data. Since flux is initially absent from the sample during zero-field-cooling, the observed diamagnetism after applying a field represents flux **exclusion** from the sample. This differs fundamentally from the field-cooled flux **expulsion** which characterizes the true Meissner effect (see IIA-3). Ideally, if the applied field is less than the effective lower critical field, the zero-field-cooled susceptibility of a bulk superconductor should approach $-1/4\pi$.

It is worth emphasizing that if a **surface** or shell of the sample is superconducting but the interior is not, the zero-field cooled moment can still show full flux exclusion from the **entire** volume. Thus, it is clear that the size of the zero-field-cooled susceptibility, or, by extension, of the ac susceptibility, does **not** in general measure the volume fraction of bulk superconductivity. While such measurements are widely quoted in the literature for confirming the presence of superconductivity, it should always be remembered that the results may only reflect the properties of a surface sheath. In fact, Kittel et al.[72] have emphasized that at sufficiently low fields, complete screening can occur even when the thickness of a surface superconducting layer is much less than the penetration depth.

In reality, few samples show complete screening, even after proper correction for sample shape demagnetization. Some deviation can occur simply from London penetration at the sample surfaces, but this correction should be very small for samples large compared to the penetration depth. This effect grows if the sample is powdered, providing the basis for perhaps the most direct measurement of the temperature dependence of the penetration depth, as will be discussed further in IIG-2. Conversely, a reduced susceptibility of crystals has on occasion been used[186] as evidence for their granularity. However, to make this quantitative, one must consider "dead" or nonsuperconducting surface layers arising from oxygen out-diffusion, other sources of macroscopic inhomogeneity, and imperfect demagnetization geometries.

A study of the field-dependent susceptibility after zero-field-cooling is the standard method for determining the lower critical field. However as will be seen in IVB, the interpretation of the data is complicated by flux pinning and flux penetration at corners. Perhaps it will come as no surprise that in ceramics, because of the grain-boundary or weak-link effects to be described in more detail in IIB, flux penetration can occur at a field as low as a few milliOersteds, for example in the careful SQUID studies of LaSrCuO by Esparza, Civale et al.[110,120] By contrast, values of order hundreds of Oersteds emerge from the fairly clean onset of flux penetration in the best crystals (see IVB).

## IIA-3. Field-cooled moment: the Meissner effect

In contrast to the zero-field-cooled case, field-cooling starts with flux uniformly present in the sample above $T_c$. Therefore, any diamagnetic moment $M_{fc}$ which appears as a result of field-cooling represents a flux **expulsion**, the true expression of the Meissner effect. In general this field-cooled susceptibility is observed to be significantly smaller than the full flux expulsion $\chi = M_{fc}/H = -1/4\pi$ expected at fields below $H_{c1}$ for a perfect (and not intrinsically pinned) superconductor. Thus it is useful to define a "Meissner fraction" $-4\pi M_{fc}/H$, where H is the internal field (applied field corrected for shape demagnetization).

There are many possible reasons for incomplete flux expulsion (Meissner fraction less than unity). One is flux pinning. Another mechanism emerges from the Ebner-Stroud superconducting glass model. These mechanisms will be discussed further in III below. Here we consider a third reason, namely that the sample may not be fully superconducting. This is perhaps the least interesting from a physical point of view, but tremendously important in the intensive search for new and improved high-$T_c$ superconductors.

The simplest hypothesis is that the Meissner fraction measures the volume fraction of superconducting phase. The discrepancy with the much larger zero-field-cooled susceptibility can then be understood, as suggested by Ginley et al.,[44] if the superconductivity occurred in a surface "shell" or sheath. This is a not implausible suggestion considering the slowness of the oxygen in-diffusion required in many cases to establish high-temperature superconductivity, for then the interior of the grain or crystal could remain oxygen-starved.

The problem with interpreting the Meissner fraction as the superconducting volume fraction is, of course, that other mechanisms, such as flux-pinning or superconducting-glass effects, can also lower it. Experimentally, the problem is highlighted by the discovery of Krusin-Elbaum et al.,[97,183,273] using a non-commercial, ultra-low-field SQUID magnetometer, that the Meissner fraction is strongly field-dependent in YBaCuO, both ceramic and crystalline, and rises towards 100% at fields usually below 1 Oe. Such low fields are not easily available in many commercial magnetometers because of remanence in their superconducting solenoids.

One particularly interesting case, from recent work of D. C. Cronemeyer (private communication) on two crystals of YBaCuO, is shown in Fig. 2. The more strongly oxygenated material shows the Meissner fraction rising at low fields to almost complete flux expulsion, while the other sample saturates at the lowest fields with values considerably less than 100%. It seems reasonable to suppose that this low-field saturation measures the true fraction of superconducting material. The essentially complete zero-field-cooled flux exclusion in Fig. 2 points, then, as Ginley et al. suggested, to a kind of superconducting shell and the possibility of inadequately oxygenated material inside bulk crystals. Duan et al.[138] confirmed this directly by mechanically stripping away the surface of their YBaCuO crystals.

In ceramics, Maletta, Cronemeyer, Malozemoff et al.[97,108,274] found a somewhat weaker dependence of the Meissner fraction on field. More recently, Celani et al.[292] found that the strength of the dependence increases with the density and quality of the sample, suggesting that the intergrain coupling strength controls this effect.

In summary, it is risky to interpret a Meissner fraction measured in several tens of Oersteds as the superconducting volume fraction. Nevertheless it is probably correct to treat it as a lower limit because, as far as is known, other effects can only decrease the fraction. These ideas underlie and confirm the early work of Cava et al.[14,18] establishing the bulk nature of superconductivity LaSrCuO and YBaCuO ceramics. Meissner fractions have been reported in countless other works on all the major materials families. Perez-Ramirez et al.[123] and Malozemoff et al.[97] first reported essentially 100% Meissner fractions in their ultra-low field measurements in YBaCuO ceramics and crystals respectively. In the more recent work of Sunshine et al.[89] and Lin et al.[95] on BiSrCaCuO crystals, Meissner fractions up to 75% have been attained even without measuring in the sub-Oersted range, and Liu et al.[300] observed 40% in TlBaCaCuO. This suggests that these materials are highly reversible (see IIIC-2) compared to YBaCuO, although very large critical currents have been reported recently, particularly by Itozaki et al.,[265] in both BiSrCaCuO and TlBaCaCuO films.

The peculiar field-dependence of the Meissner fraction is a clue to understanding the flux pinning or glassy nature of the superconductivity; it will be discussed further in IIIB and C below.

IIA-4. Remanent moment

Finally let us turn to the remanent moment $M_{rem}$, obtained by turning off the field after field-cooling and measuring as a function of increasing temperature.[16,35,57,67,97,247,274] Cronemeyer et al.[274] found that in YBaCuO ceramics at low fields, this effect follows a simple rule all the way up to $T_c$:

$$M_{rem} = M_{fc} - M_{zfc} \qquad (2)$$

Here $M_{fc}$ and $M_{zfc}$ are the field-cooled and zero-field-cooled moments measured at the same field. Malozemoff et al.[97] verified Eq. 2 in both ceramic and bulk crystal YBaCuO over the same range of fields where the Meissner fraction $M_{fc}/H$ was changing strongly (e. g. see Fig. 2). This result is important in the comparison of glassy and flux-pinning models and will be discussed in III below.

Studying the temperature-dependent remanent moment in YBaCuO ceramics at somewhat higher fields of order 500 Oe, Tuominen et al.[67] and Cronemeyer et al.[274] found that Eq. 2 broke down. This is of particular importance in the region near $T_c$ where remanence was observed to persist **above** the temperature where the field-cooled and zero-field-cooled moments merged. If that merging is interpreted as reversibility, then one would not expect remanent moment to persist. This raises questions about the meaningfulness of the irreversibility line determined from these dc magnetic measurements, a problem which needs to be studied more thoroughly.

The remanent moment has been related[50,113] to the presence of persistent supercurrents around the perimeter of ceramic YBaCuO rings. Several authors[223,232,328] have measured the field directly inside ceramic toroids of YBaCuO and TlBaCaCuO and confirmed the presence of persistent currents, though typically only at very low current densities, consistent with the ceramic nature of the materials used. Grader et al.[232] detected a slow decay of the persistent current indicating flux creep (IIIC).

Leiderer and Feile[255] pointed out that in such experiments, a remanent moment could arise simply from trapped flux without any true persistent current circulating around the ring. They developed a local magnetic probe to scan across the ring and observed the flux distribution, allowing them to distinguish between trapped flux and the fields arising from a true persistent current. They observed that this current only occurred at fields below a few tenths of an Oersted. Similarly Renard et al.[160] measured a quadrupolar moment after compensating the small field cooled moment in YBaCuO ceramics to demonstrate the difference between surface screening currents and internal trapped flux. The measurements of Gyorgy et al.[223] with a local probe inside a hollow tube should also be sensitive to the true persistent current.

IIB. Hysteresis loops and critical currents: the weak-link picture

IIB-1. Experimental hysteresis loops

Closely related to the dc measurements of magnetization versus temperature are hysteresis loop measurements of magnetization versus field, with field first increasing from zero to a maximum value, then decreasing, reversing in sign, and so forth. The results are complex but have formed a basis for the granular or weak-link picture of the high-$T_c$ superconductors.

One of the most extensive studies on ceramics has been performed by Oussena, Senoussi et al.[6,52,245,294] Data of theirs on YBaCuO ceramic is shown in Figs. 3 and 4. For field excursions as small as an Oersted, a small hysteresis loop opens up, then closes at somewhat higher fields in the range of tens of Oersteds, an effect seen in all the different high-$T_c$ ceramic systems.[46,82,114,123,124,128,133,139,166,200,246,287] This effect is absent in bulk crystals or single-grain powders, and Wong et al.[231] directly demonstrated that it arises from the intergranular coupling by powdering their YBaCuO ceramics and watching this low-field loop disappear.

At larger field excursions (e.g. of order 1 Tesla), the loop takes on a new, flatter shape similar to that reported by Dinger et al.[19,142] and Senoussi et al.[83] in bulk high-$T_c$ crystals as well as that seen earlier in low temperature superconductors. Such high-field loops in an YBaCuO ceramic are shown in Fig. 4. The diamagnetism first increases linearly with applied field (shown as negative in Fig. 4), with a slope indicating a high degree of flux exclusion (after correction for shape demagnetization). At a field corresponding roughly to the bulk lower critical field $H_{c1}$, the magnetization begins to deviate away from linearity, although the exact point is hard to pick out exactly, an issue we will return to in IVB below. Above a field $H^{\star}$, corresponding to a maximum in the magnitude of the diamagnetism, a plateau or gradual decrease sets in. When the field is now lowered from a high value (typically several Tesla), the magnetism changes sign and usually attains a new plateau with, at low temperatures at least, a small maximum near zero field (see Fig. 4).

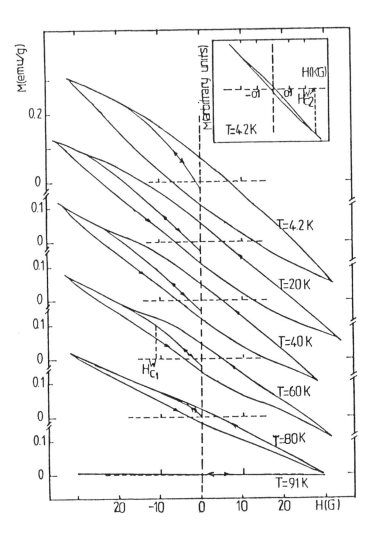

Fig. 3. Low-field hysteresis loops of YBaCuO ceramic showing a small loop which closes at higher field and a gradual collapse to zero with increasing temperature. (After S. Senoussi et al., Ref. 245.)

At higher temperatures the amplitude of the M-H loop shrinks rapidly, and, as shown by Senoussi et al.,[83] develops a characteristic dip near the zero-field crossings, giving the loop an overall "butterfly" shape.[148] Near $T_c$, the hysteresis entirely disappears at high fields, leaving an equilibrium diamagnetism. Highly reversible loops have also been observed in BiSrCaCuO, for example by Kumakura et al.,[228] confirming the tendency towards large shifts in the irreversibility line to low temperatures (IIA-1). Kwak et al.[268] found similar highly reversible loops in TlBaCaCuO films in spite of large measured critical currents.

As an aside, it is important to mention that the loops are not always smooth. In big YBaCuO crystals, Tholence et al.[188] found regular jumps in the magnetization, giving a sawtooth pattern as a function of field. Gough et al.[197] also reported jumps of many tens of flux quanta in hysteresis loops of YBaCuO ceramic. Datta et al.[161] claimed chaotic behavior in a certain temperature range.

IIB-2. The Bean critical-state model: the "magnetic" $J_c$

Many features of these high-field hysteresis loops can be understood on the basis of the simple Bean critical-state model.[331] The basic idea of this approach is that in a homogeneous though defected Type II superconductor, the hysteretic magnetization is related to the critical current density $J_c$ through a modified Ampere's law curl $h = (4\pi/c)J_c$, where h is now the locally averaged magnetic induction inside the superconductor. Thus the hysteretic magnetization can be used to derive $J_c$. This assumes, of course, that the superconductor is homogeneous and not a collection of weakly coupled grains.

Bean pictured the penetration of flux from both sides of the sample (here taken to be a slab of thickness D with external field H parallel to the slab plane) in terms of a flux gradient set by Ampere's law and by the boundary condition h=H at the surfaces of the sample. As the field is increased, the flux fronts penetrate further into the sample, until, at an applied field $H^{\star} = 2\pi D J_c/c$, they meet in the center. The magnetization is the difference of the applied field H and the average flux density $B = D^{-1}\int dx\, h(x)$ in a slab of thickness D. For $H < H^{\star}$, this gives

$$4\pi M = (cH^2/4\pi J_c D) - H \ . \tag{3}$$

For $H > H^{\star}$, and assuming a $J_c$ independent of field, a similar simple calculation shows that M becomes constant at $M^{+} = -J_c D/4c$.

As field is reversed, the flux contour changes slope at the surface, giving the bird-wing patterns of h(x) shown in classic texts.[335] Once the field has been reduced by $2H^{\star}$, a constant $M^{-}$ is again achieved, with the same magnitude as before but opposite sign. This leads to the widely used relationship for the "magnetically de-

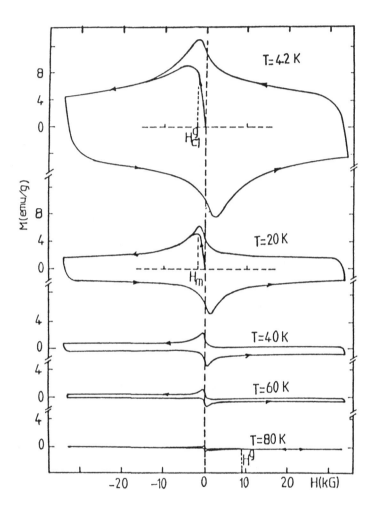

Fig. 4.    High-field hysteresis loops of YBaCuO ceramic with field H along the c-direction.  The breakaway from linearity of the virgin curve (after zero-field-cooling) indicates the lower critical field $H_{c1}$.  (After S. Senoussi et al., Ref. 245.)

termined $J_c$ ," written here for a superconducting slab of thickness D with field applied in the slab plane, in practical units (A/cm$^2$, Gauss and cm):

$$J_c = 20| -M^+ + M^- |/D \simeq 40M_{rem}/D \ . \tag{4}$$

If $J_c$ has a field dependence, then this formula is best applied at a field value sufficiently large to insure a relatively small variation in flux density across the sample. But since the observed field dependence of the hysteretic magnetization in bulk high-$T_c$ crystals is usually not large, at least at low temperature,[19,46,83] and since the plateaus are rather symmetric (see Fig. 4), it is often adequate to simply use double the high-field remanence $M_{rem}$ at zero applied field in Eq. 4, instead of $|M^+ - M^-|$. Eq. 4 is valid only if the field is cycled up to several times $H^{\star}$, which is of order 1 Tesla for bulk crystals. $M_{rem}$ resulting from this process is obviously different from the field-cooled remanence discussed in IIA-4 earlier.

Atzmony et al.[246] and Grover et al.[287] have used these ideas to describe the observed high field loops semiquantitatively, although as discussed in IIB-3 next, D should be taken as the grain size rather than the total sample size in the case of ceramics. Near $T_c$, Stancil et al.[239] have pointed out that critical-state hysteresis becomes weaker, and that therefore one must take more careful account of the reversible diamagnetism of the Type II superconductor (see IIG).

IIB-3. The weak-link model of ceramics

In ceramic high-$T_c$ superconductors, the blind application of Eq. 4 to the high-field remanence, and using the full sample size for D, has yielded rather modest "magnetic" $J_c$ values, though usually larger than values from direct transport measurements, as shown by many authors.[82,189,203,209,212,242,243,247,299] This suggests that transport currents are blocked at grain boundaries while the dominant contributions to the high-field remanence are coming, not from a continuous flux gradient across the entire sample, but rather from the much more localized contributions of individual grains.

To quantify this possibility, let us consider a critical state to exist only within the grains, with an average grain size a and an internal critical current density $J_{c,g}$. Then, from Eq. 4, the remanence is $aJ_{c,g}/40$ . Using this formula, many authors[52,54,82,220] deduced values of up to $J_{c,g} \simeq 10^7$A/cm.$^2$ at low temperature. This value is in rough agreement with evaluations for bulk crystals,[13,19,57] in which the full dimensions of the crystal were used in the analysis. This interpretation has been further confirmed by Küpfer al. al.[256] who have powdered their ceramics and demonstrated that the magnetic hysteresis at high fields remains the same.

Further evidence for this idea, and for a Josephson weak-link model for the grain-boundary behavior, has come from the dramatic discrepancy in the field dependence of the magnetic and transport critical current densities in ceramics. In typical data of Kwak et al.[82] on YBaCuO ceramic, the transport $J_c$ drops off initially linearly with field, extrapolating to zero by a few tens of Oersteds, although there is a tail to higher fields. This contrasts with the much slower suppression of the magnetic $J_c$ with field, again suggesting that the magnetic data are dominated by intragranular flux.

A number of authors[82,84,149] have interpreted the field-suppression of the transport-$J_c$ in terms of the well-known effect of a magnetic field on a Josephson junction: The vector potential within the junction modulates the current sinusoidally and creates a "diffraction" pattern as a function of position along the junction. The net critical current obeys

$$J_c(H) = J_{c0} | \sin(\pi H/H_0)/(\pi H/H_0)| \ , \tag{5}$$

with $H_0$ equaling the flux quantum $\Phi_0$ divided by an area corresponding, for a transverse field, to the length of the junction times twice the bulk penetration depth $\lambda$ (ignoring the actual junction thickness). For example, a 10 $\mu$m grain size assumed equal to the junction length, and $\lambda \simeq 0.1$ $\mu$m, give $H_0 \simeq 10$ Oe, in rough agreement with experiment. Petersen and Ekin,[84] Kwak et al.[82] and Zhao et al.[149] also suggested simulating the high-field tail by superimposing a distribution of junction sizes.

These ideas lead to a natural interpretation of the low-field hysteresis loop of ceramics. For example, Senoussi et al.[52,83] identify intergranular magnetic penetration with an effective lower critical field

$$H_{c1,i} \simeq 2\pi D J_{c,i}/c \ , \tag{6}$$

where $J_{c,i}$ is now the intergranular current density measured in a transport measurement. This is consistent with the more complete expression derived by Clem[144] using a picture analogous to Josephson penetration between the grains. Typical estimates give values below an Oersted, and indeed, in LaSrCuO ceramics, Raboutou et al.[8] and Civale et al.[120] have found $H_{c1,i}$ to be smaller than a few mOe!

Raboutou et al.[8] have suggested an alternative expression

$$H_{c1,i} \simeq (2\Phi_0 J_c/ca)^{1/2} \ , \tag{7}$$

which is presumably appropriate when the Josephson penetration depth

$$\lambda_J = [c\Phi_0/16\pi^2 \lambda J_c]^{1/2} \ , \tag{8}$$

is much larger than the grain size a. Here $\lambda$ is the London penetration depth, and junction thickness is neglected. For ceramics with 10 $\mu$m grains and $J_c$ of order 1000 A/cm$^2$, the Josephson penetration depth is comparable to the grain size. Eq. 7 also predicts very low values of the effective $H_{c1,i}$. The actual values for the effective lower critical field and intergrain coupling depend sensitively, of course, on the processing and density of the material.[165]

Once the applied field exceeds the characteristic Josephson diffraction field $H_0$ of Eq. 5, the grains are effectively decoupled, and reversibility is reestablished. This explains the disappearance of hysteresis at intermediate fields, as shown in the insert to Fig. 3. The detailed field dependence of intergranular flux penetration is a tricky issue because of possible frustration effects, to be discussed in IIIB. However Civale et al.[120] found a rather simple behavior: flux increases quadratically with field, as predicted by Eq. 3, suggesting that a kind of critical-state model can be applied to the intergranular material.

In summary, it would appear that the general picture of weakly coupled grains with Josephson links is well-established for ceramics, along with the idea of a "grain-decoupling field" above which **intra**granular effects dominate the hysteresis. In addition to evidence from the magnetic data and the related neutron depolarization data of Papoular and Collin[87], observation of flux interference effects in the transport properties of polycrystalline SQUIDs or patterned loops or even random ceramics as a function of field[2,191,210,214,218,280] demonstrate the Josephson weak-link behavior directly. Furthermore, studies by Chaudhari, Dimos et al.[25,29] on individual grain boundaries in thin films grown epitaxially on bicrystals have now confirmed directly that these boundaries act as barriers to current flow, and that they show Josephson interference effects. However it is still an issue whether these Josephson elements are tunnel junctions or point contacts.

IIB-4. Weak links in crystals?

There is yet another, still controversial and very important aspect to the magnetic $J_c$ data. This concerns the possibility of a weak-link hierarchy in these materials: even crystals or crystalline grains may be composed of smaller, "subgranular" structures. Twin boundaries, prevalent in bulk crystals of YBaCuO, with less than 1000 Å spacings in some cases, are a possible origin of this substructure, as will be discussed further in IIIB.

A natural incentive for considering such a possibility is the kind of discrepancy between transport and magnetic critical current densities observed in ceramics. Unfortunately, this kind of data is generally lacking at this time, simply because of the difficulty of electrically contacting small powders or of putting large enough currents through bulk crystals. Some data exist on epitaxial YBaCuO films by Oh et al.[219]

and Mannhart et al.[315] showing rough agreement in the absolute values of the magnetic and transport $J_c$'s at low temperatures.

Another incentive would be the failure of the size-scaling predicted in Eq. 4 for the Bean critical-state model. Experiments by Oh et al.[219] to subdivide the films confirmed the expected linear dependence of the remanence. But the grain-size dependence of powders has been less conclusive. Early results of Braginski[271] and more recent work of Shimizu and Ito[264] have shown remanence growing linearly with powder size as expected from the critical state model. But a number of other studies using a somewhat different technique (the Campbell method, see IIE-3) have tended to give an apparent $J_c$ increasing with decreasing grain size. These studies, by Daeumling et al.[224], Küpfer et al.[275], and Hibbs et al.,[281] point to a subgranular structure within the grains. Perhaps the most dramatic result was that of Küpfer et al.[167,275] who ground up a bulk crystal of YBaCuO and also found an increasing $J_c$.

Furthermore, Küpfer et al.[167,275] reported measurements of the effective $J_c$ as a function of dc field strength, showing the same features in ceramics, whether random or textured, **and** in a bulk crystal: $J_c$ first drops steeply with increasing field, but then saturates or even shows a secondary maximum at fields of order several Tesla. This maximum becomes more pronounced and shifts to lower fields with increasing temperature. Küpfer et al. interpreted this behavior in terms of a granular substructure of the crystals, suggesting that fields of order 1 Tesla were required to decouple the internal boundaries; at higher fields the characteristic $J_c$ of the subgrains dominated the magnetic response. Polturak et al.[329] also reported a related effect in measuring the penetration depth of YBaCuO powders of ever smaller dimensions. They also found evidence for twins on a scale of a few hundred Angstroms, which they suggested as the origin of the internal boundaries.

There is an apparent discrepancy here with the monotonic and only weak dependence of the remanence with field at low temperatures, in the hysteresis loops of Dinger et al.,[19] Kes,[174] and Crabtree et al.,[54] among others. The main concern in comparing these different and apparently contradictory results is the quality and control of the material. The starting crystal of the Küpfer experiment had a rather low $T_c \simeq 82K$ and a wide (8K) inductive transition. Similarly in the powder work, one must consider the possibility that as the grain size gets smaller, the surface-to-volume ratio increases and the surface defects increase the critical current.

In another recent study, Sulpice et al.[13] summarized various data in the literature, including their own, on the size-dependence of the low temperature remanence of bulk YBaCuO crystals. While there is inevitable scatter, the highest values for $H \parallel c$ increase roughly linearly with crystal dimension up to about 0.1 cm and then roll over to reach a size-independent limiting value of 2300 emu/cm$^3$ . Thus suggests that there are barriers or weak links on the scale of 0.1 cm which limit the spatial scale of supercurrents. This is an altogether different limit from that considered above, and the authors suggest, not twin boundaries, but regions of possible secondary decomposition of the off-stoichiometric crystals into regions of high and low oxygen concentration.

In summary, there is contradictory evidence on the size scaling of the magnetic remanence, some pointing to a Bean critical-state model, some to a subgranular picture. Uncertainties about the materials still cloud the results, and more conclusive experiments on high quality bulk crystals need to be done. Other experiments which bear on the crystal granularity issue will be summarized in III below.

IIB-5. Temperature, field and defect-dependent critical current

Assuming a homogeneous picture of the bulk crystal superconductors, what has been learned about the critical current density? A first result was the observation of large anisotropy in the magnetic $J_c$ along the principal axes of YBaCuO crystals, by Dinger et al.[19] and Crabtree et al.,[54] with values higher by more than an order of magnitude for $H \parallel c$ . Even more extreme differences are found in BiSrCaCuO crystals.[89,95,240] Recent results on YBaCuO crystals, however, show much smaller anisotropy (D. C. Cronemeyer and T. R. McGuire, private communication).

Another feature, first pointed out by Senoussi et al.[83] and verified by many others,[174,181,249,315] is that the magnetic $J_c$ decreases linearly and rapidly with increasing temperature, developing a long tail approaching $T_c$ . The data in some cases approximately fit an exponential form $J_c \propto \exp(-T/T_0)$ , where $T_0$ is a fitting parameter, typically 20 or 30 K in YBaCuO crystals.

An interesting example is shown in Fig. 5 from the work of Mannhart et al.,[315] who measured both the magnetic and the transport $J_c$ in the same epitaxial YBaCuO film by the following technique: After magnetic hysteresis measurements on large circular regions, the films was patterned by laser ablation into bridges tens of microns wide within single grains. With the large contact pads and the concentration of current into these bridges, it was possible to measure the critical current density throughout the temperature range. Measurements on bulk crystals or larger areas of epitaxial films have usually been limited by heating at the contacts; this is why there are surprisingly few studies of the full temperature dependence of the transport $J_c$ in the literature.[176,236] Results are still rather controversial; for example the reported $J_c(T)$ in epitaxial YBaCuO films near $T_c$ varies in one case as $(1 - T/T_c)^{3/2}$, in another as $(1 - T/T_c)^3$.

Fig. 5 reveals another remarkable effect, observed only in single-grain epitaxial films or crystals (and **not** in ceramics where the opposite trend occurs as discussed

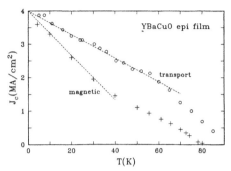

Fig. 5.    Magnetic (+'s) and transport (o's) critical currents as a function of temperature in the same YBaCuO epitaxial film on SrTiO₃ The transport data
were taken on a single large grain patterned into a microbridge by laser
ablation.  Dashed lines suggest a linear dependence at low temperature,
differing for the two techniques.  (After J. D. Mannhart, private communication. See also Ref. 315.)

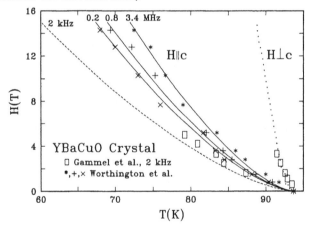

Fig. 6.    Irreversibility or melting lines in the H-T plane, in YBaCuO crystals. The
data with dots (H⊥c) and stars, crosses and x's (H∥c) are obtained from
the ac susceptibility anomaly after Worthington et al.[293]  For H⊥c, the
frequency dependence is weak, but for H∥c, there is a strong frequency
dependence indicated by data at 0.2, 0.8 and 3.4 MHz.  Originally interpreted as $H_{c2}$ , these data are now believed to indicate some kind of dynamical vortex phenomenon, involving the onset of irreversibility.  The
dashed line is an extrapolation of the data to 2 kHz using Eq. 27.  The
boxes are data from Gammel et al.[33] obtained from measurements with a
2 kHz mechanical oscillator and originally interpreted as a melting line.
The temperatures are shifted to match the $T_c$ of the susceptibility data.

above). While the low temperature values of the magnetic and transport $J_c$'s are in reasonable agreement, their temperature dependences seem different, with the initial slope with temperature being weaker in the transport. That is, the magnetic $J_c$ data lie **below** the transport data at finite temperature. Mannhart et al. compared the transport temperature dependence to a flux creep model, as will be discussed further in IIIC. Since the transport data are taken within a single grain, the difference with the magnetic data is presumably not due to grain-boundary weak links. In BiSrCaCuO crystals, van Dover et al.[227] also reported a systematic difference between magnetic and transport $J_c$, although in this case the low temperature $J_c(T)$ slopes were about the same, and the difference developed at higher temperatures.

Mannhart et al.[315] also measured $J_c(T)$ across single grain boundaries, using a technique to be described more fully in IIIA below. They found a flatter behavior at low temperatures, consistent with an earlier study of a patterned polycrystalline film with multiple grain boundaries by de Vries et al.[176] The transport $J_c$ temperature dependence has been compared[176,315] to predictions of weak-link SNS and SIS models,[335] such as the Ambegaokar-Baratoff expression for an SIS junction of normal state resistance $R_n$

$$J_c(T) = [\pi\Delta(T)/eR_n] \tanh [\Delta(T)/2kT] .  \qquad (9)$$

Here $\Delta(T)$ is the temperature-dependent superconducting energy gap.

The field dependence of $J_c$ also shows a significant fall-off in crystals, though much less than in ceramics.[217] Senoussi et al.[83] and Grover et al.[287] have suggested an exponential form in field, analogous to that in temperature. On the other hand, Atzmony et al.[246] have used a form $J_c \propto 1/(B + B_0)$, while Kes[174] found $1/\sqrt{B}$. Recently the latter dependence has been confirmed in the transport $J_c$ measurements of Mannhart et al.[315] By contrast, as mentioned in the previous section, Küpfer et al.[275] reported a high-field maximum in $J_c(H)$, in ceramics and also in crystals or individual grains. The proper form is far from clear at this time and may differ from crystal to crystal, depending on material quality.

Another classic way of studying the field-dependence of $J_c$ at fixed temperature is to plot the so-called "flux-pinning force" $F_p \propto HJ_c$. In crystals and epitaxial films, Kes[174] and Horie et al.[263] found an initial sharp rise in $F_p(H)$, then a maximum and a tail towards zero at $H_{c2}$. However Kes et al.[326] found that like other determinations of $H_{c2}$ (see IIE-2 and IIF-1), this method could be affected by flux creep.

Whether in the granular or flux pinning pictures, the issue of material quality and its correlation to $J_c$ is of great interest for applications. In classic low temperature superconductors like NbTi, optimization of defects was essential for achieving high critical current density because, as will be discussed in IIIC, $J_c$ is determined by the strength of the flux-line pinning. Some evidence for the role of defects in high-$T_c$

materials comes from the fact that much higher critical currents have generally been observed in epitaxial thin films than in bulk crystals, presumably because of interfacial or surface defects in the films. The huge literature on thin film $J_c$'s will not be reviewed here, but values of order $10^6$ A/cm$^2$ at 77 K are widely reported in epitaxial YBaCuO, and more recently[265] even in BiSrCaCuO and TlBaCaCuO films. Bulk crystal values at this temperature are much lower.[19,54,174] Polycrystalline films also have higher $J_c$'s than their ceramic counterparts, even in low-field transport measurements. One might have expected the transport $J_c$ to be limited by the grain boundaries, and diffusion along exposed thin-film grain boundaries to be higher than in ceramics, leading to weaker coupling. Many studies have been directed towards the problem of understanding the materials science of grain boundaries, but they will not be reviewed here.

More systematic studies are beginning to introduce controlled numbers of defects into the bulk. Using neutron irradiation on YBaCuO crystals, for example, Umezawa et al.[61] found a factor of two increase in the magnetic $J_c$ while $T_c$ was only decreased by 2K. Similarly, in YBaCuO ceramics, Wisniewski et al.[122,163,164] found an increase in the magnetic $J_c$ by a factor of almost 20 with no decrease in $T_c$ for a dose of up to $10^{17}$ n/cm$^2$, while Willis et al.[328] observed a factor of 3 increase. Wördenweber et al.[173] observed that chemical doping can also change the critical current density, though often simply because of changes in the microstructure of the ceramics. Schneemeyer et al.[64] found that different magnetic rare earths in the 123 crystals do not change $J_c$ significantly. Crabtree et al.[54] changed $J_c$ in YBaCuO crystals by as much as two orders of magnitude by annealing.

In summary, the peculiar field and temperature dependences of the bulk critical currents are a challenge for theory, and the work to identify the defects responsible for flux pinning and $J_c$ is just beginning.

IIC. Magnetic relaxation

One of the characteristic features of the magnetic behavior of these superconductors is their large and non-exponential time-relaxation. First reported by Müller et al.[16], this effect was studied extensively in LaBaCuO and LaSrCuO ceramics by Mota et al.,[53,147] who found the magnetic relaxation after application and then removal of a field to be approximately logarithmic in time t. The logarithmic slope dM/dln(t) increased roughly linearly with temperature at low fields, although at the lowest temperatures there appeared to be a finite (tunneling?) limit. It also increased as a high power - 3 or 4 - of the applied magnetic field.

Strong, roughly logarithmic, relaxation has also been observed in either the zero-field-cooled or remanent magnetization of YBaCuO ceramics[67,151,152,153,162,247] and thin films.[150] The temperature dependence of dM/dln(t) shows a peak at relatively low temperatures (20-30 K). As pointed out first by Tuominen et al.[67], even the relaxation $(1/M_0)dM/d\ln(t)$, normalized by the initial magnetization $M_0$, drops to zero approaching $T_c$. This has been a point of controversy because the simplest flux-creep model gives a monotonically increasing dependence, as will be discussed further in IIIC-1. Foldeaki et al.[151] and Yeshurun et al.[107] found that the field dependence of $dM/d\ln(t)$ also showed a maximum. Related observations of power law or logarithmic decay of trapped flux in ceramic YBaCuO tubes have been made by Tjukanov et al.[62] and Mohamed et al.[76]. Norling et al.[152] found in YBaCuO ceramics that at higher fields the relaxation deviates from logarithmic and is better described by a power law $t^{-a}$ in time t. Related torque measurements of Giovannella et al.[3] and Hagen et al.[156] indicated a Kohlrausch or stretched exponential dependence $\exp[-(t/t_0)^{\beta}]$ of the torque with time (see IID-1).

The field-cooled magnetization shows much less relaxation. However a small effect was detected by Norling et al.[152] and Yeshurun and Malozemoff,[28] indicating that the field-cooled magnetization (see IIA-3) may not be in equilibrium, as originally supposed.[16] This effect is important in the comparison of glassy and flux-pinning theories (IIIC).

Perhaps the large relaxation in ceramics was not surprising in view of the weak coupling between grains. It was more surprising when Yeshurun and Malozemoff[28,107,249] reported comparable relaxation in YBaCuO **crystals**, suggesting either that the crystals themselves were granular or that some other mechanism, like a giant flux creep, was occurring. They found that the temperature dependence of the zero-field-cooled $dM/d\ln(t)$ showed a peak around 30 K, with larger amplitude for field along the c-axis. Similarly, Tuominen et al.[157] confirmed a peak in $dM/d\ln(t)$ vs temperature, measured from the remanent moment. Yeshurun et al.[276] found similar effects in BiSrCaCuO 85 K crystals, with a peak at even lower temperatures and an even stronger anisotropy. That ceramics can show the superimposed relaxation effects of both intergranular and intragranular trapped flux at different time scales has been shown by Civale et al.[284] in careful studies of LaSrCuO systems with varying oxygen content.

A remarkable ageing effect in the magnetic relaxation has recently been observed by Rossel et al.[324] in an YBaCuO crystal. After cooling the sample in a field to about 75 K and waiting for a time $t_w$, they increased the field and watched the approximately logarithmic relaxation exhibit a step at a time $t_w$ **after** the field step. This "echo" tracked as a function of $t_w$. Such effects are well-known in the phenomenology of various kinds of glasses and give a compelling indication for glassy behavior in the new materials.

All these magnetic relaxation data, and the related torque relaxation (IID), will be a starting point in deciphering the magnetic behavior of the high-$T_c$ superconductors in Section III, since comparably large effects are not generally observed in low-temperature superconductors.

## IID. Torque magnetometry; motion and levitation in a field

### IID-1. Torque magnetometry

Torque magnetometry and a variety of other experiments relying on motion and levitation of the superconducting samples in a field, have added further evidence for the unusual magnetic effects such as time-logarithmic relaxation and an irreversibility or melting line.

Giovannella et al.[3] first reported magnetic torque measurements on ceramic LaSrCuO samples, showing a nonexponential time-decay of the torque after the sample was turned through some small angle. Depending on the field and temperature, they found either algebraic or stretched exponential (Kohlrausch) decays, much as in the closely related magnetization studies described in IIC above. Hagen et al.[156] also found Kohlrausch decays in YBaCuO ceramics and showed how how one can extract the critical current density from the torque data. Fruchter et al.[185,290-1] reported torque measurements on YBaCuO crystals and deduced the anisotropic $J_c$'s. Comparing with results from hysteresis loops, they found significant discrepancies which they associated with a distribution of pinning strengths (see IIIC). They also showed how torque measurements could detect flux penetration and thus determine the lower critical fields (see IVB).

Recently Farrell et al.[327] studied the torque on c-axis-aligned grains of YBaCuO and TlBaCaCuO as a function of angle in the ac plane. The skewed experimental curve fit a theory by Kogan,[104] based on the "intrinsic torque" arising from the anistropy in the Abrikosov vortex lattice.[91] The data fit determined the upper critical field anisotropy ratio $H_{c2}^{\perp}/H_{c2}^{\parallel} \simeq 5$, which will be compared to other anisotropy measurements in IVB below.

Related to the torque studies are the measurements of Wolfus et al.[75] in YBaCuO ceramics, in which the magnetization M was measured as a function of sample angle $\theta$ in a magnetic field. The irreversible part of the magnetization is expected to rotate as a rigid entity, giving a sinusoidal dependence of $M(\theta)$ measured along the original field direction. Wolfus et al.[75] observed the sine amplitude decreasing with increasing field strength. But at higher fields and temperatures, $M(\theta)$ becomes highly asymmetric, suggesting a glassy-like relaxation. Niobium, by contrast, shows a field-independent sinusoid up to a critical field above which plateaus develop, cutting off the sinusoid extrema. These results highlight the qualitatively different magnetic properties in low- and high-temperature superconductors.

IID-2.  Mechanical oscillator experiments

The vibrating reed experiment of Esquinazi et al.[196] is a probe involving motion of the superconducting sample in a magnetic field. The mechanical resonance frequency and damping are sensitive to flux penetration and pinning. Esquinazi et al. found that in LaSrCuO and EuBaCuO ceramics, low-temperature flux penetration began at about 80 or 200 Oe respectively. But in the related magnetic measurements of Civale et al.,[120] flux penetration was detected at far lower fields. This suggests that the direct magnetic measurement is sensitive to grain-boundary flux penetration, while the vibrating read detects the **bulk** penetration and thus can be used to measure the bulk $H_{c1}$. However, because of the large and special sample shape required, the method is not easily applicable to bulk crystals, which are needed to find the anisotropy of $H_{c1}$.

Gammel et al.[33] recently reported a related experiment probing higher fields. They mounted their YBaCuO and BiSrCaCuO crystals on a silicon mechanical oscillator excited capacitively into a resonant oscillation at about 2 kHz. In sufficiently large applied fields perpendicular to the direction of motion, they observed a small frequency shift and a peak in the dissipation as a function of temperature. Repeating the experiment in different fields, they determined lines in the H-T plane for the two orientations, as shown in Fig. 6 by the boxes. While the results for H$\perp$c approximately reproduced the $H_{c2}$ data reported earlier from ac susceptibility measurements[20] (see IIE), results for H$\parallel$c lay below the corresponding $H_{c2}$. The data fit a straight line lying roughly parallel to $H_{c2}(T)$ and extrapolating to a zero-field temperature a few degrees below $T_c$. Similar data on BiSrCaCuO extrapolated to 45 K below $T_c$.

Gammel et al.[33] interpreted their results in terms of flux lattice melting and a vortex liquid phase in the temperature range between their lines and $H_{c2}(T)$. But leaving interpretation aside, it is obviously of interest to establish the empirical relationship of this phenomenon to the dc (IIA-1) and ac (IIE-2) irreversibility lines, which do not form a straight line but curve smoothly to $T_c$ at zero field. Although the experiments have been done on different crystals, there appears to be remarkable coincidence between the results as we shall now see.

First it is necessary to recognize (see IIE-2) that what was initially thought to be $H_{c2}(T)$ in the ac susceptibility measurements is most likely also some kind of dynamical transition involving vortices. Secondly, this transition is frequency-dependent, so that to compare results, we need to extrapolate the ac susceptibility measurements, performed in the megaHertz range, to the 2 kHz of the mechanical oscillator experiment. As shown in Fig. 6, the H$\parallel$c data from the two techniques now overlap reasonably well. The frequency dependence is much weaker for H$\perp$c, explaining why no shift is needed to achieve agreement in this case. Furthermore, the mechanical oscillator experiment so far requires fields above about 1 Tesla; so the proposed linear extrapolation to zero field, which gives the main discrepancy with the ac irreversibility line, has not been confirmed. The BiSrCaCuO dc irreversibility lines reported by Yeshurun et al.[276] drop dramatically to lower temperatures with increasing field, rising to 1 Tesla at about 25 K, and again this temperature is consistent within a few degrees with the mechanical oscillator experiments.

In short, a plausible hypothesis is that the ac susceptibility and mechanical oscillator measurements are both observing the same irreversibility phenomenon. Its origin, whether melting, depinning or a superconducting-glass effect, is a separate question, to be considered in III below.

For completeness, we mention another mechanical measurement: the magnetic field dependence of the ultrasound velocity in ceramic LaSrCuO by Higgins et al.,[320] who observed a strong field-dependent stiffening of the sound velocity and thereby determined the irreversibility line in a novel way.

IID-3. Levitation and sample orientation

Several other effects involve translation or rotational forces on the superconductor. Perhaps the most widely known, especially in the popular media, is magnetic levitation. Typically, a small SmCo permanent magnet is placed on top of an YBaCuO ceramic disk. Cooled through the transition temperature to 77 K, the magnet rises up, usually a few millimeters off the disk. As discussed by Hellman et al.[244] and Williams and Matey,[225] the effect can be understood in terms of the Meissner expulsion of magnetic field from the sample. The resulting field gradient balances the magnet's gravitational force. The effect can also be understood in terms of a magnetic repulsion arising from the "image" of the magnet "reflected" in the superconducting surface.

In fact a quantitative treatment[244] requires consideration of the very significant flux penetration and trapping. The trapped flux gives rise to forces which hold the magnet in place: When the levitating magnet is jostled gently to one side, it tends to come back to its original position, an effect studied quantitatively by Williams and Matey[225] and related to the hysteresis loops by Brandt.[237] Similarly Moon et al.[226] found the vertical levitation force to be hysteretic in cycling the superconductor-magnet distance up and down. Trapped flux undoubtedly also explains the remarkable discovery by Huang et al.[283] and Harter et al.[235] that a magnet can even be suspended **beneath** the surface of the superconductor.

Single grains can be oriented in a magnetic field, both below and above $T_c$. Above $T_c$, Farrell et al.[55,73] aligned powders in a field of several Tesla. YBaCuO orients with its c-axis parallel to the applied field, because of the anisotropy in the normal-state susceptibility. Using the differing anisotropic susceptibilities of various magnetic rare earth ions, Arendt et al.[272] could obtain either a-axis or c-axis alignment. Aligned samples have played an important role in studies probing anisotropy in cases where bulk crystals have not been of adequate size. One particular example is the study of the reversible magnetization near $T_c$ to determine the true upper critical field[73,180] (see IIG below).

Solin et al.[23] magnetically aligned YBaCuO particles **below** $T_c$ , finding a preferred $c \perp H$ orientation and even an indication of some untwinned particles aligned with $b \parallel H$. Kogan[100] explained the $c \perp H$ tendency in terms of the anisotropic free energy of the superconductor in a field. Since $H_{c2}^{\perp}$ is larger than $H_{c2}^{\parallel}$ at a given temperature, and since superconductivity represents a state of lowered free energy, a sample in an intermediate field could only reduce its free energy by aligning with $c \perp H$.

The magnetic forces have also been exploited to screen superconducting from non-superconducting powders, and to identify the highest transition temperatures. Barsoum et al.[221] used levitation forces to separate powders on the basis of $T_c$ and size. Riley et al.[68] wiggled a magnet under such a mixture and saw particles oscillating in response to the field. By gradually raising the temperature, they claimed to be able to identify the highest $T_c$ particle, but their claim of 300 K superconductivity has not been verified.

IIE. Ac susceptibility

IIE-1. $\chi(T)$ in ceramics

Measurements of the real (inductive or $\chi'$) and imaginary (lossy or $\chi''$) components of the ac susceptibility have been among the most important and most common magnetic measurements of high-$T_c$ superconductors. An ac susceptibility-vs.-temperature measurement offers a quick identification of the critical temperature, and was used, for example, in the first discovery of superconductivity above 90 K in YBaCuO by Wu et al.[15]

This measurement is in some sense analogous to the zero-field-cooled dc measurement described earlier in IIA-1 because at any given temperature, the field is changed; so the signal is related to the flux penetration with field. A superconducting surface sheath is thus sufficient to give a signal corresponding to $\chi \simeq -1/4\pi$. This is why $\chi_a(T)$ measurements are so effective in screening samples for possible superconductivity and determining $T_c$. Conversely, they do not easily reveal whether the superconductivity is a bulk phenomenon or not.

Nevertheless, detailed ac susceptibility measurements can offer a great deal more, such as a determination of the flux penetration profile using the susceptibility change as a function of ac-field amplitude (Campbell's method[337]). Furthermore ac measurements have identified the frequency-dependent irreversibility line and have recently spurred a major re-evaluation of earlier upper critical field measurements.

$\chi'$ and $\chi''$, measured as a function of temperature in ceramic high Tc superconductors,[138,168,201,207,208,211,262,285] are complex. Typically, $\chi'$ shows upward steps as a function of increasing temperature, from values near $-1/4\pi$ at low temperatures to zero (ignoring the small normal state paramagnetic susceptibility) above $T_c$. At the same time, $\chi''$ shows either single or multiple peaks in the range where the $\chi'$ steps are steepest. For example, Goldfarb et al.[285] saw two peaks, associating the lower peak with intergranular and the upper peak with intragranular material in their ceramic YBaCuO samples, because powdering the sample caused the lower peak to disappear.

IIE-2. $\chi(T)$ in crystals: a reinterpretation of $H_{c2}$

Couach et al.[86,187] showed that oxygen concentration gradients can cause multiple $\chi''(T)$ peaks in YBaCuO crystals. However the best crystals exhibit only a single sharp peak, along with a single sharp step in $\chi'(T)$. For example, Worthington et al.[20,142,293] found features only 0.1 K wide, allowing precise definition of the characteristic temperature. Some of their measurements of this temperature as a function of field and frequency in an YBaCuO crystal are plotted in Fig. 6, showing an upward curvature which follows the power law $1 - T/T_c \propto H^{2/3}$ for H $\parallel$ c. Actually, the exponents vary with orientation, frequency and material, with values ranging from $1/2$ to $3/4$. These power laws have been found to hold over an impressively wide field range from rather small fields of only a few hundred Oe all the way to 20 T. This information will be important in IIIB-2 in setting the defect size-scale in the superconducting-glass and other models.

The frequency dependence of the transition was first recognized by Malozemoff, Worthington et al.[103,293] and is shown in Fig. 6 for an YBaCuO crystal. Measurements from 100 to $10^8$ Hz showed the transition line shifting to higher temperatures, roughly logarithmically with frequency. The frequency dependence $dT/d\ln(\omega)$ increases with field. A much stronger frequency shift has been observed in BiSrCaCuO crystals by Yeshurun et al.[276] and van den Berg et al.;[297] the temperature increases 30 K over four orders of magnitude in frequency at 1 kOe. This huge effect raises a warning flag for future studies of the BiSrCaCuO system: the frequency of the measurement must be taken into account to permit consistent results.

Further evidence for the unusual nature of the transition is its dependence on the amplitude $H_\omega$ of the ac field. In YBaCuO ceramics, Goldfarb et al.[285] observed a linear shift in temperature of the $\chi(T)$ anomaly with $H_\omega$ (although their initial inter-pretation of these anomalies as $H_{c1}$ does not seem plausible). Barbara et al.[12] also reported a downward temperature shift in an YBaCuO crystal, but with an $H_\omega^{2/3}$ dependence, crossing over to a stronger dependence of the same power above about 10 Oe. On the other hand, Kes et al.[296] saw no dependence in measurements on a BiSrCaCuO crystal. This ac-field dependence will be discussed further in IIE-3 and IIIC.

Before the discovery of the frequency dependence, the ac susceptibility anomalies were interpreted as a measure of the upper critical field $H_{c2}$. But since $H_{c2}$ is a thermodynamic transition with very little expected frequency dependence, the ob-servation of large logarithmic frequency shifts seems to exclude this interpretation. More likely we are dealing here with some sort of dynamic transition involving the vortices, which are known to contribute significantly to inductance and loss in superconductors. The nature of this transition will be discussed at length in Section III, and is a central question of this review. The ac susceptibility anomalies are also closely connected to the mechanical oscillator anomalies described in IID above, and to an activated tail in the resistivity-vs.-temperature curves in a field discussed in IIF below. Thus, all these studies are probably **not** measuring the upper critical field, a rather shocking suggestion in view of the huge literature on $H_{c2}$ which has accumu-lated in the last two years, and which will be reviewed further in IIF.

IIE-3. $\chi(H_\omega)$: Campbell's method

Another group of measurements exploiting the ac susceptibility are those based on Campbell's method.[337] The idea is simply to study the change in the diamagnetic response as a function of the amplitude $H_\omega$ of an ac field superimposed on a dc field $H_{dc}$. In the simplest critical-state picture, the response is determined by the change in flux penetration given by Eq. 3 and quadratic in $H_\omega$. From this equation, the de-rivative of the magnetization with respect to $H_\omega$ is proportional to $H_\omega/J_c$, which is in turn proportional to the depth of flux penetration. Campbell[337] showed that a plot of $H_\omega$ versus depth gives the flux density profile in the material, and the slope of this plot gives the local critical current density. This result is quite general for critical current densities varying as a function of depth, provided only that $H_\omega << H_{dc}$.

Daeumling et al.[154] applied this method to DyBaCuO ceramics and found that at low temperatures and above a small intergranular penetration field, the magnetic response was indeed quadratic $H_\omega$ and thus could be interpreted in terms of a simple critical state model within the grains. At 77 K, however, the bulk flux gradient was much smaller, and was masked by an extremely large surface contribution. A more surprising result was obtained by comparing grains of different sizes: The apparent critical currents increased with decreasing grain size. This has been discussed earlier in IIB-4; it gives evidence for the "subgranular" interpretation of the grains. Küpfer et al.[256] confirmed this kind of granularity in BiSrCaCuO and TlBaCaCuO ceramics.

For completeness, one other interesting measurement based on the ac susceptibility is mentioned here. A study of the low temperature tail of $\chi''$ in ceramics has been argued by Polturak and Fisher[59] to depend on the Josephson current of the grain boundaries. Assuming the Ambegaokar-Baratoff temperature dependence of Eq. 9 for $J_c$, they extracted the energy gap versus temperature from the ac loss, obtaining $2\Delta/kT$ of about 5, consistent with many direct tunneling measurements.

IIF. Resistivity versus temperature and field

IIF-1. Determination of $H_{c2}$?

While not directly a magnetic measurement, the study of the temperature dependent resistivity in a magnetic field has been the principal method for determining the upper critical field $H_{c2}$, and so is closely related to the ac susceptibility studies reviewed in IIE-1 above. Although a huge number of studies have been performed with this method, our conclusion here is perhaps the most unsettling in this review, namely that they are all most likely complicated by flux flow and so appear to be measurements of an irreversibility line, not of $H_{c2}$.

Early studies by Wu et al.,[15] Sun et al.[17] and many others on all the major families of oxide superconductors revealed a remarkable broadening of the resistivity as a function of temperature with increasing field. Initially there was the plausible suggestion that this effect came somehow from the anisotropy of the material and the random orientation of the grains[48], or from a hypothetical percolation effect across the grains.[149] But more detailed measurements, particularly on crystals of YBaCuO by Iye et al.[259] and Oda et al.,[85] and on crystals of 85 K BiSrCaCuO by Palstra et al.[96] and Juang et al.[99] showed that this broadening persists and is characteristically anisotropic, with greater broadening for a field applied along the c-axis. An example of the data of Iye et al.[259] is shown in Fig. 7.

In conventional homogeneous superconductors, a magnetic field shifts the resistive transition to lower temperatures, with the attendant broadening usually less than the shift; so the upper critical field remains well defined. In this context, the peculiar broadening in the high-$T_c$ superconductors has been a puzzle. Uncertainty about where to define the temperature corresponding to $H_{c2}$ has led to a variety of criteria, but usually the point chosen is where the resistivity drops to some small

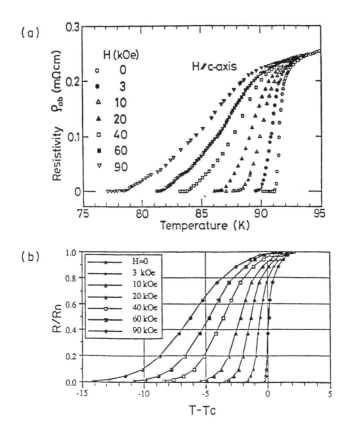

Fig. 7.    a) Resistive transition as a function of temperature in various fields, from Iye et al.[259]  b) Theoretical prediction based on Eq. 28, using a flux flow model by Tinkham.[31]  The broadening in $\rho(T)$ is suggested to be a flux flow phenomenon, which raises questions about studies using the foot in the resistive transition to determine $H_{c2}$.  (After Tinkham, Ref. 31.)

fraction, say 10%, of its normal state value, or where it drops to the sensitivity level of the measurement apparatus.

An incredible number of papers have appeared over the last two years claiming to measure $H_{c2}$ on this basis in all the major materials families: LaBaCuO and LaSrCuO ceramic[15,36,37,38,109–111,116,143,216] and crystal,[143] YBaCuO and REBaCuO ceramic[4,7,17,18,41,49,58,190,204,269] and crystal,[65,69,78,85,112,177–8,181–2,258–60] BiSrCaCuO ceramic[101,137,170] and crystal,[96,99,219,316] and TlBaCaCuO crystal.[241] Here, polycrystalline films are referenced along with ceramics and epitaxial films along with crystals.

Accepting for a moment the initial interpretation in terms of $H_{c2}$, the results generally suggest very high slopes of $H_{c2}(T)$ near $T_c$. But the reported field dependences are inconsistent, sometimes linear, sometimes curved according to $(1 - T/T_c) \propto H^{2/3}$, much as in the dc and ac susceptibility data. Exceedingly high values for the low temperature upper critical fields have been extrapolated using classic formulas, with values over 1 MOe being common for $H \perp c$. Using pulsed magnetic fields, Tajima et al.[79] and Yamagishi et al.[178] have succeeded in tracking $H_{c2} \| c$ all the way down to low temperatures in EuBaCuO crystals; they found $H_{c2}(0) = 27.5 \pm 2.5$ T at 4.2 K. They noted that their values for $dH_{c2}/dT$ near $T_c$ in this orientation differed considerably from those of other work[258–260,267] and attributed this difference to the different normal-state resistivities. Boon et al.[195] also observed peculiar variations in $H_{c2}(T)$ with annealing in YBaCuO ceramics.

The anisotropy has been of particular interest. Reported critical field slopes for $H \perp c$ invariably exceed those for $H \| c$, with typical ratios of 6-8 for YBaCuO, 15 for BiSrCaCuO(85 K), and over 70 for TlBaCaCuO (100 K). Typical values for YBaCuO are $dH_{c2}/dT \simeq 4$ T/K for $H \perp c$ and $\simeq 0.5$ T/K for $H \| c$. Similarly for BiSrCaCuO (85 K), $dH_{c2}/dT \simeq 8$ T/K for $H \perp c$ and $\simeq 0.5$ T/K for $H \| c$. Values as high as $dH_{c2}/dT \simeq 70$ T/K have been reported by Kang et al.[241] for TlBaCaCuO oriented films with $H \perp c$. But these values wander from study to study, by a factor of at least 2 and the results depend strongly on the criterion used for defining the point of zero resistance.

These troubling inconsistencies have been highlighted by recent more careful studies of the small tail which typically appears at the bottom of the $\rho(T)$ curves. Palstra et al.[32] found that the resistivity in a BiSrCaCuO crystal could be fit to the form $\rho(T,H) = \rho_0 \exp [-U(H)/kT]$, suggesting activated behavior. The likely culprits here are the flux lines pinned at defects in the material, with the resistive tail representing a flux flow resistivity. This hypothesis has been given added credence by Tinkham,[31] who reproduced the general form of the resistivity broadening in field, as shown in Fig. 7, using a flux flow model to be discussed further in IIIC.

The problems in determining the point of zero resistivity and the likelihood that flux flow dominates the resistive tail suggests using a different criterion to identify $H_{c2}$. Some authors[69,78] used the point where the resistance drops to only 90% of its normal state value. As is obvious from Fig. 7, this criterion significantly increases the apparent $dH_{c2}/dT$. But questions remain: why 90% and not 50% or 100%? Moodera et al.[63] used the field dependence of the onset, or even the field dependence of the fluctuation conductivity fit to the Aslamasov-Larkin form, but this is also of questionable accuracy, especially when applied to highly anisotropic randomly oriented crystallites in a ceramic.

## IIF-2. Nonlinearities and hysteresis effects

For completeness several other interesting transport phenomena involving magnetic effects are mentioned here. At the superconducting transition, Dubson et al.,[24] using YBaCuO ceramics, found strong nonlinearities in the I-V characteristics. Meszaros et al.[206] found these nonlinearities to be sensitive to applied fields and frozen-in flux. It is tempting to interpret these effects in terms of the granular nature of the material. Surprisingly, Yeh and Tsuei[321] found similar effects, though on a much narrower temperature scale, in an YBaCuO crystal. Many interpretations have been given for these nonlinearities, which show an increasingly high power $\alpha$ ($V \propto I^{\alpha}$) with decreasing temperature. These include Kosterlitz-Thouless behavior and superconducting-glass phenomena. Flux creep should also be considered because of the nonlinearities to be discussed in IIIC below. This is a new topic which is just beginning to be explored.

Various authors[11,118,206] reported field and temperature hysteresis in the resistivity of YBaCuO ceramics. Carolan et al.[118] in particular have used this hysteresis to define an irreversibility line, obtaining the classic 2/3 law (Eq. 1).

Another test of the flux-flow picture is the effect of magnetic field orientation of the critical current. Since the force on the vortices is $J_c \times B$, the force should be weaker and flux flow resistivity smaller when field is applied parallel rather than perpendicular to the current flow. This has indeed been confirmed by Wördenweber et al.[173] Another possible flux-flow effect has been observed by Kapustin et al.[308] in pulsed magnetic field measurements on LaSrCuO ceramic. The resistivity shows a peculiar minimum at the middle of the magnetic field pulse; the surrounding maxima are proposed to be generated by flux-flow resistivity.

## IIG. Reversible dc magnetization

## IIG-1. Determination of the thermodynamic $H_{c2}$

Finnemore et al.[34] first emphasized that, because the magnetic data become reversible near $T_c$ or $H_{c2}$, study of the reversible magnetization as a function of field and temperature in this region permits, in principle, a determination of the thermodynamic free energy and the upper critical field. Reversibility can also be achieved at sufficiently low fields throughout the temperature range; this is the region of London penetration at the surface, and is of interest for determining the size and temperature dependence of the penetration depth.

First we discuss the high-field, high-temperature region. According to the Abrikosov theory,[335] the reversible magnetization near $T_c$ goes as

$$\frac{4\pi M}{H} = -\frac{1}{(2\kappa^2 - 1)\beta_A}[\frac{H_{c2}}{H} - 1] , \qquad (10)$$

where $\kappa$ is the ratio of the penetration depth to the coherence length and $\beta_A$ is a number of order 1. Since $H_{c2}$ is linear in reduced temperature, Eq. 10 predicts the magnetization versus temperature or field to be linear and to go to zero at $H_{c2}$. For large $\kappa$, most certainly appropriate for these materials, the slope is exceedingly low. This is the essential challenge in the magnetic measurement: to achieve a high enough signal-to-noise ratio to pick out a break from small slope to zero slope. Bulk crystals have so far not been big enough to give conclusive results.

The many such measurements[58,104,115,117,118,124,125,198] on ceramics have given slopes of $dH_{c2}/dT$ ranging from 2 to 5 T/K, but the meaningfulness of these results is compromised by the overlapping contributions of the anisotropic randomly oriented grains. Early results on magnetically aligned powders of YBaCuO and HoBaCuO (see IID-3) were obtained by Farrell et al.,[55] Fang et al.[73] and Schwartzkopf et al.[180] While the low values of Farrell et al. have not been reproduced, the results of the latter two papers show a quite unexpected behavior, $H_{c2}^{\parallel} \propto (1 - T/T_c)^{1/2}$. This square-root rather than linear form was initially interpreted in terms of twin-boundaries limiting the spacial extent of the vortices in one direction and causing the standard Ginzburg-Landau expression $H_{c2} = \Phi_0/2\pi\xi^2$ to be replaced by $H_{c2} \propto \Phi_0/\xi D$ where D is the spacing between the twins. We would then have only a single power of $\xi$ in the denominator, and the conventional Ginzburg-Landau dependence $\xi \propto (1 - T/T_c)^{-1/2}$ would lead to the observed dependence of $H_{c2}$.

More recently, Athreya, Fang et al.[317-8] found that within experimental error, $H_{c2}^{\parallel}(T)$ rises vertically in the H-T plane. This result cannot easily be interpreted in the context of standard Ginzburg-Landau theory, especially in view of the relatively modest values (27 T) for the low temperature critical field $H_{c2}^{\parallel}$ determined by Tajima, Yamagishi et al.[79,178]

Athreya, Fang et al.[317,318] have investigated the thermodynamic properties near $T_c$ in more detail. A Gibbs free energy difference can be derived by integrating the moment at a given temperature T over field

$$G_0 - G_H = \int_0^H M(H')dH' ,$$

and can be related to the difference of the zero-field and field-dependent specific heat according to

$$C_0 - C_H = - Td^2(G_0 - G_H)/dT^2 .$$

These authors carried out this procedure for both YBaCuO and TlBaCaCuO aligned powders and found agreement with the corresponding specific heat measurements of Salamon et al.[88] and Fisher et al.[319] But the results remain most peculiar. The specific heat broadens rapidly with field, with the upper end of the anomaly apparently not moving downward with field. This is another expression of the apparently almost vertical $H_{c2}(T)$ .

In summary, because of the flux creep and flux flow problems in the ac susceptibility (IIE-2) and resistivity (IIF-1) measurements of $H_{c2}$, the reversible magnetization measurements would appear to be preferable. Since measurements on bulk crystals have not yet shown adequate signal-to-noise ratio to give convincing results, the data on aligned particles are the best we now have for $H_{c2}$, and the observed vertical slope has so far defied interpretation. This is clearly one of the major problems for the future and a fundamental issue in establishing the mechanism of high temperature superconductivity.

IIG-2. Low-field measurements of the penetration depth

Interpreting low-field reversible magnetization data for ceramic YBaCuO, Farrell et al.[43] assumed London penetration of the grains and, with a simple "two-fluid" dependence $\lambda(T) = \lambda(0)[1 - (T/T_c)^4]^{-1/2}$, extracted an angle-averaged $\lambda(0)$ of 1000-2000 Å. More detailed studies of YBaCuO powders by Cooper et al.[71] and Monod et al.,[192] as well as related microwave work on films, for example by Drabeck et al.,[322] have given comparable $\lambda(0)$ but have revealed a $T^2$ contribution to $\lambda(T)$ . More recent measurements by Krusin-Elbaum et al.[314] on YBaCuO crystal platelets rule out the $T^2$ effect and seem consistent with conventional singlet BCS pairing, as will be discussed in IVC below.

IIH. Magnetic decoration.

Magnetic decoration experiments have been essential in understanding the magnetic properties of the high temperature superconductors because, in contrast to all the other macroscopic measurements we have discussed so far, they give a direct picture of the vortex distribution. This has settled many controversial issues, including 1) determining the size of the flux quantum in these materials to be hc/2e, 2) demonstrating that twin boundaries pin flux and that these boundaries, like grain boundaries, are regions of weakened - rather than enhanced - superconductivity, 3) indicating the presence of yet another, more local, mechanism for flux pinning, 4) ruling out a superconducting-glass picture in crystals at low temperatures on certain size scales, and 5) suggesting the possibility of a new kind of vortex glass behavior.

Typically the decoration has been accomplished by evaporating iron or nickel particles onto the superconductor cooled to low temperature. Delicate control of the filament heating and ambient gas pressure is required to obtain the required particle sizes, which must be magnetic and single-domained. The particles are attracted to the locations of strongest field just above the vortices. After decoration at low temperature, the sample is removed from the cryostat and examined with scanning electron microscopy, or, for sufficiently dilute (low-field) arrays, with optical microscopy.

The first experiments by Gammel et al.[22] and Vinnikov et al.,[135,215,304] convincingly revealed the vortex array for H∥c at 4.2 K. A typical optical micrograph is shown in Fig. 8 from the work of Vinnikov et al.[215] for an YBaCuO crystal cooled in 10 Oe with H∥c. The micrograph shows mounds of magnetic particles which identify the vortex locations. With crossed polarizers, the optical contrast is sensitive to twin boundaries which lie perpendicular to the ab-plane and which can be correlated with the vortex positions.

Attempts to decorate at 77 K have been unsuccessful, possibly because the vortices are "melted" at this temperature or because the growth of the penetration depth with temperature makes the vortices magnetically less well-defined. Recently, experiments on BiSrCaCuO crystals at 15 K showed worm-shaped decoration patterns, suggesting vortex motion, while YBaCuO decorations at 15 K remained round (D. Bishop, private communication). This low temperature for BiSrCaCuO flux motion recalls the low temperature of the irreversibility line in this material (IID-2). For unknown reasons, the decorations have been unsuccessful in the interesting H⊥c direction (private communication, L. Ya. Vinnikov), which should show an anisotropic vortex lattice[91,171,174] and give direct information about the anisotropy of the penetration depth.

Fig. 8 reveals several important effects. First, the local configuration of vortices is on average hexagonal, as predicted by theory.[335] Secondly, in these experiments the samples were cooled in fields between a few and several hundred Oersteds, which is sufficient to trap most of the flux (see IIA-3). Under these conditions, the average spacing scales with applied field B according to $\sqrt{\Phi_0/B}$ , where $\Phi_0$ is the standard flux quantum hc/2e $= 2 \times 10^{-7}$ Gcm$^2$ determined by electron **pairing** This was not

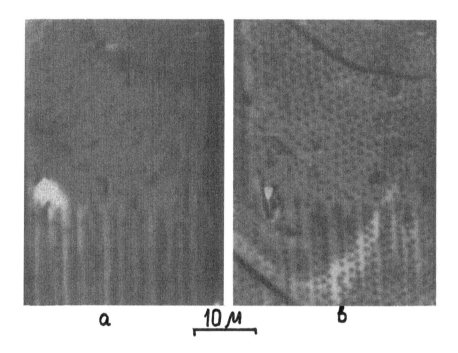

Fig. 8. Magnetic particle decoration of an YBaCuO crystal ab-face, observed in optical microscopy a) with and b) without polarized light. In a) the twin structure is visible, showing a heavily twinned region and a relatively twin-free region. In b) the corresponding vortex array is observed, showing clear trapping at the twins, as well as disorder even in the twin-free regions suggesting some other pinning mechanism. The sample was cooled in 10 Oe. (After Vinnikov et al., Ref. 215, 304.)

a trivial point at the time of the first work because of some novel theories involving nonstandard values for the flux quantum.

Another feature of Fig. 8 is that the vortex array lacks long range order, even in twin-free regions. This implies the existence of some highly dense and localized pinning centers, invisible in the optical micrograph. It also suggests the idea of a vortex glass[313] (see IIID). More recent work of Dolan et al.[323] has shown more perfect arrays, both in YBaCuO and BiSrCaCuO crystals, but lack of response to low-temperature field changes confirms that even these very perfect arrays are strongly pinned at low temperature. A possible origin for this pinning, suggested by Tinkham,[277] is oxygen nonstoichiometry or some other atomistic chemical effect. That pinning at this atomistic level is conceivable follows from the extremely short coherence lengths, as will be discussed further in IIIA. In fact, as suggested by D. Pines (private communication), intrinsic pinning arising from the discreteness of the atomic lattice is also a possibility.

Fig. 8 also shows vortex pinning by twin boundaries. There is still ambiguity as to whether or not the vortices can move easily **along** a twin boundary, but at least it is clear from this figure and others that vortex motion perpendicular to the boundary is constrained. The more recent work of Dolan et al.[323] also shows vortices hanging on to twins as they penetrate the otherwise vortex-free regions of the sample at the edge. This, coupled with observations like Fig. 8, is strong evidence for the fact that twin boundaries are planes of weakened rather than enhanced superconductivity because vortices are attracted to weakened regions. This conclusion is of considerable importance, since there had been evidence in low temperature superconductors for twin-enhanced superconductivity, and various proposals have been made[23,73,254,307] for such enhancement in the new materials.

Thus there appears to be a hierarchy of pinning mechanisms in these superconductors, one on a microscopic, almost atomic scale, and at least one other - the twin boundaries - on a more macroscopic scale. Magnetic measurements average over all these effects, and initial interpretations have by and large, and probably mistakenly, assumed only a single pinning mechanism.

Another important implication of these studies is the **lack** of evidence for structures of the type expected in the superconducting-glass model, that is, of supercurrents and their associated fields fixed by the position of weak links in the sample, as will be discussed in IIIB below. The behavior of ceramics is of particular interest in this connection, since here there is more concrete evidence for the importance of weak links. Initial efforts to decorate ceramics, however, have not been entirely convincing.[66,127,222]

II-I. Muon spin rotation

In a muon spin rotation ($\mu$SR) experiment, spin-polarized positive muons are implanted into a sample, with a magnetic field typically applied perpendicular to the initial muon polarization direction. In the local magnetic field, the muon spins precess at the Larmor frequency 135.5 MHz/T ($=\gamma_\mu/2\pi$, where $\gamma_\mu$ is the gyromagnetic ratio of the muon). The precession is detected via positrons, which are emitted preferentially along the muon spin direction when the muon decays after a mean lifetime of 2.2 $\mu$ sec.

Thus the measured precession frequency permits a determination of the local magnetic field B at the muon, and the depolarization rate gives a measure of the local field distribution.[9] In particular, for a Gaussian time-relaxation function $\exp(-\sigma^2 t^2)$, the second moment of the field distribution is given by

$$[\Delta B^2]_{ave} = 2\sigma^2/\gamma_\mu^2 = (H\Phi_0/4\pi\lambda^2)/[1 + (4\pi^2\lambda^2 H/\Phi_0)] , \qquad (11)$$

where H, the applied field, is assumed to lie between $H_{c1}$ and $H_{c2}$ with a square Abrikosov lattice. While a complete review of muon data is beyond the scope of this article, we include those results which bear on the internal field distribution arising from the vortex array.

Both the muon precession frequency $\nu_\mu$ and the depolarization rate $\sigma$ reveal features analogous to the magnetization measurements, as first reported for ceramic LaSrCuO by Aeppli et al.[39] and Kossler et al.,[40] and as studied more completely on ceramic YBaCuO by Keller et al.[148] In 20 mT experiments on ceramic YBaCuO, $\nu_\mu$ is low in the zero-field-cooled state, as expected if flux is excluded from the samples, and $\sigma$ is large, suggesting a non-uniform flux penetration near the edges of the grains. With increasing temperature, $\nu_\mu$ climbs and $\sigma$ drops as flux penetrates and spreads more uniformly through the sample, and also as the penetration depth increases, smoothing out the local variations in the flux profile. Finally, above $T_c$ these quantities become constant. Field cooling causes only a slight decrease of $\nu_\mu$ and only a slight increase in $\sigma$, exactly analogous to the magnetization data. Indeed, one can also identify an irreversibility temperature which coincides with an earlier resistive determination[118] and gives a $(1 - T/T_c) \propto H^{2/3}$ dependence. The depolarization rate even tracks the magnetization hysteresis loop, though with an as yet unexplained offset.[148]

Perhaps the principal focus of muon measurements on the new superconductors has been the determination of the temperature-dependent penetration depth. Gygax et al.,[5] Kossler et al.[40] and Aeppli et al.,[39] working with ceramic LaSrCuO, Harshman et al.,[47] working with ceramic YBaCuO, and Uemura et al.,[90] working with oxygen-depleted YBaCuO, all used Eq. 11 to analyze the $\mu$SR relaxation rate versus temperature to extract $\lambda(T)$. They found an angle-averaged $\lambda(0)$ of about 2000 and 1400 Å respectively in the LaSrCuO and YBaCuO systems respectively, with a temperature dependence consistent with a BCS singlet pairing in the two-fluid model (see IVC).  Cooke et al.[80] found 1550 to 1900 in two ceramic GdBaCuO samples, also with a two-fluid BCS-type temperature dependence.  In an oriented specimen, Schenk[175] reported $\lambda_c$ of 6000 Å or greater. There are many assumptions in this treatment, such as ignoring local demagnetizing fields and averaging over different orientations and anisotropic flux line lattices. But the principal result has recently been confirmed by the first measurements on a single crystal by Harshman et al.[325]  They found a conventional singlet temperature dependence for $\lambda(T)$ with $\lambda_{ab} = 1400$Å and $\lambda_c = 7000$Å.

There is still considerable controversy about the reliability of the muon determination of $\lambda$. Brandt[74] has pointed out that the distribution of local fields in a vortex array is not Gaussian, as assumed in Eq. 11, but asymmetric. Furthermore, pinning shifts the vortex positions (see Fig. 8), which adds to the measured muon relaxation rate. This extrinsic contribution could mask the intrinsic thermal contribution in the determination of $\lambda(T)$ . Barford and Gunn[172] showed that for H⊥c, one must take the anisotropy of the vortex lattice into account. Thus direct magnetic measurements of $\lambda(T)$ (IIG-2) have been important in confirming the muon results, as will be discussed in IVC.

## III. INTERPRETATION

### IIIA. General remarks

The underlying reason for many of the unusual magnetic properties of the high temperature superconducting oxides, as emphasized particularly by Deutscher and Müller,[21,140] is surely the unusually small coherence length $\xi$. Why it is so small is suggested by the clean-limit BCS formula

$$\xi = \hbar v_F / \pi \Delta , \qquad (12)$$

where $v_F$ is the Fermi velocity and $2\Delta$ is the superconducting energy gap. $v_F$ is expected to be smaller than in typical low temperature superconductors, because in the simplest picture the carriers are presumably just the small number of holes determined by chemical valence considerations. On the other hand the gap, measured for example by tunneling or far-infrared absorption, appears to be considerably larger. Rough estimates of $\xi$ give values of order tens of Angstroms, comparable to the size of the unit cell in these complex oxides. As will be discussed in IVA, there is at present no reliable experimental determination of $\xi$, although upper limits can be set and they are surely in this range.

The small coherence length makes possible suppression of the superconductivity over exceedingly short distances.  For example, a planar defect such as a grain boundary or even a twin boundary, which in a conventional low-temperature superconductor would act merely as a scattering center, might now form a Josephson barrier.

The existence of such Josephson elements at grain boundaries has been confirmed, first by Esteve et al.[2] and Koch et al.,[218] through the observation of Josephson interference effects in SQUIDs made of polycrystalline YBaCuO films. An even more direct measurement was made by Dimos et al.,[29] who grew YBaCuO films on polished faces of bicrystals formed by sintering two $SrTiO_3$ crystals together at fixed angles. The epitaxial film reproduced the grain boundary of the underlying substrate. Laser-ablation patterning of microbridges across the boundary or inside a single grain permitted direct electrical characterization of the boundary in comparison to the surrounding grains. The boundary $J_c$ was found to be an order of magnitude or two lower than in the grains, confirming directly that the boundary was acting as a barrier to electrical transport. Furthermore, a SQUID made of two such microbridges across one boundary exhibited the periodic dependence of voltage on applied field characteristic of Josephson interference.

While the evidence for Josephson elements at grain boundaries is overwhelming, the possibility of Josephson effects within bulk crystals is less clear. It is useful in this context to summarize a number of results, all found in the best crystals and reviewed in the previous sections:

1. an irreversibility line $(1 - T/T_c) \propto H^{2/3}$ from dc M(T) data,

2. a field-cooled "Meissner fraction" growing with decreasing field,

3. a remanent moment equal to the difference of field-cooled and zero-field-cooled moments at low temperatures and fields,

4. a roughly rectangular high field hysteresis loop at low temperature,

5. a critical current decreasing linearly with temperature at low temperatures,

6. a magnetically determined transport current lying **below** the transport critical current at finite temperatures,

7. a large time-logarithmic magnetic relaxation of zero-field-cooled and remanent magnetizations, with the logarithmic relaxation rate showing a peak as a function of both temperature and field,

8. an ageing effect in the magnetic relaxation at high temperatures,

9. a frequency-dependent ac susceptibility anomaly tracking as $(1 - T/T_c) \propto H^{2/3}$ up to 20 T, with the temperature shifting logarithmically upward with increasing frequency,

10. a well-defined vortex lattice with a density proportional to applied field and exhibiting pinning.

Since many of these properties are unconventional in the context of low temperature superconductivity, it is natural to ask, as Müller et al.[16] did in their seminal paper, could these properties arise somehow from the unique propensity of the new materials to form weak links? This suggestion is reinforced by the obvious similarity in the properties of crystals to those of ceramics, where the weak-link interpretation is already well established. A disordered array of weak links coupling together small superconducting grains is the basis of the "superconducting-glass" model, developed principally by Ebner and Stroud[332] and gives, as we shall see in IIIB, an appealing explanation of many of the experimental results.

But what is the physical origin of the weak links? While twin boundaries have been a candidate, they cannot explain the similar behavior of twin-free YBaCuO crystals and of the BiSrCaCuO and TlBaCaCuO systems, which also have low twin density. This has led a number of other authors, including Scheidt et al.,[169] Ravi Kumar and Chaddah[288] and Yeshurun, Malozemoff et al.[28,279] to suggest an alternative picture based on more conventional flux pinning. Yeshurun and Malozemoff emphasized that the key new ingredient which might explain the unconventional behaviors of the new superconductors was the combination of high temperatures and low pinning energies stemming from the low coherence length, leading to an unusually high level of thermally activated flux creep and flux flow. Related proposals by Nelson,[27] Gammel et al.,[33] Houghton and Moore[26] and by now many others have involved vortex-lattice melting, and Fisher[313] has recently suggested a vortex-glass phase transition.

In the next sections, we will briefly outline the main ideas of these different pictures with a view to answering the following questions: How well do they explain the experimental results? To what extent are they complementary, or are they just alternative expressions of the same underlying physics? A recent review by Tinkham and Lobb[278] has in fact taken this point of view, namely that many physical phenomena, such as upper and lower critical fields, can be interpreted equally well in either picture.

As emphasized in Section I, the answer to these questions is basic to our understanding of high-temperature superconductivity, because of the importance of the upper and lower critical fields, and of the coherence lengths and penetration depths which we deduce from them. We must know whether these parameters characterize a twin boundary or some other planar defect, or whether we are measuring the intrinsic properties of a true bulk superconductor. In spite of the very earliest papers claiming single-phase material, the bulk nature of superconductivity in these materials has remained very much a live issue.

IIIB. The Ebner-Stroud superconducting-glass model

IIIB-1. The model and its underlying physics

While there has been a rich literature on random and ordered arrays of super-conducting grains coupled by weak links,[306,338] perhaps the most accessible treatment for the experimentalist has been the work of Clem[144] and Tinkham and Lobb[278] on the low-field properties, and of Ebner and Stroud,[146,332] extended by Morgenstern et al.[145,253] and Schneider et al.,[257] for the non-linear properties.

We start with a brief review of the model,[332] which consists of a system of superconducting grains, usually taken to be small compared to the London penetration depth, and described by a complex order parameter with phase $\phi_i$ for the ith grain. The grains i and j are coupled by an energy $J_{ij}$, with the Hamiltonian

$$\mathcal{H} = -\Sigma_{ij} J_{ij} \cos(\phi_i - \phi_j - A_{ij}) , \qquad (13)$$

where the phase factor or "gauge field" $A_{ij}$ is determined by the vector potential A according to

$$A_{ij} = (2\pi/\Phi_0) \int_i^j A \cdot dl , \qquad (14)$$

and where $\Phi_0$ is the flux quantum hc/2e.

Eq. 13 leads to a Josephson-like current between grains i and j

$$I_{ij} = (2e J_{ij}/h) \sin(\phi_i - \phi_j - A_{ij}) , \qquad (15)$$

and a cluster moment

$$\mu = (1/2c) \Sigma_{ij} X_{ij} \times I_{ij} x_{ij} , \qquad (16)$$

where $X_{ij}$ is the vector joining the origin to the midpoint between grains i and j and where $x_{ij}$ is the vector distance from grain i to grain j. In essence, this moment is the product of the current times the area of the loop linking a set of neighboring grains.

In the low-field or high-temperature limit, Tinkham and Lobb[278] pointed out the formal equivalence between this granular model and a continuum model based on standard Type II superconductivity with defects pinning flux. An effective London penetration at the surface can be interpreted as coming from Josephson penetration along the weak links, and an effective lower critical field can be defined (Eqs. 6-8). Intergranular vortices can be formed, with their field distributions spanning many grains, as discussed also by Clem[144] and Sonin.[305]

The effective area occupied by a flux quantum is given roughly by

$$S_{eff} \simeq \Phi_0/H . \qquad (17)$$

Thus at sufficiently low fields, the area will be large compared to the average individual loop size S, and the mean-field Ginzburg-Landau limit should apply. However there remain questions about random arrays because of the possibility that frustration could cause the effective lower critical field to go to zero.[274,332] But surely, as the field increases to the point where the effective area approaches the local loop size, nonlinearities in the magnetic behavior and thus significant deviations from Ginzburg-Landau theory can occur.

This is the aspect treated in the superconducting-glass model. Ebner and Stroud[332] obtained insight into the predictions of this model by considering a single loop of N grains enclosing an area S, connected by equal coupling energies J. The gauge-field phase factors $A_{ij}$ between two grains, given by $2\pi BS/N\Phi_0$, determine the current I according to Eq. 15 and the moment IS/c from Eq. 16. There are multiple states for the system, corresponding to N different choices of the phase difference $\phi_{ij} = 2\pi m/N$ between two neighboring grains, where $0 < m \leq N$ is an integer. This is the physical origin of metastability in this model.

In view of this multiplicity of possible states, whenever B is near a multiple of $\Phi_0/S$, the system can achieve lowest energy by adjusting the phase difference $\phi_{ij}$ to compensate for $A_{ij}$. The ground state for this system (see Fig. 1 of Ref. 10) thus corresponds, according to Eq. 15, to a zig-zag current pattern as a function of field, with a simple linear dependence at the lowest field, passing through zero, and with the first break from linearity occuring at $\Phi_0/2S$. Between these states, however, are energy barriers; so a zero-temperature field-excursion causes the current to follow the full sinewave of Eq. 15 beyond the nonlinearity field $\Phi_0/2S$.

This simple picture now provides an explanation for the difference between field-cooled and zero-field-cooled moments, for once the field is above $\Phi_0/2S$, the field-cooled or equilibrium moment reverses in direction, while the zero-field-cooled moment continues to rise (in a negative or diamagnetic sense). This means that the zero-field-cooled diamagnetic susceptibility remains constant while the field-cooled one changes abruptly above $\Phi_0/2S$.

Ebner and Stroud showed that with a distribution of cluster sizes, the zero-field-cooled susceptibility was constant as a function field while the field-cooled susceptibility exhibited a sharp dropoff, above a critical field determined by the largest loop size in the distribution. By the time the field has reached the value corresponding to the average loop size, the gauge-field phase factors $A_{ij}$ induce random currents of either sign, i. e. clockwise or counterclockwise around the loops, and the corresponding moments can be either positive or negative, depending on the specific size of the loop in the distribution.

As a further step towards reality, simulations of more complex two-dimensional systems of coupled random loops by Ebner and Stroud[332] and Morgenstern et al.[253] showed that above some lower critical field $H_{nl}$ which is probably determined by the size of the array and which is in any case considerably smaller than $\Phi_0/S$, the magnitude of the net array moment drops with increasing field because of increasing local moment reversal. The moment reaches essentially zero for applied fields of order $\Phi_0/2S$ where S is once again the average local loop size. At this point there are equal numbers of local moments pointing up and down, and the system can be termed completely frustrated. Morgenstern et al.[253] and Jose[309] showed that the temperature dependence of the field-cooled and zero-field-cooled moments qualitatively resemble experiment.

IIIB-2. Comparison to experiment

These results suggest an explanation for the observed field-dependent Meissner fraction discussed in IIA-3 above and for the field-independence of the zero-field-cooled moment. At this point, there appears a difficulty, namely that in the original form of this two-dimensional model, screening is ignored, that is, the moments created by the local loop currents are assumed not to contribute to the flux threading the loop and hence to setting the gauge field phase factors. It should come as no surprise, then, that the zero-field-cooled susceptibility in this model bears no relationship to $-1/4\pi$ and is usually much smaller. More recently, several authors[97,257] have added screening to remedy this problem.

Numerical simulations of the two-dimensional model, by Morgenstern et al.[253] and Jose,[309] have shown evidence for an irreversibility line: They found that the difference between zero-field-cooled and field-cooled moments disappears at some temperature below the intragranular $T_c$, and that this temperature varies with field according the irreversibility line Eq. 1 with a power $q \simeq 2/3$. Vinokur et al.[306] analytically predicted a $q=1/2$ power. An alternative approach was suggested by Schneider et al.,[257] who extended the Ebner-Stroud model to include screening and found an approximate $H^{2/3}$ suppression of the phase-locking temperature.

The superconducting-glass model also suggests an explanation for the slow non-exponential relaxations observed in experiment, since such relaxations are a characteristic feature of all glassy systems. The ageing experiments of Rossel et al.[324] also point strongly in this direction (IIC). Aksenov and Sergeenkov[301] have worked out an analytical glass model to predict the magnetic viscosity and exponent of the power relaxation law in good agreement with experiment. The field-cooled susceptibility, however, is an equilibrium property in this model and is predicted not to relax. Although a small relaxation has been observed experimentally in this case (IIC), it is indeed much smaller than the relaxation of either the zero-field-cooled or remanent moments.

In spite of these successes, the Ebner-Stroud picture has problems when a more quantitative comparison to experiment is attempted. On the one hand, the fact that the Meissner fraction drops to a very small value by 10 Oe in most YBaCuO crystals suggests a length scale of order 1 $\mu$m for the loops (see Eq. 17). Since the moment reduction is supposed to be caused by reversed polarity moments associated with the current loops, one would expect to see a granular loop structure roughly independent of field in the magnetic decoration experiments (IIH). This conflicts with the observation of a classic vortex array with an area per flux quantum scaling simply according to Eq. 17.

Another problem arises in explaining the positive remanent moment and its simple relationship to the difference of field-cooled and zero-field-cooled moments (Eq. 2). Malozemoff et al.[97] argued that the superconducting-glass model typically gives little or no remanence for fields of order $\Phi_0/2S$. A further difficulty comes from the observation by Worthington et al.[20,293] of an irreversibility line (originally called $H_{c2}$) which follows a $(1 - T/T_c) \propto H^{2/3}$ law all the way to 20 T. In the context of either the Morgenstern et al.[253] or Schneider et al.[257] calculations, this requires the loop size in Eq. 17 to be much less than 100 Å, which is implausibly small for twin spacings.

Let us, nevertheless, accept for a moment the existence of such fine-scale weak links. Of course, at this scale they would not be visible in decoration experiments. The low-field properties would then appear to be those of a standard Type II superconductor, except that the microscopic meaning of the parameters would have to be modified. Pinning would be associated with the spatial inhomogeneities of the random granular array, as suggested by Tinkham and Lobb.[278]

But there are other problems if the weak links are really on this scale. Kwak et al.[82] estimated the intragranular critical current from the Bean formula (Eq. 4) and the observed magnetic remanence after going around a high field hysteresis loop. Adapting their approach to our numbers, we take the apparent low-temperature $J_c$ deduced from the standard analysis with crystals of dimension 0.1 mm to be $10^7$ A/cm$^2$. Then using arguments much as in our weak-link treatment in IIB-3 above, it is easy to see that the same remanence would, in a granular model with 100 Å grains, imply an intragranular $J_c$ of more than $10^{11}$ A/cm$^2$! In the context of typical values quoted up to now for the depairing critical current density[339] $J_d$, this is an incredible value. Since $H_{c2}^{\parallel}(0)$ is known[79] to be of order 30 T in YBaCuO, and $J_d$ is given by $H_c/\lambda$, $\lambda(0)$ must be considerably less than 300 Å, in disagreement with all determinations so far. The effective $H_{c1}$ from Eq. 6 would be only about 10 Oe, far less than reported in most measurements (IVB).

While these problems are serious, alternative pictures to be described in IIIC are also not completely satisfactory, particularly in the high-temperature region where some of the most characteristic "glassy" phenomena like the ageing effect appear (see IIC). Furthermore the superconducting glass model is in an early state of development, based on largely two-dimensional models; a more realistic three-dimensional treatment with screening might improve the picture. Even if not appropriate to crystals, the superconducting-glass model is relevant for ceramics, where the granular weak-link network surely exists. The problem here is that most macroscopic magnetic properties will be dominated by the intragranular contributions, which will mask the glassy intergranular effects.

IIC. Flux pinning and giant flux creep

IIIC-1. Interpretation of magnetic relaxation

A more conventional flux-pinning picture and critical-state model were invoked quite early by a number of authors to explain the hysteresis loops, as we have seen in IIB-2 above. The vortex decoration experiments (IIH) seemed to confirm the standard Type II superconducting behavior underlying this approach (IIH). The equality of $M_{rem}$ to $M_{fc} - M_{zfc}$ (see IIA-4, Eq. 2) also has a natural explanation[97,274] in this picture: The zero-field-cooled moment is simply the diamagnetic screening moment $M_{zfc}$ arising from the persistent current around the perimeter of the sample. The field-cooled moment $M_{fc}$ is the screening moment counterbalanced by some positive flux (vortices) trapped within the sample. When the field is turned off, the persistent current at the edge stops, but the trapped flux remains to give a moment which is just $M_{fc} - M_{zfc}$.

But the irreversibility line as well as the field dependence of the Meissner effect initially seemed puzzling in the conventional picture and more naturally explained in the superconducting-glass model. A key to this problem emerged from studies of the time-logarithmic relaxation in YBaCuO crystals by Yeshurun and Malozemoff.[28,279] Time-logarithmic relaxation is well-known in conventional superconductors, and has been explained in the classic Anderson-Kim theory of flux creep.[334] But the early experiments of Beasley, Labusch and Webb[340] showed effects much smaller than those now being observed in the new superconductors. Yeshurun and Malozemoff recognized that in these materials, because of the high temperature and low pinning barriers, flux can creep at a rate far beyond what has been observed in low-temperature superconductors.

In the the flux creep theory of Anderson and Kim,[334] as developed by Campbell and Evetts,[333] thermal activation of vortices over an energy barrier $U_0$ is modulated by a Lorentz force on the vortices proportional to the transport current density J, which equals the critical current density $J_c$ in the critical state. Thus the simple Arrhenius relation

$$f = f_0 \exp[-(U_0/kT)(1 - J/J_{c0})] \qquad (18)$$

can be solved for the effective critical current density $J_c$ at a given measurement frequency f. $J_c$ is suppressed below its value $J_{c0}$ in the absence of thermal activation:

$$J_c = J_{c0}[1 - (kT/U_0)\ln(f_0/f)] , \qquad (19)$$

where $f_0$ is a characteristic attempt frequency of order $10^{10}$ Hz.

The Bean model (Eqs. 3 and 4 for low and high field respectively) can be used to relate $J_c$ to the magnetization (at least at low temperature, see below), and $J_c$ in Eq. 19 depends logarithmically on frequency f or time t. Thus one can directly derive a logarithmic magnetic relaxation for a slab of thickness D, valid in the limit $H > H_{c1}$ and $H > H^{\star} = 2\pi DJ_c/c$ :

$$dM/d\ln(t) = (DJ_{c0}/4c)(kT/U_0) . \qquad (20)$$

Here $U_0$ is the activation energy in the absence of a flux gradient. For $H^{\star} > H >> H_{c1}$, such that the flux does not penetrate entirely through the sample, an alternative formula can be derived (assuming $J_c$ independent of field):

$$dM/d\ln(t) = (cH^2/16\pi^2 DJ_{c0})(kT/U_0) . \qquad (21)$$

These equations predict logarithmic relaxation of the type observed experimentally and thus offer an alternative to the superconducting-glass model of the relaxation. They also describe the temperature and field dependence of $dM/d\ln(t)$ (IIIC). The observed initial linear rise with temperature is explicit in Eqs. 20-21. The roughly cubic increase with field can be understood from Eq. 21 if $J_c$ is assumed to depend inversely on B, a dependence observed at least approximately in some experiments (see IIIB-5). The maximum of $dM/d\ln(t)$ with field can be understood in terms of a crossover between Eqs. 21 and 20, that is, in terms of first incomplete and then complete flux penetration through the sample thickness.

Using the measured $J_c$ and $dM/d\ln(t)$ in Eq. 21, Yeshurun et al.[107] extracted $U_0^{\parallel} \approx 0.02$ eV and $U_0^{\perp} \approx 0.2$ eV from the anisotropic relaxation of the zero-field-cooled magnetization in several YBaCuO crystals (note: the original evaluation in Ref. 28 was in error). These values are an order of magnitude or two smaller than in typical low temperature superconductors.[340] Yeshurun et al.[276] deduced an even stronger anisotropy in the pinning energies of BiSrCaCuO crystals.

Do such small pinning energies make sense? A possible explanation is suggested by a simple scaling form of Anderson and Kim:[334] $U_0 \propto H_c^2 \xi^3/8\pi$ , where $H_c$ is the thermodynamic critical field and $\xi$ the coherence length. For conventional type II materials this usually comes out to several eV. With $\xi^3 = \xi_{ab}^2 \xi_c$ and typical published parameters for YBaCuO, one obtains of order 10 meV. Again we see the importance of the small coherence lengths for generating new physics in the high-$T_c$ superconductors. This estimate is of course complicated by the possibility of collective vortex effects, and we will use a rather different scaling form in discussing the irreversibility line below.

Because the flux creep contribution scales as $kT/U_0$ in Eq. 19, materials with low pinning energies and high superconducting transition temperatures can show "giant" flux creep, several orders of magnitude larger than in conventional strongly pinned Type II superconductors.

Just as in the superconducting-glass model (IIIB), many problems remain in comparing the flux pinning and creep picture to experiment. One problem is that the anisotropy in the pinning barriers $U_0$ described above shows the opposite trend in comparison to the critical current densities, which were initially reported[19,54] to be larger for H‖c than for H⊥c. In the simplest model, $J_c$ is related to the pinning energy by

$$J_c \simeq cU_0/BdV \ , \tag{22}$$

where B is the magnetic induction, V an activation volume and d a characteristic length scale for the pinning potential. That is, $U_0/dV$ is the pinning **force** on a flux line per unit volume. Thus the difference in $J_c$ – and $U_0$ -anisotropy must come in this picture from a hypothetical anisotropy in the length scale and activation volume. But recent results (D. C. Cronemeyer and T. R. McGuire, private communication) indicate that the $J_c$-anisotropy may not be as large as originally thought, in part because of problems in choosing the relevant Bean formula to analyze the hysteresis data.

As discussed in IIH, little is known for sure at this time about the nature of the pinning. One possibility, suggested by the strong pinning for H⊥c, is that the flux is confined in almost normal layers between the superconducting CuO sheets. Then the barriers to flux motion could be **intrinsic** in the sense that no defect is required but only the discreteness of the atomic lattice itself (D. Pines, private communication). This mechanism is possible only when the coherence length approaches atomic dimensions, which seems valid in these materials. However a simple estimate of dV from Eq. 22 gives length scales of order hundreds of Angstroms. Doubtless, cooperative flux motions are required to understand the pinning forces in detail. Kes and van den Berg et al.[174,181] have described such a cooperative mechanism, the so-called "pin avoidance" mechanism, in which strong but widely spaced defects limit flux motion via the vortex lattice stiffness. Twin boundaries which typically lie in the ac planes are likely candidates for such pinning when H‖c. This is a complex topic still at an early stage of development.[296]

Another problem arises, as first pointed out by Tuominen et al.,[67] in interpreting the observed maximum in dM/dln(t) as a function of temperature. Considering that the starting magnetization in the high-field limit goes as $DJ_c$ (Eq. 4), the normalized relaxaton rate $(1/M_0)dM/dln(t)$ from Eq. 20 goes as $kT/U_0$. Since the barrier energy must go to zero at $T_c$, one expects this rate to diverge. Experiment shows it does not (see IIC). A possible resolution to this problem emerges from the work of Dew — Hughes,[286] who pointed out that when the barrier is low and the temperature is high, both forward and backward hopping must be considered, leading to a generalized Arrhenius relation

$$f = f_0 \exp( -U_0/kT) \sinh(U_0J/kTJ_{c0}) \ . \tag{23}$$

This equation, when solved for $J = J_c$, suppresses the divergence of the simpler treatment.

Eq. 19 and the more complete Eq. 23 offer an explanation of the previously puzzling behavior of the critical current density as a function of temperature, described in IIB-5 above. Eq. 19 explicitly shows the linear dropoff with temperature. In conventional superconductors, Campbell and Evetts[333] had argued that this term was negligibly small, but the small $U_0$ in the new materials makes it significant. Yeshurun et al.[249] showed that the pinning energy required to explain the linear dropoff in the magnetic $J_c(T)$ agrees approximately with that extracted from the magnetic relaxation.

Malozemoff et al.[295] argued that the frequency dependence in Eq. 19 offers an explanation for the discrepancy in the temperature dependence of transport and magnetic $J_c$'s shown in Fig. 5. The point is that the two measurements have different effective frequencies. A dc magnetic measurement with a standard SQUID magnetometer is typically done with at least a minute per point. On the other hand the effective frequency of the transport measurement[333] is determined via Maxwell's equation $\nabla \times E = - \partial B/\partial t$ as follows: Flux hopping at a rate given by Eq. 18 or 23 implies a $\partial B/\partial t$ and hence an electric field E. If the effective critical current density is determined by some minimum measurable voltage or electric field, this in turn determines an effective frequency. Typical estimates[103] give frequencies in the range of 1 kHz to 1 MHz, significantly higher than for the dc magnetic measurement. Thus even though the frequency enters only logarithmically into Eq. 19, the many orders of magnitude difference in frequency can explain the observed factor of two difference in slopes.

Since the hopping of vortices at a rate f leads through Maxwell's equation to a measurable voltage, Eq. 23 predicts a V(I) dependence which is linear at low current and exponential at high current. Indeed nonlinear I-V characteristics are observed in high-$T_c$ superconductors, as described in IIF-2, but no attempt has yet been made to compare such a theory to experiment.

Problems remain in explaining the full temperature dependence[83] of $J_c$, particularly in the high-temperature region, with Eq. 19 or 23. Malozemoff et al.[295] introduced a distribution of energy barriers to explain the observed $J_c(T)$. McHenry et al.[247] and Geshkenbein and Larkin[302] have also used a distribution of barriers to account for the maximum in $dM/d\ln(t)$ versus temperature. The randomness inherent in these models brings one back to a kind of "glassy" picture, though perhaps on a different physical basis than the Ebner-Stroud granular model considered above.

Kes et al.[296] pointed out that because of the large flux flow at high temperature, it is necessary to consider explicitly the constraints set by the flux continuity equation

$$\frac{\partial B}{\partial t} = \frac{\partial}{\partial x} \left( D \frac{\partial B}{\partial x} \right) , \tag{24}$$

where D is a flux diffusion constant. Using this equation, they derive a stretched exponential or Kohlrausch form for the magnetic relaxation, as observed in certain experiments (see IIC). They call this limit "thermally activated flux flow" or TAFF.

An important practical consequence of giant flux creep and thermally assisted flux flow is flux noise in electronic devices such as SQUIDs (Superconducting Quantum Interference Devices). Koch and Malozemoff[266] showed that a broad distribution of pinning barriers would lead to a $1/f$ noise spectrum, in approximate agreement with measurements on SQUIDs[214] and films.[234]

IIIC-2. Irreversibility line in the flux pinning model

Scheidt et al.[169] first pointed out that the reversible magnetic behavior near $T_c$ might be explained by vortex depinning. Yeshurun and Malozemoff[28] and Worthington et al.[293] suggested more explicitly how the giant flux creep could be related to the irreversibility line determined in the ac susceptibility measurements: Too little flux penetration during an ac cycle represents simple screening. Complete penetration implies full equilibration of the flux during each cycle. Thus, maximum loss should occur at intermediate penetration, that is, at the temperature where the ac field causes flux to penetrate just to the center of the sample. If the amplitude of the ac field is $H_\omega$, and the thickness of the slab is D, this condition is reached, according to the Bean critical-state model, when $J_c(T) = cH_\omega/2\pi D$. Substituting in Eq. 23 and linearizing the sinh, one finds

$$U_0(H,T) = kT \ln \left( \frac{cU_0 H_\omega f_0}{2\pi DkTJ_{c0}f} \right) , \tag{25}$$

which implicitly determines the irreversibility line.

The field and temperature dependences of this line are determined by the field and temperature dependences of $U_0$. Yeshurun and Malozemoff[28] gave a simple scaling argument for $U_0$ as follows: If it is proportional to the condensation energy per volume $H_c^2/8\pi$ times a characteristic excitation volume, for sufficiently large fields that volume could be limited laterally by the area which a single flux quantum occupies in the flux line lattice. This area is approximately $a_0^2$, where $a_0 \simeq \sqrt{\Phi_0/B}$ is the flux line spacing in field B. In the third direction, along the applied field, the minimum possible extent of the activation volume is $\xi$, giving $U_0 \propto (H_c^2/8\pi)a_0^2\xi$. According to the Ginsburg-Landau theory, $H_c$ scales as $(1 - T/T_c)$ near $H_{c2}$ and $\xi$ scales as $(1 - T/T_c)^{-1/2}$. Then

$$U_0 \propto (1 - T/T_c)^{3/2}/B \propto J_d\Phi_0/B \ , \qquad (26)$$

where $J_d$ is the temperature-dependent depairing current density.[335] In the last part of this equation, the pinning energy is written in a different though equivalent form, as suggested by Tinkham.[31] This form suggests an alternative physical interpretation for $U_0$, namely that it is proportional to the depairing critical current along the field direction in a cross-section area equal to the unit cell of the Abrikosov flux lattice. The form is convenient because it involves only a single material parameter $J_d$ rather than both $H_c$ and $\xi$. Of course the temperature scaling is the same because in Ginzburg-Landau theory, $J_d$ scales as $(1 - T/T_c)^{3/2}$.

Substituting Eq. 26 in 25, approximating T by $T_c$, and solving for $(1 - T/T_c)$, Worthington et al.[293] found an expression for the irreversibility line

$$1 - T/T_c = p[BkT_c \ln(cU_0H_\omega f_0/2\pi DkTJ_{c0}f)]^{2/3} \ , \qquad (27)$$

where p is a constant to be determined experimentally. Thus the $B^{2/3}$ dependence is recovered, in agreement with experiment. This derivation was important in providing the first explanation of the irreversibility line in the context of conventional flux creep, and thus it offered the first credible alternative to the superconducting-glass model.

A further test of this model comes from the frequency dependence of Eq. 27. As discussed in IIE-2, the measured frequency dependence is roughly logarithmic but with a weak upward curvature in $T_{irr}$ vs. ln $\omega$ as predicted from the 2/3 power in Eq. 27. Using the irreversibility line data at fixed frequency to determine the constant p, Malozemoff, Worthington et al.[103,293] were able to predict $dT_{irr}/d$ ln $\omega$ as a function of field, in reasonable agreement with experiment. The success of this simple picture in explaining such a large amount of data was perhaps the decisive argument in discounting the original interpretation of the ac susceptibility anomaly in terms of $H_{c2}$.

Eq. 27 also shows a dependence on amplitude of the ac field $H_\omega$. As mentioned in IIE-2, some experiments have shown a dependence and others have not. In the TAFF limit (see above), Kes et al.[296] derived a formula similar to Eq. 27 but without the $H_\omega$ dependence. These authors argue that in at high temperatures, it is essential to take the continuity equation, Eq. 24, into account, but it was ignored in deriving Eq. 27. The viscosity $\eta$ of vortex motion should also be included, although the high normal-state resistivity $\rho_n$ will reduce $\eta$, according to the conventional Bardeen-Stephen formula[335] $\eta \simeq \Phi_0 H_{c2}/\rho_n c^2$. Further work is required to develop and test these theories.

The flux-creep picture of the irreversibility line clarifies several other phenomena. Krusin-Elbaum et al.[183] explained the field dependence of the Meissner fraction (IIA-3) from the assumption that during field-cooling, the flux (and hence the magnetization) are frozen in when the irreversibility line is reached. Fig. 9 shows the H-T plane with the $H_{c1}(T)$ and $H_{c2}(T)$ lines and the upward-curving irreversibility line estimated for an YBaCuO crystal. Also shown are the equilibrium lines of $4\pi M/H$ calculated from the Abrikosov formula Eq. 10. The low temperature Meissner fraction is determined by the intersection of the irreversibility line with these contours, and since the irreversibility line curves upwards, the trapped-in Meissner fraction increases with decreasing field, in a way which can be fit to the data.[183]

In this model, the unusual field dependence of the Meissner fraction is a consequence of the irreversibility line and hence of flux pinning. Large Meissner fractions for a given field imply low pinning or, equivalently, a large reversible region, and small Meissner fractions imply the opposite. A more complete theory, however, will require inclusion of critical-state effects (J. R. Clem, private communication).

Flux depinning above the irreversibility line implies the possibility of flux-flow resistivity. This explains the otherwise puzzling broadening with field of the resistivity versus temperature (IIF-1). Using an analogy to the heavily-damped current-driven Josephson junction and assuming the pinning energy of Eq. 26, Tinkham[31] derived the flux-flow resistivity $\rho$ relative to the normal-state resistivity $\rho_n$

$$\rho/\rho_n = \{I_0[A(1 - T/T_c)^{3/2}/2H]\}^{-2} ,  \qquad (28)$$

where $I_0$ is the modified Bessel function and A is a constant reflecting the pinning strength. His predicted resistance versus temperature in different fields is shown in Fig. 7, in remarkable agreement with experiment.

In summary, the flux-creep and flow model has been successful in explaining many features of the magnetic data. At this point it is considerably more quantitative

126

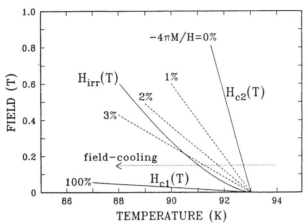

Fig. 9.   Schematic H-T phase diagram of an YBaCuO Type II superconductor, showing the Meissner fractions in the equilibrium Abrikosov theory of Eq. 10, with a superimposed irreversibility line. Field cooling corresponds to a horizontal line whose intersection with the irreversibility line determines the trapped-in flux and hence the Meissner fraction.   (After Krusin-Elbaum et al., Ref. 183.)

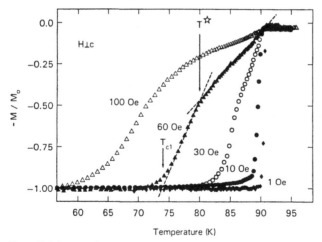

Fig. 10.   Zero-field-cooled susceptibility of an YBaCuO crystal versus increasing temperature, for a set of fields perpendicular to the c-axis. The flux penetration moves to consistently lower temperatures with increasing field. The linear extrapolation to $T_{c1}$ is taken to determine the lower critical field, while the temperature $T^{\star}$ indicates the point at which the flux fronts meet in the center of the sample according to the Bean critical-state model. (After Krusin-Elbaum et al., Ref. 106.)

than the alternative superconducting glass-model, although basic questions remain, for example in explaining the ageing effect (IIC). Therefore it seems plausible to proceed, as we will in IV, to interpret $H_{c1}$ and $H_{c2}$ data in terms of bulk Type II superconductivity.

IIID. Flux-lattice melting and anisotropy

An interesting alternative to the flux-creep and flow interpretation of the irreversibility line is the possibility of flux-lattice melting, suggested by Nelson,[27] Gammel et al.,[33] Houghton and Moore,[26,94,310] and Houghton and Pelcovitz.[311] The basic idea is that because of the high temperatures and low coherence lengths, entropy causes large random displacements of the vortices in the Abrikosov lattice. According to the Lindemann criterion, the lattice melts if the mean-square displacement reaches of order 10% of the lattice parameter. Houghton and Pelcovitz,[311] for example, predict a melting contour starting at $T_c$ and H=0 with an upward curvature in the H-T plane which can be fit approximately to the observed irreversibility line. This interpretation, at least as developed so far, is defect-independent and implies that the irreversibility line should be universal in a given material. Yet the line shifts from sample to sample, suggesting the importance of defect-pinning. More conclusive experiments are needed here with systematic correlation of the irreversibility line to defect density.

Several theories propose other kinds of entropic effects on flux lines. Nelson[27] argues that in fields slightly above $H_{c1}$, temperature-induced flux-line "wandering" leads to a novel entangled vortex state which acts to give stiffness to a vortex array which might otherwise be melted. Feigel'man and Vinokur[312] show that thermal fluctuations of flux lines can strongly reduce the effective pinning and hence also the critical current density in the mixed state. Neither of these theories has yet been tested by experiment.

Most recently, Fisher[313] argues that random pinning acts like a random field on the vortices, destroying long range order. He proposes a new thermodynamic phase in the mixed state, called a vortex glass superconductor, separated from a melted vortex phase by a sharp equilibrium phase boundary. The dynamic properties will be glassy in nature. This suggests that the frequency-dependent irreversibility line will have a finite temperature limit at zero frequency, although this limit may be hard to achieve kinetically, as in many other glassy phase transitions.

Kogan, Campbell et al.,[91,100,105,330] Schopohl and Baratoff,[171] Kes[174] and Klemm[98] have explored the anisotropic properties of vortex lattices, particularly for angles of applied field off the principal axes. New phenomena can arise, such as moments perpendicular to the applied field.

## IV. FUNDAMENTAL PARAMETERS

IVA. Upper critical field and coherence length

One the principal interests in magnetic measurements, as mentioned in the introduction, is to determine important fundamental parameters of the new superconductors: the anisotropic coherence lengths $\xi$ from the upper critical fields $H_{c2}$, and the anisotropic and temperature-dependent penetration depths $\lambda(T)$ from the lower critical fields $H_{c1}$. As suggested by Eq. 12, the coherence lengths give information about the electron pairing, and the small values reported have spurred interesting though still highly speculative real-space-pairing models or planar models with Josephson couplings between the planes. The penetration depths, interpreted in the context of the standard London formula

$$\lambda = (mc^2/4\pi n_s e^2)^{1/2} , \tag{29}$$

give information about the conduction-electron (or hole) density $n_s$ and the enhanced band mass m. Their temperature dependences are controlled by the symmetry of the electron pairing, with the conventional BCS singlet pairing giving an exponentially small variation at low temperatures, but with p-wave and d-wave pairing typically giving power-law contributions to $\lambda(T)$. Since many recent theories are based on unconventional pairing, data on $\lambda(T)$ are of particular interest.

Let us first summarize the situation for $H_{c2}$ and $\xi$ . In spite of a huge number of articles purporting to determine these quantities, it is likely that flux creep and flow effects dominate the experiments (IIE-2, IIF-1); so these transition lines in the H-T plane are more likely some kind of vortex depinning or melting transition, as discussed in IIIC-D above. If this is true, then the coherence lengths determined from the classic mean-field relation

$$H_{c2}(T) = \Phi_0/2\pi\xi^2(T) \tag{30}$$

are not reliable and the widely quoted anisotropy may rather reflect the anisotropy of pinning in these materials.

This conclusion may be hard to accept, considering the sheer number of articles advancing the $H_{c2}$-interpretation. It is particularly frustrating to those theories[119,135] which have been developed precisely to account for the upward curvature in the data. One might argue that if these lines are claimed not to be $H_{c2}$ but rather irreversibility lines, then why should they be so reproducible from sample to sample? For when a consistent criterion is used, the values wander by factors of only two or three, and convincing correlation with defect concentration has not yet been demonstrated. On the other hand, the microscopic origin of the irreversibility line is not yet established; it is possible that the dynamic transition could be dominated by intrinsic bulk parameters, though in a very different way than predicted in Eq. 30. This would be the case, for example, if the irreversibility line were a flux lattice melting phenomenon, if it were determined by intrinsic pinning or if a pin avoidance mechanism were operative (see IIIC and D).

A more plausible method for determining $H_{c2}$ is the reversible magnetization measurement (IIG-1). Nevertheless most of these measurements, done on randomly oriented ceramics, are hard to interpret because they do not unravel the anisotropy. The only such measurements on oriented materials give a result which is quite puzzling in the context of conventional theory: an essentially vertical slope $dH_{c2}/dT$ near $T_c$. The first explanation in terms of a twin boundary enhancement effect[73] is elegant but unlikely in view of the evidence for large twin-free regions in the YBaCuO crystals (see IIH).

Do the irreversibility lines at least set a lower bound on $H_{c2}$? If so, the coherence lengths would be even smaller than those quoted in the literature so far (from 15-30 Å in the plane and from 7 to 1.5 Å perpendicular). But as the estimated $\xi$ gets smaller, the Ginzburg criterion for the size of the critical region (e. g. see Deutscher[140]) grows rapidly (as $\xi^{-6}$). Thus the peculiar experimental results may reflect non-mean-field critical phenomena. It is also conceivable that in finite field, there is no well-defined phase transition at all and hence no $H_{c2}$, but only some sort of crossover behavior (P. W. Anderson and M. P. A. Fisher, private communication). This is because flux-lattice melting may destroy the conventional superconducting order parameter. Then the reversible Meissner region might in some sense represent just a strong fluctuation diamagnetism.

Further speculation at this time seems premature. The experimental challenge is to measure $H_{c2}$ as reliably as possible. Meanwhile, all present theoretical conclusions based on specific values of the coherence length and its anisotropy should, in the author's view, be regarded with some skepticism. This is particularly true of the anisotropy, because, as we shall see next, the anisotropy deduced from the **lower** critical fields is a good deal smaller than that quoted from the questionable $H_{c2}$ measurements. This could have serious consequences for theories involving two-dimensionality and weak interlayer coupling.

IVB. Lower critical field and penetration depth

The experimental situation for the lower critical fields and the penetration depths has been no less confused. Values quoted for $H_{c1}$ of bulk YBaCuO, for example, have ranged from 1 to 5000 Oe! Similarly, reported penetration depths have varied from a few hundred to 10,000 Å. These problems are closely related because in the London limit, valid over a large range of the H-T diagram for high-$\kappa$ superconductors, there is a simple relationship between $H_{c1}$ and $\lambda$. In an anisotropic effective mass formalism, this relation can be written to a good approximation as[341-2]

$$H_{c1} = \frac{\phi_0}{4\pi\lambda_i\lambda_j} [\ln \frac{\lambda}{\sqrt{\xi_i\xi_j}} + 0.50], \tag{31}$$

where $\lambda_i = \lambda\sqrt{m_i}$, $\xi_i = \xi/\sqrt{m_i}$ , where the effective masses $m_i$ obey $m_a m_b m_c = 1$, and where $\Phi_0$ is the flux quantum.

It is useful in this context to emphasize some geometric considerations in determining the penetration depth. For $H \parallel c$, screening currents flow in the ab CuO planes; so $H_{c1}^{\parallel}$ is determined by $\lambda_{ab}$. Correspondingly, $H_{c1}^{\perp}$ depends on the product of $\lambda_{ab}$ and $\lambda_c$. One must be careful, however, to consider the current flow geometry for each experiment individually.[93,341-2] For example, when the field is applied in the ab plane of a thin slab, a London penetration depth experiment using the dc susceptibility (IVC) determines only $\lambda_{ab}$ because the screening currents flow predominantly along the ab planes.

Aside from techniques involving ESR,[56] torque magnetometry[185] and the vibrating read,[196] most determinations of the lower critical field of LaBaCuO,[108,111,116,117] YBaCuO[20,57,92,93,179,184,204] and BiSrCaCuO[95] have been made by measurements of the virgin hysteresis cycle, in which, as explained in IIB-1,2 above, a deviation from linearity in M(H) is supposed to signal $H_{c1}$. But this deviation is gradual and its onset is hard to pick out. It seems that the higher the experimental resolution, the lower the reported $H_{c1}$! But the low values are suspect too, even after correction for the overall demagnetization factor N, which lowers the flux penetration field to $H_{c1}(1 - N)$. The problem is that the crystals have imperfect demagnetization geometries, and flux penetration will start in the corners at even lower fields. Another problem is that because of magnetic relaxation, a virgin hysteresis curve represents, at each successive point, ever longer relaxation times, so that the curve is distorted by time effects. For these reasons the detailed results of many works using this method are not summarized here; the reader is referred to the references indicated above.

These problems led Yeshurun et al.[107] to propose a novel method for determining $H_{c1}$. They successively measured the magnetic relaxation of zero-field-cooled YBaCuO crystals at a series of different fields and found a characteristic onset in the time-logarithmic slope $S=dM/d\ln(t)$. This approach, while far more laborious than the conventional M(H) measurement, has the further advantage that there is no need to subtract a background linear diamagnetism. Another improvement was to fit the results to a relaxation theory based on the Bean critical-state model including $H_{c1}$. In this way, they could discriminate between edge and bulk penetration. This method gave results close to those from torque magnetometry,[185] namely $H_{c1}^{\parallel} \approx 900$ and $H_{c1}^{\perp} \approx 200$ Oe respectively, at 4.2 K.

Another method has recently been proposed by Krusin-Elbaum et al.[106] In YBaCuO crystals, they observed a systematic shift of the flux penetration with increasing field, as in Fig. 10. In these zero-field-cooled measurements with increasing temperature, flux penetration begins at the temperature where $H_{c1}(T)$ drops to the point where it equals the applied field. Thus a determination of the temperature for penetration permits a measurement of $H_{c1}(T)$. The authors argued that the small foot in the data below the linear region represented corner penetration, and they explained the change in slope at a somewhat higher temperature in the figure in terms of the critical-state flux fronts meeting in the center of the sample. The method only worked down to about $T_c/2$ because at lower temperature the flux penetration became progressively more rounded. An important issue in all $H_{c1}$ measurements is the possible role of surface barriers; however in these experiments crystals with different surface treatments gave consistent results. Samples with different demagnetizing factors were also measured to test the validity of the large demagnetization corrections.

The results[106] showed an almost linear behavior of $H_{c1}(T)$ essentially down to 0.5 $T_c$, substantiating many earlier reports. This result has important implications for the temperature dependence of $\lambda$, as will be discussed further below. A surprise was the rather small anisotropy ratio of about 3 for fields along the two principal directions. Extrapolations of the data, based on a weak-coupling BCS form[335] for $\lambda$ gave $H_{c1}^{\parallel} \approx 550$ and $H_{c1}^{\perp} \approx 200$ Oe, respectively, in rough agreement with the relaxation measurements.[107] Using Eq. 31, these results imply upper limits for the anisotropic penetration depths of $\lambda_{ab} \approx 1300$ and $\lambda_c \approx 4500$ Å respectively. Recent torque measurements of Farrell et al.[327] in the reversible regime gave a ratio of 5 between $H_{c2}^{\perp}$ and $H_{c2}^{\parallel}$. This can be directly compared to the $H_{c1}$ anisotropy since in anisotropic Ginzburg-Landau theory, the ratios of the anisotropic upper and lower critical fields are approximately the inverse of each other.

The anisotropy ratio of 3-5 is significantly smaller than the ratio of 6-10 commonly quoted from the more questionable resistive $H_{c2}$ measurements. Thus YBaCuO appears to be less anisotropic than previously thought. This conclusion is problematic for many theories of high-temperature superconductivity based on two-dimensional sheets with only weak interplanar coupling, since these theories assume high anisotropy.

An interesting technique for measuring $H_{c1}$ has been proposed by Ravi Kumar and Chaddah,[289] exploiting low-field anomalies which are occasionally seen in the hysteresis loop near remanence. They interpret these anomalies, observed in YBaCuO ceramics, using a Bean critical-state model, and they determine a bulk

$H_{c1}$ of only about 1 Oe. While this is out-of-line with other data, an interpretation of these anomalies does seem like a useful counterpart to the other methods for determining $H_{c1}$ and deserves further study in crystals.

IVC. Temperature-dependent penetration depth

One of the principal methods for measuring the penetration depth has been muon spin relaxation, as described in II-I above. The main limitation until recently has been the directional averaging in randomly aligned ceramics. In an **aligned** YBaCuO ceramic, Schenk[175] found $\lambda_c \approx 6000$ Å. In an YBaCuO crystal, Harshman et al.[325] found 1400 and >7000 Å for $\lambda_{ab}$ and $\lambda_c$ respectively. $\lambda_{ab}$ is in good agreement with the values deduced above from $H_{c1}$; $\lambda_c$ still differs by a factor of about almost 2.

Uemura et al.[250] compiled muon data on LaSrCuO, YBaCuO (including oxygen depleted samples with different $T_c$'s) and TlBaCaCuO, showing a systematic and roughly linear dependence of the zero-temperature muon relaxation rate on the transition temperature. According to Eq. 11 the relaxation rate goes as $\lambda^{-2}$, and in the London penetration depth formula, Eq. 29, the carrier density $n_s$ scales in the same way (assuming constant carrier mass). These results imply that $T_c$ scales with carrier density, a feature of some recent theories.

Of even greater interest is the temperature dependence of the penetration depth. While a large number of experiments[192,193,194,270] have reported a power law contribution at low temperatures, and theories[233] were devised to explain this effect, more recent experiments give an extremely flat dependence of $\lambda(T)$ up to about 0.5 $T_c$. Since there are many possible sources of non-intrinsic flux penetration which could be enhanced with increasing temperature, it seems reasonable to consider as most reliable those experiments which show the weakest variation at low temperatures.

The muon spin relaxation measurements first showed this weak temperature dependence. The YBaCuO data were fitted to the so-called "two-fluid" form

$$\lambda(T) \propto \lambda(0)[1 - (T/T_c)^4]^{-1/2} . \tag{32}$$

As emphasized recently by Rammer,[10,60,213] this is an approximation to various strong-coupling limits of conventional singlet BCS theory. More recent results using direct magnetic measurements on epitaxial YBaCuO films by Fiory et al.[30] and on bulk crystal platelets by Krusin-Elbaum et al.[314] have also shown a weak low-temperature dependence. At higher temperature, the results fitted either the strong or weak-coupling singlet BCS forms. The $H_{c1}$ data (IVB), however, favor the weak-coupling limit.

In summary, the results for YBaCuO are consistent with conventional singlet BCS pairing and set limits on p or d-wave pairing of the type considered earlier in heavy fermions systems.

## V. CONCLUSIONS

What conclusions can we extract from this mass of seemingly contradictory experiments and interpretations? At least a few points are clear. The low coherence lengths of the high-$T_c$ superconductors give them a propensity to weak-link formation, and the behaviors of ceramics and polycrystalline films with grain boundaries are clearly dominated by the weak-links formed at those grain boundaries.

It is also clear that many magnetic and related transport properties of these materials are unconventional in the context of low temperature superconductors. These properties include the irreversibility line, the strong temperature dependence of the critical currents and the very large non-exponential time relaxations. These properties appear in both ceramics and bulk crystals.

These unusual effects suggest that even bulk crystals may be granular in nature. This hypothesis has been discussed in detail, and neither the experimental evidence nor the theories are fully consistent at this time. The interpretation in terms of conventional bulk Type II superconductivity with vortex pinning and large thermal activation seems so far to give the most detailed description of the experimental results. But this may be simply because that picture has a larger base of knowledge than the more novel superconducting-glass model.

The many studies of the upper and lower critical fields, and of the corresponding coherence lengths and penetration depths, have been evaluated in the context of the bulk picture. A surprising conclusion is that the upper critical fields and the coherence lengths derived from them are not well known. On the other hand the lower critical field has now been determined consistently by a number of authors in YBaCuO crystals shows only a modest anisotropy ratio of 3-5. The related penetration depth has a very flat temperature dependence up to $T_c/2$, consistent with conventional BCS singlet pairing.

There is a great deal of work still to do, and it is safe to predict that even within a few months our picture of the macroscopic magnetic properties of the new superconductors should be much more complete.

Finally it should be pointed out that many phenomena described in this review bear on the possible practical applications of high-$T_c$ superconductors.[339] The irreversibility line limits the temperature range for useful superconductivity in a way which depends on field and frequency. Flux line hopping gives an important mech-

134

anism for noise in electronic devices. The penetration depths set lower limits on the practical linewidths of superconducting interconnects. And the weak links limit the transport current density. All of these effects will have to be taken into account before truly practical applications emerge.

Whatever the prospect for applications, interest in high-temperature superconductivity is sure to remain high, both in the search for new materials and in the search for the underlying mechanism. In both respects, macroscopic magnetic measurements will continue to play a major role.

The author thanks his colleagues at IBM Yorktown Heights: Y. Yeshurun (on leave from Bar-Ilan University, Israel), T. Worthington, L. Krusin-Elbaum, D. C. Cronemeyer, F. Holtzberg, T. Dinger, T. R. McGuire, R. L. Greene, C. Tsuei, J. D. Mannhart, P. Chaudhari and many other others, and also his colleagues at IBM Zürich: K. A. Müller, J. G. Bednorz, I. Morgenstern and C. Rossel, for the close cowork which led to the many ideas in this review. The author also thanks many other colleagues, especially those mentioned with private communications throughout the text, for valuable insights and unpublished information.

## REFERENCES

1. J. G. Bednorz, M. Takashige and K. A. Müller, EUROPHYS. LETT. **3**, 379 (1987).

2. D. Esteve, J. M. Martinis, C. Urbina, M. H. Devoret, G. Collin, P. Monod, M. Ribault and A. Revcolevschi, ibid., p. 1237.

3. C. Giovannella, G. Collin, P. Rouault and I. A. Campbell, EUROPHYS. LETT. **4**, 109 (1987).

4. A. Junod, A. Bezinge, T. Graf, J. L. Jorda, J. Muller, L. Antognazza, D. Cattani, J. Cors, M. Decroux, O. Fischer, M. Banovski, P. Genoud, L. Hoffman, A. A. Manuel, M. Peter, E. Walker, M. Francois and K. Yvon, ibid., p. 247.

5. F. N. Gygax, B. Hitti, E. Lippelt, A. Schenck, D. Cattani, J. Cors, M. Decroux, O. Fischer and S. Barth, ibid., p. 473.

6. M. Oussena, S. Senoussi and G. Collin, ibid., p. 625.

7. J. C. Ousset, M. F. Ravet, M. Maurer, J. Durand, J. P. Ulmet, H. Rakoto and S. Askenazy, ibid., p. 743.

8. A. Raboutou, P. Peyral, J. Rosenblatt, C. Lebeau, O. Pena, A. Perrin, C. Perrin and M. Sergent, ibid., p. 1321.

9. G. Waysand, EUROPHYS. LETT. **5**, 73 (1988).

10. J. Rammer, ibid., p. 77.

11. S. Sun, Y. Zhao, G. Pan, D. Yu, H. Zhang, Z. Chen Q. Yitai, W. Kuan and Q. Zhang, EUROPHYS. LETT. **6**, 359 (1988).

12. B. Barbara, A. F. Khoder, M. Couach and J. Y. Henry, ibid., p. 621.

13. A. Sulpice, P. Lejay, R. Tournier and J. Chaussy, EUROPHYS. LETT. 7, 365 (1988).

14. R. J. Cava, R. B. van Dover, B. Batlogg and E. A. Rietman, PHYS. REV. LETT., **58**, 408 (1987).

15. M. K. Wu, J. R. Ashburn, C. J. Torng, P. H. Hor, R. L. Meng, L. Gao, Z. J. Huang, Y. Q. Wang and C. W. Chu, ibid., p. 908.

16. K. A. Müller, M. Takashige and J. G. Bednorz, ibid., p. 1143.

17. J. Z. Sun, D. J. Webb, M. Naito, K. Char, M. R. Hahn, J. W. P. Hsu, A. D. Kent, D. B. Mitzi, B. Oh, M. R. Beasley, R. H. Hammond and A. Kapitulnik, ibid., p. 1574.

18. R. J. Cava, B. Batlogg, R. B. van Dover, D. W. Murphy, S. Sunshine, T. Siegrist, J. P. Remeika, E. A. Rietman, S. Zahurak and G. P. Espinosa, ibid., p. 1676.

19. T. R. Dinger, T. K. Worthington, W. J. Gallagher and R. L. Sandstrom, ibid., p. 2687.

20. T. K. Worthington, W. J. Gallagher and T. R. Dinger, PHYS. REV. LETT. **59**, 1160 (1987).

21. G. Deutscher and K. A. Müller, ibid., p. 1745.

22. P. L. Gammel, D. J. Bishop, G. J. Dolan, J. R. Kwo, C. A. Murray, L. F. Schneeneyer and J. V. Waszczak, ibid., p. 2592.

23. S. A. Solin, N. Garcia, S. Vieira and M. Hortal, PHYS. REV. LETT. **60**, 744 (1988).

24. M. A. Dubson, S. T. Herbert, J. J. Calabrese, D. C. Harris, B. R. Patton and J. C. Garland, ibid. p. 1061.

25. P. Chaudhari, J. Mannhart, D. Dimos, C. C. Tsuei, C. C. Chi, M. M. Oprysko and M. Scheuermann, ibid., p. 1653.

26. A. Houghton and M. A. Moore, ibid., p. 1207.

27. D. Nelson, ibid., 1973.

28. Y. Yeshurun and A. P. Malozemoff, ibid., p. 2202. Note: the evaluation of the pinning energies $U_0$ is in error; see Ref. 107 for corrected values.

29. D. Dimos, P. Chaudhari, J. Mannhart and F. K. LeGoues, PHYS. REV. LETT. **61**, 219 (1988).

30. A. T. Fiory, A. F. Hebard, P. M. Mankiewich and R. E. Howard, ibid., p. 1419.

31. M. Tinkham, ibid., p. 1658.

32. T. T. M. Palstra, B. Batlogg, L. F. Schneemeyer and J. V. Waszczak, ibid., p. 1662.

33. P. L. Gammel, L. F. Schneemeyer, J. V. Wazczak and D. J. Bishop, ibid., p. 1666.

34. D. K. Finnemore, R. N. Shelton, J. R. Clem, R. W. McCallum, H. C. Ku, R. E. McCarley, S. C. Chen, P. Klavins and V. Kogan, PHYS. REV. B **35**, 5319.

35. F. S. Razavi, F. P. Koffyberg and B. Mitrovic, ibid., p. 5323.

36. B. Batlogg, A. P. Ramirez, R. J. Cava, R. B. van Dover and E. A. Rietman, ibid., p. 5340.

37. W. K. Kwok, G. W. Crabtree, D. G. Hinks, D. W. Capone, J. D. Jorgensen and K. Zhang, ibid., 5343.

38. T. P. Orlando, K. A. Delin, S. Foner, E. J. McNiff, Jr., J. M. Tarascon, L. H. Greene, W. R. McKinnon and G. W. Hull, ibid., p. 5347.

39. G. Aeppli, R. J. Cava, E. J. Ansaldo, J. H. Brewer, S. R. Kreitzmann, G. M. Luke, D. R. Noakes and R. F. Kiefl, ibid., p. 7129.

40. W. J. Kossler, J. R. Kempton, X. H. Yu, H. E. Schone, Y. J. Uemura, A. R. Moodenbaugh, M. Suenaga and C. E. Stronach, ibid., 7133.

41. T. P. Orlando, K. A. Delin, S. Foner, E. J. McNiff, Jr., J. M. Tarascon, L. H. Greene, W. R. McKinnon and G. W. Hull, ibid., p. 7249.

42. C. Allgeier, J. S. Schilling and E. Amberger, ibid., p. 8791.

43. D. E. Farrell, M. R. DeGuire, B. S. Chandrasekhar, S. A. Alterovitz, P. R. Aron and R. L. Fagaly, ibid., p. 8797.

44. D. S. Ginley, E. L. Venturini, J. F. Kwak, R. J. Baughman, B. Morosin and J. E. Shirber, PHYS. REV. B **36**, 829 (1987).

45. Y. Yeshurun, I. Felner and H. Sompolinsky, ibid., p. 840.

46. G. Xiao, F. H. Streitz, A. Gavrin, M. Z. Cieplak, J. Childress, M. Lu, A. Zwicker and C. L. Chien, ibid., p. 2382.

47. D. R. Harshman, G. Aeppli, E. J. Ansaldo, B. Batlogg, J. H. Brewer, J. F. Carolan, R. J. Cava, M. Celio, A. C. D. Chaklader, W. N. Hardy, S. R. Kreitzman, G. M. Luke, D. R. Noakes and M. Senba, ibid., p. 2386.

48. D. O. Welch, M. Suenaga and T. Asano, ibid., p. 2390.

49. T. P. Orlando, K. A. Delin, S. Foner, E. J. McNiff, Jr., J. M. Tarascon, L. H. Greene, W. R. McKinnon and G. W. Hull, ibid., p. 2394.

50. W. J. Yeh, L. Chen, F. Xu, B. Bi and P. Yang, ibid., 2414.

51. F. Zuo, B. R. Patton, D. L. Cox, S. I. Lee, Y. Song, J. P. Golben, X. D. Chen, S. Y. Lee, Y. Cao, Y. Lu, J. R. Gaines and J. C. Garland, ibid., p. 3603.

52. S. Senoussi, M. Oussena, M. Ribault and G. Collin, ibid., p. 4003.

53. A. C. Mota, A. Pollini, P. Visani, K. A. Müller and J. G. Bednorz, ibid., p. 4011.

54. G. W. Crabtree, J. Z. Liu, A. Umezawa, W. K. Kwok, C. H. Sowers, S. K. Malik, B. W. Veal, D. J. Lam, M. B. Brodsky and J. W. Downey, ibid., p. 4021.

55. D. E. Farrell, B. S. Chandrasekhar, M. R. McGuire, M. M. Fang, V. G. Kogan, J. R. Clem and D. K. Finnemore, ibid., p. 4025.

56. C. Rettori, D. Davidov, I. Belaish and I. Felner, ibid., p. 4028.

57. T. R. McGuire, T. R. Dinger, P. J. P. Freitas, W. J. Gallagher, T. S. Plaskett, R. L. Sandstrom and T. M. Shaw, ibid., p. 4032.

58. P. J. M. van Bentum, H. van Kempen, L. E. C. van de Leemput, J. A. A. J. Perenboom, L. W. M. Schreurs and P. A. A. Teunissen, ibid., p. 5279.

59. E. Polturak and B. Fisher, ibid., p. 5586.

60. J. Rammer, ibid., p. 5665.

61. A. Umezawa, G. W. Crabtree, J. Z. Liu, H. Weber, W. K. Kwok, L. H. Nunez, T. J. Moran and C. H. Sowers, ibid., p. 7151.

62. E. Tjukanov, R. W. Cline, R. Krahn, M. Hayden, M. W. Reynolds, W. N. Hardy, J. F. Carolan and R. C. Thompson, ibid., p. 7244.

63. J. S. Moodera, P. M. Tedrow and J. E. Tkaczyk, ibid., p. 8329.

64. L. F. Schneemeyer, E. M. Gyorgy and J. V. Waszczak, ibid., p. 8804.

65. P. Chaudhari, R. T. Collins, P. Freitas, R. J. Gambino, J. R. Kirtley, R. H. Koch, R. B. Laibowitz, F. K. LeGoues, T. R. McGuire, T. Penney, Z. Schlesinger, A. P. Segmüller, S. Foner and E. J. McNiff, Jr., ibid., p. 8903.

66. A. Ourmazd, J. A. Rentschler, W. J. Skocpol and D. W. Johnson, Jr., ibid., p. 8914.

67. M. Tuominen, A. M. Goldman and M. L. Mecartney, PHYS. REV. B 37, 548 (1988).

68. J. F. Riley, W. S. Sampath, K. Y. Lee, N. Mate and J. W. Blake, ibid., p. 559.

69. J. S. Moodera, R. Meservey, J. E. Tkaczyk, C. X. Hao, G. A. Gibson and P. M. Tedrow, ibid., p. 619.

70. M. E. McHenry, J. McKittrick, S. Sasayama, V. Kwapong, R. C. O'Handley and G. Kalonji, ibid., p. 623.

71. J. R. Cooper, C. T. Chu, L. W. Zhou, B. Dunn and G. Grüner, ibid., p. 638.

72. C. Kittel, S. Fahy and S. G. Louie, ibid., p. 642.

73. M. M. Fang, V. G. Kogan, D. K. Finnemore, J. R. Clem, L. S. Chumbley and D. E. Farrell, ibid., p. 2334.

74. E. H. Brandt, ibid., p. 2349.

75. Y. Wolfus, Y. Yeshurun and I. Felner, ibid., p. 3667.

76. M. A.-K. Mohamed, W. A. Miner, j. Jung, J. P. Franck and S. B. Woods, ibid., p. 5834.

77. T. Datta, C. P. Poole, Jr., H. A. Farach, C. Almasan, J. Estrada, D. U. Gubser and S. A. Wolf, ibid., p. 7843.

78. B. Oh, K. Char, A. D. Kent, M. Naito, M. R. Beasley, T. H. Geballe, R. H. Hammond, A. Kapitulnik and J. M. Graybeal, ibid., p. 7861.

79. Y. Tajima, M. Hikita, T. Ishii, H. Fuke, K. Sugiyama, M. Date, A. Yamagishi, A. Katsui, Y. Hidaka, T. Iwata and S. Tsurumi, ibid., 7956.

80. D. W. Cooke, R. L. Hutson, R. S. Kwok, M. Maez, H. Rempp, M. E. Scillaci, J. L. Smith, J. O. Willis, R. L. Lichti, K.-C. B. Chan, C. Boekema, S. P. Weathersby, J. A. Flint and J. Oostens, ibid., p. 9401.

81. A. M. Hermann, Z. Z. Sheng, D. C. Vier, S. Schultz and S. B. Oseroff, ibid., p. 9742.

82. J. F. Kwak, E. L. Venturini, P. J. Nigrey and D. S. Ginley, ibid., p. 9749.

83. S. Senoussi, M. Oussena, G. Collin and I. A. Campbell, ibid., p. 9792.

84. R. L. Petersen and J. W. Ekin, ibid., p. 9848.

85. M. Oda, Y. Hidaka, M. Suzuki and T. Murakami, PHYS. REV. B **38**, 252 (1988).

86. M. Couach, A. F. Khoder, F. Monnier, B. Barbara and J. Y. Henry, ibid., p. 748.

87. R. J. Papoular and G. Collin, ibid., p. 768.

88. M. B. Salamon, S. E. Inderhees, J. P. Rice, B. G. Pazol, D. M. Ginsberg and N. Goldenfeld, ibid., p. 885.

89. S. A. Sunshine, T. Siegrist, L. F. Schneemeyer, D. W. Murphy, R. J. Cava, B. Batlogg, R. B. van Dover, R. M. Fleming, S. H. Glarum, S. Nakahara, R. Farrow, J. J. Krajewski, S. M. Zahurak, J. V. Waszczak, J. H. Marshall, P. Marsh, L. W. Rupp, Jr., and W. F. Peck, ibid., p. 893.

90. Y. J. Uemura, V. J. Emery, A. R. Moodenbaugh, M. Suenaga, D. C. Johnston, A. J. Jacobson, J. T. Lewandowski, J. H. Brewer, R. F. Kiefl, S. R. Kreitzman, G. M. Luke, T. Riseman, C. E. Stronach, W. J. Kossler, J. R. Kempton, X. H. Yu, D. Opie and H. Schone, ibid., 909.

91. L. J. Campbell, M. M. Doria and V. G. Kogan, ibid., 2439.

92. X. Obradors, A. Labarta, J. Tajada, M. Vallet and J. M. Gonzalez-Calbet, ibid., 2455.

93. A. Umezawa, G. W. Crabtree, J. Z. Liu, T. J. Moran, S. K. Malik, L. H. Nunez, W. L. Kwok and C. H. Sowers, ibid., p. 2843.

94. A. Houghton and M. A. Moore, ibid., p. 5045.

95. J. J. Lin, E. L. Benitez, S. J. Poon, M. A. Subramanian, J. Gopalkrishnan and A. W. Sleight, ibid., p. 5095.

96. T. T. M. Palstra, B. Batlogg, L. F. Schneemeyer, R. B. van Dover and J. V. Waszczak, ibid., 5102.

139

97. A. P. Malozemoff, L. Krusin-Elbaum, D. C. Cronemeyer, Y. Yeshurun and F. Holtzberg, ibid., p. 6490.

98. R. A. Klemm, ibid., p. 6641.

99. J. Y. Juang, J. A. Cutro, D. A. Rudman, R. B. van Dover, L. F. Schneemeyer and J. V. Waszczak, ibid., p. 7045.

100. V. G. Kogan, ibid., p. 7049.

101. S. S. P. Parkin, E. M. Engler, V. Y. Lee, A. I. Nazzal, Y. Tokura, J. B. Torrance and P. M. Grant, ibid., p. 7101.

102. Y. R. Wang and M. J. Rice, ibid., p. 7163.

103. A. P. Malozemoff, T. K. Worthington, Y. Yeshurun and F. Holtzberg, ibid., p. 7203.

104. D. H. Wu, C. A. Schiffman and S. Sridhar, ibid., to be published.

105. V. G. Kogan, M. M. Fang and S. Mitra, ibid., to be published.

106. L. Krusin-Elbaum, A. P. Malozemoff, Y. Yeshurun and F. Holtzberg, ibid., to be published.

107. Y. Yeshurun, A. P. Malozemoff, F. Holtzberg and T. R. Dinger, ibid., Dec., to be published.

108. H. Maletta, A. P. Malozemoff, D. C. Cronemeyer, C. C. Tsuei, R. L. Greene, J. G. Bednorz and K. A. Müller, SOLID STATE COMMUN. **62**, 323 (1987).

109. M. J. Naughton, P. M. Chaikin, C. W. Chu, P. H. Hor and R. L. Meng, ibid., 531.

110. D. A. Esparza, C. A. D'Ovidio, J. Guimpel, E. Osquiguil, L. Civale and F. de la Cruz, SOLID STATE COMMUN. **63**, 137 (1987).

111. E. Zirngiebl, J. O. Willis, J. D. Thompson, C. Y. Huang, J. L. Smith, Z. Fisk, P. H. Hor, R. L. Meng, C. W. Chu and M. K. Wu, ibid., p. 721.

112. H. Noel, P. Gougeon, J. Padiou, J. C. Levet, M. Potel, O. Laborde and P. Monceau, ibid. p. 915.

113. F. J. Kedves, S. Meszaros, K. Vad, G. Halasz, B. Keszei and L. Mihaly, ibid., p. 991.

114. C. Lin, G. Lu, Z. Liu, Y. Sun, J. Lan, G. Zhong, S. Feng, C. Wei and Z. Gan, ibid., p. 1129.

115. A. Bezinge, J. L. Jorda, A. Junod and J. Muller, SOLID STATE COMMUN. **64**, 79 (1987).

116. F. Zuo, B. R. Patton, T. W. Noh, S.-I. Lee, Y. Song, J. P. Golben, X.-D. Chen, S. Y. Lee, J. R. Gaines and J. C. Garland, ibid., p. 83.

117. C. Ayache, B. Barbara, E. Bonjour, H. Calemczuk, M. Couach, J. H. Henry and J. Rossat-Mignod, ibid., p. 247.

118. J. F. Carolan, W. N. Hardy, R. Krahn, J. H. Brewer, R. C. Thompson and A. C. D. Chaklader, ibid., p. 717.

119. J. Xu, ibid., p. 893.

120. L. Civale, H. Safar, F. de la Cruz, D. A. Esparza and C. A. D'Ovidio, SOLID STATE COMMUN. **65**, 129 (1988).

121. X. Cao, G. Han, T. Zhang, H. Wen, Y. Wang, J. Wang, X. Mao and Y. Zhang, ibid., p. 359.

122. A. Wisniewski, M. Baran, P. Przyslupski, H. Szymczak, A. Pajaczkowska, B. Pytel and K. Pytel, ibid., p. 577.

123. J. G. Perez-Ramirez, K. Baberschke and W. G. Clark, ibid., p. 845.

124. P. N. Arberg, F. S. Razavi, F. P. Koffyberg and B. Mitrovic, ibid., 849.

125. X.-L. Chen and S.-G. Chen, ibid., 873.

126. Z. H. Mai and X. Chu, ibid., 877.

127. H. Zhang, S. S. Yan, H. Ma, J. L. Peng, Y. X. Sun, G. Z. Li, Q. Z. Wen and W. B. Zhang, ibid., 1125.

128. J. L. Tholence, H. Noel, J. C. Levet, M. Potel and P. Gougeon, ibid., 1131.

129. Z. Przelozny, ibid., 1171.

130. F. Gömöry and P. Lobotka, SOLID STATE COMMUN. **66**, 645 (1988).

131. P. Pureur, J. Schaf, J. V. Kunzler and E. R. Fraga, ibid., p. 931.

132. J. H. P. Emmen, W. J. N. de Jonge, C. v. d. Steen, J. H. J. Dalderop, P. M. A. Geppaart and V. A. Brabers, ibid., p. 1089.

133. G. J. Nieuwenhuys, T. A. Friedmann, J. P. Rice, P. M. Gehring, M. B. Salamon and D. M. Ginsberg, SOLID STATE COMMUN. **67**, 253 (1988).

134. B. Ya. Shapiro, ibid., p. 409.

135. L. Ya. Vinnikov, L. A. Gurevich, G. A. Yemelchenko and Yu. A. Ossipyan, ibid., p. 421.

136. O. Laborde, P. Monceau, M. Potel, P. Gougeon, J. Padiou, J. C. Levet and H. Noel, ibid., p. 609.

137. M. Avirovic, Ch. Neumann, P. Ziemann, J. Geerk and H. C. Li, ibid., p. 795.

138. H. Duan, L. Lu and D. Zhang, ibid., p. 809.

139. H. Zhang, S. Yan, W. Zhang, Z. Shen, Y. Sun, G. Li and K. Wu, ibid., p. 1183.

140. G. Deutscher, PHYSICA C **153-155**, 15 (1988).

141. Y. Iye, T. Tamegai, T. Sakakibara, T. Goto, N. Miura, H. Takeya and H. Takei, ibid., p. 26.

142. T. K. Worthington, W. J. Gallagher, D. L. Kaiser, F. H. Holtzberg and T. R. Dinger, ibid., p. 32.

143. M. Sato, ibid., p. 38.

144. J. Clem, ibid., p. 50.

145. I. Morgenstern, K. A. Müller and J. G. Bednorz, ibid., p. 59.

146. D. Stroud and C. Ebner, ibid., p. 63.

147. A. C. Mota, A. Pollini, P. Visani, K. A. Müller and J. G. Bednorz, ibid., p. 67.

148. H. Keller, B. Pümpin, W. Kündig, W. Odermatt, B. D. Patterson, J. W. Schneider, H. Simmler, S. Connell, K. A. Müller, J. G. Bednorz, K. W. Blazey, I. Morgenstern, C. Rossel and I. M. Savic, ibid., p. 71.

149. Y. Zhao, S. Sun. Z. Su, W. Kuan, H. Zhang, Z. Chen and Q. Zhang, ibid., p. 304.

150. C. Rossel and P. Chaudhari, ibid., p. 306.

151. M. E. McHenry, M. Foldeaki, J. McKittrick, R. C. O'Handley and G. Kalonji, ibid., p. 310; M. Foldeaki, M. E. McHenry R. C. O'Handley, submitted.

152. P. Norling, P. Svedlindh, P. Nordblad, L. Lundgren and P. Przyslupsky, ibid., p. 314.

153. S. B. Nam, ibid., p. 316.

154. M. Daeumling, J. Seuntjens and D. C. Larbalestier, ibid., p. 318.

155. I. Kostadinov, M. Mateev, I. V. Petrov, P. Vassilev and J. Tihov, ibid., p. 320.

156. C. W. Hagen, M. R. Bom, R. Griessen, B. Dam and H. Veringa, ibid., p. 322.

157. M. Tuominen, A. M. Goldman and M. L. Mecartney, ibid., p. 324.

158. C. Giovannella, L. Fruchter and C. Chappert, ibid., p. 326.

159. C. Giovannella, ibid., p. 328.

160. J. P. Renard, C. Giovannella, L. Fruchter and C. Chappert, ibid., p. 330.

161. T. Datta, C. Almasan, D. U. Gubser, S. A. Wolf and L. E. Toth, ibid., p. 332.

162. E. Agostinelli, G. Balestrino, S. Barbanera, P. Filaci, D. Fiorani and A. M. Testa, ibid., p. 334.

163. P. Przyslupski, A. Wisniewski, S. Kolesnik, W. Dobrowolski A. Pajaczkowska, K. Pytel and B. Pytel, ibid., p. 345.

164. K. V. Rao, R. Puzniak, D.-X. Chen, N. Karpe, M. Baran, A. Wisniewski, K. Pytel, H. Szymczak, K. Dyrbye and J. Bottiger, ibid., p. 347.

165. L. E. Wenger, W. Win, J. T. Chen, J. Obien, M. Wali, M. Bhullar, E. M. Logothetis, R. E. Soltis and D. Ager, ibid., p. 353.

166. J. S. Abell, M. S. Colclough, E. M. Forgan, C. E. Gough, C. M. Muirhead, D. A. O'Connor, W. F. Vinen and F. Wellhofer, ibid., p. 359.

167. H. Küpfer, I. Apfelstedt, R. Flükiger, R. Meier-Hirmer, W. Schauer, T. Wolf and H. Wühl, ibid., p. 367.

168. X. Obradors, C. Rillo, M. Vallet, A. Labarta, J. Fontcuberta, J. Gonzalez-Calbet and F. Lera, ibid., p. 389.

169. E.-W. Scheidt, M. Schaefer, H. Riesmayer and K. Luders, ibid., p. 391.

170. J.-P. Locquet, M. d'Halle, J. Vanacken, W. Boon, I. K. Schuller, C. Van Haesendonck and Y. Bruynseraede, ibid., p. 631.

171. N. Schopohl and A. Baratoff, ibid., p. 689.

172. W. Barford and J. M. F. Gunn, ibid., p. 691.

173. R. Wördenweber, K. Heinemann and H. C. Freyhardt, ibid., p. 870.

174. P. Kes, ibid., p. 1121.

175. A. Schenk, ibid., p. 1127.

176. J. W. C. de Vries, M. A. M. Gijs and G. M. Stollman, ibid., p. 1437.

177. R. Buder, M. Boujida, J. C. Bruyere, C. Escribe-Filippini, J. Marcu, P. L. Reydet and C. Schlenker, ibid., p. 1441.

178. A. Yamagishi, H. Fuke, K. Sugiyama, M. Date, Y. Tajima, M. Hikita, T. Ishii, A. Katsui, Y. Hidaka, T. Iwata and S. Tsurumi, ibid., p. 1459.

179. A. Umezawa, G. W. Crabtree and J. Z. Liu, ibid., p. 1461.

180. L. A. Schwartzkopf, M. M. Fang, L. S. Chumbley and D. K. Finnemore, ibid., p. 1463.

181. J. van den Berg, C. J. van der Beek, P. Kes, J. A. Mydosh, A. A. Menovsky and M. J. V. Menken, ibid., p. 1465.

182. O. Laborde, P. Monceau, M. Potel, P. Gougeon, J. Padiou, J. C. Levet and H. Noel, ibid., p. 1467.

183. L. Krusin-Elbaum, A. P. Malozemoff, Y. Yeshurun, D. C. Cronemeyer and F. Holtzberg, ibid., p. 1469.

184. Y. Isikawa, K. Mori, K. Kobayashi and K. Sato, ibid., p. 1471.

185. C. Giovannella, L. Fruchter, G. Collin and I. A. Campbell, ibid., p. 1473.

186. V. A. M. Brabers, W. J. M. de Jonge, A. T. A. M. de Waele, J. H. P. M. Emmen and P. M. A. Geppaart, ibid., p. 1475.

187. A. F. Khoder, M. Couach and B. Barbara, ibid., p. 1477.

188. J. L. Tholence, H. Noel, J. C. Levet, M. Potel, P. Gougeon, G. Chouteau and M. Guillot, ibid., p. 1479.

189. S. E. Male, J. Chilton and A. D. Caplin, ibid., p. 1483.

190. P. J. M. van Bentum, H. van Kempen, L. E. C. van de Leemput, J. A. A. J. Perenboim, L. W. M. Schreurs, E. van der Steen and P. A. A. Teunissen, ibid., p. 1485.

191. R. Steinman, P. Lejay, J. Chaussy and B. Pannetier, ibid., p. 1487.

192. P. Monod, B. Dubois and P. Odier, ibid., p. 1489.

193. J. R. Cooper, M. Petravic, D. Drobac, B. Korin-Hamzic, N. Brnicevic, M. Paljevic and G. Collin, ibid., p. 1491.

194. P. Peyral, J. Rosenblatt, A. Raboutou, C. Lebeau, C. Perrin, O. Pena, A. Perrin and M. Sergent, ibid., p. 1493.

195. W. Boon, M. d'Halle, J.-P. Locquet, M. Bruggeman, H. Strauven, O. B. Verbeke, J. Vanacken, C. van Haesendonck and Y. Bruynseraede, ibid., p. 1497.

196. P. Esquinazi and C. Duran, ibid., p. 1499.

197. C. E. Gough, M. N. Keene, C. Mee, A. I. M. Rae and S. J. Abell, ibid., p. 1501.

198. X. Obradors, A. Labarta, M. Vallet and J. Gonzalez-Calbet, ibid., p. 1503.

199. W. Andrä, H. Danan, R. Hergt, E. Jäger, T. Klupsch and H. Pfeiffer, ibid., p. 1505.

200. A. Ding, Z. Yu, K. Shi and J. Yan, ibid., p. 1509.

201. E. Babic, Z. Marohnic, D. Drobac, M. Prester and N. Brnicevic, ibid., p. 1511.

202. J. Cornelis, J. Vansummeren, H. OOms, R. Gilissen, J. Cooymans, N. Maene, F. Biermans and A. van den Bosch, ibid., p. 1519.

203. C. S. Jacobsen, N. E. Dam, K. F. Nielsen, I. Johannsen and A. S. Petersen, ibid., p. 1521.

204. N. Kobayashi, H. Iwasaki, S. Terada, K. Noto, A. Tokiwa, M. Kikuchi, Y. Syono and Y. Muto, ibid., p. 1525.

205. E. Leyarovski, L. Leyarovska, N. Leyarovska, Chr. Popov and M. Kirov, ibid., p. 1527.

206. S. Meszaros, G. Halasz, F. J. Kedves and Sz. Balanyi, ibid., p. 1529.

207. J. S. Munoz, A. Sanchez, T. Puig, D.-X. Chen and K. V. Rao, ibid., p. 1531.

208. C. Rillo, F. Lera, J. Garcia, J. Bartolome, R. Navarro, D. Gonzalez, M. A. Alario-Franco, D. Beltran, D. H. A. Blank, J. Gonzalez-Calbet, J. Flokstra, R. Ibanez, E. Moran, J. S. Munoz, X. Obradors, A. Sanchez and M. Vallet, ibid., p. 1533.

209. K. Kwasnitza, V. Plotzner, M. Waldmann and E. Widmer, ibid., p. 1565.

210. C. E. Gough, ibid., p. 1569.

211. R. Flükiger, T. Müller, W. Goldacker, T. Wolf, E. Seibt, I. Apfelstedt, H. Küpfer, W. Schauer, ibid., p. 1574.

212. D. C. Larbalestier, S. E. Babcock, X. Cai, M. Daeumling, D. P. Hampshire, T. F. Kelly, L. A. Lavanier, P. J. Lee and J. Seuntjens, ibid., p. 1580.

213. J. Rammer, ibid., p. 1625.

144

214. R. H. Koch, C. P. Umbach, M. M. Oprysko, J. D. Mannhart, B. Bumble, G. J. Clark, W. J. Gallagher, A. Gupta, A. Kleinsasser, R. B. Laibowitz, R. B. Sandstrom and M. R. Scheuermann, ibid., p. 1685.

215. L. Ya. Vinnikov, G. A. Emelchenko, P. A. Kononovich, Yu. A. Ossipyan and I. F. Shchegolev, ibid., p. 1359.

216. D. W. Capone, II, D. G. Hinks, J. D. Jorgensen and K. Zhang, APPL. PHYS. LETT. **50**, 543 (1987).

217. A. J. Panson, G. R. Wagner, A. I. Braginski, J. R. Gavaler, M. A. Janocko, H. C. Pohl and J. Talvacchio, ibid., p. 1104.

218. R. H. Koch, C. P. Umbach, G. J. Clark, P. Chaudhari and R. B. Laibowitz, APPL. PHYS. LETT. **51**, 200 (1987).

219. B. Oh, M. Naito, S. Arnason, P. Rosental, R. Barton, M. R. Beasley, T. H. Geballe, R. H. Hammond, A. Kapitulnik and J. M. Graybeal, ibid., p. 852.

220. H. Kumakura, M. Uerhara and K. Togano, ibid., p. 1557.

221. M. Barsoum, D. Patten and S. Tyagi, ibid., p. 1954.

222. C. J. Jou, E. R. Weber, J. Washburn and W. A. Soffa, APPL. PHYS. LETT. **52**, 326 (1988).

223. E. M. Gyorgy, G. S. Grader, D. W. Johnson, Jr., L. C. Feldman, D. W. Murphy, W. W. Rhodes, R. E. Howard, P. M. Mankiewich and W. J. Skocpol, ibid., p. 328.

224. M. Daeumling, J. Seuntjens and D. C. Larbalestier, ibid., p. 590.

225. R. Williams and J. R. Matey, ibid., p. 751.

226. F. C. Moon, M. M. Yanoviak and R. Ware, ibid., p. 1534.

227. R. B. van Dover, L. F. Schneemeyer, E. M. Gyorgy and J. V. Waszczak, ibid., p. 1910.

228. H. Kumakura, K. Takahashi, D. R. Dietderich, K. Togano and H. Maeda, ibid., p. 2064.

229. J. H. Kang, R. T. Kampworth and K. E. Gray, ibid., p. 2080.

230. A. T. Fiory, A. F. Hebard, P. M. Mankiewich and R. E. Howard, ibid., 2165.

231. D. Wong, A. K. Stamper, D. D. Stancil and T. E. Schlesinger, APPL. PHYS. LETT. **53**, 240 (1988).

232. G. S. Grader, E. M. Gyorgy, L. G. Van Uitert, W. H. Grodkiewicz, T. R. Kyle and M. Eibschutz, ibid., p. 319.

233. J. Takada, Y. Bando and M. Mazaki, ibid., p. 332.

234. M. J. Ferrari, M. Johnson, F. C. Wellstood, J. Clarke, P. A. Rosenthal, R. H. Hammond and M. R. Beasley, ibid., p. 695.

235. W. G. Harter, A. M. Hermann and Z. Z. Sheng, ibid., p. 1119.

236. L. H. Allen, P. R. Broussard, J. H. Claassen and S. A. Wolf, ibid., p. 1338.

237. E. H. Brandt, ibid., p. 1554.

238. D.-X. Chen, R. B. Goldfarb, J. Nogues and K. V. Rao, APPL. PHYS. LETT., submitted.

239. D. D. Stancil, T. E. Schlesinger, A. K. Stamper and D. Wong, ibid., submitted.

240. S. Martin, A. T. Fiory, R. M. Fleming, G. P. Espinosa and A. S. Cooper, ibid., submitted.

241. J. H. Kang, K. E. Gray, R. T. Kampwirth and D. W. Day, Appl. Phys. Lett., submitted.

242. D. C. Larbalestier, M. Dauemling, X. Cai, J. Seuntjens, J. McKinnell, D. Hampshire, P. Lee, C. Meingast, T. Willis, H. Muller, R. D. Ray, R. G. Dillenburg, E. E. Hellstrom and R. Joynt, J. APPL. PHYS. **62**, 3308 (1987).

243. J. W. Ekin, A. I. Braginski, A. J. Panson, M. A. Janocko, D. W. Capone II, N. J. Zaluzec, B. Flandermeyer, O. F. de Lima, M. Hong, J. Kwo and S. H. Liou, ibid., 4821. T. R. McGuire, F. Holtzberg,

244. F. Hellman, E. M. Gyorgy, D. W. Johnson, Jr., H. M. O'Bryan and R. C. Sherwood, J. Appl. Phys. **63**, 447 (1988).

245. S. Senoussi, M. Oussena and S. Hadjoudj, ibid., p. 4176.

246. U. Atzmony, R. D. Schull, C. K. Chiang, L. J. Swartzendruber, L. H. Bennett, ibid., p. 4179.

247. M. E. McHenry, J. McKittrick, S. Sasayama, V. Kwapong, R. C. O'Handley and G. Kalonji, ibid., p. 4229.

248. C. W. Yuan, B. R. Zhao, Y. Z. Zhang, Y. Y. Zhao, Y. Lu, H. S. Wang, Y. H. Shi, J. Gao and L. Li, J. APPL. PHYS. **64**, 4091 (1988).

249. Y. Yeshurun, A. P. Malozemoff and F. Holtzberg, J. APPL. PHYS., to be published.

250. Y. J. Uemura, B. J. Sternlieb, D. E. Cox, V. J. Emery, A. Moodenbaugh, M. Suenaga, J. H. Brewer, J. F. Carolan, W. Hardy, R. Kadono, J. R. Kempten, R. F. Kiefl, S. R. Kreitzman, G. M. Luke, P. Mulhern, T. Riseman, D. L. Williams, B. X. Yang, W. J. Kossler, X. H. Yu, H. Schone, C. E. Stronach, J. Gopalakrishnan, M. A. Subramanian, A. W. Sleight, H. Hart, K. W. Lay, H. Takagi, S. Uchida, Y. Hidaka, T. Murakami, S. Etemad, P. Barboux, D. Keane, V. Lee and D. C. Johnston, ibid., to be published.

251. B. Renker, I. Apfelstedt, H. Küpfer, C. Politis, H. Rietschel, W. Scheuer, H. Wuhl, U. Gottwick, H. Kneissel, U. Rauchschwalbe, H. Spille and F. Steglich, ZEITSCHRIFT. PHYS. B **67**, 1 (1987).

252. F. Hulliger and H. R. Ott, ibid., p. 291.

253. I. Morgenstern, K. A. Müller and J. G. Bednorz, ZEITSCHRIFT. PHYS. B **69**, 33 (1987).

254. N. Garcia, S. Vieira, A. M. Baro, J. Tornero, M. Pazos, L. Vazquez, J. Gomez, A. Aguilo, S. Bourgeal, A. Buendia, M. Hortal, M. A. Lopez de la Torre, M. A. Ramos, R. Villar, K. V. Rao, D.-X. Chen, J. Nogues and N. Karpe, ZEITSCHRIFT. PHYS. B. **70**, 9 (1988).

255. P. Leiderer and R. Feile, ibid., p. 141.

256. H. Küpfer, S. M. Green, C Jiang, Y. Mei, H. L. Luo, R. Meier-Hirmer and C. Politis, ZEITSCHRIFT. PHYS. B **71**, 63 (1988).

257. T. Schneider, D. Würtz and R. Hetzel, ZEITSCHRIFT. PHYS. B **72**, 1 (1988).

258. Y. Hidaka, Y. Enomoto, M. Suzuki, M. Oda, A. Katsui and T. Murakami, JPN. J. APPL. PHYS. **26**, L726 (1987).

259. Y. Iye, T. Tamegai, H. Takeya and H. Takei, ibid., p. L1057.

260. T. Sakakibara, T. Goto, Y. Iye, N. Miura, H. Takeya and H. Takei, ibid., p. L1892.

261. K. Shiraki, T. Fukami and S. Mase, JPN. J. APPL. PHYS. **27**, L1895 (1988).

262. K. Okuda and S. Noguchi, **Superconducting Materials**, ed. S. Nakajima and H. Fukuyama (Japanese Journal of Applied Physics, Tokyo, 1988), p. 51.

263. Y. Horie, T. Yasuda, A. A. A. Youssef and R. Kondo, JPN. J. APPL. PHYS., to be published.

264. E. Shimizu and D. Ito, Proceedings of ISS '88, Nagoya, Japan, Aug. 29-31, 1988 (Springer, Tokyo, 1988), to be published.

265. H. Itozaki, K. Higaki, K. Harada, S. Tanaka, N. Fujimori and S. Yazu, ibid., to be published.

266. R. H. Koch and A. P. Malozemoff, ibid., to be published.

267. Y. Hidaka, M. Oda, M. Suzuki, A. Katsui, T. Murakami, N. Kobayashi and Y. Muto, in **Proceedings of the Eighteenth Yamada Conference on Superconductivity in Highly Correlated Fermion Systems**, Physica B **148**, 329 (1987).

268. J. F. Kwak, E. L. Venturini, R. J. Baughman, B. Morosin and D. S. Ginley, Physica C, to be published.

269. A. P. Ramirez, B. Batlogg, R. J. Cava, L. Schneemeyer, R. B. van Dover, E. A. Rietman and J. V. Waszczak, in **Novel Superconductivity**, ed. S. A. Wolf and V. Z. Kresin (Plenum, New York, 1987), p. 689.

270. R. R. Corderman, H. Wiesmann, M. W. Ruckman and M. Strongin, ibid., p. 759.

271. A. I. Braginski, ibid., p. 935.

272. R. H. Arendt, A. R. Gaddipati, M. F. Garbauskas, E. L. Hall, H. R. Hart, Jr., K. W. Lay, J. D. Livingston, F. E. Luborsky and L. L Schilling, in **High Tem-**

perature Superconductors, ed. M. B. Brodsky, R. C. Dynes, K. Kitazawa and H. L. Tuller (Materials Research Society Pittsburgh PA, 1988), Vol 99, p. 203.

273. L. Krusin-Elbaum, A. P. Malozemoff and Y. Yeshurun, ibid., p. 221. The temperature calibration of these measurements was in error; $T_c$ should have been given as 89 K.

274. D. C. Cronemeyer, A. P. Malozemoff and T. R. McGuire, ibid., p. 837.

275. H. Küpfer, I. Apfelstedt, R. Flükiger, C. Keller, R. Meier-Hirmer, B. Runtsch, A. Turowski, U. Wiech and T. Wulf, International Conference on Critical Currents in High Temperature Superconductors, Aug. 16-19, 1988, Snowmass Village, Colorado USA, to be published.

276. Y. Yeshurun, A. P. Malozemoff, T. K. Worthington, R. M. Yandrofski, L. Krusin-Elbaum, F. Holtzberg, T. R. Dinger and G. V. Chandrasekhar, ibid., to be published.

277. M. Tinkham, Helv. Phys. Acta 61, 443 (1988).

278. M. Tinkham and C. J. Lobb, Solid State Physics 42, to be published.

279. A. P. Malozemoff, Y. Yeshurun, L. Krusin-Elbaum, T. K. Worthington, D. C. Cronemeyer, T. Dinger, F. Holtzberg, T. R. McGuire and P. Kes, Proceedings of the LACHTS Conference, Rio de Janeiro, Brazil, May 4-6, 1988 (World Scientific, Singapore), to be published.

280. C. E. Gough, M. S. Colclough, E. M. Forgan, R. G. Jordan, M. Keene, C. M. Muirhead, A. I. M. Rae, N. Thomas, J. S. Abell and S. Sulton, Nature 326, 855 (1987). (1987).

281. A. D. Hibbs and A. M. Campbell, Proceedings of the Applied Superconductivity Conference, San Francisco CA, Aug. 1988.

282. J. O. Willis, M. E. McHenry, M. P. Maley and H. Sheinberg, ibid., to be published.

283. C. Y. Huang, Y. Shapira, E. J. McNiff, P. N. Peters, B. B. Schwartz, M. K. Wu, R. D. Shull and C. K. Chiang, Modern Phys. Lett. B 869, (1988).

284. L. Civale, H. Safar and F. de la Cruz, Modern Phys Lett. B, submitted.

285. R. B. Goldfarb, A. F. Clark, A. I. Braginski and A. J. Panson, Cryogenics 27, 476 (1987).

286. D. Dew-Hughes, ibid., to be published.

287. A. K. Grover, C. Radhakrishnamurty, P. Chaddah, G. Ravi Kumar and G. V. Subba Rao, Pramana 30, 569 (1985).

288. G. Ravi Kumar and P. Chaddah, Pramana 31, 1141 (1988).

289. G. Ravi Kumar and P. Chaddah, Pramana, submitted.

290. L. Fruchter, C. Giovannella, M. Oussena, S. Senoussi and I. A. Campbell, J. de Physique, to be published.

291. L. Fruchter, C. Giovannella, G. Collin and I. A. Campbell, ibid., to be published.

292. A. Celani, R. Messi, N. Sparvieri, S. Pace, A. Saggese, C. Giovannella, L. Fruchter, C. Chappert and I. A. Campbell, ibid., to be published.

293. T. K. Worthington, Y. Yeshurun, A. P. Malozemoff, R. Yandrofski, F. Holtzberg and T. Dinger, ibid., to be published.

294. S. Senoussi, P. V. P. S. S. Shastry, J. V. Yakhmi and I. A. Campbell, ibid., to be published.

295. A. P. Malozemoff, T. K. Worthington, R. M. Yandrofski and Y. Yeshurun, in **Towards the Theoretical Understanding of High Temperature Superconductors**, ed. S. Lundqvist, E. Tosatti and Yu Lu (World Scientific, Singapore, 1988), to be published.

296. P. H. Kes, J. Aarts, J. van den Berg, C. J. van der Beek and J. A. Mydosh, Superconductor Science and Technology, submitted.

297. J. van den Berg, C. J. van der Beek, P. H. Kes and J. A. Mydosh, ibid., submitted.

298. J. Hammann, M. Ocio, E. Vincent, A. Bertinotti, D. Luzet, A. Revcolevschi and J. Jegoudez, Physica C **148B**, 325 (1987).

299. M. Polak, F. Hanic, I. Hlasnik, M. Majoros, F. Chovanec, I. Horvath, L. Krempsky, P. Kottman, M. Kedrova and L. Galikova, Physica C **156**, 79 (1988).

300. J. Z. Liu, K. G. Vandervoort, H. Claus, G. W. Crabtree and G. J. Lam, Physica C, submitted.

301. V. L. Aksenov and S. A. Sergeenkov, ibid., submitted.

302. V. B. Geshkenbein and A. I. Larkin, submitted.

303. M. B. Kartsovnik, V. A. Larkin, V. V. Ryazanov, N. S. Siborov and I. F. Shchegolev, JETP Lett. **47**, 595 (1988) (in Russian).

304. L. Ya. Vinnikov, L. A. Gurevich, G. A. Emelchenko and Yu. A. Ossipyan, Sov. Phys. JETP Lett. **47**, 109 (1988).

305. E.B. Sonin, ibid., p. 415 (in Russian).

306. V. M. Vinokur, L. B. Ioffe, A. I. Larkin and M. V. Feigel'man, JETP 93, 343 (1987) (in Russian).

307. A. A. Abrikosov, A. I. Buzdin, M. L. Kulic and D. A. Kuptsov, submitted.

308. G. Kapustin, E. Meilikhov, S. Oleinik and V. Shapiro, in **Superconductivity**, ed. V. Ozhogin (Kurchatov Atomic Energy Institute, Moscow, 1988), Vol 1, p. 33.

309. J. V. Jose, in **High Temperature Superconductivity**, ed. J. Heiras et al. (World Scientific Pub., Singapore, 1988).

310. M. A. Moore, Phys. Rev. B, to be published.

311. A. Houghton and R. A. Pelcovits, submitted.

312. M. V. Feigel'man and V. M. Vinokur, Phys. Rev. Lett., submitted.

313. M. P. A. Fisher, submitted.

314. L. Krusin-Elbaum, R. L. Greene, F. Holtzberg, A. P. Malozemoff and Y. Yeshurun, submitted

315. J. Mannhart, P. Chaudhari, D. Dimos, C. C. Tsuei and T. R. McGuire, Phys. Rev. Lett., to be published.

316. M. J. Naughton, R. C. Yu, P. K. Davies, J. E. Fischer, R. V. Chamberlin, Z. Z. Wang, T. W. Jing, N. P. Ong and P. M. Chaikin, Phys. Rev. B **37**, to be published.

317. K. Athreya, O. B. Hyun, J. E. Ostenson, J. R. Clem and D. K. Finnemore, Phys. Rev. B, to be published.

318. M. M. Fang, J. E. Ostenson, D. K. Finnemore, D. E. Farrell and N. P. Bansal, Phys. Rev. B, to be published.

319. R. A. Fisher, S. Kim, S. E. Lacy, N. E. Phillips, D. E. Morris, A. G. Markelz, J. Y. T. Wei and D. S. Ginley, submitted.

320. M. J. Higgins, D. P. Goshorn, S. Bhattacharya and D. C. Johnston, submitted.

321. N.-C. Yeh and C. C. Tsuei, submitted.

322. L. Drabeck, J. P. Carini, G. Güner, T. Hylton, K. Char and M. R. Beasley, Phys. Rev. B, submitted.

323. G. J. Dolan, G. V. Chandrasekhar, T. R. Dinger, C. Feild and F. Holtzberg, submitted.

324. C. Rossel, Y. Maeno and I. Morgenstern, Phys. Rev. Lett., submitted.

325. D. R. Harshman, L. F. Schneemeyer, J. V. Waszczak, G. Aeppli, R. J. Cava, B. Batlogg, L. W. Rupp, E. J. Ansaldo, R. F. Kiefl, G. M. Luke, T. M. Riseman and D. Ll. Williams, submitted.

326. P. H. Kes, J. van den Berg, C. J. van der Beek, J. A. Mydosh, L. W. Roeland, A. A. Menovsky, K. Kadowski and F. R. de Boer, Proceedings of the LACHTS Conference, Rio de Janeiro, Brazil, May 4-6, 1988 (World Scientific, Singapore), to be published.

327. D. E. Farrell, C. M. Williams, S. A. Wolf, N. P. Bansal and V. G. Kogan, submitted.

328. J. O. Willis, D. W. Cooke, R. D. Brown, J. R. Cost, J. F. Smith, J. L. Smith, R. M. Aikin and M. Maez, submitted.

329. E. Polturak, D. Cohen and A. Brokman, submitted.

330. V. G. Kogan and L. J. Campbell, submitted.

331. C. P. Bean, Phys. Rev. Lett. **8**, 250 (1962); Rev. Mod. Phys. **36**, 31 (1964).

332. C. Ebner and D. Stroud, Phys. Rev. B **31**, 165 (1987).

333. A. M. Campbell and J. E. Evetts, Adv. Phys. **21**, 199 (1972).

334. P. W. Anderson, Phys. Rev. Lett. **9**, 309 (1962); Y. B. Kim, Rev. Mod. Phys. **36**, 39 (1964).

335. M. Tinkham, **Introduction to Superconductivity** (McGraw-Hill Inc., New York, 1975).

336. J. R. L. de Almeida and D. J. Thouless, J. Phys. **A11**, 983 (1978); **C11**, L871 (1978).

337. A. M. Campbell, J. Phys. **C2**, 1492 (1969).

338. S. John and T. C. Lubensky, Phys. Rev. Lett. **55**, 1014 (1985); Phys. Rev. B **34**, 4815 (1986).

339. A. P. Malozemoff, W. J. Gallagher and R. E. Schwall, ACS Symposium Series, Vol 351, ed. D. L. Nelson, M. S. Wittingham and T. F. George (American Chemical Society, Washington DC 1987), Chapter 27, pp. 280-306.

340. M. R. Beasley, R. Labusch and W. W. Webb, Phys. Rev. **181**, 682 (1969).

341. V. G. Kogan, Phys. Rev. B **24**, 1572 (1981).

342. R. A. Klemm and J. R. Clem, Phys. Rev. B **21**, 1868 (1980).

# 4
# NEUTRON SCATTERING STUDIES OF STRUCTURAL AND MAGNETIC EXCITATIONS IN LAMELLAR COPPER OXIDES — A REVIEW

Robert J. Birgeneau
*Department of Physics, Massachusetts Institute of Technology*
*Cambridge, MA 02139, USA*

and

Gen Shirane
*Department of Physics, Brookhaven National Laboratory*
*Upton, NY 11973, USA*

I.    INTRODUCTION

The discovery of superconductivity in doped $La_2CuO_4$ by Bednorz and Müller[1] has stimulated an enormous number of studies of this and related lamellar copper oxide materials.  It is broadly viewed that in order to understand the novel superconductivity it will be necessary to elucidate fully the behavior across the phase diagram as a function of doping, temperature, etc.  It has already been found that the phase diagrams are remarkably rich with structural phase transitions,[2-5] three dimensional (3D) Néel states,[6-9] 2D antiferromagnetic quantum spin fluctuations,[10,11] spin glassbehavior,[12-19] insulator-metal transitions[5,20] and, of course, the superconductivity transitions.[1, 21-23]

In order to unravel the essential physics of these materials one requires extensive and accurate experimental information obtained with the full panoply of probes.  In this book, data obtained so far using bulk thermodynamic and transport techniques, optical and light scattering probes, x-ray and neutron scattering, NMR and NQR and many other techniques are reviewed.  In the vast majority of models for the superconductivity, the spin degrees of freedom of the $Cu^{2+}$ ions are essential and indeed, in many models, the magnetism is the progenitor of the high temperature superconductivity.[24]  For these aspects of the problem, neutron scattering plays a unique role since it is able to provide detailed microscopic information on the spatial and temporal variation of the spin fluctuations.  As we shall discuss in this review, the magnetic behavior is unexpectedly rich and complicated in both the insulating and metallic states.

Neutron scattering also is able to probe both the structures and lattice dynamical excitations.[25]  Of course, in conventional superconductors the phonons mediate the pairing.[26]  Isotope effect measurements suggest that the phonons play a minor role in the $La_2CuO_4$ materials[27] and have virtually no observable effect on $T_c$ in compounds such as $Y_1Ba_2Cu_3O_7$ where $T_c$ is near 90 K.[28]  It is of interest, nevertheless, to characterize the phonon excitations in order to elucidate any possible anomalous behavior.

For both the lattice dynamical and spin fluctuation measurements, meaningful information can only be obtained from experiments utilizing large, high quality homogeneous single crystals. Thus these measurements place great demands on crystal growth techniques. Remarkable success has been achieved in the growth of single crystals of $La_{2-x}Sr_xCuO_4$[29,30] and, most recently, $YBa_2Cu_3O_{6+x}$.[31] As far as we are aware, no single crystals of the Bi or Tl lamellar $CuO_2$ materials have been grown so far which are suitable for inelastic magnetic neutron scattering experiments. Thus, this article will be limited to a review of experiments in $La_{2-x}Sr_xCuO_4$ and $YBa_2Cu_3O_{6+x}$ and related compounds. Furthermore, to-date, successful inelastic scattering experiments have been performed almost exclusively at Brookhaven National Laboratory. Thus this review will by necessity emphasize the Brookhaven experiments.

In brief, extensive elastic, quasielastic and inelastic neutron scattering studies have been performed on the lamellar $CuO_2$ materials $La_{2-x}Sr_xCuO_4$, and $YBa_2Cu_3O_{6+x}$ as well as related compounds. In this chapter we review the essential information obtained in these neutron experiments. For $x = 0$, both systems are antiferromagnetic insulators with quite high ordering temperatures. Further the spins in the $CuO_2$ sheets are well described by the 2D $S = 1/2$ square lattice Heisenberg model - a system of fundamental interest in Statistical Physics. With increasing x the Néel state is destroyed and, in $La_{2-x}Sr_xCuO_4$ at least, is replaced at low temperatures by a slowly fluctuating 3D incommensurate spin fluid state. The spin-spin correlation length in the doped $CuO_2$ sheets equals the average separation between the holes which ultimately carry the supercurrent. The exact connection between this novel magnetic state and the superconductivity remains to be elucidated. Lattice dynamical studies of $La_{2-x}Sr_xCuO_4$ reveal classic soft mode behavior at the tetragonal-orthorhombic transition but the lattice excitations do not seem to be central to the high $T_c$ superconductivity.

The format of this paper is as follows. In Sec. II, we review measurements on the crystal structure, structural phase transition and low lying phonon modes in $La_{2-x}Sr_xCuO_4$. In Sec. III, magnetic structure measurements on pure and lightly doped $La_2CuO_4$ are discussed. Section IV presents neutron studies of the instantaneous spin correlations and magnetic excitations in pure $La_2CuO_4$. In the next section (V), measurements of the static and dynamic spin correlations in $La_{2-x}Sr_xCuO_4$ in both insulating and metallic samples are reviewed. Analogous experiments in the isomorphous compounds $La_2NiO_4$ and $La_2CoO_4$ are discussed in Sec. VI. The much more limited data available for 123-type materials are reviewed in Sec. VII. Finally our conclusions are given in Sec. VIII.

II.    $La_{2-x}Sr_xCuO_4$ STRUCTURE AND LATTICE DYNAMICS

The crystal structure of $La_2CuO_4$ is shown in Figure 1. At high temperatures, the structure is body-centered tetragonal, space group I4/mmm, with one $La_2CuO_4$ formula unit per primitive cell. At a temperature $T_0$, which is ~ 530 K in stoichiometric material,

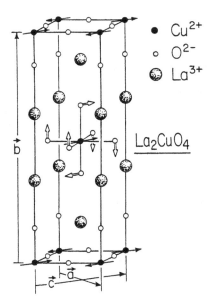

FIG. 1. Crystal and magnetic structure of $La_2CuO_4$; the arrows associated with the center oxygens indicate the direction of rotation in the orthorhombic phase.

the crystal exhibits a second order transition to the orthorhombic phase, space group Cmca, which has two formula units per unit cell.[2,3] The structural transition involves primarily a staggered rotation of the $CuO_6$ octahedra about either a $(1\ 1\ 0)_T$ or $(1\ \bar{1}\ 0)_T$ axis; here the subscript T implies that one is using the tetragonal unit cell. The identical structural transition occurs in the Sr-doped material $La_{2-x}Sr_xCuO_4$ but with $T_0$ decreasing continuously to 0 K as x increases to ~ 0.2.[5,32,33,34] This is illustrated in the phase diagram, Fig. 2, for $La_{2-x}Sr_xCuO_4$.

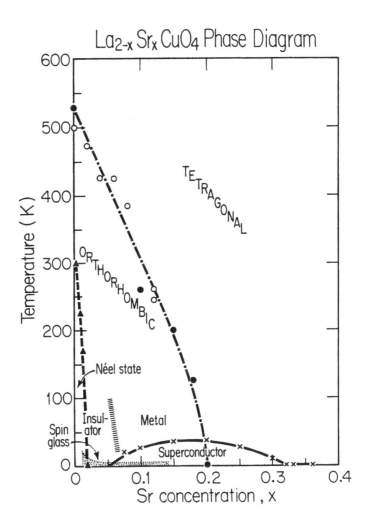

FIG. 2. Phase diagram of La$_{2-x}$Sr$_x$CuO$_4$. See Ref. 5, 13-18, 52 and 87.

Measurements of the superlattice peak intensities in the orthorhombic phase as well as the strain for one sample[3,35] are shown in Fig. 3. The solid lines are the results of fits to a single power law $I \sim (T_0 - T)^{2\beta}$; for these two samples $\beta = 0.28\pm0.03$; an average over all samples studied to-date gives $\beta = 0.30\pm0.03$.

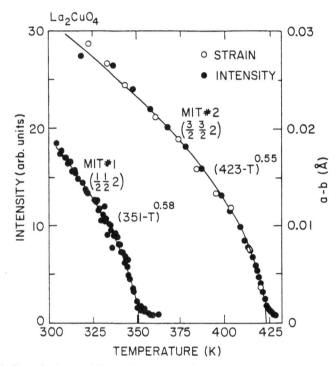

FIG. 3. Superlattice peak intensities in $La_{2-x}Sr_xCuO_4$. The solid lines are the results of fits to the power laws indicated in the figure. Note the excellent scaling between the orthorhombic distortion and Bragg intensity for sample MIT#2. Both samples have $\sim$ 5% Li replacing the Cu; MIT#1 also has 0.03 Sr replacing the La. Figure from Ref. 35.

As discussed in Ref. 3, the atomic displacements involved in the transformation are described by a doubly degenerate (n = 2) order parameter $\{Q_x\} = (Q_\alpha, Q_\beta)$ with X-point wavevectors $\{\vec{q}_x\} = (\vec{q}_\alpha, \vec{q}_\beta)$, where $\vec{q}_\alpha = (\frac{1}{2}\frac{1}{2}0)_T$ and $\vec{q}_\beta = (\frac{1}{2}\frac{1}{2}0)_T$. The primitive lattice vectors can be written as $\vec{b}_1 = (0\ 1\ 1)_T$, $\vec{b}_2 = (1\ \bar{1}\ 0)_T$, and $\vec{b}_3 = (1\ 0\ \bar{1})_T$ in terms of the conventional tetragonal lattice vectors and thus $-\vec{q}_\alpha = \vec{q}_\alpha - \vec{b}_1 - \vec{b}_3 \equiv \vec{q}_\alpha$, since $\vec{q}_\alpha$ is defined modulo a reciprocal lattice vector. Similarly $-\vec{q}_\beta \equiv \vec{q}_\beta$ and therefore both amplitudes $(Q_\alpha, Q_\beta)$ are real. The

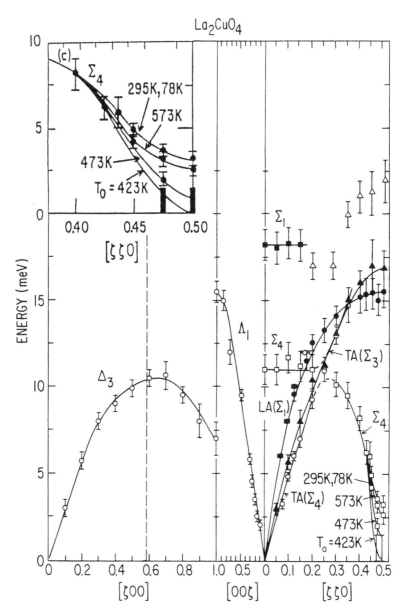

FIG. 4. Summary of the low-lying phonon branches in $La_2CuO_4$. The lines are guides to the eye. The modes are labelled according to Ref. 38. The dispersion curves are only weakly temperature dependent with the exception of the TO phonons near the X point (see inset (c)). Figure adapted from Ref. 35.

critical behavior is determined by the nature of the symmetry invariants of the
Landau-Wilson Hamiltonian. Direct inspection shows that $Q_\alpha \rightarrow \pm Q_\alpha$ or $Q_\beta \rightarrow \pm Q_\beta$ for
all of the tetragonal symmetry operations. Therefore the lowest order symmetry invariant
combinations of $(Q_\alpha, Q_\beta)$ can be chosen as $(Q_\alpha^2 + Q_\beta^2)$, $(Q_\alpha^2 + Q_\beta^2)^2$ and $(Q_\alpha^4 + Q_\beta^4)$, which
is identical to the well known $n = 2$ XY-model with cubic anisotropy.[36] For three spatial
dimensions $(d = 3)$ the stable (isotropic) fixed point for the model gives $2\beta \approx 0.70$.
However, the asymptotic critical region may be small due to the nearby cubic fixed point.
It is therefore likely that the measurements were largely outside the true critical region and
we do not regard the disagreement between measured and theoretical values of $\beta$ as
significant. Within this same Landau theory, the strain should be proportional to the
order parameter squared and therefore to the superlattice peak intensity. This is clearly
confirmed by the data for MIT-2 shown in Fig. 3.

It was explicitly confirmed in Ref. 3 and by Boni *et al.*[35] that the structural phase
transition is driven by the softening of the $\Sigma_4$ X-point phonon. The dispersion relations
of the soft phonons together with other low lying branches is shown in Fig. 4. It is
evident that the T-O transition is a classic soft mode phase transition.[37] Further, there is
no evidence for the breathing mode instability of the LA mode near the zone boundary
which was predicted by the earliest lattice dynamical calculations.[38]

According to the model discussed above the soft mode is doubly degenerate in the
tetragonal phase and this degeneracy should be lifted in the orthorhombic phase. In the
first neutron experiments discussed above only one of the two modes was observed.[3,35]
However, a second mode with $A_g$ symmetry was observed at higher energies via Raman
scattering.[39] Following the Raman work, Thurston *et al.*[33] have mapped out the
temperature dependencies of the two modes using neutrons in samples with $x = 0, 0.08$
and $0.14$; the latter two samples are, at least in part, superconducting. The results are
shown in Fig. 5. As noted above, the upper mode is the $A_g$ zone center mode; the lower
mode is, in fact, at the zone boundary of the orthorhombic cell. Condensation of that
mode would lead to a transition to the new tetragonal structure with space group $P4_2/ncm$
which has a unit cell $\sqrt{2} \times \sqrt{2}$ of that of the high temperature I4/mmm tetragonal phase.
Such a condensation in fact occurs at low temperatures in $La_{2-x}Ba_xCuO_4$ for a restricted
range of $x$[40] and apparently in $La_2CoO_4$[41] as will be discussed in Sec. VI.

In summary, $La_{2-x}Sr_xCuO_4$ exhibits a classic soft mode phase transition similar
to that observed in many perovskites; further there is then a residual soft optic mode
which persists down to low temperatures. However, there is no obvious connection
between these rotary phonon modes and the superconductivity. There also are no other

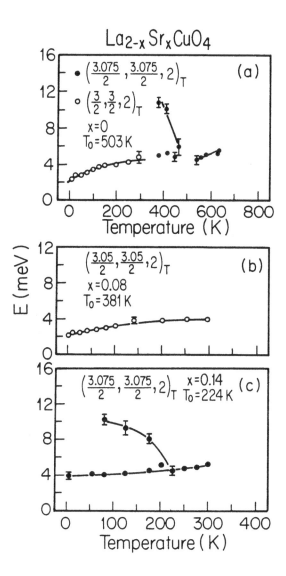

FIG. 5. Summary of the soft phonon energies in $La_{2-x}Sr_xCuO_4$ for $x = 0$, 0.08 and 0.14. The lines are guides to the eye. Figure from Ref. 33.

apparent pathologies in the lattice dynamics; at least for energies up to ~ 30 meV. It is evident, therefore, that the superconducting pairing in $La_{2-x}Sr_xCuO_4$ must originate from some other mechanism. A primary candidate for the origin of the pairing is the magnetism. We therefore proceed to discuss the magnetic properties of $La_{2-x}Sr_xCuO_4$.

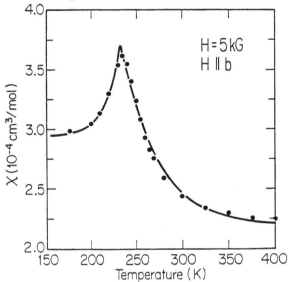

FIG. 6. Susceptibility as a function of temperature in a single crystal of $La_2CuO_4$ corrected for core susceptibility. The solid curve is a fit of the theory[51] as discussed in the text. Figure from Ref. 51.

III.     MAGNETIC ORDERING IN $La_2CuO_4$

A.     Stoichiometric Material

One-electron band structure calculations predict that $La_2CuO_4$ should be a non-magnetic metal.[42] On the other hand, models for the electronic structure which begin from the fully correlated ionic limit would lead one to expect a quasi two dimensional antiferromagnetic insulating state at low temperatures.[12,24,43-45] Specifically, from simple valence considerations one expects the Cu ion to be in the 2+ : $3d^9$ configuration. Crystal field considerations then dictate that this corresponds to a single hole in the $d_{x^2-y^2}$ orbital. Further, one expects from superexchange considerations that there will be a strong antiferromagnetic exchange coupling

$$H = \sum_{\delta(nn)} J_{nn} \vec{S_i} \cdot \vec{S}_{i+\delta} \tag{1}$$

between nearest neighbors. Thus, to first order, in this picture $La_2CuO_4$ would correspond to a stack of 2D S = 1/2 square lattice Heisenberg antiferromagnets. This is a model which in itself is of fundamental interest in Statistical Physics.[46-49] As we shall discuss below, there will then be many small correction terms to Eq. (1) which will control the final symmetry[50,51] and may make possible a transition to true long range order (LRO). There is, in fact, direct evidence for an antiferromagnetic phase transition in $La_2CuO_4$ from bulk susceptibility measurements.[13,51,52] Fig. 6 shows representative susceptibility data in a nearly stoichiometric single crystal; the peak at 240 K strongly suggests a transition to LRO. As we shall discuss below, the detailed shape is, however, anomalous and it requires a quite detailed theoretical analysis.[51]

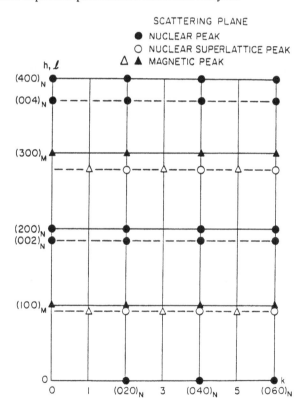

FIG. 7. Reciprocal $\vec{a}^*$ ($\vec{c}^*$) – $\vec{b}^*$ planes superposed by the twinned structure in the orthorhombic phase. The open and closed circles denote nuclear and nuclear superlattice reflections respectively. Open and closed triangles denote magnetic reflections with the propagation direction $\vec{\tau}$ ‖ [1 0 0] and the spin direction $\vec{S}$ ‖ [0 0 1].

The occurrence of three dimensional antiferromagnetic LRO can only be unambiguously tested by magnetic neutron or x-ray scattering. The first experimental confirmation of the antiferromagnetic long range order was a neutron powder diffraction study[6]. In order to discuss the neutron measurements, it is useful first to review the relevant reciprocal space geometry. All of the interesting magnetic behavior occurs in the orthorhombic Cmca phase. Accordingly, from this point on in this review we shall use exclusively the Cmca orthorhombic unit cell. Representative lattice constants at 5 K are $a = 5.339$ Å, $b = 13.100$ Å and $c = 5.422$ Å. We show in Fig. 7 superposition of the $\vec{a}^* - \vec{b}^*$ and $\vec{c}^* - \vec{b}^*$ reciprocal lattice planes; because of twinning both are observed simultaneously. This is, in fact, very often a convenience in single crystal studies.

Vaknin et al.[6] observed peaks at the (1 0 0), (0 1 1), (0 3 1), (1 2 0) and (3 0 0) positions whose onset temperature coincided with that expected from consideration of the susceptibility. From an analysis of the intensities they deduced the structure shown in Fig. 1; the spin $\vec{S}$ is along [0 0 1] while the antiferromagnetic modulation $\vec{\tau}$ is along [1 0 0]. This basic magnetic structure has since been confirmed by a number of workers using single crystal neutron techniques.[7,10,11] The ordered moment is about $0.5 \pm 0.15$ $\mu_B$, much less than the expected saturated value for $Cu^{2+}$ of $\sim 1.1$ $\mu_B$. This apparent discrepancy arises primarily from zero point fluctuation effects.[46,48] Specifically, for a 2D, S = 1/2 near-Heisenberg magnet the zero point fluctuations reduce the moment by $\sim 0.62$, that is, to 0.68 $\mu_B$. The residual discrepancy probably arises from transfer of the $Cu^{2+}$ moment to the oxygen ion and from some residual effects of disorder.

The magnetic structure illustrated in Fig. 1 is not quite correct. Recently, Thio et al.[51] as well as other groups,[53] discovered that there is a field-driven magnetic transition below $T_N$ in which a ferromagnetic moment of $\sim 2.1 \pm 0.2$ x $10^{-3}$ $\mu_B$ per Cu atom develops at high fields. A detailed theory for this is developed in Ref. 51. Due to the rotation of the $CuO_6$ octahedra, the correct nearest neighbor (nn) exchange Hamiltonian has the form

$$H = \sum_{<nn>} \vec{S}_i \cdot \overline{\overline{J}}_{nn} \cdot \vec{S}_{i+\delta} \text{ with } \overline{\overline{J}}_{nn} = \begin{pmatrix} J^{aa} & 0 & 0 \\ 0 & J^{bb} & J^{bc} \\ 0 & -J^{bc} & J^{cc} \end{pmatrix} \qquad (2)$$

The antisymmetric exchange term $J^{bc}(S_i^b \, S_{i+\delta}^c - S_i^c \, S_{i+\delta}^b)$ leads to a uniform canting of the spins within each layer by an amount $\theta = J^{bc}/2J_{nn}$ where $J_{nn} = 1/3(J^{aa} + J^{bb} + J^{cc})$. For small $J^{aa} - J^{cc}$ this also leads to the prediction that the spin will lie in the $\vec{b} - \vec{c}$ plane slightly rotated away from $\vec{c}$. This 2D ferromagnetic moment in each $CuO_2$ sheet then is ordered antiferromagnetically along $\vec{b}$. The field-driven transition thus corresponds to a transition from antiferromagnetic to ferromagnetic alignment of the 2D canted moments so that the system is then a macroscopic ferromagnet. The transition occurs when the field energy equals the interplanar exchange energy.

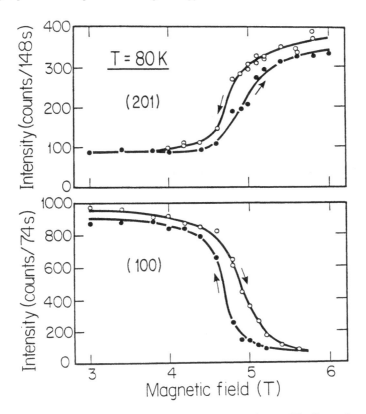

FIG. 8. Magnetic field dependence of the structure factor. The Bragg intensity at (100) decreases with H (lower panel), and that at (201) increases with H (upper panel) because of the transition from antiferromagnetic ordering of the canted component of the spins to ferromagnetic ordering. Figure from Ref. 54.

This model makes a simple prediction for the presumed magnetic structure. At low fields the structure is as in Fig. 1. At high fields the body-centered spin should reverse direction thus changing the propagation direction $\vec{\tau}$ from [1 0 0] to [0 0 1]. This results in the interchange of $\vec{a}^*$ and $\vec{c}^*$ for the magnetic peaks in Fig. 7. We show in Fig. 8 data from Kastner *et al.*[54] at T = 80 K. At ~ 4.5 T, the field at which a transition is observed in the uniform moment, the (1 0 0) peak indeed vanishes while a peak at (2 0 1) appears as expected. The large hysteresis reflects the first order nature of the transition. From a detailed mean field theory in which the 2D correlations (to be discussed in Sec. IV) are treated "exactly", the authors of Ref. 51 are able to calculate a wide range of magnetic properties including the bulk susceptibility as shown in Fig. 7. The relevant parameters deduced from this and related experiments[50] are $J_{nn} = 116$ meV, $J^{bc} = 0.55$ meV, $J^{bb} - J^{cc} = 0.004$ meV and the net between-plane coupling $J^{\perp} = 0.002$ meV. As we shall discuss in the next section it is actually the between-plane coupling $J^{\perp}$ which drives the phase transition to 3D LRO.

It is evident that the bulk magnetic properties of pure $La_2CuO_4$ are now quite well understood. One might ask whether the ansatz of a localized $Cu^{2+}$ $3d^9$ (S = 1/2) moment itself has been justified. The moment eigenfunction is accessible through the q-dependence of the neutron magnetic form factor. The magnetic form factor has been measured in both the antiferromagnetic[55] and paramagnetic states[56] through the staggered and field-induced uniform moments respectively. In both cases, the measured form factors agree well with the calculated free ion $Cu^{2+}$ $3d^9$ form factor. The authors of Ref. 56 conclude that the in-plane oxygen $P_\sigma$ contribution to the moment must be very small. Thus the data support strongly the ionic picture which is at the basis of a number of models for the superconductivity in the doped samples.

B.      Lightly Doped $La_2CuO_4$

One of the earliest and most dramatic discoveries in this field was the observation of the extreme sensitivity of the Néel state to any deviation from ideality.[7,11-18,52] In order to place these results in proper perspective it is useful to recall that for the square lattice the nearest neighbor site percolation threshold is x = 0.41. Further it has been explicitly confirmed in the isostructural materials $Rb_2Mn_{1-x}Mg_xF_4$ and $Rb_2Co_{1-x}Mg_xF_4$ that the Néel state vanishes for x = 0.41±0.01.[57] By contrast, the Néel state in $La_2CuO_4$ is extraordinarily sensitive to doping. The doping may be effected by varying the oxygen concentration in $La_2CuO_{4-\delta}$, by substitution of $La^{3+}$ with a metal with valence 2+ such as $Ca^{2+}$, $Ba^{2+}$ or $Sr^{2+}$, or by replacement of the $Cu^{2+}$ by $Li^+$.

The extreme sensitivity of the Néel state to oxygen concentration was demonstrated by use of neutron scattering techniques independently by Freltoft *et al.*[58] and Yamada *et al.*[7] We show in Fig. 9 results from Ref. 7 on three samples subject to different oxygenation conditions. In essence, as oxygen is added both the Néel temperature and the long range ordered moment are progressively diminished. The actual oxygen concentration is varied by $\lesssim 0.03$; we assume that the maximum $T_N$ of 300 K occurs for stoichiometric material although, given the extreme sensitivity of the Néel state to doping, this is difficult to prove unambiguously.

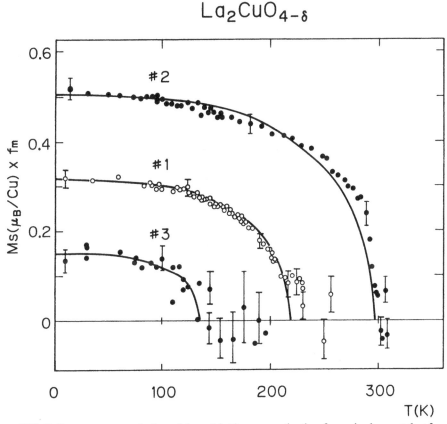

FIG. 9. Temperature evolution of the sublattice magnetization from single crystals of pure $La_2CuO_4$ with different heat treatments, #1 as grown, #2 Ar annealed (deoxygenated) and #3 $O_2$ annealed (oxygenated). Lines indicate Brillouin functions with S = 1/2. Figure from Ref. 7.

This system has been studied using muon spin resonance (μSR) techniques by Uemura and co-workers.[18,59] They have also observed an extreme dependence of the antiferromagnetism on the oxygen content. However there are some essential differences in the neutron and muon results which are intimately connected to the fact that the latter is a local probe. Most importantly, in Ref. 59 it is shown that the ordered moment barely changes with $\delta$ even for samples with $T_N$ as low as 15 K. This shows that all of the spins are indeed freezing at low temperatures. However, the neutron results then require that for the samples with a low $T_N$ only a small fraction of all the $Cu^{2+}$ spins are participating in the 3D LRO state. We will discuss a model[12] for this at the end of this section.

A more extensive neutron study of an oxygenated single crystal of $La_2CuO_{4+\delta}$ has been performed by Endoh et al.[11] They find one new and important feature beyond that exhibited in Fig. 9. In a sample with $T_N \sim 100$ K they find that the 3D sublattice magnetization rises smoothly down to $\sim 30$ K and then starts to *decrease* with further decrease in temperature. As will be discussed in more detail in the next section this decrease is accompanied by the appearance of intense 2D quasielastic scattering. This is closely analogous to the re-entrant behavior observed in certain spin glass systems such as $Eu_{1-x}Sr_xS$.[60]

A second method for doping $La_2CuO_4$ is to substitute $Li^+$ for $Cu^{2+}$. This was done inadvertently by the MIT group in their first $La_2CuO_4$ single crystal studies.[3] Specifically, the first large single crystals were grown using an $Li_4B_2O_5$ flux. It turns out that crystals grown from the flux incorporate $\sim 5\%$ $Li^+$ in place of the $Cu^{2+}$. In close analogy to the $Sr^{2+}$-doped crystals to be discussed in Sec. V, 5% $Li^+$ completely destroys the Néel state and instead a 2D quantum spin fluid state (QSF) with a correlation length of $\sim 40$ Å is observed.[11]

Many groups have studied the magnetic properties of $La_{2-x}M_xCuO_4$ with $M = Ca^{2+}$, $Ba^{2+}$ and $Sr^{2+}$ using each of NMR, NQR,[14,15] muon precession[16,18] and bulk susceptibility.[13,52] All techniques show that the Néel temperature drops precipitously with doping as illustrated in Fig. 2. Indeed, neutron experiments[32] confirm that for x = 0.02 the LRO is completely destroyed and the Néel state is again replaced by a 2D QSF state with a correlation length of $\sim 35$ Å. Again these results in the intermediate doping regime will be discussed in more detail in Sec. V.

A phenomenological model for this unusual behavior has been proposed by Aharony et al.[12] From our comments at the beginning of this sub-section it is clear that a classical percolation model[57] will be inadequate to explain the destruction of the Néel state for x ~ 0.02 and that one must consider explicitly the effects of the holes introduced by the doping. Transport is carried by the electronic holes, which reside on the oxygen ions.[61] For concentrations $x \leq 0.05$, the holes are localized. The argument of Ref. 12 is as follows. Consider first an instantaneous configuration, with a single hole on one $O^-$ ion. The spin of the hole, $\vec{\sigma}$, will have strong exchange interactions with the two neighboring Cu spins $\vec{S}_1$ and $\vec{S}_2$. Writing

$$H_\sigma = -J_\sigma \; \vec{\sigma} \cdot ( \vec{S}_1 + \vec{S}_2 ), \tag{3}$$

it is intuitively clear that, regardless of the sign of $J_\sigma$, the ground state of $H_\sigma$ prefers $\vec{S}_1 \parallel \vec{S}_2$. Quantum mechanically, the exact ground state of $H_\sigma$ indeed has $S_{12} = 1$ (where $\vec{S}_{12} = \vec{S}_1 + \vec{S}_2$, that is, $< \vec{S}_1 \cdot \vec{S}_2 > = 1/4$ ). It is thus reasonable to replace $H_\sigma$ by a ferromagnetic (F) interaction, $\bar{H}_\sigma = - K \; \vec{S}_1 \cdot \vec{S}_2$, where $K = O \, (|J_\sigma|) >> |J_{nn}|$. Here $J_{nn} \sim 1300 \, °K \sim 0.12$ eV is the AF exchange interaction between neighboring Cu spins in the $CuO_2$ plane, and $K >> |J_{nn}|$ because the Cu-Cu distance is twice that of Cu-O. The replacement of $H_\sigma$ by $\bar{H}_\sigma$ is exact for classical spins at low temperatures.

Since a strong F bond in the $CuO_2$ plane destroys the local AF order, it also influences the coupling to the neighboring planes. The Cu spins thus feel competing AF and F interactions. In the extremely localized case, the concentration of the F bonds would be x. As x increases, the localization length $l_0$ of each hole increases, and this will increase the effective concentration of F bonds.

Competing AF and F interactions are known to yield a sharp decrease in the Néel temperature $T_N$, a spin glass (SG) phase[62] and a re-entrance from the AF to the SG phase[63] upon cooling, because of frozen random local moments. This yields the magnetic parts of Fig. 2. In the isostructural $K_2Cu_{1-x}Mn_xF_4$, the Cu ferromagnetism is lost at $x \simeq 0.2$ corresponding to a concentration 0.36 of the very weak Cu-Mn and Mn-Mn AF bonds.[64] As recently shown by Vannimenus et al.,[65] a large ratio $K/|J|$ brings the threshold concentration down. The fact that $(l_0/a) \gtrsim 3$ also renormalizes the threshold. Furthermore, quantum fluctuations also seem to lower the threshold, as indicated by preliminary Monte Carlo simulations.[66] All of these can plausibly explain why in doped $La_2CuO_4$ the Néel state vanishes at the low concentration $x \simeq 0.02$ of holes.

168

IV.    2D STATIC AND DYNAMIC SPIN CORRELATIONS IN $La_2CuO_4$

It is evident that in $La_2CuO_4$ one expects to observe strong 2D spin fluctuations above $T_N$. These were observed by Shirane et al.[10] essentially immediately after single crystals of adequate size and quality became available. Before reviewing the measurements, it will prove useful to discuss the various correlation functions measured in magnetic neutron scattering experiments.[25]    Generally, one has

$$\frac{\partial^2 \sigma}{\partial \Omega_f \partial E_f} \sim \sum_\alpha (1 - \hat{Q}_\alpha^2) S^{\alpha\alpha} (\vec{Q},\omega), \tag{4}$$

where

$$S^{\alpha\alpha} (\vec{Q},\omega) = \frac{1}{2\pi N} \int_{-\infty}^{\infty} e^{i\omega t} < S^\alpha (-\vec{Q},0) S^\alpha (\vec{Q},t) > dt \tag{5}$$

and

$$S (\vec{Q},t) = \sum_R e^{i\vec{Q}\cdot\vec{R}} S(\vec{R},t). \tag{6}$$

Here $\vec{Q} = \vec{k_i} - \vec{k_f}$ where $\vec{k}_i$ and $\vec{k_f}$ are the incoming and outgoing neutron wave vectors, respectively. In an experiment in which one integrates over the energy at fixed $\vec{Q}$, one measures the *instantaneous* correlation function

$$S^{\alpha\alpha}(\vec{Q}) = \int d\omega S^{\alpha\alpha} (\vec{Q},\omega)$$

$$= \frac{1}{N} <S^\alpha (-\vec{Q},0) S^\alpha (\vec{Q},0)> \tag{7}$$

As we shall discuss extensively below, because of the large energy scale for the magnetic fluctuations in $La_2CuO_4$, the experimental scattering geometry must be chosen carefully in order to carry out the energy integration in Eq. (7) properly. For a system with long-range order the Bragg scattering is given by the $\omega = 0$ response in Eq. (4), that is,

$$\frac{\partial^2 \sigma}{\partial \Omega_f \partial E_f}\bigg|_{Bragg} \simeq \sum_\alpha (1 - \hat{G}_\alpha^2) \frac{1}{N} <S^\alpha (\vec{G},t)>^2 \delta(\omega) \tag{8}$$

where $\vec{G}$ is a magnetic reciprocal-lattice vector. The directional term $(1 - \hat{G}_\alpha^2)$ enables one to determine the ordered spin direction.

As discussed above, if the energy integration is carried out correctly at fixed $\vec{Q}$, one measures the instantaneous response function $S^{\alpha\alpha} (\vec{Q})$. It was discovered in the original critical scattering experiments in $K_2NiF_4$ that this energy integration may be carried out rather elegantly in 2D systems.[67] The appropriate geometry is shown in Fig. 10. If the outgoing neutron wave vector $\vec{k_f}$ is along $-\vec{b}^*$, that is, perpendicular to the $CuO_2$ sheets, then the in-plane momentum transfer $\vec{Q}_{2D}$ is a constant independent of the magnitude of $\vec{k_f}$. If all such neutrons are detected then one integrates over E from $-E$ (cutoff) to $\hbar^2 k_i^2/2m$ where E (cutoff) is determined by the $Cu^{2+}$ form factor. For these experiments $- E$ (cutoff) $\simeq - 70$ meV which is much larger than thermal energy at room temperature and therefore is essentially infinite. The integration is also modified by the kinematical factor $k_f/k_i = (1 - E/E_i)^{1/2}$; this will typically have a quantitative rather than qualitative effect.

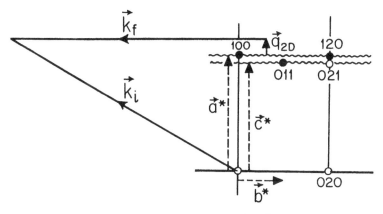

FIG. 10. Illustration of scattering geometry with $\vec{k_f} \parallel -\vec{b}^*$ for $E_i = 13.7$ meV neutrons.

We show in Fig. 11 both two-axis and three-axis scans at 200 K in a crystal labelled NTT-2 which has $T_N = 195$ K; these measurements utilized neutrons of energy, 13.7 meV and the spectrometer had collimation 40'-10'-10'-(40'). At (100), 3D magnetic Bragg scattering is observed; the 3D Bragg peak is still present at 200 K because of the

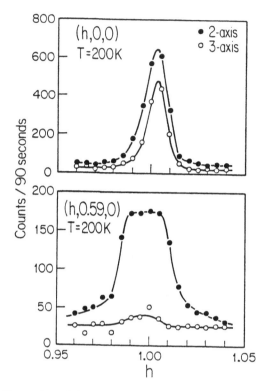

FIG. 11. Two-axis and three-axis scans across the 3D (100) magnetic peak (upper) and 2D rod (lower) in NTT-2. The two-axis Q vector corresponds to the E = 0 position. The incoming wave vector was 2.57 Å$^{-1}$ and the collimators were 40′-10′-10′. Figure from Ref. 10.

rounding of the transition. For both the energy integrating (two-axis) and $\omega = 0$ (three axis) measurements a sharp resolution-limited peak is observed at (100). The difference in intensity of the two-axis and three-axis scans evident in Fig. 11 is due to the 70% reflectivity of the graphite analyzer. Quite different behavior occurs for scans across the 2D rods; for $k_i = 2.57$ Å$^{-1}$, a simple calculation gives $\vec{k_f} \parallel - \vec{b}^*$, when $\vec{Q}_\perp = 0.59\,\vec{b}^*$. For the two axis scan along the in-plane wave vector $\vec{Q}_{2D}$ with $\vec{Q}_\perp = 0.59\,\vec{b}^*$ a flat-top peak with very sharp edges is observed. This flat-top peak is actually a superposition of the $(1\zeta0)$ and $(0\zeta1)$ rods of scattering (cf. Fig. 10) which are slightly separated in $\vec{Q}$ as a result of the orthorhombic distortion. In this case, however, hardly any signal occurs in the three-axis scan. Thus, the response function comes overwhelmingly from fluctuations which are at energies much greater than the three-axis energy window of $\simeq 1$ meV.

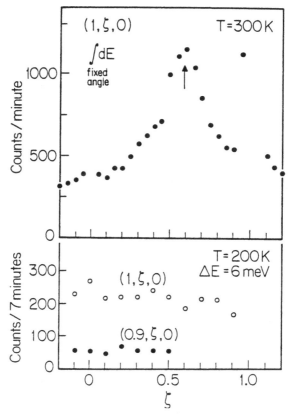

FIG. 12. Top: scan along the (1 $\zeta$ 0) rod at T = 300 K in NTT-2. The neutron wave vector is 2.57Å$^{-1}$. The arrow gives the position at which $\vec{k_f} \parallel -\vec{b}^*$ so that the energy integration for the in-plane fluctuations is carried out properly. Bottom: scans along $\vec{b}^*$ at a fixed energy transfer of 6 meV in NTT-2. Figure from Ref. 10.

This large inelasticity is demonstrated in another fashion in Fig. 12. At the top of this figure is shown a two-axis scan at 300 K, again with $k_i$ = 2.57 Å$^{-1}$, with the in-plane wave vector $\vec{Q}_{2D}$ fixed at $\vec{a}^*$ and $\vec{Q}_\perp = \zeta \vec{b}^*$ varied. It is evident that a sharp peak occurs for $\zeta \simeq 0.6$, that is, for $\vec{k_f} \parallel - \vec{b}^*$. No corresponding peak occurs for $\zeta = -0.6$ so that this effect does not have the symmetry of the reciprocal lattice and hence cannot be intrinsic to La$_2$CuO$_4$. Thus when $k_i$ = 2.57 Å$^{-1}$ and $\vec{Q}_\perp = +0.59\vec{b}^*$ one measures properly the 2D correlation function $<S^\alpha (-\vec{Q}_{2D},0) \cdot S^\alpha (\vec{Q}_{2D},0)>$. In general, the inelasticity involved in the 2D spin fluctuations is so large that for positions along the rod away from $\vec{Q}_\perp = 0.59\,\vec{b}^*$, the in-plane wave vector $Q_{2D}$ is varied significantly in the process of integrating over the energy thus smearing the peak and lowering the measured intensity. In order to verify the 2D nature of the scattering, a scan along the rod at 200 K was performed by use of a triple-axis spectrometer set for an energy transfer of 6 meV. The results are shown at the bottom of Fig. 12. Significant scattering occurs at $(1\,\zeta\,0)$ whereas for $(0.9\,\zeta\,0)$ the intensity is at the background level. As expected, the scattering for E = 6 meV at $(1\,\zeta\,0)$ is, except for trivial geometrical factors, independent of $\zeta$, thus confirming that these dynamic spin fluctuations are purely two dimensional in character.

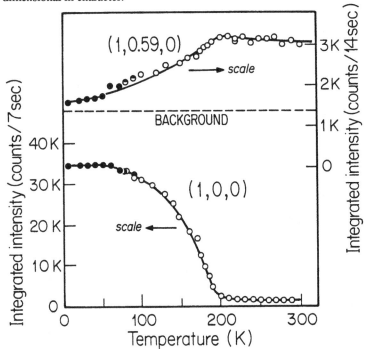

FIG. 13. Integrated intensities of the (100) 3D antiferromagnetic Bragg peak and the (1 0.59 0) 2D rod in NTT-2. The open and closed circles represent separate experiments which were normalized in the overlap region. Figure from Ref. 10.

Even without quantitative analysis, two important qualitative features of the spin correlations in $La_2CuO_4$ may be deduced from the two-axis scan shown at the bottom of Fig. 11. First from the flat-top shape one concludes that there are equal intensity contributions from the $\vec{a}^* - \vec{b}^*$ and $\vec{c}^* - \vec{b}^*$ planes. From Eqs. (4) and (7) it follows that, at 200 K, $S^{aa}(\vec{Q}_{2D}) = S^{cc}(\vec{Q}_{2D})$, that is, the fluctuations have at least XY symmetry. This in turn means that at this temperature the in-plane anisotropy is not playing an important role. Second, from the sharp edges one may deduce that at 200 K the correlation length must exceed 200 Å.

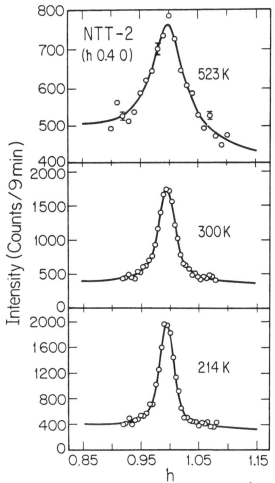

FIG. 14. Two-axis scans across the 2D rod in NTT-2 for $\vec{Q}_\perp = 0.4b$ at a series of temperatures. The incoming neutron energy was 30.5 meV and the collimation was 40'-10'-10'. The solid lines are the results of fits to the superposition of two Lorentzians, one centered about $\vec{Q}_{2D} = \vec{a}^*$ and the other about $\vec{Q}_{2D} = \vec{c}^*$, convoluted with the instrumental resolution function. Figure from Ref. 11.

The integrated intensity in the two-axis scan across the rod for NTT-2 is shown as a function of temperature at the top of Fig. 13. The rod intensity increases slightly below 300 K with decreasing temperature. However, at $T_N = 195$ K the 2D scattering intensity begins to decrease. This is consistent with the heuristic notion that the 2D dynamic scattering is converted into 3D Bragg scattering at the transition to 3D long-range order. It should be emphasized, however, that in $La_2CuO_4$ this process

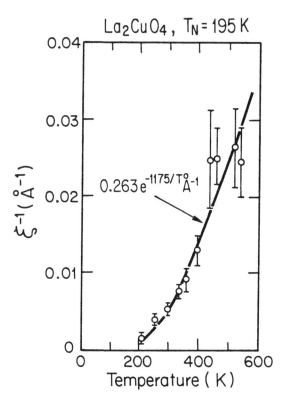

FIG. 15. Inverse correlation length versus temperature in $La_2CuO_4$ (NTT-2) with $T_N = 195$ K. The solid line is the theoretical prediction of Ref. 47.

occurs very gradually. By contrast, in $K_2NiF_4$, due to the 2D Ising nature of the transition, the transfer of intensity from the rod (which in this case is nearly elastic) to the 3D Bragg peak occurs within 2% of $T_N$.[67] This reflects a fundamental difference between the 3D phase transition in $La_2CuO_4$ and those in previously studied planar antiferromagnets. In the latter, the transitions to LRO are essentially 2D in character with

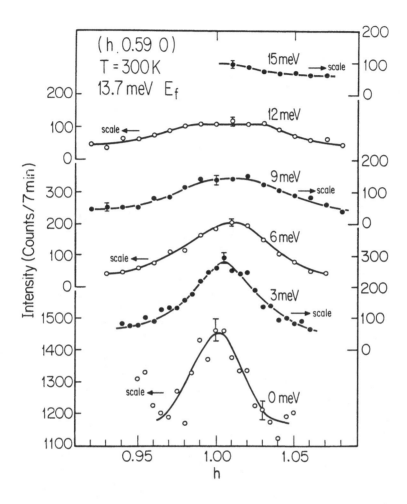

FIG. 16. Inelastic constant energy scans across the 2D rod in NTT-2 at T = 300 K. The outgoing neutron energy was held fixed at 13.7 meV and the collimators were 40'-40'-40'-40'; here $\vec{Q}_{2D} = h\,\vec{a}^*$. The solid lines are guides to the eye. Figure from Ref. 11.

the 3D ordering following parasitically. In $La_2CuO_4$ the 3D coupling drives the transition.[10,11] As we shall discuss in Sec. VI, the materials $La_2NiO_4$[68] and $La_2CoO_4$[49] which have S = 1 and S = 3/2 respectively, exhibit 2D Ising phase transitions to LRO in exact analogy to $K_2NiF_4$. Thus the S = 1/2 Heisenberg character of the $Cu^{2+}$ ion in $La_2CuO_4$ seems to give unique new behavior.

In order to characterize the 2D correlations quantitatively, measurements were carried out using 30.5 meV neutrons and a tight collimator configuration.[11] Representative data are shown in Fig. 14. The flat top seen in Fig. 11 is not visible in Fig. 14 because of reduced resolution. The solid lines in Fig. 14 are the results of fits to two isotropic Lorentzians $A/(\kappa^2 + q_{2D}^2)$, one centered about $\vec{Q}_{2D} = \vec{a}^*$ and one other about $\vec{Q}_{2D} = \vec{c}^*$, both convoluted with the instrumental resolution

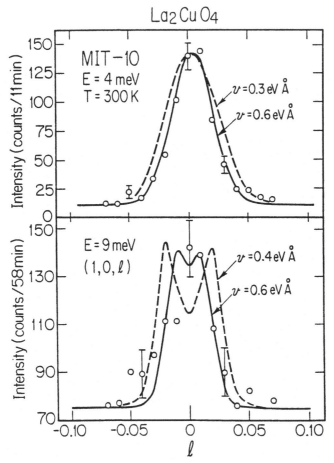

FIG. 17. Constant energy transverse scans through the 2D peak position(1 0 $l$) at T = 300 K in MIT-10 ($T_N$ = 235 K). The lines are the calculated profiles for presumed linear excitations with the indicated velocities and fixed energy widths of 1 meV. The background and peak intensities were chosen to best fit the data. The incoming neutron energy was 14.7 meV and the collimation was 20'-20'-20'-40'. Figure from Ref. 32.

function. Here $q_{2D} = |\vec{Q}_{2D} - \vec{a}^*|$ or $|\vec{Q}_{2D} - \vec{c}^*|$; the fits also included a sloping background. The results for the inverse correlation length $\kappa$ are shown in Fig. 15. The correlation length evolves from about 40 Å at 500 K to about 400 Å at $T_N$. The solid line in Fig. 15 is the prediction of the renormalized classical model of Chakravarty, Halperin and Nelson[47] which evidently is quite successful.

Initial measurements of the dynamics in $La_2CuO_4$ determined that the excitation dispersion relation had an extremely steep slope.[10,11] Because of this it was necessary to carry out scans with the energy transfer fixed and the momentum transfer $\vec{Q}_{2D}$ varied. Representative results for $T = 300$ K are shown in Figs. 16. We remind the reader that at 300 K the spin correlations are purely 2D with a correlation length of $\sim$ 200 Å. For energies up to at least 12 meV the peak is centered about $\vec{Q}_{2D} = \vec{a}^*$ (or $\vec{c}^*$), that is $h = 1$ implying that one cannot separate the $-\vec{q}_{2D}$ and $+\vec{q}_{2D}$ branches.

Measurements at higher resolution which have been carried more recently are reported in Ref. 32. Results for $E = 4$ meV and 9 meV are shown in Fig. 17. Again even with very high resolution it is not possible to separate the $-\vec{q}_{2D}$ and $+\vec{q}_{2D}$ branches. The dashed and solid lines are calculated profiles assuming a linear dispersion with the indicated velocities and widths. Only the overall heights were adjusted. From these and similar calculations the authors of Ref. 32 conclude that at $T = 300$ K the measurements are consistent with underdamped 2D spin excitations with $v \geq 0.6$ eVÅ. From their analysis Thio et al.[51] deduce a value for $J_{nn}$ which corresponds to a spin wave velocity of 0.6 eVÅ. The two-magnon Raman experiments of Lyons et al.,[69] if simply interpreted, suggest a magnon velocity of $\sim$ 0.7 evÅ. We conclude therefore that all of the measurements are at least consistent with each other. Of course, this very high velocity and concomitant large value for $J_{nn}$ is an essential ingredient in most magnetic pairing models for the superconductivity.[24]

Various models for the QSF state predict that $S(\vec{Q},\omega)$ should have a non-trivial geometry in reciprocal space. Accordingly, the authors of Ref. 32 carried out a broad survey of the inelastic magnetic scattering at $T = 300$ K. They find that at fixed $\omega$, $S(\vec{Q},\omega)$ is circularly symmetric about (100) and (001) and they find no evidence for significant low energy fluctuations about any other high symmetry positions such as (1/2,0,1/2) in the Brillouin zone. This is consistent with a classical spin wave description of the excitations and it precludes many Fermi-surface based models for the QSF state.

It must be said, on the other hand, that a full test of models for the dynamics above $T_N$ has not yet been completed. Specifically, a number of authors have argued that the spin dynamics in the paramagnetic state of $La_2CuO_4$ should be understandable using

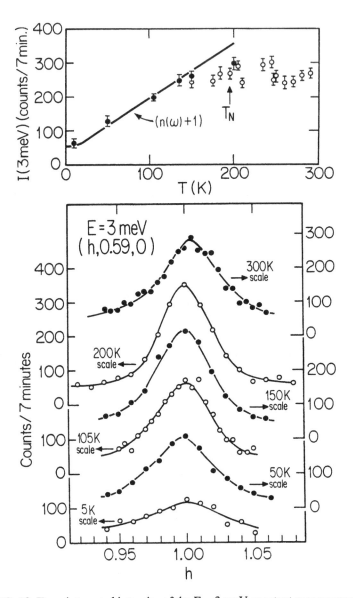

FIG. 18. Top: integrated intensity of the E = 3 meV constant energy scans vs temperature in NTT-2. $T_N$ is the Néel temperature in this sample. The solid line is the prediction for the temperature dependence of the intensity assuming Bose statistics I ~ $(e^{\hbar\omega/kT} -1)^{-1} +1$. Bottom: constant energy scans at E = 3 meV in NTT-2. The outgoing neutron energy is 13.7 meV and the collimation 40'-40'-40'-40'. Figure from Ref. 11.

conventional dynamical scaling ideas without involving exotic quantum effects.[47,49] Since the dynamic scaling exponent Z is 1 for the 2D Heisenberg model[47] the energy scale for the fluctuations can remain quite large even for correlation lengths as large as 100 lattice constants. It is hoped that definitive experiments addressing this issue will be completed in the near future.

We finish this section with a review of the behavior of the spin excitations as T is lowered through $T_N$.[11] Representative results are shown in Fig. 18. In the bottom panel we show the E = 3 meV scan in NTT-2 ($T_N \simeq 195$ K) as a function of temperature between 5 and 300 K. The peak intensity versus temperature is shown in the top panel. The intensity is approximately independent of temperature between 300 and ~ 200 K and drops off considerably in intensity below 150 K. From 5 up to 150 K the intensity

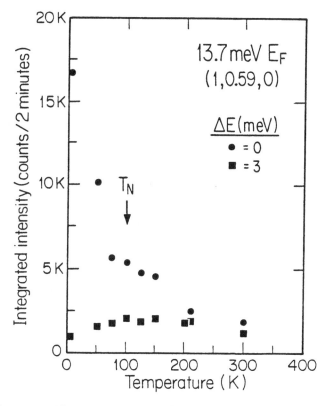

FIG. 19. Integrated intensity of constant energy scans vs temperature at 0 and 3 meV in NTT-3. $T_N$ is the Néel temperature. The outgoing neutron energy was held fixed at 13.7 meV and the collimation was 40'-40'-40'-40'. Figure from Ref. 11.

follows the Bose statistics curve quite well. We conclude, therefore, that the quantum spin fluid excitations become simple spin waves in the Néel state, as one might have expected.

In Fig. 19, we show the integrated intensity in NTT-3 ($T_N \simeq 100$ K) at energy transfers of 0 and 3 meV. The behavior at 3 meV is closely similar to that in NTT-2. However, for E = 0 the intensity increases gradually through $T_N$ and rises dramatically at low temperatures. This rapid rise of the 2D quasielastic scattering at low temperatures mirrors the reentrant diminution of the 3D Bragg intensity discussed in III.[11] As discussed there, this may be interpreted as indicating reentrant spin-glass behavior. Finally, the low energy excitations in NTT-2 at 80 K have been studied by Peters et al.[50] Representative data from their study are shown in Fig. 20. Above $T_N = 195$ K the scattering has a half-width of ~ 3 meV and the profile depends only weakly on temperature. As emphasized previously, there is no evidence for critical slowing down in the 2D scattering as the 3D Néel temperature is approached; this is in marked contrast to

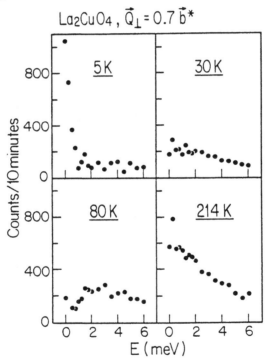

FIG. 20. Inelastic scans with background removed in NTT-2 at (1,0.7,0) at a series of temperatures. $E_f$ was held fixed at 14.7 meV and all collimators were 40'. Figure from Ref. 50.

the behavior in the isomorphous $S = 1$ and $S = 3/2$ systems $La_2NiO_4$ and $La_2CoO_4$ respectively as will be discussed in Sec. VI. Below $T_N$ two gaps open up; at 80 K the authors of Ref. 50 find gaps of $1.0\pm0.25$ eV and $2.5\pm0.5$ meV for motion of the spins, respectively, in and out of the plane. These lead to the anisotropy energies quoted in Sec. III(i). Finally, as is evident in Fig. 20 there is an increase in the scattering at $E = 0$ at temperatures $T < 30$ K similar to that shown for NTT-3 in Fig. 19 but of much smaller magnitude.

In summary, pure $La_2CuO_4$ seems to be a close approximation to the idealized 2D $S = 1/2$ Heisenberg antiferromagnet. In general, the properties of the stoichiometric material are readily explainable using classical magnetic ideas although the detailed dynamics above $T_N$ remain to be elucidated. We have already seen from the extreme sensitivity of $T_N$ to doping and the unusual "re-entrant" effects that the magnetic behavior of $La_2CuO_4$ with holes - either localized or itinerant - is unexpectedly rich. Of course, this is actually quite encouraging since the material evolves with increased doping into a superconductor with a remarkably high $T_c$ and one might presume that the two phenomena are related. We therefore now proceed to discuss the spin fluctuations in the insulating, metallic and superconducting regimes in $La_{2-x}Sr_xCuO_4$.

## V.    SPIN CORRELATIONS IN INSULATING, METALLIC AND SUPERCONDUCTING $La_{2-x}Sr_xCuO_4$

As discussed extensively in Sec. III, addition of holes to $La_2CuO_4$ has a drastic effect on the magnetic properties. We are, of course, especially interested in the magnetism in the superconducting regime. It is clear, however, that in order to understand the basic physics of the high-$T_c$ materials it will be necessary to understand the evolution of the properties as the number of holes is varied. As indicated in Fig. 2, $La_{2-x}Sr_xCuO_4$ undergoes an insulator-metal transition for $x \simeq 0.05$. In the regime $x \lesssim 0.05$ for $T \lesssim 100$ K one observes typically variable range hopping conductivity, $\ln \sigma \sim (T_o/T)^{1/4}$, although in different samples the exponent may vary from $1/4$ to $1/2$.[3,20] In this regime the carriers are localized due to disorder. For $x > 0.05$ the holes de-localize and the material is then a metal at high temperatures and a superconductor at low temperatures. The superconducting state itself is very sensitive to the perfection of the material and indeed for many doped single crystals often only a small fraction ($\sim 10\%$) will be superconducting.

Shortly after the Bednorz-Müller discovery,[1] samples of $La_{2-x}M_xCuO_4$ with $M = Ca^{2+}$, $Ba^{2+}$ or $Sr^{2+}$ were studied with a wide range of probes including bulk

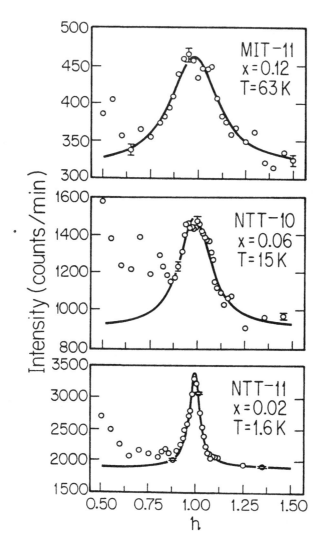

FIG 21. Two-axis scans across the rod along (h,ζ,0) in NTT-11 (x = 0.02) at
T = 1.6 K; NTT-10 (x = 0.06), T = 15 K; and MIT-11 (x=0.12), T = 63 K. In these
scans $Q_\perp = \zeta \vec{b}^*$ varied from $0.1\vec{b}^*$ to $1.1\vec{b}^*$ as h varied from 0.5 to 1.5. The
incoming neutron energy was 14.7 meV and the collimation was 40'-40'-40'-open. The
solid lines are the results of fits to a single Lorentzian. The scattering intensity is
normalized to the volumes of the crystals; the relative volumes were determined by
measuring intensities of the transverse-acoustic phonons. Figure from Ref. 32.

susceptibility,[13,52] NMR, NQR,[14,15] muon precession[16,18] and neutron scattering.[10,11,32] As discussed in III, all probes indicate the destruction of magnetic LRO for $x \gtrsim 0.02$. However for samples with $0.02 < x < 0.05$ both NQR and μSR indicate some type of phase transition at low temperatures, $T \lesssim 10$ K. Indeed, the muon experiments indicate a freezing of the spins into a Néel-like state with a local moment of $\sim 0.5$ $\mu_B$, close to that in pure $La_2CuO_4$ in the Néel state.[16-18] The explicit nature of the spin correlations in this state can only be determined using neutron scattering techniques.

An extensive set of neutron experiments has now been performed at Brookhaven in large single crystals of $La_{2-x}Sr_xCuO_4$ with x varying between 0.02 and 0.14.[32,70] However, only some of the crystals with $x > 0.05$ are actually superconducting at low temperatures and then the Meissner fraction varies from a few % to 80%. The explicit microscopic origin of this drastic variation in Meissner fraction is not yet fully understood. As we shall discuss below, the neutron experiments show first that the full $Cu^{2+}$ moment is retained in the superconducting samples. Thus the magnetism coexists synergistically with the superconductivity. Second, the character of the spin fluctuations varies smoothly across the insulator-metal boundary. This is difficult to understand using current theoretical models. It should be noted that this smooth variation of the spin correlations through $x \simeq 0.05$ was already evident in the bulk susceptibility as x varied through 0.05 although the importance of this result was not generally recognized.[13,52] Indeed, many workers misinterpreted the 2D $Cu^{2+}$ high temperature susceptibility as the Pauli susceptibility of the itinerant holes.

We now discuss the quasielastic and inelastic neutron experiments. The experimental techniques were identical to those discussed in III for the pure and lightly doped materials. One minor difference is that for the two-axis scans because of the breadth in Q-space it was necessary to carry out the scans such that $\vec{k_f}$ was always $\parallel \vec{b}^*$ (see Fig. 10); thus both $\vec{q}_{2D}$ and $\vec{Q}_\perp = \zeta \vec{b}^*$ had to be changed simultaneously. Before discussing the actual measurements we first note two general features of the data. First, for all samples the two axis scans at 300 K with $\vec{q}_{2D}$ fixed at 0 and $\vec{Q}_\perp$ varied exhibit a sharp peak at the position when $\vec{k_f}$ is along $\vec{b}^*$, as in Fig. 12. This shows that the fluctuations are predominantly 2D and dynamical at 300 K. This was confirmed by direct inelastic measurements for some of the samples. Second, the integrated intensity for all samples is independent of Sr concentration and is identical to that in pure $La_2CuO_4$ to an overall accuracy of $\sim 20\%$. This implies that the full $Cu^{2+}$ moment is retained in the doped samples. Of course, the neutrons are sensitive to all spins in the system, those of $Cu^{2+}$ and those of $O^-$. The above observation must be qualified by the fact that the spin fluctuations extend up to about 0.27 eV and the high energy fluctuations are not rigorously included in the integration; however, the part of $S(\vec{Q})$ near $\vec{q}_{2D} = (100)$ and $(001)$ is dominated by the low energy fluctuations which are properly included in the integration.

Representative data are shown in Fig. 21. As is evident in Fig. 21 the scattering is centered about $h = 1$ as in pure $La_2CuO_4$ and it becomes progressively broader as the $Sr^{2+}$ concentration, $x$, is increased. Further these profiles are, within the statistics, independent of temperature below 300 K. There is also some indication of a two-peaked character in the response for $x = 0.11$, especially in the scan at 63 K.

Most recently, an extensive set of experiments[70] has been carried out on two crystals, both with $x \simeq 0.11$, $T_c = 10$ K and Meissner fractions of $\sim 80\%$. These crystals which are both large, 1 - 2 cm$^3$ in volume, and of high perfection represent a significant accomplishment by Y. Hidaka and coworkers at NTT. It was discovered in these experiments that there was a striking thermal evolution in the distribution in energy of the scattering so that it was essential to separate the quasielastic ($|\Delta E| < 0.5$ meV) and integrated inelastic ($|\Delta E| > 0.5$ meV) contributions to $S(\vec{Q})$. Accordingly, the spectrometer was set up in the triple axis mode and all scans were carried out twice, first detecting neutrons scattered off the PG analyzer so that $|\Delta E| < 0.5$ meV and second detecting neutrons passing straight through the analyzer. The effective reflectivity of the analyzer was measured to be 78% so that by subtracting 29% of the first scan from the second scan, one obtains precisely in the latter scan the intensity integrated over all energies with $|\Delta E| > 0.5$ meV since the absorption by the analyzer is negligible. These latter data turn out to be particularly clean, with little contamination scattering.

Representative results are shown in Fig. 22. This shows the integrated inelastic ($|\Delta E| > 0.5$ meV) and fitted total cross sections for the $(h, h - 0.45, 0)$ scans across the 2D ridge at a sequence of temperatures. Two features are immediately evident. First the scattering is broad and flat-topped with some indication of a two peaked structure. This incommensurate two-peaked structure was suggested in previous experiments[32] as noted above. It was also discovered independently by Yoshizawa et al.[71] Second, the total cross section as measured in this particular cut through reciprocal space varies only weakly with temperature from 350 K to 12 K. However, the spin fluctuations change from being predominantly inelastic at 350 K to predominantly quasielastic at 12 K. Figure 23 shows pure two-axis scans along $(h, h - 0.4, 0)$ across the ridge at

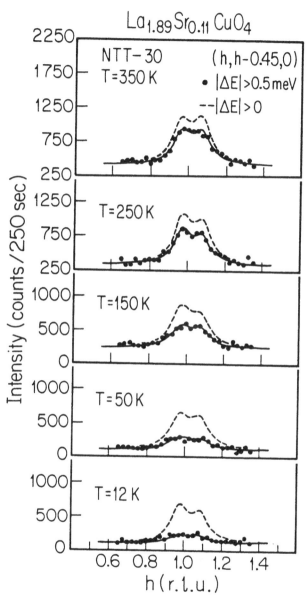

FIG. 22. Integrated inelastic ($|\Delta E| > 0.5$ meV) scattering for scans across the magnetic ridge along (h, h $-$ 0.45, 0); $E_i$ = 14.7 meV and the collimator configuration is 40'-40'-40'-80'. The solid lines are the results of fits to two displaced 2D Lorentzians as discussed in the text. The dashed lines are the results of the best fits to the total scattering, elastic plus inelastic. Figure from Ref. 70.

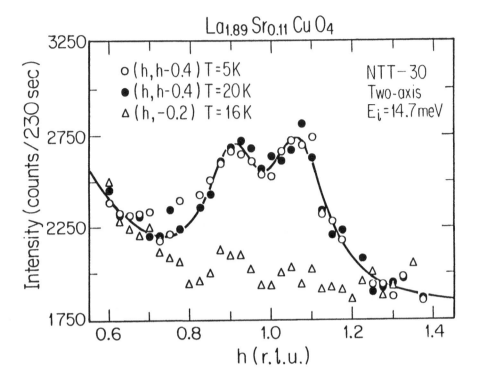

FIG. 23. Pure two-axis scans along (h, h − 0.4, 0) at T = 5 K and 20 K and along (h, − 0.2, 0) at T = 16 K. The spectrometer had 1 PG filter and no analyzer. The solid line is the result of fits to two displaced 2D Lorentzians together with a background function determined from the (h, − 0.2, 0) scan. Figure from Ref. 70.

T = 20 K (normal) and T = 5 K (superconducting) together with a background scan along (h, − 0.2, 0). It is evident that the static structure factor, $S(\vec{Q})$, is unchanged in going from the normal to the superconducting state. Further no measurable change is found in the quasielastic and integrated inelastic components separately.

Figure 24 shows both quasielastic and integrated inelastic scans at 12 K and 350 K in which $\vec{Q}_{2D}$ is held fixed at the peak position, 1.05 $\vec{a}^*$, and the momentum transfer ⊥ the CuO$_2$ planes, $\vec{Q}_\perp = k\vec{b}^*$, is varied. The integrated inelastic scan at 350 K shows a peak when $\vec{k_f} \parallel -\vec{b}^*$ as expected. More surprisingly, the quasielastic peak intensity exhibits a *sinusoidal modulation* ⊥ the CuO$_2$ sheets at both 12 K and 350 K.

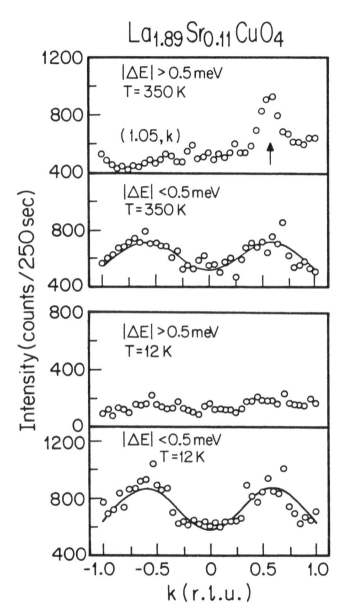

FIG. 24. Elastic ($|\Delta E| < 0.5$ meV) and integrated inelastic ($|\Delta E| > 0.5$ meV) scans $\perp$ the CuO$_2$ sheets for $\vec{Q}_{2D} = 1.05\,\vec{a}^*$. The arrow gives the position at which the outgoing neutron wavevector $\vec{k_f}$ is $\perp$ the CuO$_2$ sheets, that is along, $-\vec{b}^*$. Figure from Ref. 70.

The period of the modulation is about 2 $La_2CuO_4$ unit cells. Thus, even at temperatures as high as 350 K the low energy spin fluctuations in $La_{1.89}Sr_{0.11}CuO_4$ are fully three dimensional in character. This contrasts markedly with the fluctuations in pure $La_2CuO_4$ which are essentially 2D above the Néel temperature as discussed in Sec. IV. Thus at low temperatures the $Cu^{2+}$ structure factor in superconducting $La_{1.89}Sr_{0.11}CuO_4$ corresponds to a slowly fluctuating ($< 10^{-11}$ sec) 3D modulated spin fluid. We note that recent µSR studies on a sample prepared identically to the neutron sample indicate that the entire volume freezes magnetically below T ~ 4 K.[18]

The solid lines in Figs. 22 and 23 are the results of fits to two displaced 2D Lorentzians. Clearly this simple model works well although it certainly is not unique. For the total scattering, the peak positions, intensities and width as well as the background were all varied. For the quasielastic and integrated inelastic components the peak positions and widths were fixed at the values determined from the fits to the total scattering and only the intensities and background were allowed to vary. The quasielastic and integrated inelastic components turn out to be well-described separately by the parameters characterizing the total cross section. The 2D instantaneous spin-spin correlation length is of order 18±6 Å independent of temperature from 350 K to 5 K. The 2D incommensurability from these fits is of order 0.05 Å$^{-1}$ (Fig. 22) to 0.08 Å$^{-1}$ (Fig. 23) although larger values are obtained from quasielastic scans ⊥ the rod so the exact value of the incommensurability should be treated cautiously. AT 350 K the total cross section, which corresponds to an integral from ~ kT to + 14.7 meV, is predominantly (~ 75%) inelastic while at 12 K the $|\Delta E| < 0.5$ meV component accounts for ~ 75% of the observed scattering.

We will discuss the dynamical behavior in more detail below. We consider first the instantaneous correlations in the samples studied to-date. In Ref. 32 the two-axis scans were analyzed using a single Lorentzian profile since the data did not justify a more elaborate lineshape. However, the more recent experiments discussed above show clearly that the 2D correlations are incommensurate. Accordingly, all of the data presented in Ref. 32 have now been re-analyzed using a two Lorentzian line-shape.[72] In all cases the goodness-of-fit parameter $\chi^2$ is improved, albeit at the expense of two additional adjustable parameters. The correlation lengths so-deduced are shown in Fig. 25. The solid line, $3.8/\sqrt{x}$ Å, is the average separation between the O$^-$ holes in the $CuO_2$ planes. Remarkably, this agrees with the measured correlation lengths quite well. It is evident therefore that the holes have an extraordinarily disruptive effect on the $Cu^{2+}$ - $Cu^{2+}$ antiferromagnetic state. This may be understood at least in part with the

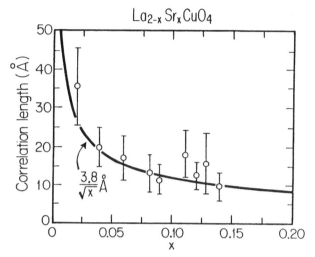

FIG. 25. Instantaneous spin correlation length vs temperature in $La_{2-x}Sr_xCuO_4$. The lengths are deduced from fits of two Lorentzians with identical widths symmetrically displaced about $h = 1$. The solid line is the function $3.8/\sqrt{x}$ Å which is just the average separation between the holes introduced by the $Sr^{2+}$ doping.

frustration model of Ref. 12. However, that model was constructed for the localized regime; it is not clear within the context of that model why the $Cu^{2+}$ spins should follow the $O^-$ hole spin so faithfully in the metallic state. Thus a more elaborate theory is most certainly required.

One final qualitative observation should be made on the instantaneous spin correlations. It is evident that profiles for the $x = 0.12$ sample (labelled MIT-11) shown at the top of Fig. 21 and those for the $x = 0.11$ sample shown in Fig. 23 (labelled NTT-30) are closely similar, if not identical. Indeed the fitted widths and incommensurabilities agree to within the errors. However, MIT-11 exhibits metallic resistivity with an upturn at the lowest temperatures whereas, as discussed above, NTT-30 exhibits a superconducting transition at 10 K. Thus the magnetic state is quite robust, depending primarily on the $Sr^{2+}$ concentration, whereas the superconductivity itself is apparently quite delicate.

We consider finally the momentum and temperature dependence of the spin excitations themselves. So far there are only limited inelastic measurements in the doped samples. We show in Fig. 26, excitation *creation* scattering processes in NTT-10, $x = 0.06$ at $T = 120$ K for $E = 3,6$ and 9 meV. The width in momentum space is identical to that obtained from the two axis scans (Fig. 21) while the peak intensity depends only weakly on energy. Studies as a function of temperature show, quite remarkably, that the profile is independent of temperature between 300 K and 5 K.

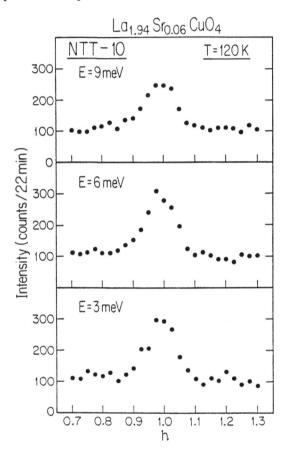

FIG. 26. Constant-energy scans across the 2D rod at T = 120 K in NTT-10 (x = 0.06). The outgoing neutron energy was 14.7 meV and the collimation was 40'-40'-40'-40'. The momentum transfer ⊥ the $CuO_2$ planes was held fixed at 0.6 $\vec{b}^*$. Figure from Ref. 32.

The excitation *annihilation* scattering is simply related to that shown in Fig. 26 by the detailed balance factor, $e^{-E/kT}$, as expected.[73] It was also explicitly verified that the inelastic response function was independent of $\vec{Q}_\perp$, that is, the excitations are 2D in character.[73]

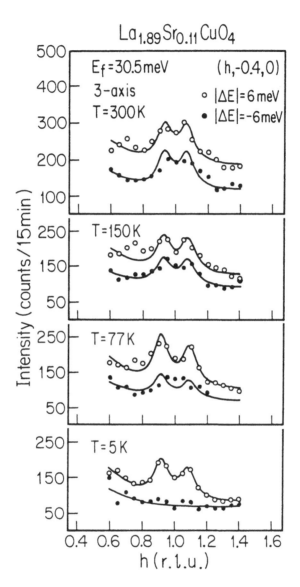

FIG. 27. Constant energy scans across the 2D rod along (h,−0.4,0) in NTT-35, $La_{1.89}Sr_{0.11}CuO_4$. The outgoing neutron energy was 30.5 meV and the collimation was 40′-40′-40′-80′. The lines for the +6 meV data are the results of fits to two symmetrically displaced Lorentzians. The lines for -6 meV are calculated from the +6 meV fits assuming detailed balance. The open circles represent excitation creation and the closed circles excitation annihilation scattering processes.

Figure 27 shows a series of scans in the superconducting sample NTT-35, $La_{1.89}Sr_{0.11}CuO_4$ for E = ±6 meV. The lineshape for excitation creation closely mirrors the integrated response (Figs. 22, 23). Thus these represent excitations out of the slowly fluctuating modulated ground state. Again, the excitation creation intensity is independent of temperature between 300 K and 5 K; this spans the region kT >> $\hbar\omega$ to kT << $\hbar\omega$. The excitation annihilation cross section is related to the creation process by the detailed balance factor $e^{-E/kt}$ as is required by time reversal symmetry.[25]  It has been explicitly verified[73] that there is no dispersion in the $\vec{b}$ direction at 6 meV, that is, the excitations are confined to the $CuO_2$ planes. This contrasts with the fluctuations with |E| < 0.5 meV which exhibit 3D sinusoidal correlations as shown in Fig. 24.

If the spin excitations were *bosons* as in pure $La_2CuO_4$ below $T_N$ (cf Fig. 18) then the intensity for E = +6 meV would have changed by a factor of 5 between 300 K and 5 K. On the other hand the temperature dependence of the excitation intensity is only consistent with *Fermi* statistics if the chemical potential is much larger than 25 meV. Thus the excitation statistics remain to be understood.

Limited measurements have also been performed[73] for energies varying between 4 meV and 18 meV. The lower limit of 4 meV is set by background considerations while above 18 meV phonon scattering processes dominate. The data at this point are incomplete but they do show that the excitation intensity depends only weakly on energy. Because of the broad distribution in energy the scattering at any given energy is quite weak. The count rate in Fig. 27 at 6 meV is 6 counts per minute; further, NTT-35 is a high quality single crystal 2 $cm^3$ in volume. It is not surprising, therefore, that similar inelastic neutron scattering experiments have not yet been successfully executed in other high temperature superconductors.

Finally, we consider the relationship between the dynamic (Fig. 27) and instantaneous spin correlations (Figs. 22, 23 and 24). The integrated inelastic measurements cover the energy range from $-$ kT to $+ E_i$, with the center 1 meV excluded. At 350 K this corresponds to $\sim -30$ meV to $+ 14.7$ meV. Further the energy gain part is amplified by the kinematical factor $k_f/k_i = (1 - \hbar\omega/E_i)^{1/2}$. At 12 K the energy integration range is from $\sim 0.5$ meV to 14.7 meV. Thus the diminution of the integrated inelastic scattering with decreasing temperature evident in Fig. 25 is simply a manifestation of detailed balance combined with the weak energy and temperature dependence of $S(\vec{Q},\omega)$ at $+ \omega$. Heuristically, the thermal spin excitations seem to condense out yielding a slowly fluctuating, modulated, spin fluid response. Further, the slow part of $S(\vec{Q},\omega)$ is correlated three dimensionally.

Clearly, the spin fluctuations in superconducting $La_{2-x}Sr_xCuO_4$ are quite complicated. There does not appear at present to be any theory which can account for these exotic results. It is clear that obtaining an understanding of the influence of the $O^-$ holes on the $Cu^{2+}$ magnetism is a necessary pre-requisite for any model of the superconductivity. We close this section with the empirical observation that the $Cu^{2+}$ - $Cu^{2+}$ spin correlation lengths $\xi_{\parallel}$ and $\xi_{\perp}$ of $18\pm6$ Å and $5.5\pm2$ Å agree with typical estimates of the superconducting coherence lengths in the $CuO_2$ superconductors.

VI.    NEUTRON SCATTERING STUDIES OF $La_2NiO_4$ AND $La_2CoO_4$

As discussed extensively in the previous sections, pure and doped $La_2CuO_4$ exhibit a variety of new, unexpected features in their static and dynamic magnetic correlations. Clearly one would like to understand which features reflect the physics of the 2D, S = 1/2 square lattice Heisenberg antiferromagnet and which follow from more general considerations. Fortunately, these questions may be answered in part by studies of the isomorphous materials $La_2NiO_4$[68] and $La_2CoO_4$.[41] From the ionic model discussed in III one would expect the $Ni^{2+}$ ion to have holes in the $d_{x^2-y^2}$ and $d_{3z^2-r^2}$ orbitals while the $Co^{2+}$ ion would have, in addition, a hole in the $d_{xy}$ orbital. This presupposes that the intra-atomic exchange energy is larger than the crystal field. Thus the $Ni^{2+}$ and $Co^{2+}$ ions should have S = 1 and S = 3/2 respectively while the orbital moment should be quenched.

So far only limited, largely qualitative, data are available on $La_2NiO_4$[68] and $La_2CoO_4$.[41] However, these first generation experiments do provide some important information. We summarize the essential results before reviewing the measurements in more detail. First the measurements confirm that the moments for $Ni^{2+}$ and $Co^{2+}$ are consistent with S = 1 and S = 3/2 respectively. Second, the magnetic state is very sensitive to the stoichiometry and specifically, in $La_2NiO_{4+\delta}$ the Néel temperature varies rapidly with $\delta$.[68] The frustration model of Ref. 12 which is based on classical ideas should actually be more realistic for these higher spin systems and we presume that it describes properly the essential physics of the sensitivity of $T_N$ to doping. Third, the exchange energy is quite large, again as in $La_2CuO_4$. The phase transition behavior is however, fundamentally different. Both $La_2NiO_4$[68] and $La_2CoO_4$[41] show 2D Ising critical behavior as in $K_2NiF_4$.[67] Further, it has been explicitly demonstrated in $La_2NiO_4$ and $La_2CoO_4$ that the energy scale for the dominant 2D critical fluctuations just above $T_N$ is orders of magnitude smaller than that in $La_2CuO_4$.[68] Thus the $S = 1/2$, Heisenberg character of $La_2CuO_4$ is indeed of fundamental importance.

We begin with a discussion of the ordering in $La_2NiO_{4+\delta}$. In all, three crystals of $La_2NiO_{4+\delta}$ have been studied.[68] For convenience we label them Ni-1, Ni-2 and Ni-3. For Ni-1 $\delta \sim 0$ and for Ni-2 $\delta \sim 0.05$ while for Ni-3 $\delta$ was not precisely known at the time of this writing. At low temperatures each has the Cmca crystal structure shown in Fig. 1. For Ni-1 both the T-O transition and a transition to a 3D Néel state occur above 300 K. For Ni-2 $T_0 = 240.5$ K on heating and 232.2 K on cooling while $T_N = 68$ K. Sample Ni-3 orders magnetically at $\sim 45$ K. For all three crystals the magnetic structure is identical to that shown in Fig. 1 except that the spins are oriented along $\vec{a}$, that is, $\vec{\tau} \parallel [100]$ and $\vec{S} \parallel [100]$. It is evident therefore, that in $La_2NiO_{4+\delta}$ the 3D Néel ordering is extremely sensitive to disorder as in the $La_2CuO_4$ system. The microscopic mechanism is presumably common to both systems.[12]

The spin dynamics near $T_N$ in Ni-2 have recently been studied[68] in some detail. It is found that near $T_N$ the 2D critical scattering has an energy width of $< 0.1$ meV. This contrasts with the behavior in $La_2CuO_4$ where, as discussed in III, the corresponding energy width (see Fig. 20) is of order 3 meV. In addition this quasielastic critical scattering shows a sharp peak at $T_N$. This is suggestive of 2D Ising critical behavior as in $K_2NiF_4$.[67]

The spin excitations in the ordered state have been studied by Aeppli and Buttrey.[68] In Ni-1 the excitation spectra are always resolution-limited implying a spin wave velocity of $> 0.3$ eVÅ, comparable to that in $La_2CuO_4$. In Ni-2 ($\delta = 0.05$) it is possible to resolve the spin waves for energies greater than $\sim 10$ meV and one is thereby able to measure the spin wave velocity $v = 0.13$ eVÅ. This should be contrasted with the situation in $La_{1.94}Sr_{0.06}CuO_4$ (Fig. 26) where there is no indication of any splitting of the $+ \vec{q}_{2D}$ and $- \vec{q}_{2D}$ excitations up to 9 meV so that $v > 0.3$ eVÅ. The lowest spin wave gap in Ni-2 at 10 K is $\sim 2.5$ meV compared with 1.1 meV in $La_2CuO_4$. Simple spin wave theory then gives that the ratio of the in-plane anisotropy field to the effective exchange field in Ni-2 is at least $10^2$ times that in $La_2CuO_4$. It is therefore not surprising that the 2D in-plane Ising anisotropy drives the transition to 3D LRO in $La_2NiO_{4.05}$ whereas the 3D coupling initiates the transition to LRO in $La_2CuO_{4-\delta}$.

We now discuss of the ordering in $La_2CoO_4$. At intermediate temperatures $La_2CoO_4$ has the orthorhombic Cmca structure shown in Fig. 1. At 135 K it transforms to a tetragonal unit cell with an in-plane lattice constant of 5.53 Å. The unit cell thus has in-plane dimensions $\sqrt{2} \times \sqrt{2}$ of those of the high temperature I4/mmm tetragonal phase.

The space group for the new low temperature tetragonal structure is not known although the measurements of Ref. 41 suggest P4$_2$/ncm. As noted by Axe *et al.*[40] this structure may be generated by condensation of the low energy phonon mode shown for La$_2$CuO$_4$ in Fig. 5.

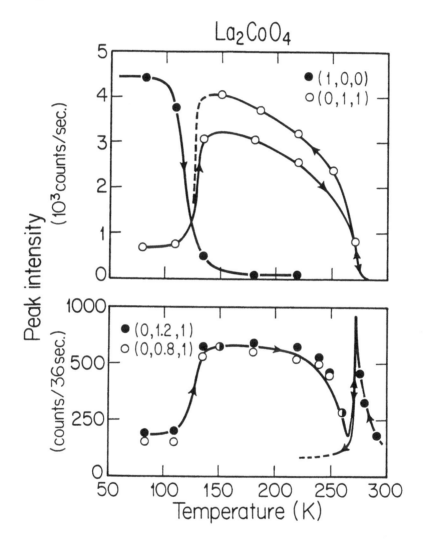

FIG. 28. Temperature dependences of the (100) and (011) peak intensities (a) and those at the "off Bragg" points of (0,1.2,1) and (1,0.8,1) on the 2D magnetic ridge (0,K,1) in La$_2$CoO$_4$. Figure from Ref. 41.

Results of magnetic neutron scattering measurements are shown in Figs. 28 and 29. In Fig. 28 the overall temperature dependences of the magnetic Bragg peak intensities at the (100) and (011) reciprocal lattice positions (see Fig. 7) and those at "off Bragg" points along the 2D magnetic ridge of (0,K,1) are shown. Fig. 28(a) clearly shows the successive phase transitions in this system. Above $T_N$, where there is no 3D magnetic order, magnetic quasi-elastic scattering corresponding to the 2D spin correlations in the $CoO_2$ planes is observed; this is, of course, similar to that observed previously in many 2D systems.[67,75] Between $T_2$ and $T_N$ (hereafter we call this region phase-I and the region $T < T_2$, phase-II), the spin structure is found to be identical to that of $La_2NiO_4$ with the propagation vector $\vec{\tau} \parallel [100]$ and the spin direction $\vec{S} \parallel [100]$. Around $T_2$, a crossover of the intensities of (011) and (100) takes place. In phase-II, the magnetic Bragg intensities for peaks which are uncontaminated by nuclear superlattice scattering can be explained by assuming an $La_2CuO_4$-type spin structure $\vec{\tau} \parallel [100]$, $\vec{S} \parallel [001]$. It is noted that the Cu-type spin structure can be constructed from the Ni-type by a 90° spin rotation or a flip of the spin on half of the $CoO_2$ planes. Because of the tetragonal symmetry, more general non-collinear structures derived from the $La_2CuO_4$ structure are also allowed by the data. The magnetic moments of the $Co^{2+}$ ions are identical in both phases within the experimental resolution. The moment is 2.9±0.1 $\mu_B$ which corresponds to that of $Co^{2+}$ in the high spin state ($S = 3/2$) assuming $g \simeq 2$. We note that this moment is reduced from its bare value by zero-point effects. The true value is probably closer to 3.3±0.2 $\mu_B$, a value somewhat larger than 3 $\mu_B$ and consistent with the XY-like symmetry of the $Co^{2+}$ spin. The Q-dependence of the magnetic Bragg intensities also follows the magnetic form factor of $Co^{2+}$ ion determined by CoO. The Ni-type spin structure persists in the residual orthorhombic structure in phase-II. On the other hand, in the residual tetragonal phase in phase-I, there appears to be no 3D magnetic long range order.

In phase-I, a dramatic hysteresis in the magnetic intensities is observed when one reenters this phase by heating the crystal from phase-II. The peak intensities of the magnetic Bragg reflections are considerably reduced compared with those obtained on cooling. This missing intensity appears as strong diffuse scattering along the $\vec{b}^*$-direction, that is, in the form of a 2D rod. The huge difference of the scattered neutron spectra between heating and cooling processes is shown in Fig. 28(b). This thermal hysteresis entirely disappears in the vicinity of $T_N$. This hysteresis undoubtedly derives from stacking faults in the $La_2CO_4$ magnetic structure which are generated at the first

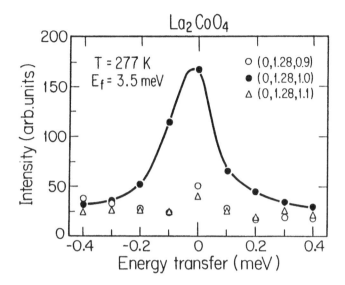

FIG. 29. High resolution energy scans at 277 K around the 2D magnetic ridge
(0,K,1) in the constant Q mode in La$_2$CoO$_4$. The final energy E$_f$ was 3.5 meV and the
collimation 60'-40'-60'-40'-80'. Figure from Ref. 41.

order transition on heating from phase-II to phase-I. These stacking faults then anneal
out near T$_N$.

The phase transition behavior at the upper transition in La$_2$CoO$_4$ is fundamentally
different from that in La$_2$CuO$_4$.[10,11] From the temperature dependence of the (011)
Bragg intensity the Néel temperature T$_N$ was determined to be 274.7±0.6 K. As noted
above, for T > T$_N$, the system exhibits 2D fluctuations. Along the 2D magnetic ridge
(0 K 1), significant 2D intensity is observed with little indication of 3D short range order
effects. The latter become observable only very near T$_N$. For scans across the ridge, the
authors of Ref. 41 report well-defined peaks corresponding to a long 2D spin correlation
length as discussed in IV. In La$_2$CuO$_4$, an unusually high inelasticity is associated with
these 2D spin fluctuations.[10,11] No such effect is observed in La$_2$CoO$_4$. Indeed, as
shown in Fig. 29, the energy spectrum of the constant Q scan on the ridge is very sharply
peaked around $\Delta E = 0$. La$_2$CoO$_4$ also exhibits a clear contrast to La$_2$CuO$_4$ in the
temperature dependence of the instantaneous 2D correlation function (S($\vec{Q}_{2D}$)). As
shown in Fig. 28(b), the peak intensity, S($\vec{Q}_{2D} = 0$) on the ridge shows a very sharp
peak at T$_N$ in good accord with the observed critical behavior in the typical 2D Ising
systems such as K$_2$NiF$_4$.[67] The 2D Ising character also manifests itself in the intensity

profiles of the scans across the ridge. For small K, the peak is centered around the (0 K 1) ridge but not the (1 K 0) ridge; for the latter one can not detect the longitudinal component of the spin correlations because the spin direction is parallel to the scattering vector. This demonstrates that the spin correlations in this compound are dominated by the *longitudinal Ising component.*

Finally, there is to-date only limited information on the spin Hamiltonian relevant to the $Co^{2+}$ ions in $La_2CoO_4$. One may, nevertheless, deduce a number of basic features from the existing neutron scattering experiments together with results from preliminary single crystal susceptibility measurements. First, the bulk susceptibility between 300 K and 440 K is characteristically 2D with the in-plane value about 50% larger than that out-of-plane. This suggests that the spins predominantly lie in the planes and, in the absence of in-plane anisotropy, would be described by an XY model. The in-plane symmetry is broken by the rotation of the $CoO_6$ octahedra as in $La_2CuO_4$. However in the Co case the spin is parallel to the rotation axis and hence also parallel to the antisymmetric exchange vector $\vec{D}$.[51] This requires that either effective crystal field or anisotropic exchange effects control the in-plane anisotropy. This intraplanar anisotropy clearly predominates over the interplanar exchange interaction, accounting for the 2D Ising character of the phase transition.

## VII. ANTIFERROMAGNETISM IN $YBa_2Cu_3O_{6+x}$

So far there is less information on the magnetism in the 123 materials compared with $La_{2-x}Sr_xCuO_4$. However, the results to-date indicate some striking similarities in the two systems. The oxygen content of $YBa_2Cu_3O_{6+x}$ can be varied continuously, with x varying from 0 to 1. For $x \geq 0.4$, the material is orthorhombic and superconducting, reaching a maximum of $T_c$ of ~ 90 K at x = 1. Below x ~ 0.4, the structure becomes tetragonal and the material is an insulator at low temperatures. By analogy with $La_2CuO_{4-\delta}$ one would expect the insulating phase to be antiferromagnetic. The first evidence for magnetic ordering came from μSR experiments by Nishida *et al.*[8] Soon after that superlattice reflections indicative of long-range 3D antiferromagnetic order were observed by Tranquada *et al.*[9] in neutron powder diffraction measurements on x = 0 and 0.15 samples. The spin structure which was deduced from the peak intensities has since been corroborated by several other groups.[76] Work has continued on powder samples with a range of oxygen concentrations, leading to a magnetic phase diagram.[77] Single crystal studies[78,79] have demonstrated that a second type of ordering can occur when sufficient oxygen is present in the system. Most recently, sufficiently large single crystals have been grown[31] that inelastic scattering from spin waves can be studied. The results of these experiments will be discussed in some detail in this section.

The structure of the copper lattice in $YBa_2Cu_3O_6$ is shown in Fig. 30(a). Within each chemical unit cell there are two $CuO_2$ layers, labelled A and C; the B layer contains

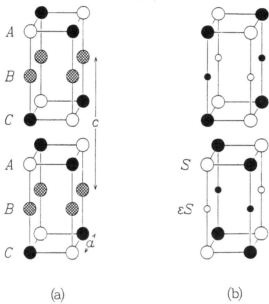

(a)                                      (b)

FIG. 30. (a) Magnetic spin structure for $YBa_2Cu_3O_{6+x}$ with x near zero, from Ref. 9. Only copper atoms are shown for clarity; cross-hatched circles represent non-magnetic $Cu^{1+}$ ions, while black and white circles indicate antiparallel spins at $Cu^{2+}$ sites. Solid lines connect pairs of sites bridged by oxygen atoms. (b) Second type of spin structure observed for larger x, from Ref. 78. The average spin at a B-layer site is a fraction $\varepsilon$ of the spin on a $CuO_2$ layer.

copper atoms with no bridging oxygens. X-ray absorption measurements[80] clearly show that the two-fold coordinated Cu atoms in the B layers have a 1+ valence, and hence those atoms are nonmagnetic (as indicated by cross-hatching in the figure). The spin structure observed[9] for the $Cu^{2+}$ atoms in the A and C layers is indicated by the black and white shading which represents antiparallel spins. Within a $CuO_2$ layer there is a simple Néel ordering with the spins lying in the plane, as in $La_2CuO_{4-\delta}$.[6] Nearest-neighbor A-C layers have no bridging oxygens between them, so that the coupling is presumably dipolar, leading to antiferromagnetic order. The superexchange coupling between next-nearest-neighbor A-C planes also results in antiferromagnetic pairing, so that the magnetic and chemical unit cells have the same height.

Neutron diffraction measurements have been performed[77] on a series of powder samples with varying oxygen contents, resulting in the phase diagram shown in Fig. 31.

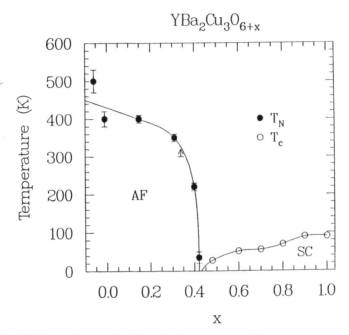

FIG. 31. Phase diagram for $YBa_2Cu_3O_{6+x}$ showing Néel temperatures determined by neutron diffraction on powder samples; AF = antiferromagnet, SC = superconducting. Solid lines are guides to the eye. From Ref. 77.

The magnetic phase boundary is in reasonable agreement with $\mu$SR results.[81] Antiferromagnetic order is observed throughout the insulating tetragonal phase, with a maximum Néel temperature, $T_N$, of ~ 500 K. With increasing oxygen content, $T_N$ decreases gradually until the tetragonal-orthorhombic phase boundary is approached, where it drops quickly to zero. This behavior is quite different from that observed in $La_2CuO_{4-\delta}$, where a subtle variation in $\delta$ causes $T_N$ to drop from ~ 300 K to zero. In that case, increasing the oxygen content introduces O 2p holes which destroy the LRO as discussed in Sec. III. In $YBa_2Cu_3O_{6+x}$, x-ray absorption measurements[80] have shown that the main effect of adding oxygen is to convert $Cu^{1+}$ atoms to $Cu^{2+}$. The presence of the extra magnetic moments has several important effects on the order which will be discussed next.

A $Cu^{2+}$ atom in a B layer will tend to frustrate the antiferromagnetic pairing of its two Cu neighbors in the A and C layers. The 3d hole of a $Cu^{2+}$ atom in the B layer is orthogonal to the 3d holes in the A and C layers, and it follows that the A-B-C coupling is ferromagnetic.[79] Such fluctuations in the local order may explain the decrease in $T_N$ with increasing oxygen content, they should also cause a decrease in the average ordered moment, again in analogy with $La_2CuO_4$. The maximum ordered moment is approximately 0.66 $\mu_B$ per magnetic Cu atom at x ~ 0. This is extremely close to the value predicted by spin-wave theory[46,48] for a quadratic-layer, spin-$\frac{1}{2}$, Heisenberg model as discussed in Sec. IV. With increasing oxygen content the average 3D LRO moment is significantly reduced, while the $\mu$SR measurements[81] indicate the size of the moments which order locally does not vary.

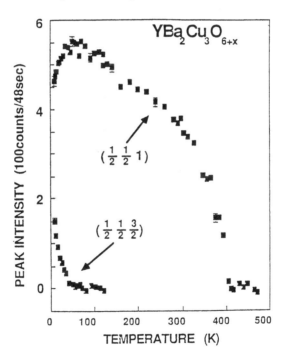

FIG. 32. Peak intensities of $(\frac{1}{2}\frac{1}{2}1)$ and $(\frac{1}{2}\frac{1}{2}\frac{3}{2})$ magnetic reflections as a function of temperature in $YBa_2Cu_3O_{6.35}$. Background (~ 90 counts/48 sec) has been subtracted. Figure from Ref. 78.

When a sufficiently large fraction of the B-layer Cu sites has become magnetic it is possible for the long-range magnetic order to change. Ferromagnetic coupling A-B-C layers would result in a magnetic unit cell with twice the height of the chemical one, as shown in Fig. 30(b). Evidence for such a phase was first observed at low temperature in a single crystal of $YBa_2Cu_3O_{6+x}$ with x ~ 0.35 by Kadowaki et al.[78] At high temperatures, only superlattice peaks of the type $(\frac{1}{2}\frac{1}{2}l)$ with $l$ an integer, characteristic of the structure shown in Fig. 30(a), were seen. However, below T ~ 40 K new peaks appeared at $(\frac{1}{2}\frac{1}{2}l)$ with $l$ a half-integer. Both types of peaks coexist down to the lowest temperature reached (5 K). These relevant peak intensities are shown in Fig. 32. The coexistence may be interpreted[78] either as two different phases present in different parts of the crystal at the same time, or as a single, homogeneous phase in which the spin direction rotates from plane to plane along the c-axis. The average ordered moment in the B layer is a small fraction $\varepsilon$ of the moment in the A and C layers.

In contrast to the Y system, a neutron diffraction study by Moudden et al.[79] of a $NdBa_2Cu_3O_{6+x}$ single crystal with x ~ 0.1 found the $(\frac{1}{2}\frac{1}{2}\frac{l}{2})$ peaks to dominate all the way up to the Néel temperature of 385 K. Only weak diffuse scattering was observed in the regions of $(\frac{1}{2}\frac{1}{2}l)$ reflections. No significant coexistence of superlattice peaks was observed down to 80 K, and the spin structure corresponds to that shown in Fig. 30(b) with $<S> = 0.4\ \mu_B$ and $\varepsilon = 0.1$. More recent measurements on a second $NdBa_2Cu_3O_{6+x}$ crystal show $(\frac{1}{2}\frac{1}{2}\frac{3}{2})$ and $(\frac{1}{2}\frac{1}{2}2)$ peaks with comparable intensities from 300 K up to $T_N$.

More experimental work will be required to determine the phase diagram for the second type of ordering. It seems clear that the ordering behavior in Y and Nd systems is quite different. It has been verified in both powder and single crystal samples that only the first type of ordering occurs at temperatures above 5 K for x $\leq$ 0.2 in the Y compounds. With Nd, the ferromagnetic coupling due to $Cu^{2+}$ in the B layers appears to be much stronger than in the Y system. This difference could be related to lattice parameters; however, some "hidden" variable such as chemical stacking faults or short-range correlations of oxygen atoms in the B layers might also be important. The precise determination of oxygen content in single crystal samples is also a problem when one attempts to compare results for different samples.

The overall picture of magnetic behavior in $RBa_2Cu_3O_{6+x}$ presented here is not universally accepted. Lynn et al.[82] have just recently reported a neutron diffraction study on two antiferromagnetic $NdBa_2Cu_3O_{6+x}$ crystals in which a different type of low temperature spin structure was found. In contrast to the second type of structure reported by Kadowaki *et al.*[78] and Moudden *et al.*,[79] Lynn *et al.*[82] find antiferromagnetic coupling between A and B (and B and C) layers and, in one crystal, ordered magnetic moments of 1.0 $\mu_B$ and 0.5 $\mu_B$ in the A and B planes, respectively. The large ordered moments and the antiferromagnetic A-B-C coupling are inconsistent with results obtained at Brookhaven; these claims also are in disagreement with theory for the moment on the

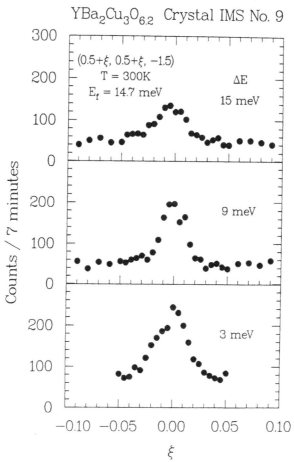

FIG. 33. Several constant $\Delta E$ scans across the 2D magnetic ridge measured in a large single crystal of $YBa_2Cu_3O_{6+x}$ with x ~ 0.2. The zone center for 2D spin-waves is along $(\frac{1}{2}\frac{1}{2}l)$. From Ref. 84.

planes (0.6 to 0.7 $\mu_B$) and for the sign of the chain-plane exchange (ferromagnetic by symmetry). Finally, Lynn *et al.*[82] suggest that their spin structure with large ordered moments in the B layers is the ground state in $RBa_2Cu_3O_{6+x}$ with $x \sim 0$. As discussed above, neutron diffraction and x-ray absorption measurements have demonstrated that for Y compounds with $x \sim 0$ essentially all of the B-layer Cu atoms are 1+ and nonmagnetic. Measurements by Kitaoka et al.[83] using nuclear quadrupole resonance confirm the absence of ordered moments in the B layers of Y compounds with $x < 0.2$. While the ground state of the Nd system with $x \sim 0$ may possibly correspond to the second type of spin structure, the average ordered moment in the B layers must go to zero as x approaches zero.

With some understanding of the magnetic order, the next step is to determine the magnetic interaction strengths through inelastic neutron scattering measurements of spin waves. Very recently Shamoto, Hosoya, and Sato[31] have succeeded in growing crystals which are large enough for such work. Initial measurements have been performed[84] on an antiferromagnetic crystal of $YBa_2Cu_3O_{6.2}$ with $T_N = 370$ K. As in $La_2CuO_{4-y}$, superexchange is expected to be quite strong within $CuO_2$ layers but relatively weak between planes. As a results, spin waves should show relatively little dispersion along the $(\frac{1}{2}\frac{1}{2}l)$ ridge but very steep dispersion perpendicular to that direction. Figure 33 shows scans across the ridge at constant energy transfers of 3, 9, and 15 meV. The dispersion is so great the the $+|q|_{2D}$ and $-|q|_{2D}$ spin-wave branches are not resolved as in $La_2CuO_4$. Taking into account the spectrometer resolution, a lower limit for the spin-wave velocity in the plane can be set at 0.5 eV Å, in good agreement with the results of an inelastic light scattering study.[85] The corresponding nearest-neighbor exchange coupling within a $CuO_2$ plane is $\geq 1000$ K.

Scanning along the $(\frac{1}{2}\frac{1}{2}l)$ ridge at a constant energy transfer, it is observed that the intensity is modulated by the factor $\sin^2(\pi z l)$, where $z = 0.28$ is the fraction of the unit-cell height c by which the nearest neighbor A and C planes are separated. The modulation is due to the fact that the $Cu^{2+}$ ions do not form a simple Bravais lattice. The modulation is still present at least 100°C above the Néel temperature, indicating that correlations still exist within bilayers.

The strong spin-wave dispersion within the $CuO_2$ planes and the weak dispersion perpendicular to the planes indicate the dominantly 2D nature of the magnetic interactions in antiferromagnetic $YBa_2Cu_3O_{6+x}$. Further evidence of the two-dimensionality comes from susceptibility measurements.[86] For samples with low oxygen content, the susceptibility peaks at high temperature and decreases smoothly with decreasing temperature. This behavior is the characteristic signature of 2D antiferromagnetic correlations. With increasing oxygen content the susceptibility changes very gradually, approaching a temperature independent form at x ~ 1. The gradual change suggests that antiferromagnetic correlations survive into the orthorhombic superconducting phase again as in $La_{2-x}Sr_xCuO_4$. So far there is only limited information on the magnetism in the superconducting phase. However, some polarized powder neutron diffraction measurements by Mezei and coworkers[19] shows evidence for low energy magnetic fluctuations in $YBa_2Cu_3O_{6.6}$. Studies on single crystals are required to substantiate these results.

## VIII.   CONCLUSIONS

It is evident that the structural and magnetic properties of the lamellar $CuO_2$ materials are remarkably subtle and rich. Studies of the phonons have been carried out primarily on $La_{2-x}Sr_xCuO_4$. Classic soft mode behavior is observed at the tetragonal-orthorhombic structural phase transition. However, no direct connections have been made between these soft phonons and the superconductivity. In particular, the doping with $Sr^{2+}$ varies $T_0$ but otherwise does not seem to alter the character of the transition and the associated soft modes. Further, these soft modes are not present in the other $CuO_2$ superconductors and hence they cannot be central to the high-$T_c$ phenomenon.

The $CuO_2$ magnetism on the other hand is quite remarkable. Most complete information is available for $La_{2-x}Sr_xCuO_4$ primarily because of the success achieved to-date in growing large, high quality single crystals. However, analogous behavior is seen in the $YBa_2Cu_3O_{6+\delta}$ system. Recently, antiferromagnetic ordering has been observed using $\mu SR$[18] in $Bi_2Sr_2YCu_2O_8$ and we are optimistic that experiments varying the hole concentration systematically in the Bi materials will yield behavior analogous to that already observed in the two systems discussed above.

$La_2CuO_4$ itself represents a close realization of the 2D, $S = 1/2$ square lattice Heisenberg antiferromagnet. However, the magnetic correlations in the $CuO_2$ sheets are remarkably sensitive to the presence of holes on the oxygen ions. The underlying mechanism for this appears to be the strong coupling of the $O^-$ hole spin to the $Cu^{2+}$ spins on either side. This yields a net $Cu^{2+}$ - $Cu^{2+}$ ferromagnetic coupling which frustrates the $CuO_2$ antiferromagnetic order. Since the $Cu^{2+}$ - $Cu^{2+}$ exchange integral is ~ 0.12 eV in magnitude the $Cu^{2+}$ - $O^-$ exchange integral may well exceed 0.5 eV. The instantaneous spin-spin correlation length in the doped $CuO_2$ sheets equals the average separation between the holes. The spin fluctuations at room temperature in heavily doped samples are primarily dynamic. However, even at 350 K there is significant weight in the low energy, $|E| < 0.5$ meV, component and these low energy fluctuations are correlated three dimensionally. As the temperature is lowered the weight in the low energy component grows progressively and $\mu$SR experiments indicate a freezing of all spins in the whole sample below ~4 K. Thus the superconductivity occurs in the presence of a slowly fluctuating 3D incommensurate spin fluid. Further, the progenitors of this novel magnetic state are the holes which carry the supercurrent themselves.

It is clear that the novel magnetism and the phenomenon of high-$T_c$ superconductivity are inextricably bound up together. At the minimum, the magnetism provides the large energy scale necessary for the high $T_c$'s. Theories are still at an early stage of development.[24] Manifestly, any realistic model must explain both the magnetism and the superconductivity. Indeed, at the most basic empirical level, one wonders if it is a coincidence that the 3D magnetic and superconducting coherence lengths are so similar in magnitude.

It is clear that many more neutron experiments remain to be done on these systems. First it is imperative that high resolution measurements be carried out on the spin excitations at energies comparable to the superconducting gap energy. Second, it is obviously important that experiments as detailed as those in $La_{2-x}Sr_xCu_2O_4$ be carried out on the other $CuO_2$ superconductors. As we have seen in Sec. V, such experiments place great demands on crystal growth techniques. We recall that in a high quality crystal of superconducting $La_{1.89}Sr_{0.11}CuO_4$ of volume 2 $cm^3$ the signal rate for 6 meV excitations was 6 counts per minute. If the crystal were, say 0.5 $cm^3$ in volume, the experiment would not have been doable with current neutron sources. It seems likely, nevertheless, that because of the unprecedented importance of this problem the requisite advances in crystal growth will indeed be achieved. We anticipate, therefore, extensive neutron studies of these systems for the indefinite future.

ACKNOWLEDGEMENTS

We are extremely grateful to all of our collaborators in these experiments including especially J. D. Axe, Y. Endoh, Y. Hidaka, M. A. Kastner, B. Keimer, P. Picone, M. Sato, S. K. Sinha, T. R. Thurston, J. Tranquada, and K. Yamada. We are also very grateful to many physicists including in particular A. Aharony and V. J. Emery for valuable discussions of the theoretical issues. The work at MIT was supported by the National Science Foundation - Low Temperature Physics Program under contract no. DMR 85-01856 and the NSF Materials Research Laboratory under contract no. DMR 84-18718. Research at Brookhaven is supported by the Division of Materials Science, US Department of Energy under contract DE-AC02-76CH00016.

Much of the research reviewed here was supported by the US - Japan Cooperative Neutron Scattering Program.

References

1.  Bednorz, J. G. and Müller, K. A., Z. Phys. B64, 189 (1986).

2.  Grande, V. B., Müller-Buschbaum, Hk., and Schweizer, M., Z. Anorg. Allg. Chem. 428, 120 (1977).

3.  Birgeneau, R. J., *et al.*, Phys. Rev. Lett. 59, 1329 (1987).

4.  Onoda, M., *et al.*, Jpn. J. Appl. Phys. 26, L363 (1987).

5.  Fleming, R. M., *et al.*, Phys. Rev. B35, 7191 (1987); van Dover, R. B., Cava, R. J., Batlogg, B., and Rietman, E. A., *ibid.* B35, 5337 (1987).

6.  Vaknin, D., *et al.*, Phys. Rev. Lett. 58, 2802 (1987).

7.  Yamada, K., *et al.*, Solid State Commun. 64, 753 (1987).

8.  Nishida, N., *et al.*, Jpn J. Appl. Phys. 26, L1856 (1987).

9.  Tranquada, J. M., *et al.*, Phys. Rev. Lett. 60, 156 (1988).

10. Shirane, G., *et al.*, Phys. Rev. Lett. 59, 1613 (1987).

11. Endoh, Y., *et al.*, Phys. Rev. B37, 7443 (1988).

12. Aharony, A., *et al.*, Phys. Rev. Lett. 60, 1330 (1988); Birgeneau, R. J., Kastner, M. A., and Aharony, A., Z. Phys. B71, 57 (1988).

13. Fujita, T., *et al.*, Jpn J. Appl. Phys. 26, L402 (1987).

14. Kumagai, K., *et al.*, Physica B148, 480 (1987).

15. Kitaoka, Y., *et al.*, Physica C153-155, 733 (1988).

16. Budnick, J. L., *et al.*, Europhys. Lett. 5, 647 (1988).

17. Harshman, D. W., *et al.*, Phys. Rev. B38, 852 (1988).

18. Uemura, Y. J., *et al.*, J. de Physique (in press) and references therein.

19. Mezei, F., *et al.*, Physica C153-155, 1669 (1988).

20. Kastner, M. A., *et al.*, Phys. Rev. B37, 111 (1988).

21. Wu, M. K., *et al.*, Phys. Rev. Lett. 58, 908 (1987).

22. Maeda, M., *et al.*, Jpn. J. Appl. Phys. Lett. 27, L209 (1988); Hazen, R. M., *et al.*, Phys. Rev. Lett. 60, 1174 (1988).

23. Parkin, S. S. P., *et al.*, Phys. Rev. Lett 60, 2539 (1988).

24. See, for example, Anderson, P. W., Science 235, 1196 (1987); Anderson, P. W., *et al.*, Phys. Rev. Lett. 58, 2790 (1987); Emery, V. J., *ibid*, 58, 2794 (1987); Lee, P. A. and Read, M., *ibid.* 58, 2691 (1987); Hirsch, J. E., *ibid.* 59, 228 (1987); Kivelson, S. A., Rokhsar, D. S., and Sethna, J. P., Phys. Rev. B35, 8865 (1987); Gros, C., Joynt, R., Rice, T. M., Z. Phys. B68, 425 (1987);

Thouless, D. J., Phys. Rev. B36, 7187 (1987); Schrieffer, J. R., Wen, X.-G., and Zhang, S.-C., Phys. Rev. Lett. 60, 944 (1988).

25. For a detailed discussion of the theory of neutron scattering see Theory of Neutron Scattering by W. Marshall and S. W. Lovesey (Oxford 1971).

26. Bardeen, J., Cooper, L. N., and Schrieffer, J. R., Phys. Rev. 108, 1175 (1957).

27. Batlogg, B. G., et al., Phys. Rev. Lett. 59, 912 (1988); Faltjens, T.A., et al., ibid. 59, 915 (1988).

28. Leary, K. J., et al., Phys. Rev. Lett. 59, 1236 (1988).

29. Picone, P. J., Jenssen, H. P., and Gabbe, D. R., J. Cryst. Growth 85, 576 (1987).

30. Hidaka, Y., et al., Jpn. J. Appl. Phys. 26, L377 (1987); J. Cryst. Growth 85, 581 (1987) and private communication.

31. Shamota, S., Hosoya, S., and Sato, M., Solid State Commun. 66, 95 (1988).

32. Birgeneau, R. J., et al., Phys. Rev. B38, Oct. 1 (1988).

33. Thurston, T. R., Axe, J. D., Birgeneau, R. J., et al., (unpublished work).

34. Jorgensen, J. D., et al., Phys. Rev. Lett. 58, 1024 (1987).

35. Böni, P., et al., Phys. Rev. B38, 185 (1988).

36. Aharony, A., Phys. Rev. B8, 4270 (1973).

37. For a review see Shirane, G., Rev. Mod. Phys. 46, 437 (1974).

38. Weber, W., Phys. Rev. Lett., 58, 1371 (1987).

39. Sugai, S., Phys. Rev. B (in press).

40. Axe, J. D., Cox, D. E., Mohanty, K., et al., (unpublished work).

41. Yamada, K., Matsuda, M., Endoh, Y., et al., Phys. Rev. B (submitted).

42. Mattheiss, L. F., Phys. Rev. Lett. 58, 1028 (1987); Yu, J., Freeman, A. J., and Xu, J.-H., ibid., 1035 (1987).

43. GuO, Y., Langlois, J. M., and Goddard, W. A., III, Science 239, 896 (1988).

44. Stechel, E. B.and Jennison, D. R., Phys. Rev. B39, 4632 (1988).

45. McMahon, A. K., Martin, R. M., and Satpathy, S., Phys. Rev. B39, (1988).

46. See for example, Anderson, P. W., Phys. Rev. 86, 694 (1952); Mater. Res. Bull 8, 153 (1973).

47. Chakravarty, S., Halperin, B. I., Nelson, D. R., Phys. Rev. Lett. 60, 1057 (1988) and to be published.

48. Reger, J. D., Young, A. P., Phys. Rev. B37, 5978 (1988); Huse, D. A., Phys. Rev. B37, 2380 (1988); Huse, D. A. and Elser, V., Phys. Rev. Lett. 60, 2531 (1988).

49. Arovas, D. P. and Auerbach, A., Phys. Rev. B38, 316 (1988); Auerbach, A., and Arovas, D. P., Phys. Rev. Lett. 61, 617 (1988).

50. Peters, C. J., *et al.*, Phys. Rev. B37, 9761 (1988).

51. Thio, T., *et al.*, Phys. Rev. B38, 905 (1988).

52. Greene, R. L., *et al.*, Solid State Commun. 63, 379 (1987); Ishi, H., *et al.*, Physica 148B, 419 (1987); Johnston, D. C., *et al.*, Phys. Rev. B36, 4007 (1987).

53. Fukuda, K., *et al.*, Solid State Commun. 65, 1323 (1988); Cheong, S.-W., *et al.*, Solid State Commun. 65, 111 (1988).

54. Kastner, M. A., *et al.*, Phys. Rev. B38, Oct. 1 (1988).

55. Freltoft, T., *et al.*, Phys. B37, 137 (1988).

56. Stassis, C., Harmon, B. N., Freltoft, T., *et al.*, Phys. Rev. B (submitted).

57. Birgeneau, R. J., *et al.*, Phys. Rev. B21, 317 (1980); Cowley, R. A., *et al.*, Phys. Rev. B21, 4038 (1980).

58. Freltoft, T., *et al.*, Phys. Rev. B36, 826 (1987).

59. Uemura, Y. J., *et al.*, Phys. Rev. Lett. 59, 1045 (1987), Physica C153-155, 769 (1988).

60. For a review of neutron experiments on $Eu_xSr_{1-x}S$ and $(Fe_xMn_{1-x})_{75}P_{16}Al_3$, see G. Aeppli, *et al.*, J. Appl. Phys. 55, 1628 (1984).

61. Tranquada, J. M., *et al.*, Phys. Rev. B35, 7187 (1987); *ibid.*, B36, 5263 (1987); Nücker, N. *et al.*, Phys. Rev. B37, 5158 (1988), Shen, Z.-X., *et al*, Phys. Rev. B36, 8414 (1987).

62. For a review see Binder, K. and Young, A. P., Rev. Mod. Phys. 58, 801 (1988).

63. Aeppli, G., *et al.*, Phys. Rev. B25, 4882 (1982).

64. Kimishita, Y., *et al.*, J. Phys. Soc. Japan 55, 3574 (1986).

65. Vannimenus, J., *et al.*, (unpublished).

66. Morgenstern, I., private communication.

67. Birgeneau, R. J., Skalyo, J., Jr., and Shirane, G., Phys. Rev. B3, 1736 (1971); Birgeneau, R. J., Als-Nielsen, J., and Shirane, G., Phys. Rev. B16, 280 (1977).

68. Aeppli, G. and Buttrey, D., Phys. Rev. Lett. 61, 203 (1988); Freltoft, T., Shirane, G., Buttrey, D. J., *et al.*, (manuscript in preparation).

69. Lyons, K. B., *et al.*, Phys. Rev. B37, 2393 (1988).

70. Birgeneau, R. J., Endoh, Y., Hidaka, Y., *et al.*, Phys. Rev. Lett. (submitted).

71. Yoshizawa, H., *et al.*, J. Phys. Soc. Japan (in press).

72. Thurston, T. R., (unpublished work).

73. Thurston, T. R., *et al.*, Phys. Rev. B (submitted).

74. Oh, B., *et al.*, Phys. Rev. B37, 7861 (1988).

75. Birgeneau, R. J., Guggenheim, H. J., and Shirane, G., Phys. Rev. B1, 2211 (1970).

76. Li, W., *et al.*, Phys. Rev. B37, 9844 (1988); Burlet, P. *et al.*, Physica C153-155, 1115 (1988).

77. Tranquada, J. M., *et al.*, Phys. Rev. B38, 2477 (1988).

78. Kadowaki, H., *et al.*, Phys. Rev. B37, 7932 (1988).

79. Moudden, A. H., *et al.*, Phys. Rev.B38, Nov. 1 (1988).

80. Tranquada, J. M., *et al.*, (unpublished work).

81. Brewer, J. H., *et al.*, Phys. Rev. Lett. 60, 1073 (1988).

82. Lynn, J. W., *et al.*, Phys. Rev. Lett. 60, 2781 (1988).

83. Kitaoka, Y., *et al.*, J. de Physique (in press).

84. Sato, M., *et al.*, Phys. Rev. Lett. 61, 1317 (1988).

85. Lyons, K. B., *et al.*, Phys. Rev. Lett. 60, 732 (1988).

86. Johnston, D. C., *et al.*, Physica C153-155, 572 (1988).

87. Torrance, J. B., *et al.*, Phys. Rev. Lett. 61, 1127 (1988).

# 5

## NORMAL STATE TRANSPORT AND ELASTIC PROPERTIES OF HIGH $T_c$ MATERIALS AND RELATED COMPOUNDS

P.B. Allen
*SUNY, Stony Brook, NY 11794, USA*

Z. Fisk and A. Migliori
*Los Alamos National Laboratory*
*Los Alamos, NM 87545, USA*

Normal state experiments in high $T_c$ superconductors have given a rich and very confusing story. The confusion stems from the sensitivity to processing history and other sample-dependent effects, from the intrinsic richness of phenomena available in these systems, and from the difficulty of reconciling the unexpected phenomena with any single unifying theory. Most of the data, from pressed powder ("ceramic") samples, represent an exceedingly complicated and sample-dependent average of the different tensor components of the microscopic transport coefficients. Therefore we shall focus on single crystal results whenever possible. The experiments to be discussed are

1. Resistivity
2. Hall coefficient
3. Thermopower
4. Magnetism
5. Ultrasound
6. Thermal conductivity. This is also nicely reviewed by Fischer, Watson, and Cahill,[1] who also review specific heat and electrical resistivity;
7. Electrodynamics

## I. Resistivity

Most high $T_c$ superconductors are orthorhombic and thus have 3 independent tensor components for the resistivity, $\rho_{xx} = \rho_a$, $\rho_{yy} = \rho_b$, $\rho_{zz} = \rho_c$. We use a notation where the $CuO_2$ planes lie in the ab or xy plane, and the CuO chains (in $YBa_2Cu_3O_7$) lie in the b or y direction. However, almost all orthorhombic crystals are microtwinned, so that only an average of $\rho_a$ and $\rho_b$ (called $\rho_{ab}$) can be measured. It is suspected that $\rho_a$ and $\rho_b$ will differ by less than a factor of 2, so the in-plane average should not be far from an arithmetic mean $(\rho_a + \rho_b)/2$. By contrast, $\rho_c$ is typically larger by a factor of 50 or more.

For homogeneous single-phase samples which are superconducting,
the following general features are seen in the normal-phase
temperature dependence:

(i) $\rho_{ab}$ is "metallic" with $d\rho_{ab}/dT$ positive and comparable to
$\rho_{ab}/T$.

(ii) $\rho_c$ is larger by $\sim 10^2$ and usually "non-metallic" ($d\rho_c/dT < 0$).

(iii) $\rho_{ab}$ is surprisingly close to linear in T, and the
extrapolated value $\rho_{ab}(T{=}0)$ is close to zero.

(iv) The slope $d\rho_{ab}/dT$ is large, $\sim 0.5{-}1.0$ $\mu\Omega cm/K$, corresponding to
room-temperature resistivities of 150-300 $\mu\Omega cm$. However, variations
of a factor of 2 or more in slope are commonly seen from sample to
sample, reflecting problems such as microcracks or inhomogeneities
which diminish the effective conducting cross-sectional area,
problems with attaching leads, or possibly other factors.

These four observations will now be illustrated with data selected
from the literature. Figure I.1 shows resistivity[2] of the
non-superconducting "parent" compound $La_2CuO_4$, an antiferromagnetic

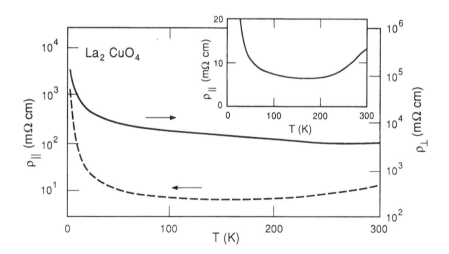

Fig. I.1   $\rho_{ab}$ (denoted $\rho_{\parallel}$) and $\rho_c$ (denoted $\rho_\perp$) of insulating
$La_2CuO_4$ from Ref. 2. The inset shows $\rho_{ab}$ on a linear
scale.

insulator. At 300 K, $\rho_{ab}$ is 1.3 x $10^4$ $\mu\Omega$cm, and $\rho_c$ is larger by a factor of 300. Both directions show a dramatic increase in $\rho(T)$ at low temperatures. This is interpreted[2] as a crossover from diffusive transport at high temperature to strongly correlated variable range hopping at low temperature. Good homogeneous single crystals of superconducting $La_{2-x}M_xCuO_4$ (where M = Ca, Sr, Ba) have been hard to grow, and fewer single-crystal resistivity data are available. Figure I.2 shows results for $\rho_{ab}(T)$ from Suzuki and Murakami[3] in a sample which is superconducting at 12 K.

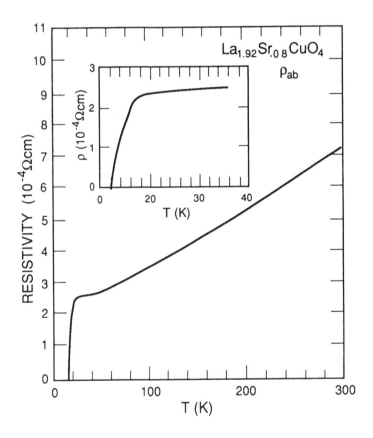

Fig. I.2    $\rho_{ab}(T)$ from Ref. 3, measured on a single-crystal thin film of $La_{1.92}Sr_{.08}CuO_4$.

Superconducting single crystals of the family $MBa_2Cu_3O_{7-\delta}$ are more common, where M = Y or various Lanthanides, and $\delta$ is most often $\sim 0$. Figure I.3 shows single crystal measurements by Penney *et al.*[4] on $YBa_2Cu_3O_7$. Notice that there are two $\rho_{ab}(T)$ curves on two different crystals. The one with the higher $T_c$ shows a lower $\rho_{ab}$, both in slope ($\sim 0.7$ $\mu\Omega cm/K$) and in extrapolated $\rho_{ab}(T = 0)$ ($\sim 20$ $\mu\Omega cm$) compared with the lower $T_c$ crystal ($\sim 1.5$ $\mu\Omega cm/K$). By contrast, consider the results of Störmer *et al.*[5] on oriented thin films, shown

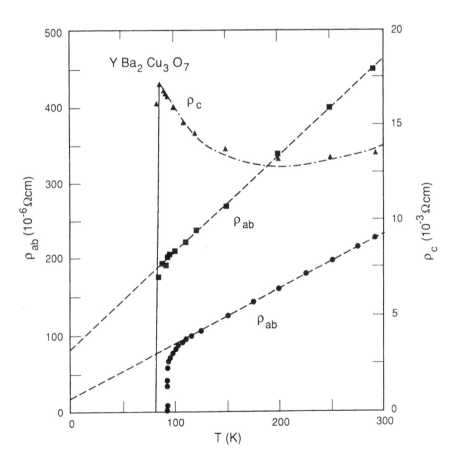

Fig. I.3     $\rho_{ab}(T)$ and $\rho_c(T)$ for $YBa_2Cu_3O_7$ from Ref. 4.   The two $\rho_{ab}(T)$ curves are from different crystals.

in Fig. I.4. Here the "best" sample (as measured by $T_c$ and critical
current density) has the <u>lowest</u> resistance ratio (3.4 for sample 1 as
opposed to 5.4 for sample 6) as well as the highest $\rho(300\ K)$
(233 $\mu\Omega$cm, similar to the better sample in Fig. I.3, for sample 1,

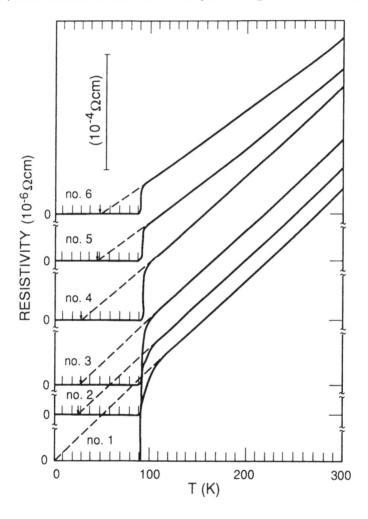

Fig. I.4    $\rho(T)$ for oriented thin films of $YBa_2Cu_3O_7$, from Ref. 5.
            Presumably the results are very similar to $\rho_{ab}(T)$ of
            single crystals. Samples numbered 1 through 6 have
            decreasing values of $T_c$ and $j_c$.

compared to 151 $\mu\Omega$cm for sample 6.) Furthermore, the best sample
(No. 1) has an extrapolated $\rho(0\ K)$ of 0 $\mu\Omega$cm, whereas the worst
sample (no. 6) has an extrapolated $\rho(0\ K)$ = -30 $\mu\Omega$cm.

More recently, single-crystal data for other layer structure
CuO-based superconductors have become available. Figure I.5 shows
results by Martin et al.[6] for an untwinned orthorhombic crystal of
$Bi_2Sr_{2.2}Ca_{0.8}Cu_2O_8$ with $T_c$ ~ 81 K. Three unusual features should be
noted: (i) The anisotropy $\rho_c/\rho_a$ and $\rho_c/\rho_b$ is extremely large, nearly
$10^5$. This reflects the large spacing of 12 A between CuO layers in
the crystal structure. (ii) The "non-metallic" behavior $d\rho_c/dT < 0$
is almost absent. (iii) $\rho_a$ and $\rho_b$ exhibit quite linear temperature
dependences, with anisotropy $\rho_a/\rho_b$ ~ 1.5-2.0 and extrapolated T = 0
intercepts $\rho_b(0)$ ~ +5 $\mu\Omega$cm and $\rho_b(0)$ ~ -15 $\mu\Omega$cm. The average slope,
$d\rho_{ab}/dT$, is 0.46 $\mu\Omega$cm/K.

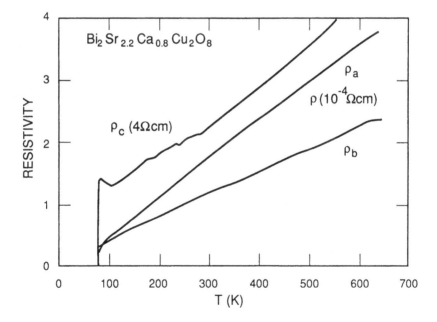

Fig. I.5    $\rho_a(T)$, $\rho_b(T)$, and $\rho_c(T)$ for $Bi_2Sr_{2.2}Ca_{0.8}Cu_2O_8$ as
measured in Ref. 6.

Perhaps the most striking aspect of these data is the surprising degree of linearity of $\rho_{ab}(T)$. The linearity persists up to 600 K as shown in Fig. I.5, and has been followed up to 1100 K in $La_{2-x}Sr_xCuO_4$ and 800 K in $YBa_2Cu_3O_7$ by Gurvitch and Fiory.[7] Of course, conventional metals also have linear $\rho(T)$ curves at higher temperature where the cross section $\sigma \propto \langle u^2 \rangle$ (for scattering degenerate-electron quasiparticles from lattice displacements u) goes as n + 1/2 (where n is the Bose occupation factor), which equals $(k_B T/\hbar\omega)$ $(1+1/12$ $(\hbar\omega/k_B T)^2 +...)$ at high temperature. Actually, it is not generally appreciated at how low a temperature the Bloch–Gruneisen resistivity begins to appear linear. This linear resistivity is partly due to the factor of 1/12 and is partly due to other factors. A good example is rhenium,[8] shown in Fig. I.6. These data show the typical weak anisotropy of an hcp metal, and the typical linearity down to T ~ 75 K, in spite of a high $\theta_D$ ~ 420 K (derived from elastic constants.) The slope $(d\rho/dT ~ 0.065 \, \mu\Omega cm/K)$ and linearly extrapolated intercepts $(\rho(0 \, K) ~ -3 \, \mu\Omega cm)$, are typical of good metals (i.e. metals with propagating electronic quasiparticles, having room-temperature mean free path $\gtrsim$ 20 A.) Therefore, it is certainly not correct to cite the linear $\rho(T)$ behavior as a priori evidence for unusual physics. In particular, $\rho_{ab}(T)$ for $La_{1.92}Sr_{0.08}CuO_4$ shown in Fig. I.2 has just the kind of deviation from linearity expected in ordinary metals, and Bloch–Gruneisen-type fits have been made[9] to the data of Figs. I.2 and I.3.

Nevertheless, there are reasons for arguing that the linear $\rho(T)$ is very unusual, and weaker but still plausible reasons for supposing that the extrapolated T = 0 value $\rho(0 \, K) ~ 0$ is in fact "intrinsic" (i.e. characteristic of the best samples). These reasons have been presented by Stormer et al.[5] based on the data of Fig. I.4 and a correspondingly unusual temperature-dependence of the Hall coefficient (to be discussed in the next section.)

A critical test would be to discover metallic samples where $\rho(T)$ can be measured to much lower temperatures, to search for curvature of $\rho(T)$, which would be expected for a conventional phonon scattering mechanism. Such a material has been studied by Xiao et al.[10] and is

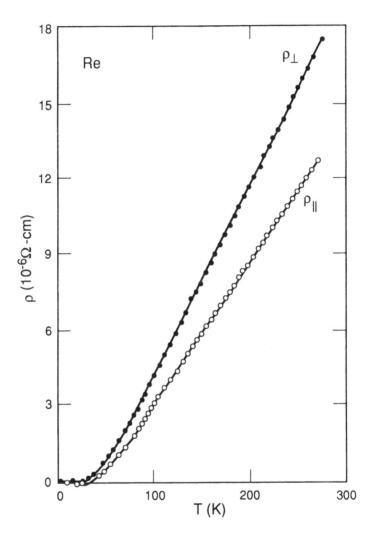

Fig. I.6    $\rho_{ab}(T)$ (denoted as $\rho_{\perp}$) and $\rho_c(T)$ (denoted as $\rho_{\parallel}$) for Re metal, from Ref. 8.

shown in Fig. I.7. It is striking that in the best superconductors (with composition near $BiSrCuO_y$) the $\rho(T)$ curve is linear down to $T_c \sim 7$ K. The absolute values of $\rho(300$ K) in these polycrystalline samples are 2–3 x $10^3$ $\mu\Omega$cm, much higher than $\rho_{ab}(300$ K) of single crystals of other superconductors, but typical for polycrystalline metals. The value of A is remarkably close to zero in good samples. However, the occurrence of negative values of A in extrapolated cases is not easy to understand unless the intrinsic $\rho(T)$ behavior is a highly nonlinear function of T at low T.

However, counter-examples can be found. For example, Torrance et al.[11] have studied various metallic but non-superconducting compounds based on CuO and find ordinary-looking curved $\rho(T)$ behavior. However, their best data were obtained from a compound $La_4BaCu_5O_{13}$ which does not have a layer structure. Less easily explained away is a measurement by Cheong et al.[12] on a single crystal of $La_{1-x}Sr_xCuO_4$ (x $\sim$ 0.02), shown in Fig. I.8. $\rho_{ab}(T)$ appears to have a small low-temperature linear slope, and a fairly ordinary upward curvature.

It has not been easy to find a theoretical explanation for an intrinsic linear behavior of $\rho(T)$. Zou and Anderson[13] gave a plausibility argument based on RVB theory, but their argument has received criticism.[14] If the behavior is intrinsic, it may not be limited to CuO systems. Gurvitch[7] has pointed out that similar behavior was seen[15] in NbO. Figure I.9 shows $\rho(T)$ for various polycrystalline samples of cubic NbO in the range 77 K $<$ T $<$ 900 K. Except for being smaller in magnitude by a factor 2–10, the $\rho(T)$ data for this low-$T_c$ oxide are very reminiscent of the best data for high-$T_c$ CuO systems. The data are also quite reminiscent of the behavior of conventional Fermi liquid metals like Re (Fig. I.6). Another low $T_c$ oxide superconductor is spinel-structure $Li_{1+x}Ti_{2-x}O_4$. Resistivity data[16] for ceramic samples are shown in Fig. I.10. Again the data are reminiscent of the high-$T_c$ CuO superconductors, except

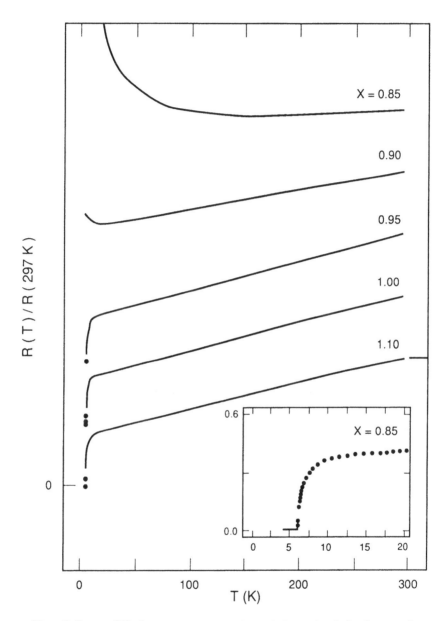

Fig. I.7    $\rho(T)$ for ceramic samples of $Bi_{2-x}Sr_xCuO_y$ for various values of x, from Ref. 10.

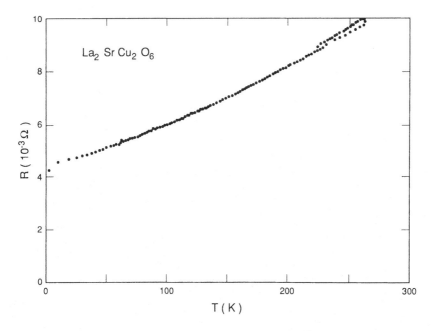

Fig. I.8    $\rho_{ab}(T)$ for $La_2SrCu_2O_6$ from Ref. 12.

that data on high-$T_c$ samples with good metallic conductivity ($\rho < 10^3$ $\mu\Omega$cm) are not available.

A second question which the single crystal $\rho(T)$ data raise is whether the rapid increase as temperature is lowered at low temperature is intrinsic, and how to explain it. Zou and Anderson[13] offered an explanation for intrinsic $\rho_c \sim 1/T$ behavior from RVB theory. Very often data fit well the form $\rho_c(T) = A/T + BT$ where the term BT is interpreted as arising from defect-related short circuits of the barriers between $CuO_2$ planes. This is illustrated in Fig. I.11, which shows data by Hagen et al.[17] plotted as $\rho_c(T)$ versus $T^2$. However, Iye et al.[18] have reported on some samples where $\rho_c(T)$ lacks completely the A/T term, and these are presumed to be the best annealed crystals. These data are shown in Fig. I.12. Ossipyan et al.[19] have also seen "fully metallic" behavior of $\rho_c(T)$.

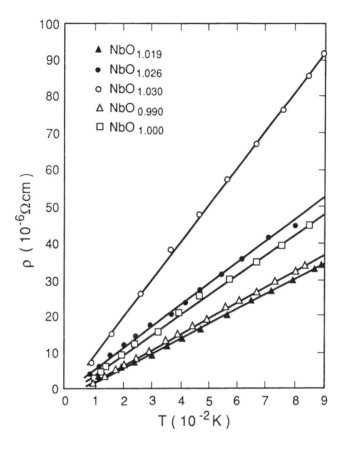

Fig. I.9    $\rho(T)$ for cubic $NbO_x$ from 77 to 900 K and various oxygen stoichiometries near $x = 1$ (from Ref. 15).

If $d\rho_c/dT < 0$ at low temperature is <u>not</u> intrinsic as Iye suggests, then this behavior may indicate incipient strong or weak localization in the c-direction.  This explanation has not found immediate theoretical support.  At $T = 0$ K, strong localization in the c-direction is inconsistent with delocalization in the ab plane. Impurity fluctuations in the hopping matrix element $t_c$ will always permit the wave function to couple better to adjacent planes at certain "hot spots," destroying strong 1D localization.

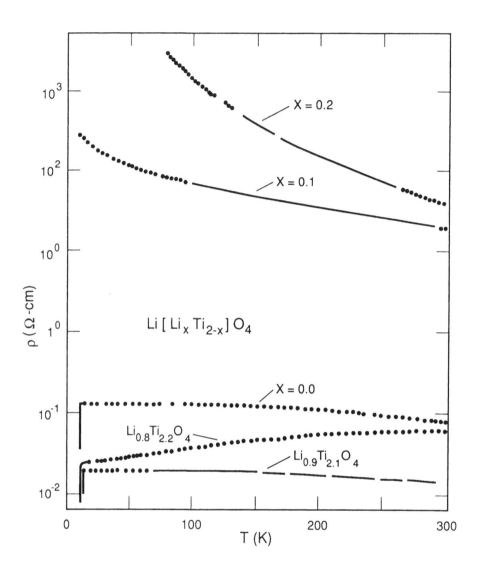

Fig. I.10   $\rho(T)$ for ceramic samples of spinel structure $Li_{1+x}Ti_{2-x}O_4$ (from Ref. 16).

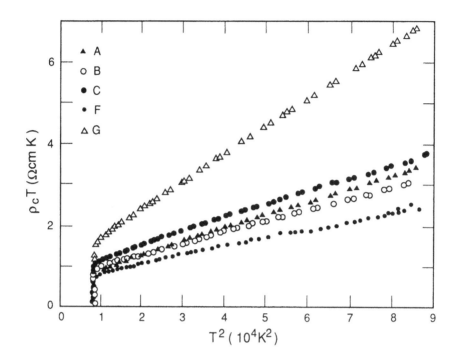

Fig. I.11    $\rho_c(T)$ for five single crystals of $YBa_2Cu_3O_7$, plotted as $\rho_c T$ versus $T^2$ (from Ref. 17).

Finally, it is worth comparing such behavior with that of a more conventional layer-structure superconductor. Single crystal data[20] for the hexagonal-structure, 2.5 K superconductor 4Hb-polytype of $TaS_2$ are shown in Fig. I.13. The metallic behavior of $\rho_{ab}$ and non-metallic $\rho_c(T)$ are quite similar to that of CuO superconductors. At T $\lesssim$ 25 K, the sharply rising $\rho_c(T)$ stops rising and seems to saturate, consistent with weak localization.

A third issue raised by the $\rho(T)$ data is the absence of "saturation" effects. Many metallic d-band compounds are known where $\rho(T)$ appears conventional (i.e. Bloch-Gruneisen-like) at low temperature, but approaches approximately a constant high temperature value, $\rho_{sat} \sim 150 \ \mu\Omega cm$. This was first emphasized by Fisk and

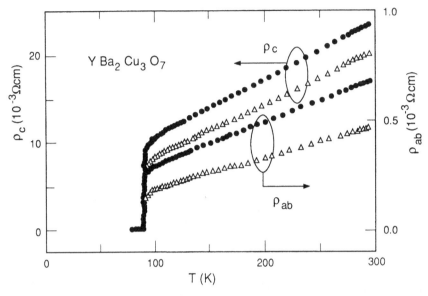

Fig. I.12  $\rho_{ab}(T)$ and $\rho_c(T)$ for a single crystal of $YBa_2Cu_3O_7$.  The two sets of data are obtained from different combinations of contacts.  Note that $\rho_c(T)$ is linear in T (from Ref. 18).

Webb,[21] and their data for A15-structure metals is shown in Fig. I.14.  The phenomenon does not depend on whether the metal is a good (like $Nb_3Sn$) or bad (like $Nb_3Sb$) superconductor, only on the fact that $\rho$ is large.[21]  It is apparent that the carriers at low tempera-ture have a fairly large mean free path, permitting a conventional Fermi liquid analysis, whereas at high temperature the scattering is so strong that the mean free path is $\lesssim$ 3 A.  In this regime, wave vectors, velocities, and mean free paths can no longer be properly defined.  The experimental observation is that $\rho(T)$ saturates in this regime, and this is often explained by the statement that the mean free path cannot become any shorter than ~3 A, so the resistivity must cease growing.  In fact, there is no good theory to explain this.[22]  The downward curvature of $\rho_{ab}$ in $TaS_2$ (Fig. I.13) is

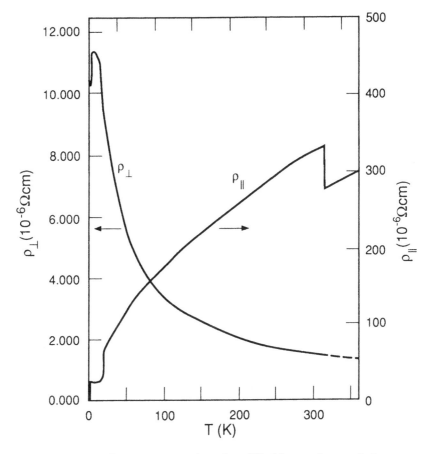

Fig. I.13   $\rho_{ab}$ (denoted as $\rho_{||}$) and $\rho_c(T)$ (denoted as $\rho_\perp$) for a single crystal of 4Hb-TaS$_2$ (from Ref. 20).

strongly suggestive of saturation with $\rho_{sat} \sim 400$ μΩcm. The question is, why are very similar effects not seen in CuO superconductors? In fact, such effects probably are seen in samples where alloying has disordered the CuO$_2$ planes. Figure I.15 shows $\rho(T)$ data[23] for ceramic samples of YBa$_2$(Cu$_{1-x}$Ga$_x$)$_3$O$_7$. As x increases to 0.07, $\rho$ has increased in magnitude by a factor of 2-3 at room temperature, and downward curvature is starting to be noticeable. The crossover point

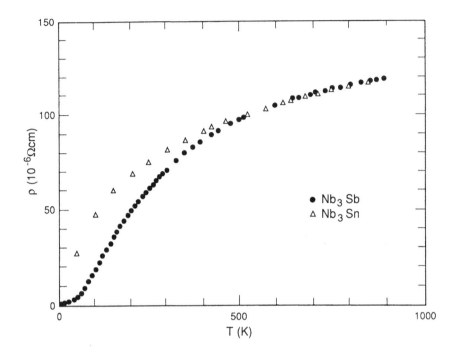

Fig. I.14   $\rho(T)$ for A15 structure metals showing high T "saturation"
(from Ref. 21).

where deviations from Bloch-Gruneisen behavior start to be seen is
typically 10 Å.  This suggests that pure $YBa_2Cu_3O_7$ has a mean free
path at 300 K of ~ 30 Å.  This is consistent with the observed[7] lack
of saturation up to 800 K in pure samples.  Such a long mean free
path is surprising in a high $T_c$ material, since most scenarios derive
superconductivity from interactions between quasiparticles, and these
interactions should cause scattering.

    If we assume a mean free path $\ell$ ~ 30 Å at 300 K in $YBa_2Cu_3O_7$, and
take quite a low value, $\rho_{ab}(300 \text{ K})$ ~ 220 µΩcm, we can estimate the
number of carriers by a crude argument.  We write $\rho = \hbar k_F/ne^2 \ell$ and
assume that the carriers are located in a single-sheeted cylindrical
Fermi surface of radius $k_F = \sqrt{2\pi nc}$ where n is the carrier density and

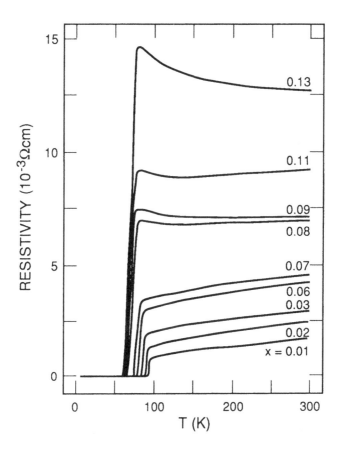

Fig. I.15   $\rho(T)$ for ceramic samples of $YBa_2(Cu_{1-x}Ga_x)_3O_7$ for various
values of x, from Ref. 23.

c is the unit cell height.  The resulting value of n corresponds to
~ 1.6 carriers per formula unit, or about 1/2 carrier per Cu atom.
Assigning a free electron value to the mass, the corresponding Fermi
velocity is $0.7 \times 10^8$ m/s and the corresponding lifetime $\tau$ is 4 ×
$10^{15}$ s.  The corresponding scattering rate $h/\tau$ is ~ 0.15 eV or about
$2\pi k_B T$ at 300 K.  In electron–phonon systems,[24] the scattering rate
corresponds to $2\pi\lambda k_B T$ where $\lambda$ is the dimensionless electron–phonon
coupling.  Thus we fit the data with $\lambda \sim 1$, a number which is not

unreasonable.  A value $\lambda \sim 1$ is much too low to account for a 90 K value of $T_c$ by a pure phonon mechanism.

This argument should not be taken very seriously, however.  For one thing, band theory does not give a single sheet in $YBa_2Cu_3O_7$. For another, the relevance of the band-picture is often doubted. However, recent photoemission measurements of List[25] show a Fermi edge.  One ought at least to allow the possibility that the carrier mass m✱ differs from the free electron value m.  Then the corresponding argument leads to $\lambda \sim 1.0$ (m/m✱).  If one wishes to increase $\lambda$ but not to make m✱/m smaller than 1, then one will have difficulty understanding how the resistivity is so large and yet does not saturate.

## II.  Hall Effect

Most of the oxide superconductors have close relatives which are semiconductors or insulators and which can be reached from the superconducting phases by compositional variations of stoichiometry. Interest of course attaches to the interrelation between the properties of the semiconducting and superconducting phases, and Hall effect measurements provide perhaps the simplest probe with which to explore this interface.  The Hall constant is by no means straightforward to interpret in general: only in the high-field limit is the meaning clear, and even then we need information as to how many bands are involved to understand the numbers.  But the variation of Hall constant with a parameter such as the concentration of a constituent in a compound can be given a clean interpretation in many cases, and this is the situation in some of the high $T_c$ oxides.

We consider first the data for $La_2CuO_4$.  The pure material is non-metallic.  The Hall coefficient[26] is large and T-dependent, roughly following the behavior of $\rho(T)$, although the mobility $\mu_H(T)$ has a broad maximum near 100 K in an antiferromagnetic sample with $T_N = 280$ K.  Data for strontium-doped systems come from Shafer et al.[26] and Ong et al.[27] and they are in substantial agreement for

these measurements on polycrystalline material. It is found that
each Sr atom substituted adds one hole, as might be expected from
simple chemical considerations. Shafer et al.[26] show that the number
of holes as determined by $R_H$ is equal to the [Cu-O]+ concentration in
the compound as determined iodometrically. This variation of hole
concentration with Sr addition continues out to a composition
$La_{1.85}Sr_{0.15}CuO_4$ beyond which there is an abrupt drop to an
essentially unmeasurably small Hall constant (Fig.II.1). This
behavior appears most cleanly in the data of Ong et al.[27] It is
important to note that the oxygen stoichiometry becomes a problem for
Sr > 0.20. Torrance et al.[28] have reported on this, and they have
shown how to maintain the stoichiometry near $O_{4.0}$ for Sr
concentrations as large as 0.40.

Band theory makes instead the following predictions. For x = 0,
there is a half-filled band with one carrier per cell. If currents
and Hall fields are in the ab plane of a single crystal, with $\vec{B} \parallel c$,
such a system will have $R_H$ small because of partial cancellations of
hole-like and electron-like contributions. The cancellation is exact
in the simplest 2D single band model. More sophisticated
calculations[29] have cancellation at x ≈ 0.20, with $R_H$ going smoothly
from positive (hole-like) at x = 0 to negative at x > 0.20. For $\vec{B} \perp$
c, $R_H$ is predicted to be negative and fairly small. The Hall data
for small x completely contradict this picture, as expected, because
$La_2CuO_4$ is actually non-metallic.

The non-metallic behavior is attributed to electron-electron
(Coulomb) correlations, and can presumably be modelled by an
appropriate Hubbard Hamiltonian. For dimension D > 1 and filling
factor deviating from 1/2, the behavior of Hubbard models is not
rigorously known. It is tempting to attribute the experimentally
observed jump in $R_H$ at x = 0.15 to a transition from a highly
correlated metal with carrier density x = number of dopants to a
conventional band Fermi-liquid with carrier density 1 - x. The
transition is presumably driven by free-carrier screening of the
repulsive Coulomb effect (Hubbard U). The difficulty is then how to
relate this transition to the occurrence of superconductivity,

234

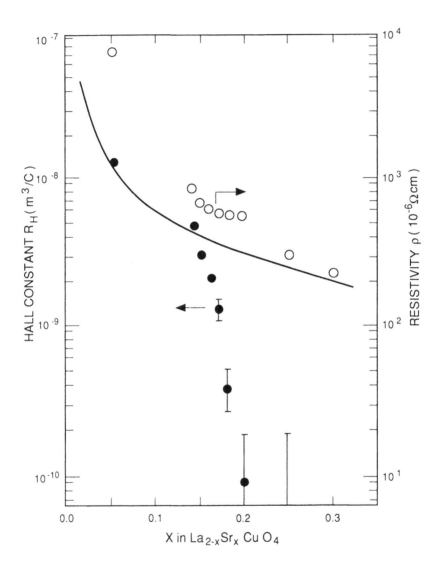

Fig. II.1    $R_H$ and resistivity at 77 K versus Sr concentration from
Ong et al. (Ref. 27).

especially since no sharp changes in superconducting properties occur at x = 0.15. Three routes are available; all invoke sample inhomogeneities to argue that the true intrinsic change in $R_H$ of a homogeneous sample would occur in a more explainable location. One route is to push the $R_H$ jump to larger x, making the superconducting region very interesting because the normal state behavior (as seen in $R_H$) is not a conventional Fermi liquid. Another route is to push the $R_H$ jump to lower x, so that the whole superconducting region is built from a conventional Fermi liquid. A third is to argue that the jump in $R_H$ is entirely an artifact, i.e., to argue that poorly made samples are always highly correlated.

Single crystals of $La_{2-x}Sr_xCuO_4$ in the superconducting range have been hard to grow. We know of only one report[3] of $R_H$ on such a single crystal, a thin film with x ≈ 0.06 and $T_c$ ~ 18 ± 2 K. Only one tensor component $R_{xyz}$ (Hall field in x or a direction, current in y or b direction, B field in z or c direction) was measured. The resulting "carrier density," $1/eR_{xyz}$ is shown in Fig II.2. At T ~ 50-100 K, their value of n corresponds to $R_H$ ~ 1.4 x $10^9$ $m^3$/C, a fairly low value, smaller by a factor of ~6 than the interpolated value at x = 0.06 in Fig II.1, and larger by a factor of ~3.5 than the band theory prediction.[29] Perhaps the most interesting aspect of Fig II.2 is that $1/R_H$ is seen to have a strong linear T-dependance A + BT above 150 K. The room temperature value of $R_H$ is even smaller, farther from the ceramic data of Fig II.1 and closer to band theory (the discrepancy of 2.4 then lies within the expected accuracy of the available solution of the Boltzmann equation[29]). However, band theory cannot easily account for the T-dependence, and the phenomenon is seen even more strikingly in $YBa_2Cu_3O_7$.

Behavior somewhat analogous to that seen in Sr doped $La_2CuO_4$ is found in $YBa_2Cu_3O_{7-x}$ as x is varied (Fig.II.3). The oxygen stoichiometry corresponding formally to pure divalent Cu is 6.5 (x = 0.5). This is exactly the region below which semiconducting behavior sets in and where we see in the polycrystalline sample a Hall constant $R_H$ with a rapid variation in nominal hole concentration[30] $1/R_H$. For x < 0.5, there is a long plateau region of $R_H$ which

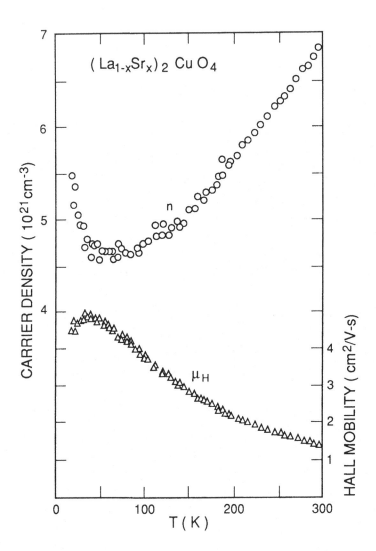

Fig. II.2   Carrier density $(1/R_H e)$ and Hall mobility for a single
crystal thin film B $\parallel$c (Ref. 3).

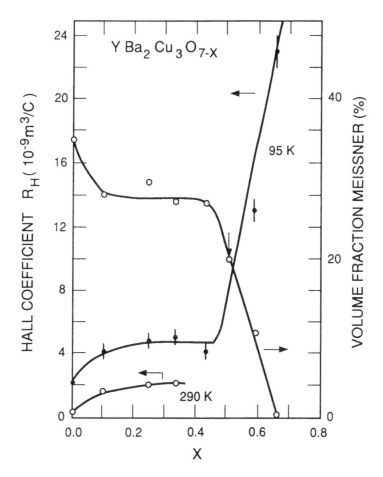

Fig. II.3    $R_H$ versus x for $YBa_2Cu_3O_7$ from Wang *et al.* (Ref. 30).
The superconducting volume fraction is also shown.

corresponds to 0.24 holes/planar Cu, close to the value found for the
Sr doped $La_2CuO_4$ 40 K superconductor.  The plateau in $R_H$ corresponds
to the plateau in $T_c$ vs. x at 55 K.  For x = 0, $R_H$ is roughly 40%
smaller than in the plateau region, and it is here that $T_c$ is 90 K.

Thus, $T_c$ correlates nicely with $1/R_H$ or with minimum carrier
density.  The most unusual feature is a strong temperature dependence

$1/R_H \sim A + BT$. Unlike $La_{2-x}Sr_xCuO_4$ (Fig. II.2), $YBa_2Cu_3O_7$ has the coefficient A close to zero. There is no generally accepted explanation.

Single-crystal data are available from several laboratories. An example from Penney $et$ $al.$[31] is given in Fig. II.4. When $\vec{B}$ is $\|c$ (so that electron orbits are in the metallic ab plane) the sign is positive (hole-like) and $1/R_H$ goes accurately like T. When $\vec{B}$ is $\perp c$, the sign is negative, and the T-variation is weak. The signs and orders of magnitude agree with the predictions of band theory,[32] but

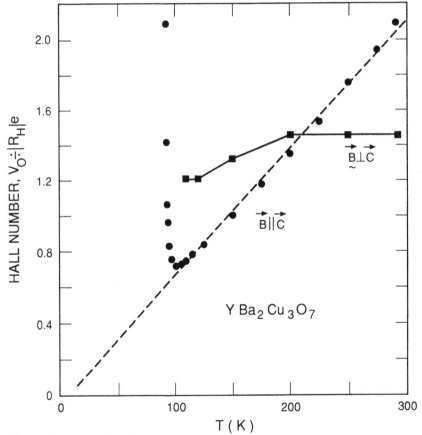

Fig. II.4   $V_o/|R_H|e$ for a single crystal of $YBa_2Cu_3O_7$ (Ref. 31). $V_o = 175$ Å$^3$ is the volume per formula unit.

band theory gives only a weak T-dependence. Forro et al.[33] report a similar experimental value of $R_H$ with $\vec{B} \perp c$ at room temperature, but find a significant T-dependence in the opposite direction($|R_H|$ decreases as T decreases). Their data with $\vec{B} \parallel c$ are smaller than those of Penney[31] but have a similar T-dependence. Stormer et al.,[5] looking at epitaxial thin films of $YBa_2Cu_3O_7$ grown on $SrTiO_3$ (100), found both the magnitude and T-dependence of $R_H$ to vary significantly with sample. Of six samples, only one (which had the highest critical current density) showed $1/R_H \sim BT$, three of the samples had somewhat similar T-dependence, and two had $R_H$ smaller and weakly T-dependent.

The Hall constant has also been reported for a single crystal sample of the Bi-based cuprates,[34] and again hole-type conduction is found in the plane. In the case of the lower $T_c$ $BaBiO_3$-based superconductors, measurements on $BaPb_{0.7}Bi_{0.3}O_3$[35] show a strong temperature variation of $R_H$ which has been interpreted as arising from the sum of activated- and temperature-independent terms, the former depending on the concentration of O-vacancies in the films. In this way it has been argued that most of the temperature dependence of the electrical resistivity arises from the temperature dependence of the carrier concentration in this material.

## III. Thermopower

Relatively little has been reported concerning the thermopower of the high $T_c$ oxides. This is presumably due in part to the fact that a simple interpretation of the data is generally difficult, the thermopower involving as it does the derivative of the density of states at the Fermi level. The measurements have the advantage that they are fairly simple and make only modest demands on sample geometry. One also determines the sign of the dominant carriers in the material.

Most of the data we are familar with are on $YBa_2Cu_3O_7$. Single crystal data of Wang and Ong[36] find that the thermopower is positive both in and out of plane. The anisotropy is considerably less than reported for the resistivity, and the temperature dependence is strong perpendicular to the Cu-O planes (Fig. III.1). There is a peak in the in-plane thermopower at $T_c$. The in-plane thermopower found by Howson et al.[37] is negative, but also peaks at $T_c$. A study by Yu et al.[38] agrees with this last result. These authors also looked at the magnetic field dependence above $T_c$ for $La_{1.85}Sr_{0.15}CuO_4$ at fields up to 30 T and found no effect. Their hope to freeze out the spin entropy contribution seems unrealistic in view of the large exchange coupling that two-magnon Raman[39] experiments appear to find. Kwok et al.[40] have reported the variation and temperature dependence of the thermopower of $YBa_2Cu_3O_x$ for $6 < x < 7$. The temperature dependence changes markedly in the vicinity of $x = 6.5$, but the metallic-like behavior for $6.1 < x < 6.5$ remains unexplained.

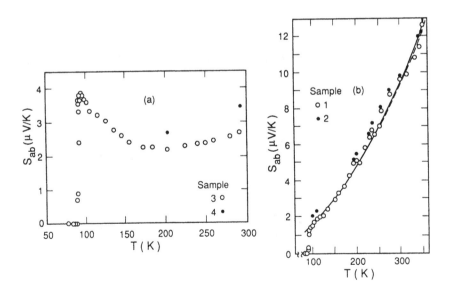

Fig. III.1  S(T) for $YBa_2Cu_3O$ from Wang and Ong (Ref. 36).

## IV.  Magnetic Properties

The nature of the underlying pairing mechanism is the primary question concerning high $T_c$ oxide superconductivity.  While there is still a slight possibility for a phonon mechanism, most theorists have looked to exciton and magnetic interactions, or their combination.  Considerable experimental attention has focussed on the striking magnetic properties of the cuprates.  There appears to be no magnetism in the $BaBiO_3$ based superconductors, and so we must conclude that either there is more than one mechanism for high $T_c$ in the oxides, or that the cuprate magnetism has nothing to do with superconductivity.  But the magnetic behavior of the cuprates is interesting in its own right, and studies of these materials are providing new arenas for the study of correlated-electron systems.

We begin with the properties of $La_2CuO_4$.  The first point to make concerns the stoichiometry of the compound.  There were early reports that large variations in La stoichiometry below 2 and oxygen stoichiometry below 4 were possible, but neutron line profile analysis now shows that La and Cu stoichiometries are within 1 atomic % of their nominal ones. However, single crystals grown from PbO fluxes were found to contain approximately 2 atomic % Pb incorporated into the crystals.[41]  Similarly, crystals grown from fluxes based on $Li_4B_2O_5$ appear to incorporate Li at similar levels.[42]  A number of the properties of $La_2CuO_4$ are now known to be quite sensitive to substitutions of this sort, and it is necessary to characterize samples whose properties are under investigation.

The oxygen stoichiometry problem is more controversial.  The usual practice has been to write the formula as $La_2CuO_{4-y}$ which carries the implicit assumption that the oxygen stoichiometry never exceeds 4.  Below we shall present arguments which suggest that the stoichiometry of oxygen is always on the high side, the lower bound of four occurring when Cu has a formal valence of +2 in the compound.  So within the error of neutron determination of the La and Cu stoichiometries, we will argue that the formula is more appropriately written $La_2CuO_{4+y}$.  Firstly, high pressure oxygen loading experiments[43] have shown that the material can be prepared with y = 0.13 in a slightly different structure. Furthermore, it is known from magnetic studies of the related compound $La_2NiO_4$ that the 3-dimensional magnetic ordering temperature is rapidly suppressed[44] when the oxygen stoichiometry deviates from 4, presumably due to the frustrated character of the magnetic ordering and the attendant defect sensitivity. In $La_2CuO_4$, inert gas or vacuum annealing is found[45] to raise the Neel temperature $T_N$ substantially, again consistent with the view that the procedure is approaching oxygen stoichiometry 4.0 from above. And we have the result that the Hall constant is positive for nominal $La_2CuO_4$, consistent with a small excess of oxygen above 4.0. Where this oxygen sits in the lattice is not yet known, and it is interesting that the other rare earth cuprates $R_2CuO_4$,[46] all of which form in a related but different structure, are n-type and cannot be made metallic via chemical substitutions.

Single crystals of $La_2CuO_4$ can be grown from excess CuO and these appear to be within 1 atomic % of having full La and Cu occupancy. The magnetic susceptibility of these crystals is quite anisotropic[47] (Fig. IV.1): the basal plane susceptibility is roughly half that perpendicular to it. There is a peak in the susceptibility near room temperature, and neutron diffraction[48] has shown this peak to correspond to the 3D ordering shown in Fig. IV.2. In the planes the spin arrangement k-vector is consistent both with what would be expected from the nested aspect of the band-structure-predicted Fermi surface and from a strongly coupled 2D Heisenberg antiferromagnet. As mentioned above, the tetragonal $K_2NiF_4$ structure has an intrinsically frustrated face-centered lattice of Ni's; the slight orthorhombic distortion present in the $La_2CuO_4$ structure lifts this frustration slightly, and the 3D magnetic structure chooses an interlayer arrangement with antiparallel spins on the nearest interlayer sites. The Neel temperature depends strongly on annealing protocol.[45] $T_N$ is generally observed in the range 240-280 K, but careful inert-gas annealing has raised $T_N$ to 308 K,[45] and a vacuum anneal, which decomposes part of the surface of a crystal, raises $T_N$ to 326 K.[45]

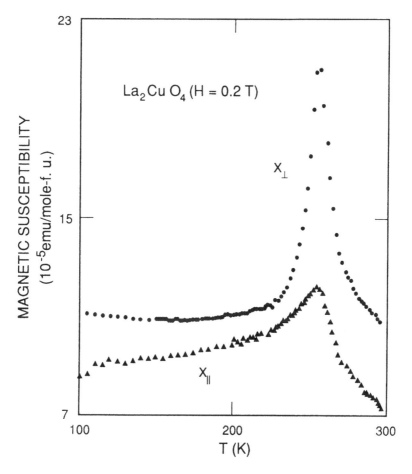

Fig. IV.1    Anisotropy of $\chi$ in $La_2CuO_4$ (Ref 47).

The chapter on neutron scattering results covers in detail the microscopic magnetic properties of $La_2CuO_4$, so our discussion will be restricted mainly to macroscopic aspects of the magnetism of the cuprates. However, we note one important result of the neutron work here. This is that there are extremely strong 2D fluctuations in the Cu-O planes, characterized by a temperature of order 1000 K.[49] These fluctuations have also been seen in two-magnon Raman scattering.[39]

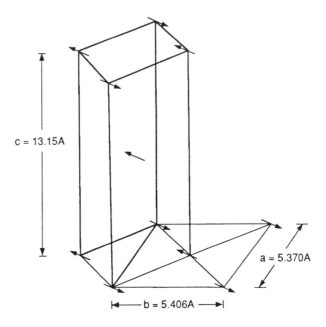

Fig. IV.2    3-D Ordering in $La_2CuO_4$ (Ref. 48)

These large interactions account for the rather flat temperature
dependence of the magnetic susceptibility, excepting the region very
near $T_N$.

Applied fields induce a weakly ferromagnetic phase below $T_N$.  This
was first detected by Cheong et al.[50] in a PbO-flux grown crystal,
which NQR[51] measurements found to be only partially ordered
magnetically, apparently from incorporated Pb in the crystal.
Subsequent neutron work[52] found that this field-induced weak
ferromagnetism arose from the preference of the Cu spins to lie in
the plane of its orthorhombically tilted (and elongated) $CuO_6$
octahedron.  This gives rise to a small net resultant moment from two
neighboring, antiferromagnetically aligned spins, but each of these
resultant moments is opposed by another anti-aligned resultant moment

in the crystal. The applied external field flips this opposed
resultant moment. We show magnetization data for this transition[53]
in Fig. IV.3. Included in the figure is the H-T phase diagram
determined by Cheong[55] which indicates that the transition field goes
discontinuously from zero to finite values at $T_N$. It has also been
found that there are large resistance effects at $T_N$, especially
perpendicular to the Cu-O planes, provided there are sufficient
carriers present (Fig. IV.4). Crystals with very low conductivity
show no conductivity anomaly at $T_N$.

The effect of Sr doping is to depress $T_N$ rapidly, and for $x > 0.02$
no long-range magnetic order is found. There has been some
suggestion that a spin-glass phase exists for $0.02 < x < 0.15$ but
this is not certain.[54] This muon experiment on Sr doped samples
indicates that the Cu moment maintains its full value in this range
of x in spite of the lack of long-range order. The superconducting
phase diagram[11] for Sr-doped $La_2CuO_4$ is shown in Fig. IV.5.
Antiferromagnetic order appears not to coexist with the
superconducting phase.

Two kinds of magnetic order have been found in the materials
related to $YBa_2Cu_3O_y$. The first is found in the materials containing
rare earth elements carrying localized f-moments.[55] These f-moments

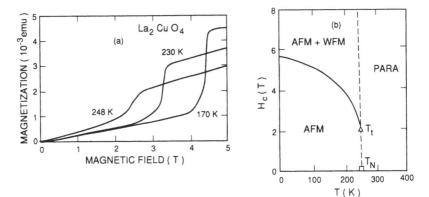

Fig. IV.3   a) M versus H and b) H-T phase diagram (Ref. 53).

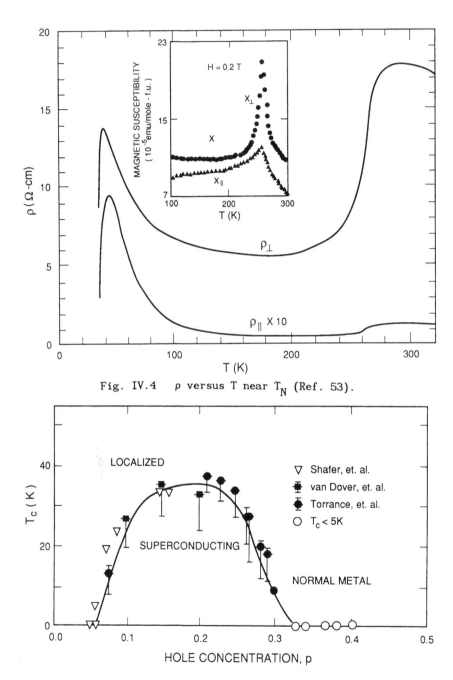

Fig. IV.4    $\rho$ versus T near $T_N$ (Ref. 53).

Fig. IV.5    Sr doped $La_2CuO_4$ phase diagram (Ref. 11).

order at low temperatures, and the ordering coexists with superconductivity. For samples of the 123-materials with y < 6.5, this ordering still occurs unchanged. There appears to be no depression of the superconducting $T_c$ by these f-moments, and the above facts show that the rare earth magnetism is essentially independent from the high $T_c$ oxide superconductivity. We give in Table IV.1 data pertinent to the rare earth ordering in the 123-materials.

Table IV.I  Néel Temperatures of $RBa_2Cu_3O_{7-x}$

| | $T_N$(K) | Ref. |
|---|---|---|
| Pr | < 0.5 | 56 |
| Nd | < 0.5 | 56 |
| Sm | 0.60 | 56 |
| Eu | 0 | |
| Gd | 2.1 | 55 |
| Tb | * | |
| Dy | 0.95 | 57 |
| Ho | 0.17 | 57 |
| Er | 0.59 | 57 |
| Tm | * | |
| Yb | * | |

*not reported

The second kind of magnetic ordering observed[58] is ordering of Cu moments in the semiconducting 123-materials with y < 6.5. This magnetic ordering does not, therefore, coexist with the superconductivity. To date, the Cu-moment ordering has been studied in Y, Nd and Er 123-materials. All these compounds show $T_N$ near 500 K for y = 6.0. $T_N$ decreases smoothly to 0 K as y → 6.5. Only the Cu's in the planes carry a moment at high T, and these are aligned antiferromagnetically along the direction perpendicular to the planes. It has also been observed in some samples that a second magnetic phase grows in and replaces the above mentioned phase at low temperature. For a Nd sample with y = 6.1, this second phase started growing in below 90 K.[59] The strange feature of this transition is that all the Cu's now carry moment, a chemically important fact indicating that, at least in these O-deficient materials, both chain and plane Cu-sites have similar oxidation states. This low temperature phase has the spins lying parallel to the Cu-O planes. It remains to be determined what conditions lead to this second ordering and how general it is.

The question of interest of course is whether or not some of these magnetic interactions survive in the materials which become superconducting. There are at present only tentative results on this question. A neutron measurement on Sr-doped $La_2CuO_4$[60] indicates that the magnetic interactions are there in force in the superconducting material, but some questions remain to be cleared up concerning the sample on which the measurements were made, principally concerning sample homogeneity. The two-magnon mode has been reported in Y 123-material, broadened but at the same energy.[61] The question here is whether or not the surface layer measured is truly indicative of the bulk. Photoemission experiments,[25] have found that oxygen is immediately lost from freshly cleaved surfaces that are not maintained near He temperature.

## V. Elastic Properties and Ultrasound

Elastic properties of the high $T_c$ superconductors and related compounds have been measured on superconducting and non-superconducting ceramics at high temperatures (above 4 K), on superconducting ceramics at very low temperatures (1 K and below), on large (2.5 mm) and superconducting single crystals for which the Meissner effect has not been measured, and on non-superconducting single crystals. In the normal state, two points immediately attract attention to the importance of the elastic constants. They are the intrinsic anisotropy of the high-temperature superconductors, and the possibility of using elastic-constant data in the high temperature, phonon-dominated regime to sort out the specific heat. Unfortunately, the most difficult measurements to interpret are those performed at high temperatures on ceramics, where no anisotropy information is available. The root of the difficulty may be inferred from the work of Blendell *et al.*[62] where ultrasound determinations of the elastic constants of sintered $YBa_2Cu_3O_y$ superconductors of various densities were corrected for porosity using the expressions of Ledbetter and Datta.[63]  Their uncorrected results are shown in Table V.I, where B is the bulk modulus, E is Young's modulus, $\mu$ is the shear modulus and $\sigma$ is Poisson's ratio.

Table V.I

| Porosity | $v_\ell$(km/s) | $v_s$(km/s) | $\sigma$ | E(Gpa) | B(Gpa) | $\mu$(Gpa) | $\rho$(gm/cc) |
|---|---|---|---|---|---|---|---|
| 12.9% | 4.87 | 2.76 | .264 | 107 | 75.5 | 42.4 | 5.56 |
| 32.6% | 3.63 | 2.06 | .263 | 46.1 | 32.4 | 18.3 | 4.30 |

As shown, they find that for their sample, with a porosity of 12.9% (theoretical density from x-rays is 6.383 gm/cc) the longitudinal sound velocity $v_\ell$ was 4.7 km/s and the transverse wave velocity $v_s$ was 2.76 km/s. For their lower-density sample, having 32.6% porosity, $v_\ell$ = 3.63 km/s and $v_s$ = 2.06 km/s. Upon correcting for porosity, the higher-density sample extrapolated to $v_\ell$ = 5.22km/s and $v_s$ = 2.92km/s, while the lower density sample extrapolated to $v_\ell$ = 4.32km/s and $v_s$ = 2.35 km/s. Furthermore, the predicted value (corrected for porosity in the denser sample) of the bulk modulus was $1.39 \times 10^{12}$ dynes/cm$^2$, an extremely low value compared to values for a single superconducting[64] crystal of $YBa_2Cu_3O_y$ measured in the b direction (long axis of the orthorhomb) of about $1.5 \times 10^{12}$ dynes/cm$^2$. Thus one must conclude that intrinsic structural differences exist between samples of differing densities. This is not surprising, as at all the frequencies for which such quantities have been measured, the wavelength of sound is much larger than the grain size. One probes,

therefore, the properties of the bonds between grains as well as the
intrinsic properties.  These structural differences, associated
perhaps with sample variations, heat treatment, and oxygen content,
can cloud any conclusions based on ceramic measurements. This is
illustrated further in $YBa_2Cu_3O_y$ superconductors by the observations
of Kim et al.,[65] who find that $v_\ell$ varies from 4.7 km/s for
y = 6.2, $\rho$ = 6.12 gm/cc to 3.95 km/s, $\rho$ = 6.37 gm/cc for y = 6.8,
opposite in density dependence to the results of Blendell, but with
no obvious fundamental processing differences except for oxygen
anneal.  More confusion exists in the interpretation of the
temperature dependence of $v_\ell$ for the purpose of understanding the
superconductivity.  It increases on cooling, as measured by Ewert
et al.[67] in $YBa_2Cu_3O_y$ ceramic, shown in Fig. V.1 but decreases on

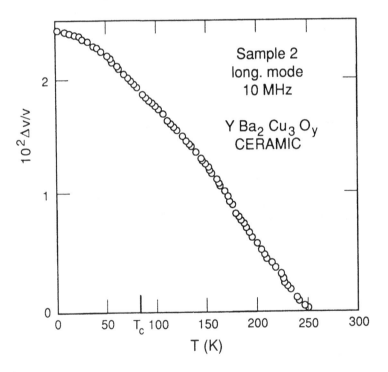

Fig. V.1    $v_\ell$ for $YBa_2Cu_3O_y$ ceramic from S. Ewert et al. showing a
normal increase in $v_\ell$ cooling below 300 K (Ref. 66).

cooling in ceramic $La_{1.85}Ba_{0.15}CuO_4$, as shown in Fig. V.2.[67]
Nevertheless, one may estimate the phonon specific heat and the Debye
temperature from the Debye-average sound velocity which is

$$v_D = \left[ \frac{1}{3} \left[ \frac{1}{v_\ell^3} + \frac{2}{v_s^3} \right] \right]^{-1/3} \tag{1}$$

We find $v_D = 3.25$ km/s if we use the corrected data of Blendell
et al. on their higher-density $YBa_2Cu_3O_y$ sample. The Debye
temperature is then a little over 280 K if we use the theoretical
density. This is to be compared with a Debye temperature of 392 K[69]
determined via specific heat for a superconducting single crystal
with a $T_c$ of 89 K. The single-crystal elastic data are consistent
with specific heat data obtained on ceramics,[69] where values nearer
to 360 K are observed. Thus the discrepancy, though suggestive of
porosity and ceramic structure effects, is certainly real because it

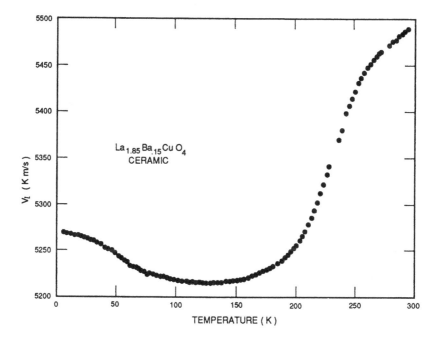

Fig. V.2  $v_\ell$ for a single-crystal $La_{1.85}Ba_{0.15}CuO_4$ ceramic showing
the decrease in sound velocity on cooling below 300 K.
From K. Fossheim et al. (Ref. 67).

is hard to imagine how this compound could have a $v_\ell$ of 7.3 km/s and a $v_s$ of 4.1 km/s, the values predicted by the specific heat data if we use $\sigma = 0.26$. A highly anisotropic structure could, however, limit the specific heat contribution from the acoustic-branch phonons.

The superconducting phase transition generates a discontinuity in compressional elastic constants which can be obtained from a thermo-dynamical treatment. Observation of such effects are relevant to the interpretation of normal-state properties because the discontinuity, $\Delta v_\ell$ in the bulk sound velocity of an isotropic superconductor, given by

$$\Delta v_\ell = -\frac{v_\ell \Delta C}{2T_c} \left[\frac{dT_c}{dP}\right]^2 \tag{2}$$

where $\Delta C$ is the discontinuity in specific heat at $T_c$ and $v_\ell$ is the bulk sound speed, is a measure of the required resolution for ultrasound measurements to shed light on the free energy changes associated with superconductivity. Saint-Paul et al.[64] find no anomaly at $T_c$ in their superconducting single crystal (nor could they measure the anisotropies in the sound velocities) within the resolution of their measurement, which is insufficient to resolve the expected drop (Fig. V.3). Neither do they find any change in slope of the sound velocity at $T_c$ in contrast to observations on ceramics by, for example Bishop et al.[70] shown in Fig. V.4. Thus, even though Saint-Paul's single crystal was probably too large to be properly oxygenated, it is unlikely that hysteretic effects or anomalous changes in $v_\ell$ or its slope are important to superconductivity. They are most likely manifestations of the ceramic nature of the samples. Even in such high-quality glasses as fused silica, the sound velocity decreases on cooling from 300 K and the ultrasonic attenuation increases[71] to peak at 40 K. In the high $T_c$ superconductors, attempts have been made to measure the discontinuity in value or in slope of the shear elastic constants expected at $T_c$. Again, the properties of the ceramics intrude. The difficulty with shear in ceramics is well discussed by Pippard.[72] Based on his arguments and

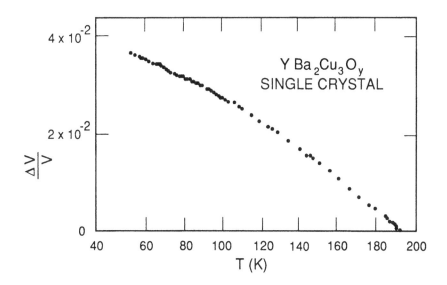

Fig. V.3    $v_\ell$ for a single-crystal YBa$_2$Cu$_3$O$_y$ with a T$_c$ of about 90 K
but with no determination of the Meissner fraction.  Note
the smooth, monotonic increase in sound speed on cooling.
The measurement was of insufficient precision to detect
discontinuities at T$_c$.    From M. Saint Paul et al. (Ref.
64).

reasonable estimates[73] for the pressure dependence of the thermodynamic critical field, we find that the expected discontinuities are
of the order of 100 ppm for the value of the bulk modulus, and,
absent the unlikely possibility of spontaneous shear occurring at T$_c$,
0 for the shear modulus; changes in slope are also small.  Thus, the
very large effects in the slope of the shear modulus such as those
observed by Ledbetter[74] or Bhattacharya[75] would overwhelm the small
thermodynamically-induced break in slope at T$_c$ at a temperature only
1 K away from T$_c$.  The implications are obvious.  It is interesting
to note that only the high-resolution data of Brown et al.[73] on the
shift in bulk modulus at T$_c$ in GdBa$_2$Cu$_3$O$_y$ (Fig. V.5) provide
thermodynamic agreement, required by Eq. (2), with the specific

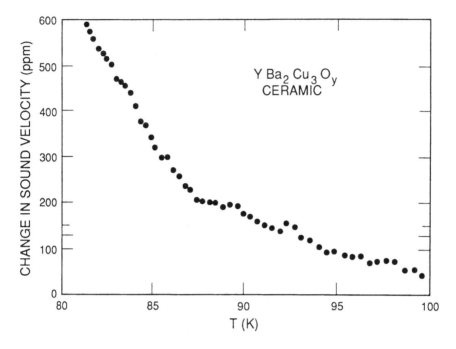

Fig. V.4    $v_\ell$ for a ceramic superconducting sample of $YBa_2Cu_3O_y$ showing a break in slope at $T_c$ which is not seen in the ceramic.  From Bishop et al. (Ref. 70).


heat data of M. E. Reeves et al.[76] and the pressure dependence of $T_c$ measured by Borges et al.[77]  Brown, however, had to cycle his pressed pellet up and down in temperature for many days before the small "earthquakes" finally subsided to the point where a 70 ppm discontinuity was unmasked.  This low-level relaxation of ceramic pellets would certainly bring into question the extremely noisy x-ray data of Horn et al.[78] where a discontinuity in the difference between the two in-plane lattice constants at $T_c$ was claimed.  On the basis of the same thermodynamic arguments which obtain discontinuities in specific heat, thermal expansion, and elastic constants at $T_c$, one finds that for the isotropic case, no discontinuity in volume is expected.  Based on similar reasoning, but including the fully

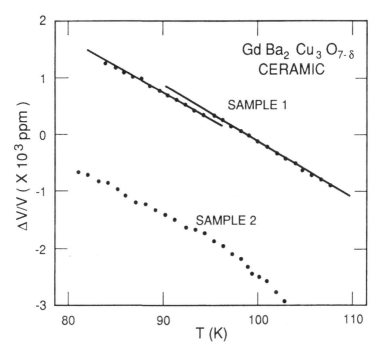

Fig. V.5    $v_\ell$ for a ceramic superconducting sample of $GdBa_2Cu_3O_y$.
Shown are the first cool-down for sample II and data for
sample I after many thermal cycles, which served to
reduce random discontinuous jumps below the level of the
expected discontinuity (shown) at $T_c$.    From S. E. Brown
*et al.* (Ref. 73).

anisotropic stress tensor, it is only by a coincidental effect that
the difference between two lattice constants could be discontinuous
for an orthorhombic solid.[72]

    The pressure dependence of elastic constants has also been studied
by using x-rays. In $La_{1.7}Ba_{0.3}CuO_{4-y}$, Moret[79] finds an anomaly in the
a-axis (short axis) compressibility:    the a-axis does not change
up to 7 kbar, while the c-axis compresses uniformly.    Above 7 kbar,
they find $da/adp = 2 \times 10^{-4}/kbar$ and $dc/cdp = 1.7 \times 10^{-4}/kbar$.
A similar anomaly is observed by Shelton[80] in $La_{1.85}Ca_{0.15}CuO_4$.

However, neither Terada nor Takahashi[81] see such effects in either $La_{1.8}Sr_{0.2}CuO_{4-y}$ or $La_{1.4}Sr_{0.6}CuO_{4-y}$ where compressibilities of both lattice constants in the former are $2.5 \pm 0.4 \times 10^{4}$/kbar and $da/adp = 1.9 \times 10^{4}$/kbar and $dc/cdp = 2.24 \times 10^{4}$/kbar for the latter. Thus the basal plane is perhaps stiffer than the direction normal to the planes, but perhaps not by as much as the error bars on the anisotropy which are about 20%.

At very low temperatures, one might expect many of the anomalous results arising from the problems with ceramics to disappear because the thermal expansion coefficients, related to the Debye temperatures of hundreds of degrees,[68] freeze out. In ultrasound measurements on $YBa_2Cu_3O_y$ from 10 K to 5 mK, Golding et al.[82] postulate a sufficiently high density of tunneling states (on "two-level systems") to fully account for the linear term observed in the specific heat[76] with a quasi-particle density as low as it would be in a fully gapped superconductor. Golding attributes the tunneling states to mobile oxygen defects.

## VI.   Thermal Conductivity

Although the thermal conductivity is a useful tool in studying superconductors, the data available for the high $T_c$ superconductors above $T_c$ is sparse. Because all the measurements we know of above 10 K have been performed on ceramics, the usual precautions must be taken in drawing conclusions. The data of Uher et al.[83] (Fig. VI.1) on $YBa_2Cu_3O_y$ ($T_c = 92$ K) is typical. Above $T_c$, the thermal conductivity increases from about 3.5W/mK at $T_c$ to about 4W/mK at 300 K. From these values and their resistivity data on the same specimen, they estimate that at least 90% of the thermal conductivity comes from phonons. This large phonon background effectively masks any features in the electronic thermal conductivity. Below $T_c$ Uher and others[84] see an increase in thermal conductivity. This is opposite the effect seen in classical superconductors, in which electronic part dominates and freezes out exponentially as the gap

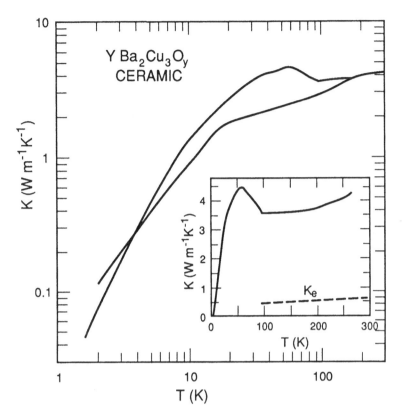

Fig. VI.1 The thermal conductivity of YBa$_2$Cu$_3$O$_y$ ceramic showing the increase below T$_c$. From C. Uher *et al*. (Ref. 83).

develops. For the high T$_c$ materials, it appears that the decrease in electron-phonon scattering below T$_c$ is more important than the freezing out of electrons. Bayot[84] concludes that this is evidence for strong electron-phonon coupling, but Uher warns that both strong and weak-coupling classical superconductors with low electronic thermal conductivity in the normal state also display this behavior.[85] [86] At very low temperatures (0.1-7 K), the thermal conductivity remains small and is similar to that observed in other sintered ceramics.[87] These data[87] overlap those of Bayot[84] and, within a factor of two, agree with them.

## VII.  Electrodynamics

Measurements of the electrical conductivity of the high $T_c$ materials in the normal state have been made at infrared frequencies and at dc.  Intermediate frequency measurements of the complex conductivity are not possible because the $\rho\epsilon$ (where $\rho$ is the resistivity and $\epsilon$ is the dielectric constant) time constant is too small in a metal.  However two sets of measurements in the microwave regime (1 GHz to 100 GHz) have been made on single crystals of $Eu_2CuO_4$ and $La_2CuO_4$.[88][89]  These materials are semiconductors with a 1.7 eV gap.  In both materials at frequencies near 10 GHz a large resonant response is observed.  Such a mode indicates collective transport, in this case with a large effective mass of many 10's of electron masses, which is the result of a highly correlated ground state. The spectral weight of the mode in the Eu compound, shown in Fig. VII.1, is activated (with an activation energy of 630 K) and is strong enough to produce a dc dielectric constant of about 1000 at 300 K.  The material had a Hall carrier density which is activated[90] (activation energy of 700 K) in agreement with the activation energy of the spectral weight.  Such an activated spectral weight is unlike that of a charge density wave (which is IR active at temperatures below $T_c$, much like the behavior of a superconductor). The $La_2CuO_4$ displays a similar mode, but with a more complicated temperature dependence:  the static dielectric constant remains high (about 70) when extrapolated to zero temperature.  This value of dielectric constant is typical of perovskites, but the collective mode is not. This mode, because of its close coupling to the dc carrier concentration, indicates that some very complicated phenomena may accompany current transport in the semiconducting copper oxides.

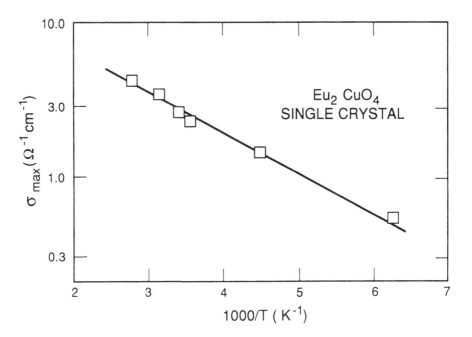

Fig. VII.1  The peak conductivity at 12 GHz of the microwave
            collective mode versus temperature for a single crystal
            of $Eu_2CuO_4$ showing the freezing out of the mode on
            cooling, from Ref. 88.

## References

1. H. E. Fischer, A. K. Watson, and D. G. Cahill, preprint.

2. S.-W. Cheong, Z. Fisk, R. S. Kwok, *et al.*, Phys. Rev. B$\underline{37}$, 5916 (1988).

3. M. Suzuki and T. Murakami, Japan. J. Appl. Phys. $\underline{26}$, L524 (1987).

4. T. Penney, S. von Molnar, D. Kaiser, *et al.*, Phys. Rev. B$\underline{38}$, 2918 (1988).

5. H. L. Stormer, A. F. J. Levi, K. W. Baldwin, *et al.*, Phys. Rev. B$\underline{38}$, 2472 (1988).

6. S. Martin, A. T. Fiory, R. M. Fleming, *et al.*, Phys. Rev. Lett. $\underline{60}$, 2194 (1988).

7. M. Gurvitch and A. T. Fiory, Phys. Rev. Lett. $\underline{59}$, 1337 (1987); also in Novel Superconductivity, ed. by S. A. Wolf and V. Z. Kresin (Plenum Press, New York, 1987) p. 663; M. Gurvitch, A. T. Fiory, L. S. Schneemeyer, *et al.*, Physica C$\underline{153-155}$, 1369 (1988).

8. N. V. Volkenshteyn *et al.*, Fiz. Met. Metalloved $\underline{45}$, 1187 (1978).

9. P. B. Allen, W. E. Pickett, and H. Krakauer, Phys. Rev. B$\underline{37}$, 7482 (1988).

10. G. Xiao, M. Z. Cieplak, D. Musser, *et al.*, preprint.

11. J. B. Torrance, Y. Tokura, A. Nazzal, *et al.*, Phys. Rev. Lett. $\underline{60}$, 542 (1988).

12. S-W. Cheong, J. D. Thompson, and Z. Fisk, unpublished data.

13. P. W. Anderson and Z. Zou, Phys. Rev. Lett. $\underline{60}$, 132 (1988).

14. C. Kallin and A. J. Berlinsky, Phys. Rev. Lett. $\underline{60}$, 2556 (1988); P. W. Anderson and Z. Zou, Phys. Rev. Lett. $\underline{60}$, 2557 (1988).

15. G. V. Chandrashekhar, J. Moyo, and J. M. Honig, J. Solid State Chem. $\underline{2}$, 528 (1970).

16. D. C. Johnston, J. Low Temp. Phys. $\underline{25}$, 145 (1976).

17. S. J. Hagen, T. W. Jing, Z. Z. Wang, *et al.*, Phys. Rev. B$\underline{37}$, 7928 (1988).

18.    Y. Iye, T. Tamegai, T. Sakakibara, *et al.*, Physica C153-155, 26 (1988).

19.    Yu. A. Ossipyan, V. B. Timofeev, and I. T. Schegolev, Physica C 153-155, 1133 (1988).

20.    W. J. Wattamaniuk, J. P. Tidman, and R. F. Frindt, Phys. Rev. Lett. 35, 62 (1975).

21.    Z. Fisk and G. W. Webb, Phys. Rev. Lett. 36, 1084 (1976).

22.    P. B. Allen, in Superconductivity in d- and f-Band Metals, edited by H. Suhl and M. B. Maple (Academic, New York, 1980) p. 215.

23.    G. Xiao, M. Z. Ciepiak, D. Messer, *et al.*, Nature 332, 238 (1988).

24.    P. B. Allen, T. P. Beaulac, F. S. Khan, *et al.*, Phys. Rev. B34, 4331 (1986).

25.    R. S. List, A. J. Arko, Z. Fisk *et al.*, Phys. Rev. B (to be published).

26.    M. W. Shafer, T. Penney and B. L. Olson, Phys. Rev. B 36, 4047 (1987).

27.    N. P. Ong, Z. Z. Wang, J. Clayhold, *et al.*, Phys. Rev. B 35, 8807 (1987)..

28.    J. B. Torrance, Y. Tokura, A. I. Nazzal, *et al.*, Phys. Rev. Lett. 61, 1127 (1988).

29.    P. B. Allen, W. E. Pickett, and H. Krakauer, Phys. Rev. B 36, 3926 (1987).

30.    Z. Z. Wang, J. Clayhold, N. P. Ong *et al.*, Phys. Rev. B 36, 7222 (1987).

31.    T. Penney, S. von Molnar, D. Kaiser, *et al.*, Phys. Rev. B 38, 2918 (1988).

32.    P. B. Allen, W. E. Pickett, and H. Krakauer, Phys. Rev. B 37, 7482 (1988).

33.    L. Forro, M. Raki, C. Ayache, *et al.*, Physica C 153-155, 1357 (1988).

34.    K. Kitazawa, H. Takagi, K. Kishio *et al.*, Physica C 153-155, 9 (1988).

35. S. Uchida, K. Kitazawa, S. Tanaka, Phase Transitions $\underline{8}$, 95 (1987).

36. Z. Z. Wang and N. P. Ong, preprint.

37. M. A. Howson, M. B. Salamon, T. A. Freidman *et al.*, prepint.

38. R. C. Yu, M. J. Naughton, X. Yan *et al.*, Phys. Rev. B $\underline{37}$, 7963 (1988).

39. K. B. Lyons, P. A. Fleury, J. P. Remeika *et al.*, Phys. Rev. B $\underline{37}$, 2353 (1988).

40. R. S. Kwok, S-W. Cheong, J. D. Thompson *et al.*, Physica C $\underline{152}$, 240 (1988).

41. M. Marezio, private communication.

42. R. J. Birgeneau, C. Y. Chen, D. R. Gabbe *et al.*, Phys. Rev. Lett. $\underline{59}$, 1329 (1987).

43. J. E. Schirber, B. Morosin, R. M. Merrill *et al.*, Physica C $\underline{152}$, 121 (1988).

44. G. Aeppli and D. J. Buttrey, Phys. Rev. Lett. $\underline{61}$, 203 (1988).

45. S-W. Cheong and J. D. Thompson, private communication.

46. S-W. Cheong, J. D. Thompson, and Z. Fisk, unpublished..

47. J. D. Thompson, S-W. Cheong, and Z. Fisk, unpublished.

48. R. J. Birgeneau, M. A. Kastner, A. Aharony *et al.*, Physica C $\underline{153-155}$, 515 (1988).

49. G. Shirane, Y. Endoh, R. J. Birgeneau *et al.*, Phys. Rev. Lett. $\underline{59}$, 1613 (1987).

50. S-W. Cheong, Z. Fisk, J. O. Willis, *et al.*, Solid State Commun. $\underline{65}$, 111 (1988).

51. M. Takigawa and P. C. Hammel, unpublished.

52. M. A. Kastner, R. J. Birgeneau, T. R. Thurston *et al.*, preprint.

53. S-W. Cheong, J. D. Thompson, and Z. Fisk, preprint.

54. D. R. Harshman, G. Aeppli, G. P. Espinosa *et al.*, Phys. Rev. B $\underline{38}$, 852 (1988).

55. See, for example, J. O. Willis, Z. Fisk, J. D. Thompson *et al.*, J. Magn. Magn. Mat. $\underline{67}$, L139 (1987).

56. A. P. Ramirez, L. F. Schneemeyer and J. V. Waszczak, Phys. Rev. B $\underline{36}$, 7145 (1987).

57.     B. D. Dunlap, M. Slaski, D. G. Hinks, *et al.* J. Magn. Magn.
        Mat. 68, L139 (1987).

58.     See, for example, J. M. Tranquada, D. E. Cox, W. Kunnmann *et*
        *al.*, Phys. Rev. Lett. 60, 156 (1988).

59.     J. W. Lynn, W-H. Li, H. A. Mook *et al.*, Phys. Rev. Lett. 60,
        278 (1988).

60.     R. J. Birgeneau, D. R. Gabbe, H. P. Jenssen, preprint.

61.     K. B. Lyons, P. A. Fleury, L. F. Schneemeyer *et al.*, Phys.
        Rev. Lett. 60, 732 (1988).

62.     J. E. Blendell, C. K. Chiang, D. C. Cranmer et al., NBS
        Gaithersburg, MD, 1987, preprint.

63.     H. M. Ledbetter, and S. K.Datta, J. Acoust. Soc. Am. 79, 239
        (1986).

64.     M. Saint-Paul, J. L. Tholence, P. Monceau, Solid State Commun.
        66, 641 (1988).

65.     T. J. Kim, B. Luthi, M. Schwartz et al., ICCF6, Frankfurt
        1988, preprint.

66.     S. Ewert, S. Guo, P. Lemmons, Solid State Commun. 64, 1153
        (1987).

67.     K. Fossheim, T. Laegried, E. Sandvold, Solid State Commun.
        (1987), to be published.

68.     S. von Molnar, A. Torressen, D. Kaiser, *et al.*, Phys. Rev. B
        37, 3762 (1988).

69.     A. P. Ramirez, P. J. Cava, G. P. Espinoza *et al.*, Proc. Mat.
        Res. Soc., Boston, MA, November 1987.

70.     D. J. Bishop, A. P. Ramirez, P. L. Gammel et al., Phys. Rev. B
        35, 8788 (1987).

71.     H. J. McSkimin, J. Appl. Phys. 24, 988 (1953).

72.     A. B. Pippard, Philos. Mag. 46, 1115 (1955).

73.     S. E. Brown, A. Migliori, Z. Fisk, Solid State Commun. 65, 483
        (1988).

74.     H. M. Ledbetter, M. W. Austin, S. A. Kim, *et al.*, J. Mater.
        Res. 2, 790 (1987).

75.     S. Bhattacharya, M. J. Higgins, D. C. Johnston, *et al.*,
        Phys. Rev. B 37, 5901 (1988).

76.  M. E. Reeves, D. Citrin, B. G. Pazol, *et al.*, Phys. Rev. B 35, 7207 (1987).

77.  H. A. Borges, R. A. Kwok, J. D. Thompson, *et al.*, Phys. Rev. B 36, 2404 (1987).

78.  P. M. Horn, D. T. Keane, G. A. Held, *et al.*, Phys. Rev. Lett. 59, 2772 (1987).

79.  R. Moret A. I. Goldman and A. Moodenbaugh, Phys. Rev. B 37, 7867 (1988).

80.  R. N. Shelton, et al., in <u>Proc. of the Special Symposium on High T$_c$ Superconductors, MRS Spring Meeting, Anaheim, California, April 1987</u>, unpublished, extended abstracts. p. 49.

81.  H. Takahashi, C. Murayama, S. Yomo et al., Jpn. J. Appl. Phys. 26, L510 (1987); N. Terada, H. Ihara, M. Hinabayashi *et al.*, Jpn. J. Appl. Phys. 26 L504 (1987).

82.  B. Golding, N. O. Birge, W. H. Haemmerle, *et al.*, Phys. Rev. B 36, 5606 (1987).

83.  C. Uher and A. B. Kaiser, Phys. Rev. B 36, 5680 (1987).

84.  V. Bayot, F. DeLannay, C. Dewitte, *et al.*, Solid State Commun. 63, 983 (1987).

85.  R. Hasegawa and L. E. Tanner, Phys. Rev. B 16, 3925 (1977).

86.  M. C. Karamargin et al., Phys. Rev. B 6, 3624 (1972).

87.  J. J. Freeman, T. A. Freidmann, D. M. Ginsberg *et al.*, University of Illinois preprint (1987)

88.  D. W. Reagor, A. Migliori, Z. Fisk, *et al.*, Phys. Rev. B 38, 5106 (1988).

89.  D. W. Reagor, A. Migliori, G. Gruner, *et al.*, to be published Phys. Rev. B.

90.  S-W. Cheong, J. D. Thompson, Z. Fisk, unpublished.

# 6

# RARE EARTH AND OTHER SUBSTITUTIONS IN HIGH TEMPERATURE OXIDE SUPERCONDUCTORS

John T. Markert, Yacine Dalichaouch, and M. Brian Maple

*Department of Physics and*
*Institute for Pure and Applied Physical Sciences*
*University of California, San Diego, La Jolla, CA 92093, USA*

## I. INTRODUCTION

Since the discovery of high temperature superconductivity below [1] and above [2] liquid nitrogen temperature in layered copper-oxide compounds, an enormous body of literature has appeared on the subject; the endeavor of such research is to assess the physical properties and, hopefully, elucidate the underlying mechanisms at work in these systems. Although many early studies were hampered by problems with the quality and characterization of these complex materials, subsequent work on pure and well-characterized specimens has permitted the emergence of a fairly consistent and comprehensive description of these remarkable compounds. In the present report, we review a plenitude of investigations into magnetic, thermodynamic, and transport properties, with an emphasis on the information provided by substitutional studies. Because of the vast quantity of literature in the field of high temperature super-conductivity, some degree of incompleteness, due to selective sampling, cannot herein be avoided; we humbly petition the reader's understanding wherever any inevitable oversights have occurred.

This review is organized according to the major system under study, in historical order; subcategories which emphasize various properties or subsystems have been established. We begin in Sec. II with a description of the so-called "40 K" or "2-1-4" superconductors, first discussing in some detail the non-isovalent partial substitutions for lanthanum in $La_2CuO_4$ which induce superconductivity; then, a brief recounting of the effects of isovalent substitution for lanthanum and copper substitution is given. The bulk of this report appears in Sec. III, where we discuss the "90 K" or "1-2-3" rare-earth substituted superconductors of the form $RBa_2Cu_3O_{7-\delta}$. Section III-A discusses critical temperatures, followed by a summary of upper critical fields and critical current densities in Sec. III-B and a potpourri of other superconducting properties in Sec. III-C. The rare-earth R substitutions are the focus of Secs. III-D and III-E; the former recounts crystalline electric field effects, while the latter reviews the antiferromagnetic ordering of $R^{3+}$ ions. The somewhat unusual case of R = La is briefly related in Sec. III-F, and the surprising non-superconducting substitution R = Pr is the subject of Sec. III-G. Other isovalent substitutions for Y (and Ba) are touched upon in Sec. III-H. Section III-I presents a fairly detailed account of substitutions for copper of the form $YBa_2(Cu_{1-x}M_x)_3O_{7-\delta}$. The discussion of the 1-2-3 compounds is concluded in Sec. III-J with a review of the effects of the non-isovalent substitutions for Y and Ba in $YBa_2Cu_3O_{7-\delta}$. Attention is then directed to the more recent bismuth and thallium copper-oxide superconductors in Sec. IV, and the non-copper-containing barium-bismuth oxide superconductors in Sec. V. Some concluding remarks appear in Sec. VI.

## II. $La_{2-x}M_xCuO_{4-\delta}$ AND RELATED COMPOUNDS

The $La_{2-x}M_xCuO_{4-\delta}$ system (M = Ba, Sr, Ca, Na, . . .) probably represents the simplest form of the copper oxide superconductors, both since its $K_2NiF_4$-type structure lacks the complicating CuO chains found in the $YBa_2Cu_3O_{7-\delta}$ compounds and because it does not possess the distorted sheets which result from size mismatch in the Bi and Tl high-temperature superconductors. For the $La_{2-x}M_xCuO_{4-\delta}$ "40 K" materials, doping of the parent compound, $La_2CuO_4$, is essential to produce superconductivity; thus, a discussion of substitutions will describe a variety of superconducting properties. We will first describe in some detail the properties of $La_{2-x}M_xCuO_{4-\delta}$ as a function of the superconductivity-inducing dopant type and dopant concentration. Then other substitutions will be considered; these include additional lanthanide ion (Ln) substitutions, i.e., $La_{2-x-y}Ln_ySr_xCuO_{4-\delta}$, and substitutions for copper of the form $La_{2-x}Sr_x(Cu_{1-y}M_y)O_{4-\delta}$, where M is generally a transition metal.

Substitution of the divalent alkaline earth metals M = Ba [1,3,4], Sr [5-9], or Ca [6], or the monovalent alkali metal M = Na [10,11] induces metallic behavior in $La_{2-x}M_xCuO_{4-\delta}$ for $x \gtrsim 0.06-0.10$; at low temperatures, superconductivity accompanies the metallic behavior. With the possible exception of M = K [11,12], no other monovalent or divalent lanthanum substituent has been found to result in a single-phased material. The transition temperature differs for the various dopants, and is a function of dopant concentration and oxygen content. Early composition-dependent studies of $La_{2-x}Sr_xCuO_{4-\delta}$ indicated that there is a strong peak [7] in the fraction of the sample which is superconducting for $x \approx 0.15$; however, the dopant concentration alone does not control superconducting properties.

The random substitution of lesser-valent ions for $La^{3+}$ affects the hole concentration, usually discussed in terms of the hole concentration per copper ion, $p$. Thus, hole-doping of the semiconducting parent compound, $La_2CuO_4$, induces the metal-insulator transition. Oxygen deficiency also regulates the hole concentration; since each missing oxygen atom would have normally accomodated two electrons, one has $p =$

$x–2\delta$ (for alkaline earth substitutions) or $p = 2x–2\delta$ (for alkali metal substitutions). Trace superconductivity is observed in what is believed to be the lanthanum-deficient material $La_{2-x}CuO_{4-\delta}$ [12–16], where one would have $p = 3x–2\delta$; in this case, the lanthanum deficiency (and superconductivity) may exist only as a filamentary grain-boundary phase. Bulk superconductivity has been observed in oxygen-enriched $La_2CuO_4$, prepared by special plasma-oxidation techniques [17,18]; it may be that the excess oxygen occupies lanthanum vacancies, in which case one would have $p = 5x$ in $La_{2-x}O_xCuO_4$. Thus it is not surprising that bulk superconductivity is obtained for much lower "dopant" concentrations in the oxygen-enriched material (~1% excess oxygen [17]), since the effect on the hole concentration can be much greater than with the metallic substitutions.

For each type of dopant M in $La_{2-x}M_xCuO_{4-\delta}$, a maximum value of the superconducting transition temperature $T_c$ is obtained as a function of dopant concentration; this value of $T_c$ generally persists for a wide range of dopant concentrations, as shown in Fig. 1 for M = Na. These values are typically $T_c \approx 39$ K, 32 K, 24 K, and 18 K for M = Sr, Ba, Ca, and Na, respectively. These values of $T_c$ are plotted in Fig. 2 as a function of the ionic radius of the substituted ion; the nine-fold

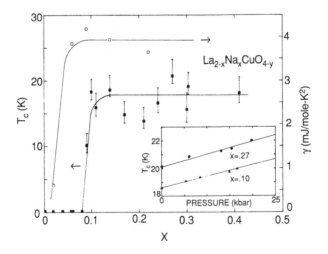

**Fig. 1.** Superconducting transition temperature $T_c$ (solid squares; left scale) and linear specific heat coefficient $\gamma$ (open squares; right scale) vs Na concentration x in $La_{2-x}Na_xCuO_{4-\delta}$. The bars represent the 10%–90% transition widths. Inset: Superconducting transition temperature $T_c$ as a function of applied pressure for two different concentrations x. After Ref. 10.

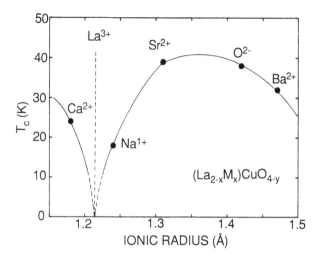

**Fig. 2.** Superconducting transition temperature $T_c$ vs dopant ionic radius for various dopant ions M in $La_{2-x}M_xCuO_{4-\delta}$. The transition temperatures shown typically occur over a range of concentrations; the ionic radii are for the nine-fold coordination appropriate to the $K_2NiF_4$-type structure. The dashed vertical line indicates the ionic radius of the host ion, $La^{3+}$. Other lines are guides to the eye.

coordinated ionic radii [19] appropriate to the $K_2NiF_4$ structure have been used. Included in Fig. 2 is the value of $T_c$ (38 K) for which the abrupt drop in resistivity for the oxygen-enriched material occurs [18]; the ionic radius of the $O^{2-}$ ion has been used. Although size effects are presumably quite secondary to the more gross electronic properties, it appears that the degree of substitutional size mismatch plays a role in determining $T_c$ in this structure. Indeed, mixed alkali substitutions [20,21] of the form $La_{2-x}(Ca,Sr,Ba)_xCuO_{4-\delta}$ result in transition temperatures intermediate between those of the singly-doped compounds. The pressure dependence of $T_c$ is shown in the inset of Fig. 1 for $La_{2-x}Na_xCuO_{4-\delta}$, where an increase of $T_c$ with applied pressure with value $dT_c/dP \approx 0.08$ K/kbar is observed; such positive values of $dT_c/dP$ are somewhat unusual among superconductors in general, but seem to be a common feature of the high $T_c$ copper oxide materials. A partially successful attempt has been made, based upon a simple model of lattice contraction, expansion, and distortion, to correlate the superconducting properties of $La_{2-x}M_xCuO_{4-\delta}$ materials with the lattice parameters for various dopants [22].

Variation of dopant and oxygen concentrations in $La_{2-x}M_xCuO_{4-\delta}$ reveals that the superconducting transition temperature increases with increasing hole concentration, at least over the range $0.06 \lesssim p \lesssim 0.25$. In such studies, the dopant concentration is known from the starting material stoichiometry and the hole concentration $p$ is measured directly by iodometric titration techniques; alternatively, the oxygen content can be determined by thermogravimetric analysis upon reaction in a reducing (usually hydrogen-argon) atmosphere, and then $p$ is calculated from charge balance. For samples prepared under typical oxygen annealing conditions (slow-cooling in 1 atm $O_2$) it is found [8] that the oxygen deficiency is quite small for low dopant concentrations, i.e., $\delta \approx 0$ for $0 \leq x \leq 0.15$. Then, at some point in the range $0.15 \leq x \leq 0.31$ (depending on the precise preparation conditions), the onset of an appreciable number of oxygen vacancies occurs; the observed lattice contraction supports the chemical evidence of oxygen depletion [9]. For samples prepared in a high pressure ($\sim 600$ bar) oxygen atmosphere, higher hole concentrations, up to $p = 0.40$, can be attained in samples without oxygen vacancies [9]. The behavior of $La_{2-x}Sr_xCuO_4$ as a function of hole concentration $p$ is shown in Fig. 3. Surprisingly, superconductivity disappears at high hole concentrations, although the metallic conductivity in the normal state improves [9].

Thus in the simple $La_{2-x}M_xCuO_4$ system, without oxygen vacancies, one observes localized holes in an antiferromagnetic material at low hole concentrations; with

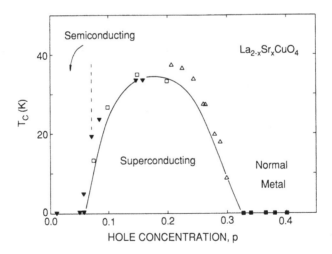

**Fig. 3.** The dependence of $T_c$ on the hole concentration per copper ion, $p$, for $La_{2-x}Sr_xCuO_4$. Open squares are from Ref. 7, solid triangles from Ref. 8, and open triangles and solid squares from Ref. 9. After Ref. 9.

increasing $p$, delocalization and superconductivity occur, with $T_c$ increasing for $0.06 \lesssim p$ $\lesssim 0.10$ and then remaining relatively constant for $0.10 \lesssim p \lesssim 0.25$. The transition temperature then decreases, and for $p \geq 0.32$, superconductivity disappears, although the normal-state conductivity becomes higher. For systems with oxygen vacancies, more complicated behavior occurs, although the dependence of superconductivity on hole concentration presumably remains much the same. Hall effect [23,24] and thermopower [24] measurements are consistent with one hole per Sr dopant ion at low concentrations; for higher concentrations, probably because of the occurrence of oxygen vacancies, there is a large drop in the Hall coefficient and a negative thermopower at high temperatures, implying the existence of two types of charge carrier, reminiscent of a compensated semiconductor. A significant result for these materials was obtained from magnetothermopower measurements [25]; these data showed that the normal-state thermopower of $La_{1.85}Sr_{0.15}CuO_{4-\delta}$ is independent of the ratio of magnetic field to temperature. Such behavior almost certainly indicates that the carriers do not have a spin degree of freedom [25]; this result supports those models in which the carriers lack spin or move in pairs with total spin zero. Thus, although the $La_{2-x}M_xCuO_{4-\delta}$ materials may be considered the simplest prototype for the high temperature superconductors, a rich variety of behaviors is observable even in this "ideal" system.

Another interesting feature of these compounds (as well as the $YBa_2Cu_3O_{7-\delta}$ 90 K superconductors) is the presence of a linear term in the specific heat at low temperatures [26–29]. Such behavior is shown for the $La_{2-x}Na_xCuO_{4-\delta}$ system in Fig. 1, where the electronic specific heat coefficient, $\gamma$, at low temperatures ($T \ll T_c$) increases from near zero at low concentrations to $\sim 4$ mJ/mole-$K^2$ in the superconducting state. The persistent presence of this feature in many studies seems to counter suggestions that it may be due to an impurity phase; it has been suggested that the nonzero $\gamma$ may reflect the presence of a large number of two-level systems, as observed in the low temperature specific heat of glasses. Alternatively, if it is due to a vanishing of the energy gap over part of the Fermi surface, such a nonzero electronic specific heat coefficient may have important implications for understanding the mechanism of high temperature superconductivity. An unusual dependence of $\gamma$ as a function of concentration has been recently reported [31] for the Ba- and Sr-doped compounds, where $\gamma$ rises from near zero at $x = 0$ to $\sim 4$ mJ/mole-$K^2$ at $x \approx 0.05$, but then drops again to $\gamma \approx 0$ at $x \approx 0.10$ before rising again at higher concentrations. Such a dip at intermediate concentrations might be attributable to an increase in oxygen deficiency near $x \approx 0.10$; the oxygen content of these samples was not reported. A similar two-peaked feature has been reported [30] for the inductively-measured superconducting transition temperature as a function of concentration x in $La_{2-x}Ba_xCuO_{4-\delta}$, where the $T_c$ vs x data display a minimum at x ≈ 0.12. A sudden drop in $T_c$ near a concentration where the onset of an appreciable number of oxygen vacancies is known to occur under typical sample preparation conditions [8] is probably due to such oxygen deficiency. Obviously, samples for doping studies need to be characterized by measurement of oxygen content, thus enabling a description in terms of net hole concentrations; a description which only includes the dopant concentration is not sufficient.

We now turn to other substitutions in this system, first considering rare-earth substitution for lanthanum. In general, the introduction of lanthanide element magnetic impurities Ln into $La_{2-x-y}Ln_ySr_xCuO_{4-\delta}$ decreases $T_c$ somewhat, although for low concentrations (y ≈ 0.1) the degradation is slight [21,31–34]. Since lanthanum is the only lanthanide element which forms the $K_2NiF_4$ structure with copper and oxygen, most lanthanide substituents can only be introduced in small amounts while maintaining a single phase, with $y_{max}$ ≈ 0.2. An exception is Ln = Pr, where $y_{max}$ ≈ 1.0–1.4 can be attained [21,32]. For the series $La_{1.8-y}Pr_ySr_{0.2}CuO_{4-\delta}$, $T_c$ decreases smoothly from $T_c$ ≈ 38 K for y = 0 to $T_c$ ≈ 15 K for y = 1.4 [32].

Substitutions of the form $La_{1.7}Ln_{0.1}Sr_{0.2}CuO_{4-\delta}$ result in the expected monotonic decrease in lattice constants as Ln is varied across the rare-earth series from Ln = La to Tm [32]. A significant feature of this series is the behavior of $T_c$, which shows a tendency to decrease from Ln = La to Gd (with the exception of Ln = Eu), and then increase from Ln = Gd to Tm [32]. These tendencies, particularly the greater $T_c$ depression for Sm than the other light rare-earth ions and for Gd than the heavier rare-earth ions, are suggestive that exchange interactions between the rare-earth magnetic ions and the conduction electrons are responsible for the depression of the superconductivity (see Fig. 22 and the accompanying discussion for a comparison of the rare-earth dependences of the exchange and dipolar interactions). That the depression of $T_c$ is noticeable at all indicates that the interactions between the rare-earth ions and the conduction electrons are probably greater in the $La_{2-x-y}Ln_ySr_xCuO_{4-\delta}$ system than in the 90 K superconductors. It has also been suggested that the $T_c$ degradation may be a volume effect [33], but this conjecture is difficult to reconcile with the observed monotonic changes in lattice constant and concurrent nonmonotonic changes in $T_c$. An upper critical field study of $La_{1.85-y}Nd_ySr_{0.15}CuO_{4-\delta}$, however, found no evidence of an exchange interaction between the magnetic moments and the superconducting electrons [34]; instead, the changes in superconducting properties were described in terms of a decreasing density of states at the Fermi energy. Clearly, careful work on well-characterized samples is necessary for a consensus to emerge.

Attempts have been made to substitute a great variety of transition metals for copper in $La_{2-x}Sr_xCuO_{4-\delta}$ [31,35–38]; these generally result in severe degradation of superconducting properties. For example, in $La_{1.85}Sr_{0.15}(Cu_{1-x}M_x)O_{4-\delta}$, superconductivity is completely suppressed by only a few (5–7) percent of magnetic impurities, M = Ni [35,36]. Suprisingly, the nonmagnetic substitution M = Zn [35] supresses superconductivity even more quickly than magnetic Ni. The $La_2CuO_4$ systems offer a simple framework for the study of these effects, since all of the copper sites are equivalent in this structure. Nonetheless, because of the much greater volume of literature to draw upon, we defer discussion and speculation concerning the various mechanisms at work until the examination of the copper-substituted $YBa_2(Cu_{1-x}M_x)_3O_{7-\delta}$ system, which appears in Sec. III-I.

## III. $RBa_2Cu_3O_{7-\delta}$ AND RELATED MATERIALS

### A. Superconducting Critical Temperatures

After the discovery [2] of superconductivity in the vicinity of 90 K in $YBa_2Cu_3O_{7-\delta}$ ($\delta \approx 0.1$), which has a layered orthorhombic perovskite-like crystal structure [39], a number of groups independently discovered superconductivity above 90 K in the isostructural lanthanide $RBa_2Cu_3O_{7-\delta}$ counterparts with R = Nd [40–43], Sm [40–44], Eu [40–44], Gd [40–43,45], Dy [40–42], Ho [40–43,46], Er [40,41,43,47], Tm [40,41], Yb [40,41,43,46,48], and Lu [40,42,43,46,49]. Shown in Figs. 4 (a) and (b) is the electrical resistivity $\rho$, normalized to its value at 120 K (80 K for R = La), vs temperature [50] for all of the superconducting $RBa_2Cu_3O_{7-\delta}$ compounds. The superconducting critical temperature $T_c$, defined as the temperature at which $\rho$ drops to 50% of it extrapolated normal state value, is typically between 90 K and 94 K, while the width of the transitions, $\Delta T_c$, taken as the difference in temperatures at which $\rho$ drops to 10% and 90% of its extrapolated normal state value, is $\leq 2$ K, except for $LaBa_2Cu_3O_{7-\delta}$ for which $T_c \approx 56$ K and $\Delta T_c \approx 10$ K. The $T_c$ of $LaBa_2Cu_3O_{7-\delta}$ can be raised to ~ 90 K using special processing techniques that are discussed in Sec. III-F. Although the Pr compound has the same layered orthorhombic perovskite-like crystal structure as the superconducting $RBa_2Cu_3O_{7-\delta}$ compounds, it is neither metallic nor superconducting

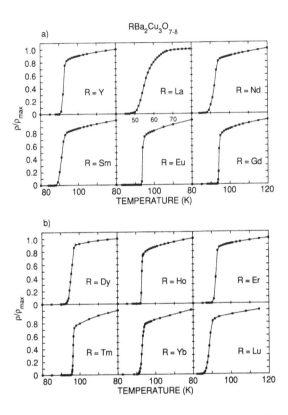

**Fig. 4.** Electrical resistivity ρ, normalized to its value at 120 K (80 K for R = La), for $RBa_2Cu_3O_{7-\delta}$ compounds with R = Y, La, Nd, Sm, Eu, Gd, Dy, Ho, Er, Tm, Yb, and Lu. From Ref. 53.

[50,51]. Attempts to prepare $RBa_2Cu_3O_{7-\delta}$ compounds with R = Ce and Tb by means of conventional solid state reaction techniques have produced mixed phase materials consisting of $BaCeO_3$ or $BaTbO_3$, CuO, and $BaCuO_2$ [52]. The compound $PmBa_2Cu_3O_{7-\delta}$ has not been prepared because of the short half-life of the radioactively unstable Pm nucleus.

There is as yet no definitive evidence that $T_c$ [53] and other superconducting properties, such as the initial slope of the upper critical field $H_{c2}$ curve, $(-dH_{c2}/dT)_{Tc}$ [54,55], and the critical current density $J_c$ [56], vary significantly with $R^{3+}$ ion in the series of $RBa_2Cu_3O_{7-\delta}$ compounds. The negligible effect of the $R^{3+}$ ions on $T_c$ suggests that the exchange interaction between the spin s of the superconducting electrons and the angular momentum **J** of the $R^{3+}$ ions is small, since $R^{3+}$ ions with partially-filled $4f$ electron shells ordinarily depress $T_c$ through pair-breaking effects at a rate that scales with the pair-breaking parameter

$$\alpha \equiv n\hbar N(E_F)\mathfrak{I}^2(g_J-1)^2J(J+1) \tag{1}$$

where $n$ is the concentration of paramagnetic $R^{3+}$ ions, $N(E_F)$ is the density of conduction electron states at the Fermi level, $\mathfrak{I}$ is the exchange interaction parameter, and $g_J$ and $J$ are, respectively, the Landé g-factor and total angular momentum of the $R^{3+}$ ion [57]. A small value of $\mathfrak{I}$ indicates minimal overlap between the $R^{3+}$ $4f$ orbital wavefunctions and the wavefunctions of the neighboring copper and oxygen atoms.

## B. Upper Critical Magnetic Fields and Critical Current Densities

In addition to their unprecedentedly high values of $T_c$, the $RBa_2Cu_3O_{7-\delta}$ compounds have enormous values of the upper critical magnetic field, $H_{c2}(T)$. Measurements of the $H_{c2}$ initial slopes, $(-dH_{c2}/dT)_{T_c}$, on polycrystalline $RBa_2Cu_3O_{7-\delta}$ superconductors have been reported for all of the rare earth elements that form the compounds except for La [50,53–55,58–63]. No systematic dependence of $(-dH_{c2}/dT)_{T_c}$ on R ion has been established, since a different effect dominates the behavior; the resistive transition curves upon which these measurements are based tend to broaden and frequently develop a "foot" in the applied magnetic field H. This broadening appears to be associated with poor intergranular conduction, which may have several causes including mismatch of crystallite orientations, unreacted material, oxygen deficiency, and impurity phases. Plotted in Fig. 5 are the magnitude of the upper critical magnetic field slope, $(-dH_{c2}/dT)_{T_c}$, and the normal state electrical resistivity just above $T_c$, $\rho(T = 100 \text{ K})$, vs the broadening of the resistive transition width in an applied magnetic field, $\Delta T_c(H = 9 \text{ T})$ $- \Delta T_c(H = 0 \text{ T})$, for fourteen $RBa_2Cu_3O_{7-\delta}$ polycrystalline specimens made from all of the rare earth elements that form the compounds except for La [54]. The figure indicates that $(-dH_{c2}/dT)_{T_c}$ tends to decrease with increasing applied-field-induced transition broadening. This broadening, in turn, tends to decrease with decreasing normal state electrical resisitivity. Figure 5 suggests the possibility of attaining initial upper critical field slopes greater than 3 T/K if values of $\Delta T_c(H = 9 \text{ T}) - \Delta T_c(H = 0 \text{ K})$ less than 7.5 K and $\rho(T = 100 \text{ K})$ less than 0.68 m$\Omega$-cm can be attained by more refined sample preparation techniques. A correlation between $(-dH_{c2}/dT)_{T_c}$ and the transition width $\Delta T_c$

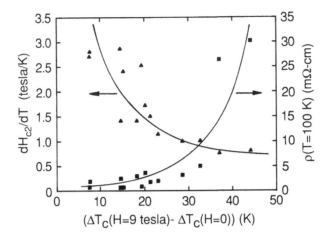

**Fig. 5.** Magnitude of the initial upper critical field slope near $T_c$ (triangles), $(dH_{c2}/dT)_{T_c}$, and the normal state electrical resistivity just above $T_c$ (squares), $\rho(T = 100$ K), vs resistive transition broadening in applied magnetic field, $\Delta T_c(H = 9$ T) $- \Delta T_c(H = 0$ T). The lines are guides to the eye. From Ref. 54.

in zero and finite magnetic field has been noted by other researchers [63].

Exemplary $H_{c2}$ vs T data (solid squares) for the compound $YbBa_2Cu_3O_{7-\delta}$, which yield $(-dH_{c2}/dT)_{T_c} = 2.70$ T/K, are shown in Fig. 6 [53]. The data were obtained from the $\rho(T)/\rho(100$ K) curves displayed in the inset of Fig. 6 by defining $T_c$ in the applied field as the temperature at which $\rho$ drops to $\sim 0.5$ of its extrapolated normal state value for $T < 100$ K. The two curves in Fig. 6, which were fitted to the data near $T_c$, are based on the standard, three dimensional, type II, dirty limit Werthamer, Helfand, Hohenberg, and Maki (WHHM) theory [64,65] and correspond to minimum ($\lambda_{so} = \infty$) and maximum ($\lambda_{so} = 0$) paramagnetic limitation, where $\lambda_{so}$ is the spin-orbit scattering parameter. The extrapolations, which do not take into account the temperature variations of the normal state electrical resistivity, yield $H_{c2}(T = 0$ K) $= 97–166$ T and $H_{c2}(T = 77$ K) $\approx 30$ T. The $H_{c2}(T)$ curves of the $RBa_2Cu_3O_{7-\delta}$ compounds can be put into perspective by comparison with the previous record for $H_{c2}(T)$ shown in Fig. 6 for the champion among the conventional superconductors, the Chevrel phase compound $PbMo_{5.1}S_6$ [66], and by noting that the highest magnetic fields that are presently available for $H_{c2}(T)$ measurements are $\sim 70$ T pulsed fields, which do not access very much of the low temperature region of the $H_{c2}(T)$ curve from which $\lambda_{so}$ can be determined. The large

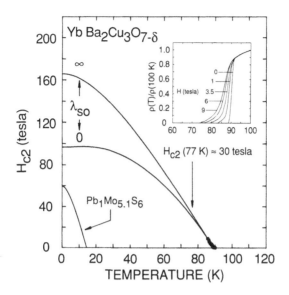

**Fig. 6.** Upper critical field $H_{c2}$ vs temperature for the compound $YbBa_2Cu_3O_{7-\delta}$. The solid lines are based on standard WHHM theory for a conventional type II superconductor in the limits of maximum ($\lambda_{SO} = 0$) and minimum ($\lambda_{SO} = \infty$) paramagnetic limitation and have been fitted to the data near $T_c$. Inset: Normalized electrical resistivity vs temperature in several magnetic fields between 0 and 9 tesla. From Ref. 53.

magnitude of the zero temperature orbital critical field $H_{c2}*(0)$ yields an estimate of the Ginzburg-Landau coherence length at $T = 0$ K, $\xi \approx 12$ Å. Such a short coherence length has important implications for microscopic models of superconductivity for these compounds and also for the performance of superconducting devices involving thin films of these materials. Measurements of $H_{c2}(T)$ on single crystal specimens of $YBa_2Cu_3O_{7-\delta}$ [67] reveal that $H_{c2}(T)$, like other physical properties, is highly anisotropic; values of $\xi$ parallel to the $c$-axis, $\xi_{/\!/} \approx 5$–7 Å, and perpendicular to the $c$-axis (within the $a$-$b$ plane), $\xi_\perp \approx 20$–30 Å, have been inferred. Somewhat smaller values of $\xi_{/\!/} = 1.4 \pm 0.2$ Å and $\xi_\perp = 14.2 \pm 1.0$ Å have been inferred from an analysis of the diamagnetic contribution to the $\chi(T)$ data from superconducting fluctuations for high purity powders of $YBa_2Cu_3O_{7-\delta}$; see Fig. 7 [68]. The data in Fig. 7 reveal that $\chi$ increases monotonically with T, with negative curvature below $\sim 200$ K that, according to the theoretical analysis, arises from the superconducting fluctuation diamagnetism. The analysis was based on a

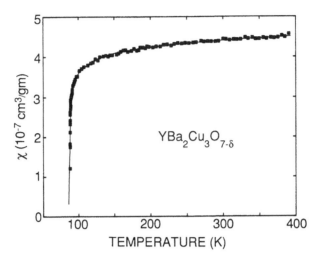

**Fig. 7.** Magnetic susceptibility $\chi$ vs temperature for $YBa_2Cu_3O_{7-\delta}$. Superconducting fluctuation diamagnetism is evident well above $T_c$. From Ref. 68.

generalization of the Lawrence-Doniach model [69] to include the angular dependence of the superconducting fluctuation diamagnetism, the results of which were then powder averaged in the 3-D and 2-D regimes for comparison with the powder $\chi(T)$ data. The $\chi(T)$ data have been fitted to the theoretical expressions for the 3-D regime between 90 K and 100 K and the 2-D regime above 100 K.

At liquid helium temperature, single crystal $RBa_2Cu_3O_{7-\delta}$ samples display high $J_c$ values, $\sim 10^6$–$10^7$ A/cm$^2$ for current flow in the $a$-$b$ plane, according to direct resistance measurements on epitaxially grown thin films [70] and indirect magnetization measurements with analyses using the Bean critical state model [71] on monolithic single crystal specimens [67] and grain-aligned pseudo-single-crystal specimens [72] (see Sec. III-D). Values of $J_c$ decrease by a factor $\sim 10$ in an applied field of $\sim 6$ T. At liquid nitrogen temperature, the values of $J_c$ drop to $\sim 10^5$–$10^6$ A/cm$^2$ in zero field and decrease by $\sim 10^2$ in a field of $\sim 6$ T. The corresponding values of $J_c$ in polycrystalline sintered $RBa_2Cu_3O_{7-\delta}$ compounds are several orders of magnitude lower and degrade more rapidly in applied fields [56,73]. This is illustrated in Fig. 8, which shows semilogarithmic $J_c$ vs H isotherms at various temperatures between about 4.2 K and about 90 K for $RBa_2Cu_3O_{7-\delta}$ compounds with R = Y, Sm, and Gd [56]. The low values, and rapid degradation with H, of $J_c$ in polycrystalline sintered materials both

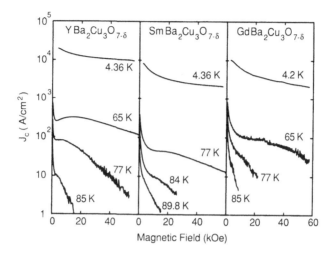

**Fig. 8.** Semilogaritmic plot of critical current density $J_c$ vs magnetic field isotherms at various temperatures between about 4.2 K and about 90 K for $RBa_2Cu_3O_{7-\delta}$ compounds with R = Y, Sm, and Gd. Peaks and shoulders are apparent at $T \geq 65$ K. From Ref. 56.

appear to be associated with the mismatch of neighboring crystallite orientations and with oxygen deficiency, impurity phases, and other imperfections at the region between the grain boundaries.

## C. Possible Clues of a Non-Phonon Electron Pairing Mechanism

The origin and nature of the high $T_c$ superconductivity of the $La_{2-x}M_xCuO_{4-\delta}$ and $RBa_2Cu_3O_{7-\delta}$ compounds, as well as the other layered copper oxide compounds, is not presently understood and constitutes a formidable challenge to experimentalists and theorists alike. One of the most intriguing possibilities is that a magnetic mechanism, rather than the electron-phonon interaction, is responsible for the formation of the superconducting electron pairs in the high $T_c$ copper oxides. The primary evidence for a magnetic pairing mechanism is the proximity of antiferromagnetism and superconductivity as the concentration of holes in the conducting $CuO_2$ planes is varied, as discussed above for the $La_{2-x}M_xCuO_{4-\delta}$ compounds and below for the $RBa_2Cu_3O_{7-\delta}$ compounds. Evidence (but by no means proof) of non-phonon mediated superconductivity is the unexpectedly high value of $T_c$ itself. Indeed, recent calculations of $T_c$ based on the electron-phonon interaction using one electron band structures [74] can account for the ~ 40 K $T_c$'s that occur in the $La_{2-x}Sr_xCuO_{4-\delta}$ system but not the ~ 95 K $T_c$'s of the $RBa_2Cu_3O_{7-\delta}$ compounds. Other evidence includes the small barium and oxygen isotope effects on $T_c$, the appearance of a T-linear term in the superconducting state specific heat (although the significance of this last effect is somewhat in question due to the complicating influence of impurity phases), and the positive pressure dependence of $T_c$. Some of these clues of non-phonon, possibly magnon, mediated pairing are briefly considered below.

Isotope effect measurements have been made on both superconducting copper oxide phases in which $^{16}O$ has been exchanged with $^{18}O$ and on $YBa_2Cu_3O_{7-\delta}$ made from $^{134}Ba$, $^{135}Ba$, and $^{139}Ba$. The isotope effect exponent $\alpha$, which appears in the relation $T_c \sim M^{-\alpha}$ where M is the isotopic mass, has been found to lie in the range ~ 0.1–0.37 for oxygen in $La_{2-x}Sr_xCuO_{4-\delta}$ [75,76] and to be ~ 0 for oxygen [77,78] and barium [79] in $YBa_2Cu_3O_{7-\delta}$. The value of $\alpha$ = 1/2 results from the attractive electron-phonon interaction; however, it is well known that $\alpha$ can be reduced by the repulsive Coulomb interaction, even to zero, as observed for some transition element superconductors such as Ru and Zr [80]. Thus, the observation of an $\alpha$-value that is close to zero is consistent with a non-phonon mechanism, but does not rule out the electron-phonon mechanism.

Two types of antiferromagnetic (AFM) order have been observed in the $RBa_2Cu_3O_{7-\delta}$ compounds, one involving $Cu^{2+}$ ions with the $3d^9$ configuration and the other $R^{3+}$ ions with partially-filled $4f$ electron shells. Investigations of both types of AFM order are of interest in their own right, and may also be relevant to the occurrence of high $T_c$ superconductivity in the layered copper-oxide materials. In the first case, there is some reason to suspect, on both experimental and theoretical grounds [81-84], that the superexchange interactions that produce the AFM order in the insulating variants of the layered high $T_c$ compounds $RBa_2Cu_3O_{7-\delta}$ and $La_{2-x}M_xCuO_{4-\delta}$ may also be responsible for the formation of the "Cooper pairs" of electrons (or holes) with oppositely oriented spins in the metallic versions of these materials; the AFM ordering of the $Cu^{2+}$ ions in the oxygen-deficient $RBa_2Cu_3O_{7-\delta}$ compounds is discussed below. In the second case, investigations of the origin and nature of the magnetic order of the $R^{3+}$ ions may yield information about the underlying electronic structure that is also responsible for the high $T_c$ superconductivity of these materials; the AFM ordering of the $R^{3+}$ ions in the $RBa_2Cu_3O_{7-\delta}$ compounds is described in Sec. III-E.

The transition from metallic and superconducting to insulating and antiferromagnetic behavior in the $RBa_2Cu_3O_{7-\delta}$ compounds has been studied for R = Y [84–89] and Nd [90], and is accomplished by increasing the oxygen vacancy concentration $\delta$ from ~ 0 to

~ 1. For example, with increasing $\delta$ in the $YBa_2Cu_3O_{7-\delta}$ system [85,86], $T_c$ decreases from ~ 94 K at $\delta \approx 0$ to ~ 60 K in the interval $0.2 \lesssim \delta \lesssim 0.5$, which appears to represent another superconducting phase, to ~ 0 K at $\delta \approx 0.6$. At slightly larger values of $\delta$, after the superconductivity has been destroyed, antiferromagnetic ordering of the $Cu^{2+}$ ions appears with a Néel temperature $T_N$ that increases with $\delta$ and reaches ~ 500 K at $\delta = 1$ [84,88,89]. The increase of $\delta$ between 0 and ~ 0.5 is also accompanied by a change of the crystal structure from orthorhombic to tetragonal symmetry at room temperature, involving the redistribution of oxygen in the CuO chain sites [39]. As mentioned in Sec. II, the proximity of antiferromagnetism and superconductivity is also found in the $La_{2-x}M_xCuO_{4-\delta}$ compounds where the antiferromagnetism in $La_2CuO_{4-\delta}$ ($T_N \approx 300$ K; $\mu \approx 0.5$ $\mu_B$) is suppressed with increasing oxygen-vacancy concentration $\delta$ or M cation concentration x. The latter also serves to drive the system metallic and superconducting, with no overlap between AFM and superconducting regions. Thus, a general feature is that increasing the number of holes in the $CuO_2$ planes, either by increasing the oxygen concentration or by cation doping, leads to the destruction of AFM order and the appearance of superconductivity, with no region of overlap between these two phenomena. However, the proximity of antiferromagnetism to superconductivity suggests that a common mechanism of magnetic origin may be responsible for the two phenomena.

Specific heat data below 50 K in the superconducting state in the form of a C/T vs T plot for the $YBa_2Cu_3O_{7-\delta}$ compound are displayed in Fig. 9 [26,50]. Since the superconducting electron contribution to C(T) should be exponentially small at these temperatures which are well below $T_c$, the data have been fitted to the sum of an electronic contribution $C_e(T) = \gamma'T$ with a "residual $\gamma$," a Debye lattice contribution with three acoustic branches and a Debye temperature $\theta_D$, and an Einstein lattice contribution with $N_E$ optical branches and an Einstein temperature $\theta_E$. The solid line labeled (a) represents a fit of the data by the sum $C_e(T) + C_D(T) + C_E(T)$, with $\gamma' = 8.2$ mJ/mole-$K^2$, $\theta_D = 161$ K, $N_E = 11.3$, and $\theta_E = 170$ K. The contribution $C_E(T)$ is depicted by solid line (b), while the sum $C_e(T) + C_D(T)$ is indicated by solid line (c).

Shown in the inset of Fig. 9 is a plot of the C/T vs $T^2$ data below 16 K which reveals the existence of the $C_e(T) = \gamma'T$ term at low temperature as well as an additional low temperature contribution $\delta C(T) \equiv C(T) - \gamma'T$ which actually exhibits a peak at 2 K with a value of ~ 20 mJ/mole-K [26,50]. The $C_e(T) = \gamma'T$ contribution below $T_c$ is similar to that which occurs in the $La_{2-x}M_xCuO_{4-\delta}$ compounds, discussed in Sec. II, and the superconducting spinel $Li_{1.1}Ti_{1.9}O_{3.95}$ [91]. Compared to the value of the electronic

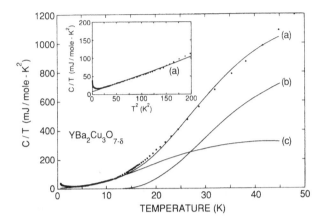

**Fig. 9.** Specific heat C divided by temperature T vs temperature for $YBa_2Cu_3O_{7-\delta}$. The solid line labeled (a) represents the sum of an electronic contribution $C_e(T) = \gamma'T$ with $\gamma' = 8.2$ mJ/mole-K$^2$, a Debye lattice contribution $C_D(T)$ with a Debye temperature $\theta_D = 161$ K, and an Einstein lattice contribution $C_E(T)$ with an Einstein temperature $\theta_E = 170$ K. The contributions $C_e(T) + C_D(T)$ are indicated by the solid line labeled (c), while the contribution $C_E(T)$ is depicted by the solid line labeled (b). Inset: C/T vs T$^2$ below 16 K. From Ref. 53.

specific heat coefficient $\gamma$ in the normal state, which has been estimated, for example, by Junod et al. [92] from the specific heat jump $\Delta C$ at $T_c$ using the BCS relation $\Delta C = 1.43$ $\gamma T_c$ to be ~ 40 mJ/mole-K$^2$, the observed value of $\gamma$ is significant. A linearly temperature dependent contribution to the specific heat in the superconducting state could be associated with an energy gap that vanishes over a portion of the Fermi surface (gapless superconductivity), which could have important theoretical implications. Unfortunately, this interpretation is complicated by the possible presence of nonsuperconducting phases in the R-Ba-Cu-O systems with nonvanishing $\gamma$'s such as $BaCuO_2$ [93].

Another interesting property of the $RBa_2Cu_3O_{7-\delta}$ compounds is the positive pressure dependence of $T_c$, at least in the low pressure region from ~ 0 to 20 kbar, where $T_c$ increases linearly with pressure at a rate $dT_c/dP$ ~ 0.1–0.2 K/kbar [94]. The application of a pressure of 149 kbar has been found to increase the onset of $T_c$ for the compound $YBa_2Cu_3O_{7-\delta}$ to about 107 K [95], the highest value of $T_c$ ever observed for any of the $RBa_2Cu_3O_{7-\delta}$ compounds, to our knowledge. Linear increases of $T_c$ with P at comparable rates appear to be a general property of the layered high $T_c$ copper oxide compounds, suggesting a common mechanism for the high $T_c$ superconductivity of these materials.

## D. Crystalline Electric Field Effects

### i. Magnetic Susceptibility

At temperatures between $T_c$ and 300 K, the magnetic susceptibility of polycrystalline specimens of the superconducting $RBa_2Cu_3O_{7-\delta}$ compounds containing $R^{3+}$ ions with partially-filled $4f$ electron shells can be described by the sum of a constant Pauli-like contribution $\chi_0$ and a Curie-Weiss component $C/(T-\theta)$ [56,96], i.e.,

$$\chi(T) = \chi_0 + C/(T-\theta). \tag{2}$$

The values of the effective magnetic moment $\mu_{eff} = (3k_BC/N)^{1/2}$, where N is the number of R ions, derived from the Curie constant C of the fit of Eq. (2) to the $\chi(T)$ data, are in reasonable agreement with the $R^{3+}$ free ion values calculated from Hund's rules, as illustrated in Fig. 10. The $\chi(T)$ data for the polycrystalline $RBa_2Cu_3O_{7-\delta}$ compounds do not reveal any obvious evidence of CEF effects, a result that can be attributed to two factors. First, the deviations of $\chi(T)$ from Curie-Weiss behavior due to CEF effects, which can be quite pronounced in single crystal specimens in the principal crystallographic directions, can average out in polycrystalline materials so that $\chi(T)$ resembles rather closely a Curie-Weiss law [97]. This has been well documented in the series of $RRh_4B_4$ compounds [98,99] which have been investigated extensively with respect to superconductivity, magnetism, the interaction between these two phenomena, and CEF effects [100]. Second, the strongest deviations of $\chi(T)$ from CEF behavior generally occur at low temperatures (below $\sim 50$ K) which are comparable to the CEF splittings between the R-ion energy levels. However, this is difficult to observe in the $RBa_2Cu_3O_{7-\delta}$ compounds since all but the Pr compound exhibit superconductivity with $T_c \approx 94$ K. The large slope of the $H_{c2}$ vs T curve near $T_c$ does not permit the superconductivity to be quenched in reasonable magnetic fields and, therefore, does not allow $\chi(T)$ to be determined at temperatures $\lesssim 90$ K without the complicating influence of the superconductivity.

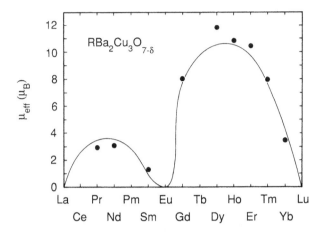

**Fig. 10.** Effective magnetic moment $\mu_{eff}$ vs rare earth ion R for $RBa_2Cu_3O_{7-\delta}$ compounds compared to the Hund's rules $R^{3+}$ free ion values (line). From Ref. 53.

## ii. Specific Heat

The first real evidence of CEF effects in the $RBa_2Cu_3O_{7-\delta}$ compounds was observed in the low temperature specific heat in the form of electronic Schottky anomalies [26,50,101]. Shown in Fig. 11 are C vs T data between ~ 0.5 K and 30 K for $RBa_2Cu_3O_{7-\delta}$ compounds with R = Ho, Tm, Yb, and, for comparison, Y, all of which do not exhibit magnetic order above ~ 0.45 K, the low temperature limit of the $^3He$ calorimeter in which the specific heat measurements were made [26,50]. The broad peak in C(T) for the Ho compound at T $\approx$ 5 K appears to be an electronic Schottky anomaly associated with CEF splitting of the $Ho^{3+}$ J = 8 Hund's rules ground state multiplet. There is no trace of a Ho nuclear Schottky anomaly in the C(T) data, as often observed for Ho compounds, which suggests that the electronic ground state of $Ho^{3+}$ in this compound is a singlet or that magnetic order occurs at lower temperature. (Actually, both $HoBa_2Cu_3O_{7-\delta}$ and $YbBa_2Cu_3O_{7-\delta}$ exhibit magnetic order at 0.17 K and 0.35 K, respectively, as discussed below.) In order to isolate the $Ho^{3+}$ electronic Schottky anomaly, the conduction electron and lattice contributions (see Fig. 9) can be approximately corrected for by subtracting the specific heat of $YBa_2Cu_3O_{7-\delta}$ from the total specific heat of $HoBa_2Cu_3O_{7-\delta}$. The difference $\Delta C \equiv C(HoBa_2Cu_3O_{7-\delta}) - C(YBa_2Cu_3O_{7-\delta})$ obtained in this manner is plotted vs T for T $\leq$ 25 K in Fig. 12 [26].

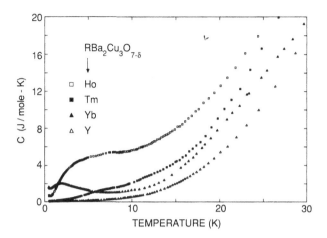

**Fig. 11.** Specific heat C vs temperature for $RBa_2Cu_3O_{7-\delta}$ compounds with R = Ho, Tm, Yb, and, for comparison, Y. From Ref. 53.

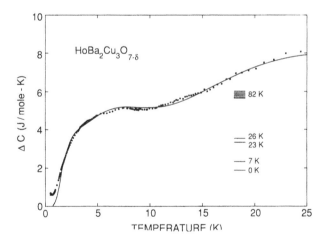

**Fig. 12.** Specific heat $\Delta C$, corrected for the $YBa_2Cu_3O_{7-\delta}$ background, vs temperature for the $HoBa_2Cu_3O_{7-\delta}$ compound. The solid line represents the calculated electronic Schottky anomaly associated with the CEF splitting of the $Ho^{3+}$ energy levels. The energies and degeneracies of the levels are shown in the figure. From Ref. 26.

The C(T) data can be described well by the solid line in the figure which is a calculated Schottky anomaly based on an energy level scheme consisting of a ground state singlet, three singlets at ~ 7, 23, and 26 K, seven levels at ~ 82 K, and higher levels at temperatures greater than 150 K that have a negligible effect on the low temperature fit [26,50].

In contrast, the C(T) data for $TmBa_2Cu_3O_{7-\delta}$ have no distinguishable peaks at low temperature, although the specific heat is several times larger than that of $YBa_2Cu_3O_{7-\delta}$ in this temperature region. The data can be described reasonably well between 4 K and 17 K by a Schottky anomaly calculated from the following energy level scheme: a quartet at 0 K, a singlet at 17 K, a triplet at 57 K, and the remaining levels at energies greater than 150 K [26,50]. The C(T) data for $YbBa_2Cu_3O_{7-\delta}$ appear to consist of an electronic Schottky anomaly with a peak near 2 K and a low temperature upturn that is presumably the high temperature tail of a lambda-type anomaly associated with the ordering of the $Yb^{3+}$ magnetic moments at ~ 0.35 K. A reasonable description of the C(T) anomaly between ~ 1.5 K and 8 K can be obtained with a Schottky anomaly based on the following possible $Yb^{3+}$ energy level scheme: doublets at 0 K, 0.2 K, 4.7 K, and 47 K [26,50].

Recently, inelastic neutron scattering experiments have been carried out on $HoBa_2Cu_3O_{7-\delta}$ by Furrer et al. [102] which have yielded an energy level spectrum consisting of a large number of well-resolved CEF transitions between 0.5 and 73 meV, from which these authors have assigned the energies of ten excited CEF energy levels and, on the basis of the observed intensities, determined the CEF parameters required for orthorhombic symmetry. The lowest lying energy levels deduced consist of singlets at 6, 21, 44, 50, 94, 125, and 135 K, in reasonable accord with the lowest lying energy levels used to describe the Schottky anomaly in the low temperature specific heat data discussed above. The CEF parameters have also been used to calculate the anisotropic magnetic susceptibility of $HoBa_2Cu_3O_{7-\delta}$, yielding an easy axis of magnetization with a cross-over from $//c$ above ~ 100 K to $\perp c$ below ~ 100 K. A change from an easy axis of magnetization $//c$ at room temperature to one $\perp c$ at 4.3 K was previously inferred [72] for $HoBa_2Cu_3O_{7-\delta}$ as recounted below in Sec. III-D(iii), which describes experiments on magnetically aligned crystallites of the $RBa_2Cu_3O_{7-\delta}$ compounds.

### iii. Paramagnetic Anisotropy

The paramagnetic anisotropy of the $RBa_2Cu_3O_{7-\delta}$ compounds has been characterized by Ferreira et al. [72] and Livingston et al. [103] during the course of investigations of the magnetic alignment of single crystal grains of these compounds.

In the experiments of Ferreira et al., the $RBa_2Cu_3O_{7-\delta}$ crystallites were mixed in epoxy at room temperature and aligned in an applied magnetic field $H_A = 18$ kOe. The epoxy was allowed to harden in the field, producing a cylindrical specimen with the cylindrical axis in the direction of $H_A$. The original motivation of the magnetic alignment of powder grains was to prepare pseudo-single-crystal specimens for research and to develop the procedure as a potential processing step for the production of high $T_c$ superconductors with higher critical current densities. The alignment of the crystallites is associated with their anisotropic paramagnetic susceptibility; they tend to align so that the direction of maximum magnetization is along the applied field direction. Magnetic alignment of single crystal grains of $YBa_2Cu_3O_{7-\delta}$ suspended in epoxy at room temperature was first demonstrated by Farrell et al. [104] who employed a very large magnetic field of 94 kOe. The $RBa_2Cu_3O_{7-\delta}$ compounds with partially-filled $4f$ electron shells, which carry localized magnetic moments which can be as large as $\mu_{eff} \approx 10.6$ $\mu_B$ for R = Dy and Ho, should be amenable to alignment in smaller and more easily accessible values of $H_A$.

X-ray analyses indicated that the $c$-axis of the $RBa_2Cu_3O_{7-\delta}$ crystallites aligns parallel to $H_A$ for R = Y, Nd, Sm, Dy, and Ho, and perpendicular to $H_A$ for R = Eu, Er, Tm, and Yb. Similar results were reported by Livingston et al. [103] for R = Y, Nd, Eu, Gd, Dy, Ho, and Er. Since the crystallites align with their maximum paramagnetic susceptibilities in the direction of $H_A$, this determines the easy axis of magnetization at room temperature. Except for Y, Sm, and Eu, the room-temperature alignment direction (easy axis of magnetization) correlates with the sign of the second order Stevens factor $\alpha_J$ of the CEF Hamiltonian, as noted by Livingston et al. [103] and Ferreira et al. [72], similar to what was previously observed for the series of $RRh_4B_4$ compounds [97-100]. The easy axes of magnetization at room temperature for the $RBa_2Cu_3O_{7-\delta}$ and $RRh_4B_4$ compounds are compared to the second order Stevens factor $\alpha_J$ in Table I. The Stevens factor is expected to determine the ground state magnetic anisotropy of both series of compounds since their magnetic properties are primarily associated with the rare earth ions whose site symmetries are similar—tetragonal in the case of the $RRh_4B_4$ compounds and orthorhombic for the $RBa_2Cu_3O_{7-\delta}$ compounds. The latter are nearly tetragonal with

**Table I.** Easy axes of magnetization of $RBa_2Cu_3O_{7-\delta}$ and $RRh_4B_4$ compounds indicated as being parallel ($\parallel c$) or perpendicular ($\perp c$) to the $c$-axis, compared to the sign of the second order Stevens factor $\alpha_J$ of the CEF Hamiltonian.

| R | Sign of $\alpha_J$ | $RBa_2Cu_3O_{7-\delta}$ | $RRh_4B_4$ |
|---|---|---|---|
| Y | 0 | $\parallel c^\dagger$ | $\parallel c^\dagger$ |
| Nd | − | $\parallel c$ | |
| Sm | + | $\parallel c$ | $\parallel c$ |
| | | | $\perp c$ (low $T$) |
| Eu | 0 | $\perp c$ | |
| Gd | 0 | $\parallel c^{\dagger *}$ | $\parallel c^\dagger$ |
| Tb | − | | $\parallel c$ |
| Dy | − | $\parallel c$ | $\parallel c$ |
| Ho | − | $\parallel c$ | $\parallel c$ |
| | | $\perp c$ (low $T$) | |
| Er | + | $\perp c$ ($\parallel b^*$) | $\perp c$ |
| Tm | + | $\perp c$ | $\perp c$ |
| Yb | + | $\perp c$ | |

$^\dagger$ CEF theory predicts no anisotropy.

$^*$ Low temperature easy axis of magnetization as inferred from the orientation of the $R^{3+}$ moments in the AFM state.

orthorhombic *a* and *b* axes that differ by only a few percent [39].

In order to determine the anisotropy of the magnetization M and estimate the maximum critical current density $J_c$ at liquid helium temperatures, magnetization M vs magnetic field H curves were measured at 4.3 K. Figure 13 shows M(H) for $RBa_2Cu_3O_{7-\delta}$ powders with R = Sm, Nd, Ho, and Dy, for which x-ray scans indicated alignment of the *c*-axis of the oriented grains parallel to the cylindrical axis of the specimen and thus to the alignment field $H_A$. In all cases, $\Delta M(/\!/) > \Delta M(\perp)$, where $\Delta M(/\!/)$ and $\Delta M(\perp)$ represent, respectively, the magnetic hysteresis parallel and perpendicular to the cylindrical axis and, in turn, the *c*-axis of the oriented grains. The result $\Delta M(H/\!/c) > \Delta M(H\perp c)$ is consistent with earlier work on field aligned powders [104] and bulk single crystals of $YBa_2Cu_3O_{7-\delta}$ [105,106]. The greater hysteresis in the M(H/\!/c) curves indicates that $J_c$ is higher for supercurrent flow in the basal *a-b* $CuO_2$ planes than for flow perpendicular to these planes. Figure 14 shows M(H) for $RBa_2Cu_3O_{7-\delta}$ powders with R = Eu, Yb, Tm, and Er, for which x-ray scans indicated alignment of the

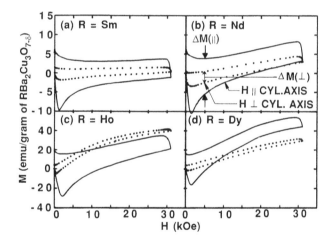

**Fig. 13.** Magnetization M vs measuring field H at 4.3 K for epoxy-embedded powders of $RBa_2Cu_3O_{7-\delta}$ with R = Sm, Nd, Ho, and Dy, all subject to an alignment field $H_A$ = 18 kOe. These specimens all displayed x-ray scans indicating that the *c*-axes of the aligned crystallites are *parallel* to the specimen's cylindrical axis. Magnetization data are for measuring field parallel to the cylindrical axis (solid lines) and perpendicular to the cylindrical axis (dotted lines). In all cases, $\Delta M(/\!/) > \Delta M(\perp)$ for these compounds. From Ref. 72.

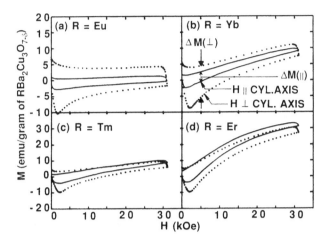

**Fig. 14.** Magnetization M vs measuring field H at 4.3 K for epoxy-embedded powders of $RBa_2Cu_3O_{7-\delta}$ with R = Eu, Yb, Tm, and Er, all subject to an alignment field $H_A$ = 18 kOe. These specimens all displayed x-ray scans indicating that the $c$-axes of the aligned crystallites are *perpendicular* to the specimen's cylindrical axis. Magnetization data are for measuring field parallel to the cylindrical axis (solid lines) and perpendicular to the cylindrical axis (dotted lines). In all cases, $\Delta M(/\!/) < \Delta M(\perp)$ for these compounds. From Ref. 72.

$c$-axes of the oriented grains perpendicular to the cylindrical axis of the specimen and thus perpendicular to the alignment field $H_A$. Opposite to the behavior shown in Fig. 13, $\Delta M(/\!/) < \Delta M(\perp)$ in Fig. 14. For H parallel to the cylindrical axis, all oriented grains have their $c$-axes perpendicular to H, so the hysteresis $\Delta M$ is at a minimum. On the other hand, assuming that the $c$-axes of grains are randomly oriented in directions perpendicular to the cylindrical axis, when H is perpendicular to the cylindrical axis there are components of H parallel to the $c$-axes of most oriented grains, so a relatively large magnetic hysteresis occurs.

The critical current density $J_c$ can be crudely estimated from the hysteresis in the magnetization $\Delta M$ from the Bean-model [71] formula $J_c(A/cm^2) = f\Delta M(emu/cm^3)/d(cm)$, where $f$ is a geometrical factor which, for a sphere of diameter $d$, is equal to 34 [56]. Using an average diameter of 5 μm, based on scanning electron microscopy studies on $HoBa_2Cu_3O_{7-\delta}$ powder, values of $J_c$(4.3 K, 5 kOe) of $10^6$–$10^7 A/cm^2$ were obtained for Dy. These values are comparable to those of bulk single crystals and field aligned grains of $YBa_2Cu_3O_{7-\delta}$ at similar temperatures and fields, but are two or three orders of magnitude higher than the corresponding values similarly estimated from $\Delta M$ and the Bean model for bulk polycrystalline specimens of $RBa_2Cu_3O_{7-\delta}$ [53,56,107,108]. This difference is consistent with present ideas that the latter materials display low $J_c$ because of poor intergranular transport, possibly associated with orientational mismatch of the anisotropically conducting grains.

Aside from the superconducting characteristics, the magnetization curves of Figs. 13 and 14 also display paramagnetic contributions due to the magnetic moments of the $R^{3+}$ ions. For $HoBa_2Cu_3O_{7-\delta}$ of Fig. 13(c), the 4.3 K paramagnetic magnetization component is larger for H perpendicular to the cylindrical axis than for H parallel to the cylindrical axis, opposite to the situation expected at room temperature where crystallites should align so that the direction of maximum paramagnetic susceptibility is along the applied field $H_A$ and hence along the cylindrical axis. That the sign change in the paramagnetic anisotropy occurs at elevated temperature is suggested by preliminary 100 K magnetization measurements which show that $M(H/\!/\text{cylindrical axis}) > M(H\perp\text{cylindrical axis})$ for $0 < H < 50$ kOe. This cross-over in the easy axis of $Ho^{3+}$ magnetization from $/\!/c$ at room temperature to $\perp c$ at liquid helium temperatures is consistent with independent conclusions based on inelastic neutron scattering data as described in Section III-E, but is contrary to expectations for *ground state* magnetic anisotropy based on the sign of the Stevens $\alpha_J$ factor.

### E. Antiferromagnetic Ordering of $R^{3+}$ Ions
### and the Anisotropic 2-D Ising Model

Evidence of magnetic ordering of the $R^{3+}$ ions in the $RBa_2Cu_3O_{7-\delta}$ compounds has emerged from specific heat measurements on the compounds with R = Nd [53,109], Sm [53,109,110], Gd [53,109-116], Dy [53,109,110,115], Ho [101], and Er [53,101,109,111,115], and Mössbauer effect measurements on the compound with R = Yb [117]. The magnetic ordering temperatures $T_M$ inferred from the sharp peaks in the C(T) contributions associated with magnetic order are 0.52 K for Nd [53], 0.61 K for Sm [53], 2.25 K for Gd [53], 0.90 K for Dy [53], 0.17 K for Ho [101], and 0.60 K for Er [53]. The onset of the hyperfine splitting for the Yb compound yields a magnetic ordering temperature of $\sim 0.35$ K. Shown in Fig. 15 are C vs T data for the $RBa_2Cu_3O_{7-\delta}$ compounds which order magnetically above $\sim 0.5$ K reported in references 53 and 109. The C(T) data for the $RBa_2Cu_3O_{7-\delta}$ compounds with R = Nd, Sm, Gd,

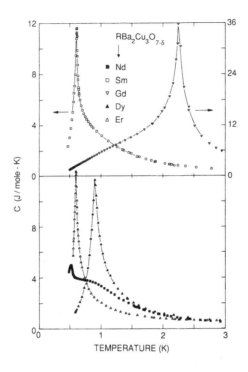

**Fig. 15.** Specific heat vs temperature between 0.5 and 3 K for $RBa_2Cu_3O_{7-\delta}$ compounds with R = Nd, Sm, Gd, Dy, and Er. From Ref. 53.

Dy, and Er have been approximately corrected for conduction electron and lattice contributions by subtracting the C(T) data for $YBa_2Cu_3O_{7-\delta}$, since Y has no $f$-electrons. The shapes of the C(T) anomalies due to magnetic ordering for the $RBa_2Cu_3O_{7-\delta}$ compounds with R = Sm, Gd, and Er have been analyzed in terms of antiferromagnetic 2-D and 3-D Ising models with isotropic exchange interactions and found to be best described by the 2-D Ising model [118-121]. The occurrence of antiferromagnetic order has been directly verified for the Nd, Gd, Dy, and Er compounds by means of neutron scattering experiments. In the cases of Nd [122], Gd [123], and Dy [124,125], the $R^{3+}$ magnetic moments are parallel to the $c$-axis, whereas in the case of Er [126] they lie in the $a$-$b$ plane and are parallel to the $b$-axis. Shown in Fig. 16 are the temperature dependences of the peak intensities of the (1/2,1/2,1/2) (open squares) and (1/2,1/2,3/2) (solid squares) magnetic Bragg reflections for $DyBa_2Cu_3O_{7-\delta}$, measured with decreasing temperature, from neutron scattering experiments by Fischer et al. [125] (the lines drawn

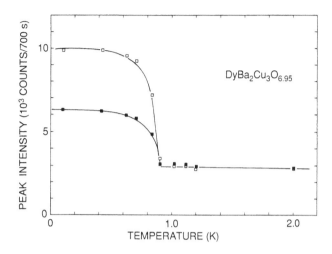

**Fig. 16.** Temperature dependence of the peak intensity of the (1/2,1/2,1/2) (open squares) and (1/2,1/2,3/2) (solid squares) magnetic reflections for $DyBa_2Cu_3O_{6.95}$ for decreasing temperature. The solid lines are guides to the eye. From Ref. 125.

through the data are guides to the eye). The data in Fig. 16 yield values for the Néel temperature $T_N$ and saturation magnetic moment $\mu_s$ of 0.90 K and 6.8 $\mu_B$, respectively. The value of $\mu_s$ exceeds the 6 $\mu_B$ value estimated for the CEF ground state by Dunlap et al. [101], but is substantially lower than the Hund's rules free-ion value of 10 $\mu_B$.

It has recently been found [122,127,128] that the C(T) data (Fig. 15) for the $RBa_2Cu_3O_{7-\delta}$ compounds with R = Nd, Sm, Dy, and Er can be described well by AFM 2-D Ising models with surprisingly *anisotropic* exchange interactions with anisotropy ratios that range from 50 for Nd to 4 for Dy. An Ising model would seem to be appropriate for these compounds, since the ground states of the $R^{3+}$ ions in the CEF are doublets [109]. Specific heat measurements on nonsuperconducting oxygen deficient $RBa_2Cu_3O_{7-\delta}$ compounds ($\delta \gtrsim 0.5$) for R = Nd, Sm, Dy, and Er have revealed that the magnetic ordering temperatures and shapes of the magnetic specific heat anomalies are dramatically different than those of the high $T_c$ superconducting oxygen-rich $RBa_2Cu_3O_{7-\delta}$ compounds ($\delta \approx 0.1$) [122,127,128].

Shown in Fig. 17 are C(T) data associated with magnetic ordering for the $DyBa_2Cu_3O_{7-\delta}$ compound between ~ 0.5 and 4 K. The solid line which has been fitted to the C(T) data represents the solution of the 2-D Ising model with a Néel temperature $T_N = 0.90$ K and exchange interaction parameters $E_1 = 0.73$ K and $E_2 = 0.18$ K. While

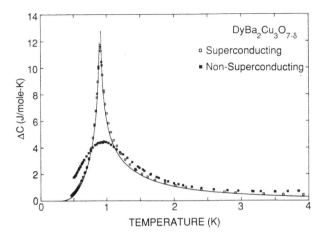

**Fig. 17.** Magnetic specific heat $\Delta C$ vs temperature for $DyBa_2Cu_3O_{7-\delta}$ in the superconducting state ($\delta \approx 0.1$; open squares) and the normal state ($\delta \approx 0.5$; solid squares). The solid line that has been fitted to the C(T) data for $\delta \approx 0.1$ represents the solution of the anisotropic two-dimensional Ising model with Néel temperature $T_N = 0.90$ K and exchange interaction parameters $E_1 = 0.73$ K and $E_2 = 0.18$ K. From Ref. 128.

the isotropic 2-D Ising model gives a reasonable description of the C(T) data for $DyBa_2Cu_3O_{7-\delta}$, the anisotropic 2-D Ising model with an anisotropy ratio $E_1/E_2 = 4$ gives a much better description, as shown in [128]. Removal of oxygen to produce nonmetallic nonsuperconducting $DyBa_2Cu_3O_{7-\delta}$ ($\delta \approx 0.5$) transforms the sharply peaked feature in C(T) due to the AFM order into a broad, Schottky-like anomaly with a maximum near 1 K that is uncharacteristic of long range magnetic order.

Shown in Fig. 18 are C(T) data due to magnetic ordering for $NdBa_2Cu_3O_{7-\delta}$ in the metallic superconducting ($\delta \approx 0.1$) and nonmetallic nonsuperconducting ($\delta \approx 0.5$) states. The C(T) data for the metallic superconducting compound can be described well by the anisotropic 2-D Ising model (solid line) with $T_N = 0.50$ K and exchange interaction parameters $E_1 = 0.85$ K and $E_2 = 0.017$ K. The anisotropy ratio $E_1/E_2 = 50$ for the Nd compound is more than an order of magnitude larger than that for the Dy compound. Removal of oxygen to produce nonmetallic nonsuperconducting $NdBa_2Cu_3O_{7-\delta}$ ($\delta \approx 0.5$) results in a marked enhancement of the magnetic ordering temperature (from 0.5 K to ~ 1.5 K) and a profound change in the shape of the C(T) anomaly.

The C(T) curves arising from magnetic ordering for the $SmBa_2Cu_3O_{7-\delta}$ compound in

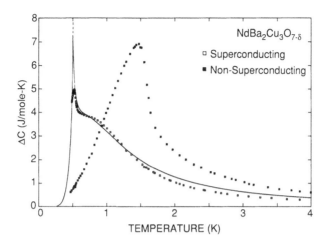

**Fig. 18.** Magnetic specific heat $\Delta C$ vs temperature for $NdBa_2Cu_3O_{7-\delta}$ in the superconducting state ($\delta \approx 0.1$; open squares) and the normal state ($\delta \approx 0.5$; solid squares). The solid line that has been fitted to the C(T) data for $\delta \approx 0.1$ represents the solution of the anisotropic two-dimensional Ising model with Néel temperature $T_N = 0.50$ K and exchange interaction parameters $E_1 = 0.85$ K and $E_2 = 0.017$ K. From Ref. 127.

the metallic superconducting ($\delta \approx 0.1$) and nonmetallic nonsuperconducting ($\delta \approx 0.5$) states are shown in Fig. 19. The solid line through the C(T) data for the metallic superconducting sample ($\delta \approx 0.1$) represents the 2-D anisotropic Ising model with $T_N = 0.61$ K, $E_1 = 0.70$ K, and $E_2 = 0.063$ K. This anisotropy ratio $E_1/E_2 = 11.1$ is intermediate between that of the Nd and Dy compounds, indicating that the anisotropy ratio decreases as R moves toward the heavier rare earths. The anisotropy ratio $E_1/E_2$ is ~ 5.4 for $ErBa_2Cu_3O_{7-\delta}$. The sharp peak at $T_N$ is absent in the C(T) data for the nonmetallic nonsuperconducting $SmBa_2Cu_3O_{7-\delta}$ compound, similar to the situation encountered for $DyBa_2Cu_3O_{7-\delta}$, indicating that there is no long range magnetic order above ~ 0.5 K. It is interesting that the solid line through the C(T) data for this compound follows from either of two distinctly different models—a 1-D Ising model with an exchange interaction parameter $E = 1.2$ K or a Schottky anomaly for a two level system with a splitting $\Delta E = 2E = 2.4$ K. A 1-D Ising model would not be appropriate, however; the oxygen deficient $RBa_2Cu_3O_{7-\delta}$ compounds are tetragonal, so there can be no anisotropy in the $a$-$b$ plane. Thus, it would appear that the $SmBa_2Cu_3O_{7-\delta}$ compounds do not exhibit magnetic order above ~ 0.5 K and that the rounded peak in

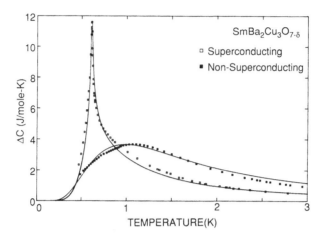

**Fig. 19.** Magnetic specific heat $\Delta C$ vs temperature for $SmBa_2Cu_3O_{7-\delta}$ in the superconducting state ($\delta \approx 0.1$; open squares) and the normal state ($\delta \approx 0.5$; solid squares). The solid line that has been fitted to the C(T) data for $\delta \approx 0.1$ represents the solution of the anisotropic two-dimensional Ising model with Néel temperature $T_N = 0.61$ K and exchange interaction parameters $E_1 = 0.70$ K and $E_2 = 0.063$ K. The solid line that has been fitted to the data for $\delta \approx 0.5$ represents the solution of the one-dimensional antiferromagnetic Ising model with exchange interaction parameter $E = 1.2$ K *or* an electronic Schottky anomaly for a two level system (presumably two doublets) separated by 2.4 K. From Ref. 127.

C(T) is associated with a ground state multiplet and a low lying excited state multiplet of equal degeneracy, presumably 2.

The Ising model analysis of the magnetic specific heat anomalies of the $RBa_2Cu_3O_{7-\delta}$ compounds suggests that the exchange interactions within the *a-b* plane are much stronger than the interactions between the planes, and that they are anisotropic. The source of this anisotropy is not readily apparent, since these materials have a rather small orthorhombic distortion and are nearly tetragonal, as noted in Section III-D. Thus, if the magnetic interactions between the $R^{3+}$ magnetic moments are mediated by electrons within the adjacent $CuO_2$ planes between which they are situated, via RKKY or superexchange interactions, $E_1$ and $E_2$ would be expected to be comparable to one another. The most likely source of anisotropy would be the CuO chains, which extend in the *b* direction and are midway between the $R^{3+}$ layers. However, the CuO chains would seem to be too far away from the planes of $R^{3+}$ ions to produce strongly anisotropic exchange interactions within the planes while maintaining only weak interactions between the $R^{3+}$ planes and thereby imbue the magnetic ordering with 2-D character. Nonetheless, the CuO chains clearly play an important role in the magnetic ordering of the $R^{3+}$ ions, since the increase of the oxygen vacancy concentration $\delta$ has a profound effect on the magnetic ordering and involves oxygen sites [39] in the CuO chains. A potential source of the exchange interaction anisotropy could be the overlap between the $R^{3+}$ doublet ground state wavefunctions and valence band states of the $CuO_2$ planes. Of course, one cannot rule out the possibility that the excellent description of the C(T) data provided by the anisotropic AFM 2-D Ising model is fortuitous and that such analysis is simply inappropriate.

The strong dependence of the magnetic ordering temperature $T_M$ on the oxygen vacancy parameter $\delta$ in $NdBa_2Cu_3O_{7-\delta}$ indicates that the coupling between the $R^{3+}$ magnetic moments involves the RKKY or superexchange interactions. This is rather surprising since C(T) measurements by Dunlap et al. [113] on $GdBa_2Cu_3O_{7-\delta}$ in the metallic superconducting ($\delta \approx 0.1$) and nonmetallic nonsuperconducting ($\delta \approx 0.85$) states yielded no change in $T_N$ nor shape of the C(T) anomaly associated with magnetic order (see Fig. 20), leading those authors to suggest that the magnetic ordering of the $R^{3+}$ ions

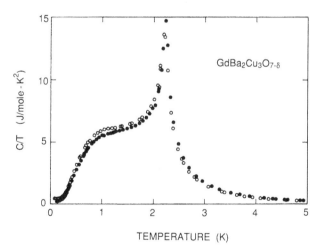

**Fig. 20.** Specific heat C divided by temperature T vs temperature for $GdBa_2Cu_3O_{7-\delta}$. The closed circles correspond to the superconducting compound ($\delta \approx 0.1$) and the open circles to the nonsuperconducting compound ($\delta \approx 0.85$). From Ref. 113.

is unchanged and due to dipolar interactions. Recent neutron scattering studies have revealed, however, that the AFM structure of $GdBa_2Cu_3O_{7-\delta}$ does depend on the oxygen vacancy concentration $\delta$ [129]. A schematic representation of the antiferromagnetic structures determined from neutron diffraction experiments on two superconducting $GdBa_2Cu_3O_{7-\delta}$ specimens — (a) a single crystal with $\delta \approx 0.5$ and $T_c \approx 40$ K [129] and (b) a powdered polycrystal with $\delta \approx 0.1$ and $T_c \approx 90$ K [123] — is displayed in Fig. 21 where only the Gd ions and the directions of their magnetic moments are shown for clarity. The magnetic moments in both structures are coupled antiferromagnetically in the $a$-$b$ plane. However, the magnetic coupling between the Gd ions along the $c$-axis is ferromagnetic for $\delta \approx 0.5$ and antiferromagnetic for $\delta \approx 0.1$. It is interesting that the temperature dependence of the sublattice magnetization in the $GdBa_2Cu_3O_{6.5}$ single crystal obtained from the intensity of the (1/2,1/2,0) reflection is consistent with 2-D behavior [129].

(a)                    (b)

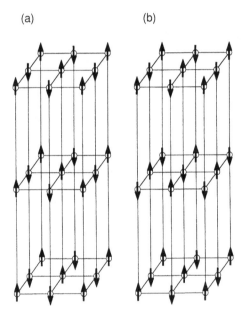

**Fig. 21.** Schematic representation of the magnetic structures of (a) superconducting $GdBa_2Cu_3O_{6.5}$ ($T_c \approx 40$ K) and (b) superconducting $GdBa_2Cu_3O_7$ ($T_c \approx 90$ K) obtained from neutron diffraction. Only the Gd ions and the directions of their moments are shown. From Ref. 129.

Evidence in favor of the RKKY interaction as the dominant mechanism for magnetic ordering in the $RBa_2Cu_3O_{7-\delta}$ compounds is the observed scaling of the magnetic ordering temperatures $T_M$ with the deGennes factor $(g_J-1)^2J(J+1)$ of the $R^{3+}$ ions [53,109,110]. This is illustrated in Fig. 22, where the values of $T_M$ for the $RBa_2Cu_3O_{7-\delta}$ compounds are plotted vs $R^{3+}$ ion and compared to the deGennes factor (the solid line) which has been normalized to the $T_M$ of the Gd compound. The discrepancies between the experimental values of $T_M$ and the deGennes scaling may be caused by CEF effects [130] and the variation in the strength of the exchange interaction between the $R^{3+}$ magnetic moments and the conduction electron spins with R [131].

## F. $LaBa_2Cu_3O_{7-\delta}$

Substitution of lanthanum for yttrium in $YBa_2Cu_3O_{7-\delta}$ has proven to be unusual among the rare-earth-subsituted high-$T_c$ compounds, since $LaBa_2Cu_3O_{7-\delta}$ is a

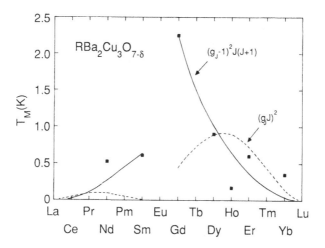

**Fig. 22.** Magnetic ordering temperatures $T_M$ vs R for $RBa_2Cu_3O_{7-\delta}$ compounds with R = Nd, Sm, Gd, Dy, and Er (Ref. 109) , Ho (Ref. 101), and Yb (Ref. 117). The solid line represents the value of $T_M$ expected from a scaling of $T_M$ for R = Gd by the deGennes factot $(g_J-1)^2J(J+1)$, where $g_J$ and J are, respectively, the Landé g factor and total angular momentum of the Hund's rule ground-state multiplet of the $R^{3+}$ ion under consideration. The dotted line represents the value of $T_M$ expected from a scaling of $T_M$ for R = Dy by the dipolar factor $(g_JJ)^2$.

superconductor but with a lower transition temperature, typically around 50 K (see Fig. 4). The decreased $T_c$ correlates with structural properties, and has been attributed to disorder on the La and Ba sites and/or oxygen vacancy disorder [132–140]. Such structural deviations are generally manifested by the degree of orthorhombic distortion, although complications due to competing phases, including superstructures of the same stoichiometry, have been identified [132]. Under restricted synthesis conditions, a superconducting onset temperature of 95 K and zero resistance temperature of at least 80 K have been attained in $LaBa_2Cu_3O_{7-\delta}$ [133].

Because the $La^{3+}$ ion is the largest of the trivalent rare-earth ions, it most readily substitutes for the larger $Ba^{2+}$ ion; such compatibility is also evident in the $Y(Ba_{1-x}La_x)_2Cu_3O_{7-\delta}$ and $La_{1+x}Ba_{2-x}Cu_3O_{7-\delta}$ compounds discussed later in Sec. III-J. The site disorder which results from La-Ba transposition probably affects superconductivity in several ways. Primarily, $La^{3+}$ ions in the $Ba^{2+}$ sites may cause occupancy of adjacent oxygen chain vacancies caused by a tendency toward local charge neutrality. This would result in a compensating depletion of oxygen chain sites elsewhere; both processes would contribute to the observed departure from orthorhombicity. Another possible effect, due to the concurrent $Ba^{2+}$ substitution at the $La^{3+}$ sites, would be the distortion of the dimpled structure of the $CuO_2$ planes [134]. As a result of the larger size and lower charge of the $Ba^{2+}$ ion, the nearby Cu-O bond angles and distances would be reduced, with possible significance for superconductivity in the $CuO_2$ planes.

Surprisingly, studies of oxygen content [134,135] show that, while increasing the oxygen stoichiometry $y = 7-\delta$ to $y \approx 7.1–7.4$ is possible in $LaBa_2Cu_3O_{7-\delta}$, such enhanced oxygen content results in lower $T_c$'s. Thus, such preparation techniques are not alleviating the oxygen chain site depletion discussed above, but, to the contrary, are probably promoting further La-Ba site disorder. The techniques which apparently do minimize such disorder involve annealing at low temperatures, typically 300°C [133,136]. Thus the order-disorder transition exhibited by $LaBa_2Cu_3O_{7-\delta}$ combined with the generally low atomic diffusivity at temperatures around 300°C indicate that extremely long low-temperature anneals are required to maximize the superconducting properties of this compound.

## G. $PrBa_2Cu_3O_{7-\delta}$ and $Y_{1-x}Pr_xBa_2Cu_3O_{7-\delta}$

Of all the $RBa_2Cu_3O_{7-\delta}$ (R = rare earth except Pm, Ce, and Tb) compounds that form in the orthorhombic crystal structure, only $PrBa_2Cu_3O_{7-\delta}$ is unexpectedly not superconducting [53]. The answer to why Pr is so different from the other rare earths in this crystal structure could illuminate key points concerning the pairing mechanism in these oxide superconductors. This has prompted various research groups to investigate the normal and superconducting properties of the $Y_{1-x}Pr_xBa_2Cu_3O_{7-\delta}$ ($0 \leq x \leq 1$) system.

Partial substitution of Pr for Y in $YBa_2Cu_3O_{7-\delta}$ is characterized by a monotonic decrease in the superconducting transition temperature $T_c$ (see Fig. 23), a change of the normal state electrical resistivity $\rho(T)$ from metallic to semiconducting behavior at $x \approx 0.4$ (see Fig. 24), and a small reduction of the orthorhombic distortion with increasing x [51,141]. The oxygen content, measured using thermogravimetric and chemical analyses [142], was found to be essentially constant throughout the series and was interpreted as evidence for unchanged ordering of the Cu-O chain sites. Recent neutron diffraction experiments [143] also led to the same conclusion. As of this writing, there are two points of view regarding the quenched superconductivity in $PrBa_2Cu_3O_{7-\delta}$; one attributes the behavior to a mixed valent state of the Pr ion in this particular structure, and another is based on hybridization effects between the Pr $4f$ and valence band states.

Normal state magnetic susceptibility $\chi(T)$ measurements were reported by several groups. The data of Ref. 51 yield an effective magnetic moment $\mu_{eff} \approx 2.7\ \mu_B/Pr$ ion as extracted from the Curie-Weiss behavior of the data. Comparison with the effective

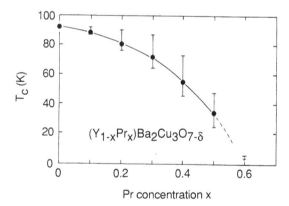

Pr concentration x

**Fig. 23.** Resistively measured superconducting transition temperature $T_c$ vs Pr concentration x for the $(Y_{1-x}Pr_x)Ba_2Cu_3O_{7-\delta}$ system. The solid line is a guide to the eye. Data points indicate transition midpoints, and the vertical bars represent the 10%–90% transition widths. From Ref. 51.

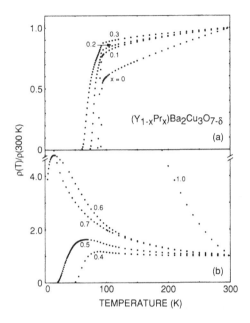

**Fig. 24.** Electrical resistivity $\rho$, normalized to its value at 300 K, vs temperature for $(Y_{1-x}Pr_x)Ba_2Cu_3O_{7-\delta}$ compounds. From Ref. 51.

moments for $Pr^{3+}$ and $Pr^{4+}$ of 3.58 $\mu_B$ and 2.54 $\mu_B$ suggests a mixed valent state for the Pr ion with an average valence $\bar{v} \approx 3.9$ throughout the whole series. Other $\chi(T)$ measurements are in fair agreement with this result, indicaing an average valence for the Pr ion which ranges from 3.5 to 4.0 [142,144–146]. Since the valence state of all other rare earths in these 90 K superconductors is +3, it appears at first sight that the higher than trivalent state of Pr in the $Y_{1-x}Pr_xBa_2Cu_3O_{7-\delta}$ system is responsible for the quenched superconductivity. It can be speculated that, in this case, the extra charge provided by $Pr^{4+}$ would cause $T_c$ to decrease by filling holes in the $CuO_2$ planes. Recent magnetic susceptibility measurements on tetragonal single crystals of $PrBa_2Cu_3O_{7-\delta}$ yielded an average valence for Pr of about 3.3, a value somewhat smaller than that obtained from polycrystalline samples.

In contrast to the magnetic susceptibility data, spectroscopic measurements such as x-ray absorption [147,148] and photoemission (PES) [149] experiments in $Y_{1-x}Pr_xBa_2Cu_3O_{7-\delta}$ yield spectra that are similar to those seen in compounds where Pr has a valence 0f +3. In x-ray absorption measurements, the spectrum has the same energy dependence as that of $Pr^{(3+)}{}_2O^{(2-)}{}_3$. The Pr 4f spectra extracted from PES are independent of x and show a large peak just below the Fermi energy $E_F$, with essentially no weight at $E_F$, contrary to what is usually expected in situations where mixed valence prevails. The broad and asymmetric 4*f* lineshapes observed in $Y_{1-x}Pr_xBa_2Cu_3O_{7-\delta}$, which are also observed in several intermetallic compounds where Pr has a valence close to +3, were taken as evidence for extensive hybridization between Pr 4*f* and the oxygen 2*p* states. Recent experiments [150] on the pressure dependence of superconductivity in this system reveal a monotonic increase of $T_c$ with pressure up to 20 kbar for $0 \leq x \leq 0.2$, then the development of a maximum at 6 kbar for x = 0.3, followed by a monotonic decrease of $T_c$ with increasing P for $x \geq 0.3$, as illustrated in Fig. 25. These results suggest an electronic transition at ~ 6 kbar for x = 0.3 that shifts to lower values of P for

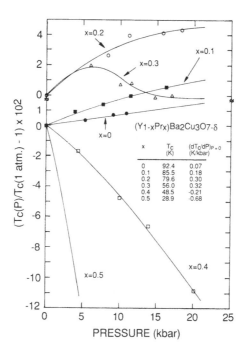

**Fig. 25.** Shift in the reduced superconducting critical temperature $T_c$ relative to its value at 1 atm, $[T_c(P)/T_c(1 \text{ atm})-1]\times 10^2$, vs pressure P between 0 and 20 kbar in $(Y_{1-x}Pr_x)Ba_2Cu_3O_{7-\delta}$ compounds. The values of $T_c$ at 1 atm and the initial rate of change of $T_c$ with P, $(dT_c/dP)_{P=0}$, for each value of x are listed in the figure. The lines drawn through the data are guides to the eye. The line for x = 0.5 passes through a point (not shown) at 5.1 kbar. From Ref. 150.

higher values of x. The $T_c(P)$ behavior was thought to be associated with hybridization between Pr $4f$ and valence band states, whereas the decrease of $T_c$ was tentatively ascribed to the strong Coulomb repulsion between electrons within the Pr $4f$ state or pair-breaking caused by the magnetic $Pr^{3+}$ ions [150].

Another possibility is that the reduced moment of Pr from that of a 3+ state, as measured through magnetic susceptibility, results from a strong $4f$-conduction band admixture due to the Kondo effect as is usually seen in intermetallic Ce and U compounds [151]. The drawback of this interpretation is that the density of states at the Fermi level in these oxide superconductors seems to be too small for this mechanism. However, inelastic neutron scattering experiments on $PrBa_2Cu_3O_{7-\delta}$ [152] indicate an unstable Pr moment and are interpreted in a strong $4f$-conduction band hybridization picture involving a nonnegligible density of conduction states at the rare earth site. In fact, recent low temperature specific heat measurements on $Y_{1-x}Pr_xBa_2Cu_3O_{7-\delta}$ compounds [153] reveal values of the electronic specific heat coefficient as high as $\gamma \approx$ 280 mJ/mole-$K^2$ (see also Fig. 26), suggesting a large density of states at the Fermi level. Measurements by other workers [154] also yield an increase of $\gamma$ with x, but the excess specific heat due to praseodymium, obtained by substracting the specific heat of $YBa_2Cu_3O_{7-\delta}$, was interpreted in terms of Schottky anomalies due to $Pr^{3+}$ and $Pr^{4+}$ ions. At the present time, it is not clear whether the enhanced values of $\gamma$ contain a magnetic contribution from the Pr ions. The specific heat data shown in Fig. 26 exhibit a rounded peak at $T \approx 16$ K, indicative of magnetic order; susceptibility measurements (inset of Fig. 26) reveal a departure from Curie-Weiss behavior below $T \approx 14$ K [127]. The sharp upturn in C/T at low temperatures in Fig. 26 appears to be the high temperature tail of a Pr nuclear Schottky anomaly.

Hall effect measurements were also performed in the $Y_{1-x}Pr_xBa_2Cu_3O_{7-\delta}$ system [142]; it was found that the Hall carrier number $(1/eR_H)$ is positive, decreases with increasing x, and is strongly temperature dependent. The linear dependence of $1/eR_H$ on T diminishes rapidly as x increases. The authors suggest that the system should be understood as a highly correlated fermion system having a large on-site Coulomb repulsion, in which case the Hall coefficient may be temperature dependent because of strong renormalization effects in a nearly half-filled Hubbard band. The superconducting temperature $T_c$ is also found to have a stronger dependence on the Hall carrier number and the Cu-O formal valence than in the oxygen depleted $YBa_2Cu_3O_{7-\delta}$ compound. From measurements of the thermoelectric power [155], it was found that the incorporation of Pr in $YBa_2Cu_3O_{7-\delta}$ was found to have a drastic effect on the magnitude,

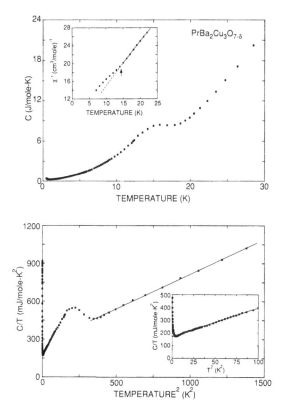

**Fig. 26.** Upper frame: Specific heat C vs temperature T for $PrBa_2Cu_3O_{7-\delta}$ between ~ 0.5 K and 20 K. Inset: Inverse magnetic susceptibility $\chi^{-1}$ vs temperature for $PrBa_2Cu_3O_{7-\delta}$ between 7 K and 22 K. The maximum in C(T) occurs at ~ 16 K and $\chi^{-1}(T)$ deviates from Curie-Weiss behavior at ~ 14 K, indicating complex antiferromagnetic order. Lower frame: C/T vs $T^2$ for $PrBa_2Cu_3O_{7-\delta}$ between ~ 0.5 and ~ 39 K and, in the inset, ~ 0.5 K and 10 K. The lines represent fits of the data to the expression $C(T) = \gamma T + \beta T^3$ in two temperature ranges — 3.5 K $\leq$ T $\leq$ 10 K with $\gamma$ = 169 mJ/mole-$K^2$ and $\beta$ = 2.25 mJ/mole-$K^4$ and 18 K $\leq$ T $\leq$ 38 K with $\gamma$ = 264 mJ/mole-$K^2$ and $\beta$ = 0.55 mJ/mole-$K^4$. From Ref. 127.

which increases, as well as the temperature dependence, which exhibits a negative slope dS/dT; together with the resistivity data, these features were interpreted as consistent with electron correlation effects in narrow bands. The decrease of $T_c$ was associated, if Pr valence is higher than +3, with a valence band filling mechanism. Another interesting observation [156] is that $T_c(x)$ curve in the $Y_{1-x}Pr_xBa_2Cu_3O_{7-\delta}$ system follows the dependence predicted the pair-breaking theory of Abrikosov and Gor'kov, seemingly indicating that Pr does not reduce $T_c$ just by filling holes in the $CuO_2$ planes, but by acting as a Cooper pair-breaker too.

It is generally agreed that the average valence of copper decreases weakly with increasing Pr concentration, reaching a value of $\sim 2.2$ for x = 0.5. Lattice parameter measurements vs x also show anomalies around x = 0.5, corresponding approximately to the transition from superconducting to nonsuperconducting states.

## H. Other Isovalent Substitutions for Y and Ba in $YBa_2Cu_3O_{7-\delta}$

Isovalent substitutions for yttrium include: 1) the rare earth R substitutions which have been thoroughly discussed above, 2) mixed rare-earth substitutions of the form $R_{1-x}R'_xBa_2Cu_3O_{7-\delta}$, where R and R' are different rare earth elements, and 3) non-rare-earth substitutions of the form $Y_{1-x}M_xBa_2Cu_3O_{7-\delta}$, where M represents various *sp* and *d* trivalent ions. Types 2) and 3) will now be discussed, as well as the isovalent substitution of strontium for barium.

Mixed rare-earth substituitions have been reported for $R_{1-x}R'_x = Y_{1-x}Eu_x$ [157], $Y_{1-x}Er_x$ [158,159], $Yb_{1-x}Er_x$ [160,161], $Yb_{1-x}Sm_x$ [157], and $Gd_{1-x}Eu_x$ [162], to name but a few of the 66 possible pairs of rare-earth elements for which both single-R-based compounds are superconducting. Generally, superconducting properties change very little for such series, with typical $T_c$ variation of 1–2 K as x varies from 0 to 1; lattice constants vary smoothly between those of the single rare-earth compounds. The transition temperature often varies systematically with x for a given series [157,160]; in other studies the deviations appear more random. In one study of $Yb_{1-x}Er_xBa_2Cu_3O_{7-\delta}$, the observed peak in the superconducting properties at x = 0.5 led to the conjecture that the two rare-earth ions order in some manner in the basal plane [161]; however, the validity of such a suggestion has not been clearly demonstrated.

Of interest among the mixed rare-earth substitutions is $Eu_{1-x}Gd_xBa_2Cu_3O_{7-\delta}$, where the dilute substitution of Gd for the nonmagnetic Eu enables determination of properties that would normally be hidden by the antiferromagnetic order occurring for pure $GdBa_2Cu_3O_{7-\delta}$. Electron spin resonance and specific heat measurements [162] enable determination of the crystal-field splitting of the $^8S_{7/2}$ ground state of the $Gd^{3+}$ ion, and a value of $\sim 1.5$ K is obtained; the nonzero value of the crystal-field interaction is probably a result of a small admixture with excited states which have nonzero orbital angular momentum. Furthermore, the derived value of the second order Stevens factor is negative, indicating the existence of an easy axis of magnetization parallel to the crystalline $c$ axis; strong evidence for such a preference is also indicated by the paramagnetic anisotropy of the $RBa_2Cu_3O_{7-\delta}$ compounds discussed in Sec. III-D(iii), which reveals an easy axis of magnetization $//c$ for Gd and by neutron diffraction measurements of magnetically ordered $GdBa_2Cu_3O_{7-\delta}$, where the ordered Gd moments lie along the $c$ axis of the orthorhombic unit cell [123].

Trivalent substitutions for yttrium include the $Y_{1-x}Sc_xBa_2Cu_3O_{7-\delta}$ system [163,165]; the solubility range of the smaller scandium ion is reported to be $0 \leq x \leq 0.5$ [164], although another study finds somewhat multiphase materials for all $x > 0$ [164]. As with any dopant, the normal state resistivity increases and $T_c$ decreases with increasing x; in this case, $T_c$ degradation is small, with $T_c \approx 89$ K for $x = 0.5$. More interesting is the magnetic behavior; a paramagnetic spin equivalent of 2.7 $\mu_B$ per Sc ion was deduced from the Curie susceptibility [163]. The contribution of impurity phases was discounted; proposed explanations include a tendency of Sc toward covalent bonding, Sc-Cu antisite disorder, and disorder-induced localization effects [163]. Other attempted trivalent substitutions for Y include Al [165], Bi, Ga, and Fe [166]; x-ray diffraction measurements indicate that such samples are multiphase for small (and perhaps all) dopant concentrations.

Isovalent substitution of the smaller Sr for Ba in $Y(Ba_{1-x}Sr_x)_2Cu_3O_{7-\delta}$ results in a very slight $T_c$ depression and a gradual decrease in the orthorhombic lattice constants with increasing x over the reported solubility range $0 \leq x \leq 0.5$ [167–170]. The rate of $T_c$ depression is only 0.16 K/at.% [170]. Under typical annealing conditions, slightly lower oxygen content is found for the doped samples than for $YBa_2Cu_3O_{7-\delta}$; however, annealing in high pressure oxygen does not improve $T_c$. In general, then, the minor alteration of superconducting properties observed in these materials highlights their relative insensitivity to isovalent chemical substitution at the Y or Ba sites.

## I. Substitutions for Copper: $YBa_2(Cu_{1-x}M_x)_3O_{7-\delta}$

Numerous investigations into the effects of the substitution of copper in $YBa_2Cu_3O_{7-\delta}$ by a great variety of metallic elements M have appeared in the literature. By far, the most extensive research performed to date, and that for which the substitutions have resulted in single-phased materials, has been for the substituent elements M = Fe [171–190], Co [172,187–189,191–197], Ni [35,172,187–189,

181–192,196,198], Zn [35,172,187–189,191–192,199–201], Ga [188,199,202], and Al [172,203,204]. Interpretation of the several varieties of behavior encountered vary, as do details of the specific features reported (solubility limits, location of structural transitions, rate of $T_c$ depression, site occupancy, etc.). However, there has emerged a fair amount of agreement on the types and extents of the gross behavior observed for a given substitution.

The rich variety of reported behavior, as well as much of the experimental disagreement, arise primarily from the preferential occupancy of the two inequivalent copper sites. The relative substitutional occupancies of the Cu1 (chain) and Cu2 (plane) sites are affected not only by the type of dopant, but also by the dopant concentration and the reaction and annealing conditions. Such substitutional site occupancy considerations are central in the interpretation of primary features such as the degree of degradation of superconducting properties, the apparent orthorhombic-tetragonal phase transformation (or phase coexistence), oxygen site occupancy, substitutional solubility, changes upon annealing, and magnetic ordering of impurities. Several mechanisms are generally invoked to explain some of these features, including electronic charge arguments (hole-filling), magnetic pair-breaking, oxygen content and coordination changes, reduction of copper valence, and structural order and disorder.

To examine some of these features, we first consider the (nominally) trivalent magnetic (Fe and Co) and nonmagnetic (Al and Ga) substitutions, and then the corresponding divalent (Ni and Zn) substitutions.

Various authors have reported that in $YBa_2(Cu_{1-x}M_x)_3O_{7-\delta}$ maximum dopant concentrations of $x \approx 0.15$–0.25 for M = Fe [171,172] and $x \approx 0.20$–0.33 for M = Co [172,195,196] can be achieved while maintaining single-phased materials. X-ray diffraction data indicate that for both Fe [172–175] and Co [172,193,195,196] an orthorhombic–tetragonal structural phase transition seems to occur near $x \approx 0.02$–0.04, although some workers [176,177] find that the orthorhombic structure persists nearly to $x = 0.15$. A common and primary feature, however, is that superconductivity persists for concentrations beyond that at which the transition occurs; $T_c$ depression is typically only about 3–5 K/at.%. Because the impurity-induced structural transition is generally attributed to Fe ions in the Cu1 (chain) sites, and since both oxygen content and the relative occupancy of Cu1 and Cu2 sites can depend on reaction and annealing conditions, the occurence of variability in the observed structural properties is not surprising. Typical structural data are shown in Fig. 27, where it is evident that oxygen depletion is *not* responsible for the observed loss of orthorhombicity.

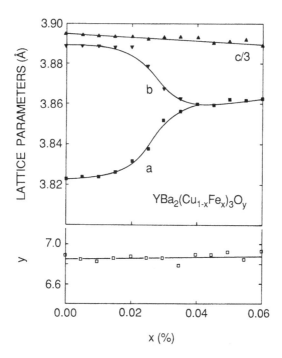

**Fig. 27.** Orthorhombic lattice parameters *a, b,* and *c* and the oxygen content y for $YBa_2(Cu_{1-x}Fe_x)_3O_y$ vs Fe concentration x from powder x-ray diffraction measurements and an inert gas fusion nondispersive IR method, respectively. An apparent orthorhombic-tetragonal structural phase transition is observed at x ≈ 0.04, while the oxygen content remains almost constant. From Ref. 175.

Mössbauer isomer shifts [178,179,181] indicate a dopant valence of $Fe^{3+}$, although $Fe^{4+}$ may be a more appropriate assignment for some of the observed doublets [184]. Neutron diffraction data [172,184,192,194] provide some of the best evidence that Fe and Co substitute preferentially (and, perhaps, in appropriately prepared samples, exclusively) on the Cu1 site, where they may more readily achieve their preferred octahedral coordination. (Due to the Cu2–Cu2 plane spacing, octahedral coordination at the Cu2 site is not possible.) Such excess impurity coordination is often presumed to cause the disordering of oxygen in the Cu1–O layer, thus causing the structural transition. However, such a picture of random oxygen ordering is probably misleading. Recent Mössbauer data [186] have been interpreted in terms of orthorhombic–tetragonal phase coexistence, where the phase mixture, which is directly dependent on the oxygen coordination of the Fe ions, is extremely sensitive to oxygen stoichiometry. A perhaps simpler and more enlightening picture emerges from the results of electron microscopy [185,197] and electron diffraction [184] studies, where it is evident that the apparent tetragonal phase is not a result of general disordering of Cu1–O chains, but, instead, is a manifestation of a microdomain structure consisting of small, ordered, orthorhombic domains with sizes of order $5 \times 5$ nm$^2$. The size of the microdomains decreases with increasing x. It is believed that the real lattice parameters in each domain do not change, and that the average tetragonal lattice parameter as observed by x-ray diffraction results from a coherence effect between adjacent microdomains. Thus, the impurity-induced structural phase transition is one in which the Cu1-O ordered chain structure persists over macroscopically small but microscopically large length scales.

Numerous Mössbauer investigations of $^{57}$Fe have been undertaken; the high temperature results vary, exhibiting either two [178,180], three [172,181–183], four [179,183], or even five [186] electric quadrupole split doublets; the relative intensity of the spectral components are observed to vary with impurity concentration and oxygen content. Attempts to derive site occupancies from these data alone are assumption-dependent, but, since there exists strong evidence from neutron diffraction [172,184,192,194] of predominant Cu1 occupancy by Fe and Co ions, one is led to associate the various doublets with different oxygen coordination numbers at the Cu1 site. Although some authors have assumed that oxygen depletion and the apparent tetragonal structure imply the possibility of smaller than 4-fold coordination [186], the usually three Mössbauer doublets are generally assigned to Fe1 coordinations of 6 (octahedral), 5 (pyramidal), and 4 (distorted tetrahedral) anions. Indeed, variation in oxygen content around $\delta = 0$ alters the relative intensity of the doublets assigned to the more common octahedral and pyramidal configurations. There exists some evidence from electron diffraction [184] that linear clustering of impurity ions occurs; such arrangements permit higher coordination numbers for a given oxygen stoichiometry. Of course, severe oxygen depletion results in oxygen loss within the microdomains, where, as in the impurity-free situation, the oxygen deficiency can vary from $\delta = 0$ to 1. Conversely, annealing the Co-doped samples in high pressure oxygen [191] improves superconductivity, probably by increasing the hole concentration and smoothing microdomain boundaries. Indeed, for impurity concentrations as high as $x = 0.05$ (~15% Co for Cu1), $T_c \approx 80$ K can be attained [191].

An interesting feature of the YBa$_2$(Cu$_{1-x}$Fe$_x$)$_3$O$_{7-\delta}$ system is the appearance of magnetic order at low temperatures. Mössbauer spectra [181–183] exhibit broadening below ~ 20 K; by 4.2 K, the clearly split spectra indicate that hyperfine fields of order 250 kOe occur. Also, specific heat measurements [190] exhibit low temperature features which are probably associated with the magnetic ordering of Fe ions. Because this coexistence of superconductivity and magnetism occurs even when Fe is substituted for only about 1% of the Cu, some authors [181,182] interpret this behavior as possibly indicative of long-range magnetic order on the Cu sublattices and suggestive that Cu spin fluctuations may be important for high-$T_c$ superconductivity. However, it is also plausible that these features arise from the clustering of Fe atoms into short chains, as discussed above [184]. In such a case of inhomogeneous magnetism, the relationship to supeconductivity would be harder to assess.

Significantly, doping with the non-magnetic trivalent substituents M = Ga or Al in $YBa_2(Cu_{1-x}M_x)_3O_{7-\delta}$ results in behavior quite similar to that which occurs with Fe or Co substitution. The apparent (as observed by diffraction techniques) orthorhombic-tetragonal structural transition again occurs, and $T_c$ suppression is slight [35,172,188, 201,202]. Thus it may be a fair assumption that all trivalent dopants with sufficiently small ionic radii primarily substitute on the Cu1 site. The suppression of $T_c$, however, is even less pronounced for Ga and Al doping than in the cases of Fe and Co substitutions. Thus, magnetic pair-breaking, while not overwhelming, may contribute to the excess $T_c$ degradation observed for M = Fe and Co. Superconducting transition temperatures as a function of dopant concentration are shown in Fig. 28 for a variety of substituent species.

In contrast to the trivalent dopants, substitution of divalent Zn for Cu results in unusually rapid depression of $T_c$, with typical values of ~ 11 K/at.%. A single-phased material can be maintained up to about x = 0.10–0.16 in $YBa_2(Cu_{1-x}Zn_x)_3O_{7-\delta}$ [172,199,201]. The structure, however, remains macroscopically orthorhombic up to the highest dopant concentrations. Some disagreement pervades the question of Zn site occupancy, with neutron diffraction measurements [192] indicating that Zn populates both sites and somewhat prefers Cu1 sites, while other authors express some conviction that Zn must substitute for Cu2 (plane) sites [172,199]. Generally, authors claiming preferential Cu2 occupancy call upon the maintained orthorhombicity [199] and the absence of oxygen-removal effects [172] as evidence that Zn does not prefer the chain sites, and upon the large $T_c$ depression as evidence that Zn does prefer the plane sites. However, no such definite conclusions can readily be drawn; since Zn often prefers tetrahedral coordination [189], the usual 4-fold coordinated Cu1 site can be occupied without changing the oxygen content [191] and without seriously altering the structure

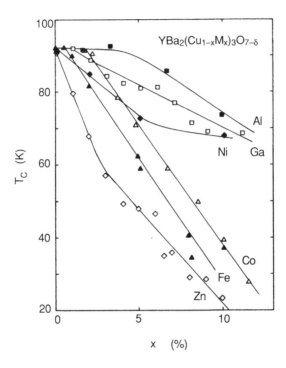

**Fig. 28.** Superconducting transition temperature $T_c$ vs dopant concentration x for YBa$_2$(Cu$_{1-x}$M$_x$)$_3$O$_{7-\delta}$. Data for M = Al (solid squares, Ref. 172), Ga (open squares, Ref. 199), Ni (solid diamonds, Ref. 191), Co (open triangles, Ref. 193), Fe (solid triangles, Ref. 173), and Zn (open diamonds, Ref. 199) are shown.

(i.e., the oxygen vacancy order) of the Cu1-O plane. Thus it is remarkable that only the small fraction (~1/3) of the Zn ions which do go to Cu2 sites [192] may be responsible for the severe degradation of superconducting properties that are observed for this dopant.

The unique behavior of the nonmagnetic M = Zn component of the YBa$_2$(Cu$_{1-x}$M$_x$)$_3$O$_{7-\delta}$ system is often attributed to its filled $d$-level electronic structure. Zinc substitution eliminates $3d$ holes [200]; such band-filling decreases the density of states at the Fermi energy. The Zn ions may also cause local distortions in the Cu2 planes [35,200]. Substitutional impurities often act as defects which are capable of inducing magnetic moments through localization effects; the development of such a moment with increasing x has been observed [35,201] in YBa$_2$(Cu$_{1-x}$Zn$_x$)$_3$O$_{7-\delta}$, and could conceivably result in magnetic pair-breaking. Alternatively, the *absence* of magnetic ions may disrupt superconductivity if the mechanism depends on antiferromagnetism or spin fluctuations. Another suggestion [189] proposes that Zn doping at the Cu2 site may induce oxygen defects in the BaO layer, which are believed to suppress $T_c$ markedly [205].

Somewhat surprising among the Cu substitutions is the behavior of $YBa_2(Cu_{1-x}M_x)_3O_{7-\delta}$ for M = Ni. Unlike the Zn-substituted species, neutron diffraction measurements [192] indicate that Ni substitutes only at the Cu2 site, even though square planar coordination is fairly common for Ni [196]. However, for this magnetic ion, superconducting properties degrade *less* quickly than in the case of nonmagnetic Zn [35,189,191]. Solubility of Ni to about x = 0.20 is observed [196] and, as expected when the Cu1 sites are undisturbed, x-ray diffraction indicates that orthorhombicity is maintained for all dopant concentrations [35,172,188,196,200]. It is remarkable that superconductivity persists even at a 10% Ni-for-Cu substitution level (i.e., 15% for Cu2), since in transition metals less than one percent of magnetic impurity ions will often destroy superconductivity. Since, here, the magnetic ion substitution apparently presents a relatively small perturbation to the Cu2 planes, such magnetic character may be intrinsic to high $T_c$ superconductivity.

## J. Non-isovalent Substitutions for Y and Ba in $YBa_2Cu_3O_{7-\delta}$

These substitutions generally fall into three categories: 1) substitutions for Y by divalent and monovalent ions, i.e., $(Y_{1-x}M_x)Ba_2Cu_3O_{7-\delta}$ for M = Ca or Na; 2) substitution for Ba by trivalent ions, especially M = La in $Y(Ba_{2-x}M_x)Cu_3O_{7-\delta}$; and 3) substitutions similar to 2), but also with Y replaced by the lanthanide element M = Ln, i.e., $Ln_{1+x}Ba_{2-x}Cu_3O_{7-\delta}$. These systems will be considered in turn. The substitution of Y by Pr, which may not be simply trivalent, has been discussed in Sec. III-G, and will not be reconsidered here.

Substitution of trivalent Y by ions M of lesser valence in $(Y_{1-x}M_x)Ba_2Cu_3O_{7-\delta}$ has been performed for M = Ca [206–208] and M = Na [51]. The limit of solubility of Ca for Y has been variously reported as x = 0.2 [207,208] and x = 0.5 [206], although it has also been claimed that a smaller, concurrent substitution of the Ca for Ba is necessary to obtain a single phase [208]. The structure remains orthorhombic over the Ca dopant range, although the orthorhombic distortion somewhat decreases [207,208]. Superconducting transition temperatures remain high. Indeed, since the Ca ion

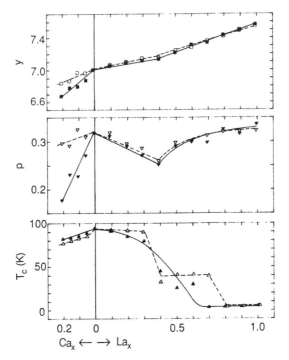

**Fig. 29.** Composition dependence of oxygen content y (squares), hole concentration per copper ion *p* (triangles), and superconducting transition temperature $T_c$ (circles) vs concentration x for $(Y_{1-x}Ca_x)Ba_2Cu_3O_{7-\delta}$ (left) and $Y(Ba_{2-x}La_x)Cu_3O_{7-\delta}$ (right). Closed symbols are for samples furnace-cooled in air; open symbols are for samples annealed at 380°C for 24 hr in $O_2$. From Ref. 207.

contributes one less electron, the doping is adding holes; as a result, there is a tendency for oxygen enrichment to *decrease* $T_c$ [207] (see Fig. 29), and even samples with oxygen deficiencies as severe as $\delta = 0.5$ can maintain $T_c \approx 80$ K [206]. Similar behaviors have been observed for isostructural compounds of the form $(Ln_{1-x}Ca_x)Ba_2Cu_3O_{7-\delta}$ for Ln = La and Nd [209]. A decrease in $T_c$ at high hole concentrations was also noted for $La_{2-x}M_xCuO_{4-\delta}$ in Sec. II (see Fig. 3).

Substitution of Na for Y has been reported to have a higher solubility limit of x = 0.5 [51]; for this $(Y_{1-x}Na_x)Ba_2Cu_3O_{7-\delta}$ system, $T_c$ degrades little up to the x = 0.3 level, where $T_c \approx 80$ K, but drops quickly thereafter to $T_c \approx 40$ K by x = 0.5. The development of a magnetic moment with increasing x was observed [51], which, if not due to impurity phases, may indicate a conversion of $Cu^{3+}$ to $Cu^{2+}$. Such behavior probably reflects a tendency toward greater oxygen deficiency with increasing dopant concentration; under typical annealing conditions, such a trend was also observed for M = Ca [207], as is evident in Fig. 29.

A primary concern of these non-isoelectronic substitutions is to elucidate the relationship between superconductivity and such charge-dependent quantities as hole concentration and copper valence. Thus, information from the above "hole-doping" experiments performed by Ca (or Na) substitution should be combined with that from the "hole-filling" substitutions, now to be described, in order to gain a consistent viewpoint of charge-regulating effects.

Substitutions of La for Ba of the form $Y(Ba_{2-x}La_x)Cu_3O_{7-\delta}$ have been reported with limiting solubilities in the range x = 0.7–1.0 [206,207,210,211]. For both air-cooled and oxygen-annealed samples, the oxygen content y = 7–$\delta$ increases to well above 7 (negative $\delta$), reaching as high as y = 7.4–7.5 for x ≈ 0.8 [207,210] (see Fig. 29). An orthorhombic-tetragonal structural transition is observed [207,210,211] near x = 0.4 in these samples, probably due to the random disordering of *excess* oxygen in the CuO chains. For air-cooled samples, the superconducting transition temperature remains relatively constant from x = 0 to 0.3, then drops rapidly for higher x. However, oxygen content can be varied by annealing in various partial pressures of oxygen, and it is found that there is a sensitive interplay of the effects of dopant charge and oxygen content in determining electronic properties; there have been attempts to sort out this interdependence. In doing so, other mechanisms, such as the effects of disorder and the possibility of exceptional oxygen occupancies are generally not considered. Typically, some relevant charge-counting parameter is identified; the common choice is often termed the hole concentration $p$ (sometimes referred to as the excess copper valence or the average [Cu-O] valence), which is directly measurable by idiometric titration techniques.

In Ref. 206, Tokura et al. have peformed a systematic variation of both the oxygen deficiency $\delta$ and the dopant concentration x for both the $(Y_{1-x}Ca_x)Ba_2Cu_3O_{7-\delta}$ and $Y(Ba_{2-x}La_x)Cu_3O_{7-\delta}$ systems. From their data, a critical insulator-superconductor phase boundary can be identified. It is roughly linear and relates the critical values of $\delta$ and $p$ approximately according to the relation: $3p = 0.5 - \delta$. This surprisingly simple relation indicates that the oxygen content changes the minimum critical hole concentration necessary for the onset of superconductivity. It is significant that no particular hole concentration is crucial for superconductivity, and that oxygen content does not simply regulate the hole concentration, as in the case of $La_{2-x}M_xCuO_{4-\delta}$ (see Sec. II). However, further extrapolations to a precise physical picture necessitate assumptions concerning considerations such as: 1) which holes are itinerant, 2) to what degree the chains and plains can accomodate localized holes, and 3) to what extent layered charge imbalances can be sustained.

Conjecture about these matters can lead to several self-consistent schemes; the authors of Ref. 206 propose that the onset of superconductivity corresponds to the first appearance of holes on the planes, and that the chains act only as a reservoir for localized charge. An alternative picture which minimizes layered charge imbalance would presume that half of the plane copper sites can accomodate localized holes. Confirmation of one such viewpoint may eventually emerge from extensive Hall effect studies of well-characterized samples as both dopant concentration and oxygen content are systematically varied. Some rather preliminary results are described below.

Other than M = La, other rare earth elements are not appreciably soluble for Ba in the yttrium-based compound $Y(Ba_{2-x}M_x)Cu_3O_{7-\delta}$ [212,213], apparently because La is closest to Ba in ionic radius. Surprisingly, however, extensive ranges of solid solution do occur in the lanthanide-based compounds. Compounds of the form $Ln(Ba_{2-x}Ln_x)Cu_3O_{7-\delta} = Ln_{1+x}Ba_{2-x}Cu_3O_{7-\delta}$ have been synthesized where Ln = La [132,205,214,215], Nd [209,216–224], Sm [212,216,224,225], Eu [216,224–226], and Gd [216]. The lanthanide-barium upper solubility limit is observed to decrease from $x \geq 0.5$ (possibly $x \approx 0.8$) for La, Nd, and Sm, to $x \approx 0.4$ for Eu, and to $x \approx 0.2$ for Gd [216]; the decreasing solubility may be attributed to the decreasing ionic radii of these ions, and thus to the increasing substitutional mismatch. For Ln = Dy, no appreciable solubility is detected [216]. These rare-earth solubility ranges are depicted in Fig. 30.

Similar to $Y(Ba_{2-x}La_x)Cu_3O_{7-\delta}$, an orthorhombic-tetragonal structural transition is observed in $Ln_{1+x}Ba_{2-x}Cu_3O_{7-\delta}$, which now occurs near $x \approx 0.2$–0.3 [216,217,221, 224,226]. This transition has been correlated with excess oxygen content [224]; it is claimed that $y = 7.10 \pm 0.01$ is the minimum oxygen occupancy required for the increase in symmetry. A very significant feature of this series of compounds is the universal behavior of $T_c$ for different lanthanide elements; this indicates that, as is the case with the rare-earth site, magnetic moments on the alkaline-earth site have little or no effect on the superconductivity [216,221,225].

Typical oxygen-regulation effects and $T_c$ behavior for this series are similar to those described for $Y(Ba_{2-x}La_x)Cu_3O_{7-\delta}$ above; oxygen content increases beyond $y = 7$ for increasing x, while $T_c$ drops [205,215,221]. The compensating oxygen ions may reside at those particular (normally vacant) chain-layer sites which are cosest to the $La^{3+}$ dopant ions, thus providing electrostatic shielding and aiding local charge balance. It has been

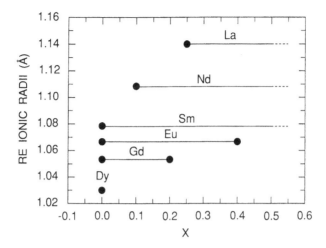

**Fig. 30.** Solid solubility regions for the $Ln(Ba_{2-x}Ln_x)Cu_3O_{7-\delta}$ systems with $Ln = La$, Nd, Sm, Eu, Gd, and Dy. The solid lines indicate the region in which a single phase was observed, while the dashed lines indicate the possible extension of the single phase into regions not studied. From Ref. 216.

further suggested [214] that such an arrangement may favor the occurrence of oxygen vacancies in the Ba-O layers, which could be responsible for the rapid decrease in $T_c$; however, the existence of such vacancies is not supported by neutron diffraction measurements [215], although such deficiencies, if extremely small, would be difficult to detect.

We now return to consideration of the role of the "hole concentration" $p$ in determining superconducting properties. As indicated above, it has been established [206] that in the substituted 1-2-3 compounds, the hole concentration alone does not determine $T_c$. In other studies [223,225], fairly successful attempts have been made to correlate $T_c$ with $p$; however, since the oxygen contents of these samples were undetermined, the inferred values of $p$ may be unreliable. Such fortuitous results probably represent a relatively constant oxygen-content cross-section of a larger family of such $T_c$ vs $p$ curves. Some insight is provided by Hall effect measurements, which provide a measure of the effective density of *itinerant* holes. Although again in a study of samples with undetermined oxygen content (and thus undetermined total hole concentration), the superconducting transition temperature has been found [222] to increase smoothly with increasing inverse Hall coefficient for $Nd_{1+x}Ba_{2-x}Cu_3O_{7-\delta}$.

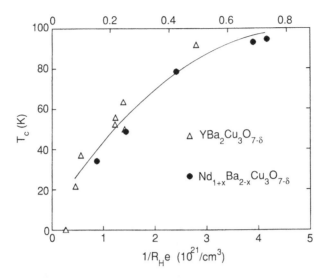

**Fig. 31.** Superconducting transition temperature $T_c$ vs the effective itinerant hole concentration $p_H = 1/R_H e$ as determined by Hall effect measurements for $YBa_2Cu_3O_{7-\delta}$ (triangles, Ref. 227) and $Nd_{1+x}Ba_{2-x}Cu_3O_{7-\delta}$ (circles, Ref. 222). Solid line is a guide to the eye. Upper abscissa is $p_H$ in units of holes per unit cell.

These data are shown in Fig. 31 along with similar data [227] for $YBa_2Cu_3O_{7-\delta}$, where the oxygen content alone governs $1/R_H$ and $T_c$. The Hall coefficients shown were measured at $T \sim 100$ K, where thermal activation of localized holes is relatively small. An understanding of this dependence of $T_c$ on itinerant hole concentration, reaching $T_c \sim$ 90 K for $3p \approx 0.5$–0.7, as well as insight into the temperature dependence of the carrier density, would provide a clearer physical picture of the conduction process. Increasingly systematic and accurate measurements of $R_H(T)$ for various $\delta$ and x in $Ln_{1+x}Ba_{2-x}Cu_3O_{7-\delta}$ and $(Y_{1-x}Ca_x)Ba_2Cu_3O_{7-\delta}$ will no doubt provide such clarification.

## IV. BISMUTH AND THALLIUM COPPER-OXIDE SUPERCONDUCTORS

The highest transition temperature superconductors to date are the members of the Bi-Sr-Ca-Cu-O and Tl-Ba-Ca-Cu-O systems. Many compounds with similar features are known. The bismuth compounds can be described by $Bi_2Sr_2Ca_{n-1}Cu_nO_{2n+4}$ and the thallium compounds similarly by $Tl_mM_2Ca_{n-1}Cu_nO_{m+2n+2}$, where $m = 1$ or 2, n = 1, 2, 3, or 4, and M = Ba or Sr. Historically, superconductivity with $T_c \approx 10$ K in $Bi_2Sr_2CuO_6$ was discovered first [228], followed by reports of much higher $T_c$'s (90–125 K) in multiphase Bi-Sr-Ca-Cu-O and Tl-Ba-Ca-Cu-O samples [229–231]. Like other high temperature superconductors, these materials possess layered structures, but apparently permit a great deal of intergrowth of multiple structures with different m and n. Additionally, site disorder is common, so that even so-called single crystals typically possess both structural and compositional disorder. For a more comprehensive review of structure, composition, and superconductivity in these materials, the reader is referred to Ref. 232.

Substitution of Pb for Bi or Tl is possible, although superconductivity is lost for high dopant levels [233]. Because of the tendency toward inhomogeneity, systematic substitutional studies are difficult. Even in the nominally ideal systems, such as $Bi_2Sr_2Ca_1Cu_2O_8$, an appreciable number of Bi (or Tl) vacancies are present, and Ca partially occupies Bi (Tl) sites. Similarly, the divalent alkaline earth sites exhibit some site disorder and appreciable solubility of, for example, additional strontium for calcium.

It is apparent from structural studies that atoms in the Bi (Tl) layers show highly correlated, large displacements from their ideal positions [234]. Such displacements arise from the size mismatch between the $CuO_2$ and BiO (TlO) layers; the copper-oxygen distances are essentially fixed, causing the bismuth-oxygen and thallium-oxygen distances to be larger than can usually be accomodated. One typical displacement which alleviates this situation entails the formation of two short and two long Bi-O (Tl-O) bonds. Such displacements generally take place in a plane perpendicluar to the c-axis; however, displacements parallel to the c-axis also occur [235].

The $CuO_2$ layers are structurally and compositionally more ideal than the other layers in the Bi and Tl systems. An interesting observation concerns the dimpling of the $CuO_2$ sheets, also observed in the other copper oxide superconductors. In the case of the $(Bi,Tl)_mM_2Ca_{n-1}Cu_nO_{m+2n+2}$ system, however, one observes a full spectrum of dimpling behavior: 1) oxygen atoms alternately above and below the copper plane, as in orthorhombic $La_2CuO_4$, 2) oxygen atoms all in one plane, but displaced from the copper plane, as in $YBa_2Cu_3O_{7-\delta}$, or 3) $CuO_2$ sheets that are essentially flat [236]. Thus the particular value of the copper-oxygen bond angle is probably not important for high temperature superconductivity, although appreciable disorder in the Cu-O bonds is believed to lower $T_c$.

The proliferation of copper oxide superconducting compounds with various numbers of $CuO_2$ sheets gave rise to some interesting speculation concerning a possible positive correlation between the number of such sheets and the transition temperature [237,238]. However, such a simple correlation probably does not apply, as there exist some series of samples which serve as counterexamples or at least indicate the saturation of such a tendency. For example, with the n = 3 and n = 4 cases, $Tl_2Ba_2Ca_2Cu_3O_{10}$ has $T_c$ = 122 K and $Tl_2Ba_2Ca_3Cu_4O_{12}$ has $T_c$ = 119 K [232].

This rich system of superconducting compounds, already including several dozen members, will no doubt grow as new substitutions are made. The Bi (Tl) compounds have established that the CuO chains in $YBa_2Cu_3O_{7-\delta}$ compounds are not essential for superconductivity at 90 K, a significant result. Also, the absence of a linear term in the specific heat has been reported for the Bi (but not Tl) compound [239], although the subtraction of several other complicating contributions makes such a determination difficult. Aside from this, the Bi (Tl) compounds do share much of the behavior of the other copper oxide superconductors: the presence of $CuO_2$ sheets, an increase in the superconducting transition temperature with applied pressure [240], and an antiferromagnetic insulating ground state at sufficiently low hole concentrations [232].

## V. BISMUTH-OXIDE SUPERCONDUCTORS: $Ba_{1-x}K_xBiO_{3-\delta}$

The discovery of superconductivity at ~ 30 K in the cubic perovskite $Ba_{1-x}A_xBiO_{3-\delta}$ system [241,242], where A = K, Rb, or Cs, is of fundamental interest; this new system is closely related to the $BaPb_xBi_{1-x}O_3$ superconductor ($T_c$ ~ 13 K at x = 0.75) discovered by Sleight et al. [243] and, most importantly, is the first copper-free compound with a $T_c$ higher than the best intermetallic superconducting compound. The system is also somewhat analogous to the $La_{2-x}M_xCuO_{4-\delta}$ materials, in which carriers are introduced in the parent $La_2CuO_4$ compound by substituting on the *inactive* La site. Similarly, substitutional doping of alkali metals, such as K or Rb, at the inactive Ba donor site of $BaBiO_3$ leads to a variation of the conduction band filling and induces superconductivity. Single phase perovskites have been prepared for $Ba_{1-x}K_xBiO_{3-\delta}$ in the range 0 < x < 0.5, with the highest $T_c$ and largest superconducting fraction for x = 0.4 [242]. Although both the bismuth-oxide and copper-oxide superconductors are based on perovskite-type structure and both have insulating parent compounds, there are several remarkable differences between them. First, the plane-like two-dimensional layered structure of the cuprates is missing; instead, the Bi-O conducting complex forms a three-dimensional network. Second, superconductivity occurs in the cuprate materials as the antiferromagnetism of the parent compound is suppressed, whereas in the bismuth compound, it occurs as the charge density wave which doubles the unit cell and opens a gap near the Fermi level in $BaBiO_3$ is suppressed [244]. Third, for the Cu-O superconductors the Cu(3$d$)-O(2$p$) bands are located at the Fermi level, whereas, for the Bi-O superconductors, the Bi(6$s$)-O(2$p$) bands are the ones near the Fermi level; the substitution of K on the Ba site has a minimal effect on the electronic states near $E_F$ and extends the metallic region of this system closer to a half-filled band [245].

Measurements of the electrical resistivity $\rho$ in the Ba-K-Bi-O system reveal a slightly negative normal-state temperature coefficient [246,247], as is also observed for Ba(Pb,Bi)O$_3$; data above 160 K can be described by $\log(\rho) \propto T^{-1/4}$ [248], suggesting phonon-assisted variable range hopping conduction. Measurements of $\rho$(T,H) reveal a positive curvature for the upper critical magnetic field and yield an initial slope as high as $-dH_{c2}/dT \approx 0.8$ T/K, of the same order as that of Ba(Pb,Bi)O$_3$ and the cuprate superconductors [247]. From extrapolation of $H_{c2}$ to T = 0, a lower limit of ~ 40 Å is deduced for the coherence length, which is much larger than the unit cell dimensions. The $^{18}$O isotope effect was determined for Ba$_{0.625}$K$_{0.375}$BiO$_{3-\delta}$ and was found to be as large ($\alpha = 0.4$) as in BaPb$_x$Bi$_{1-x}$O$_3$, indicating a probable phonon-mediated pairing mechanism [249]. The oxygen content in the K-doped BaBiO$_3$ is expected to be close to the ideal value ($\delta \approx 0$).

The question of whether or not there is a common superconducting mechanism for the bismuth-oxide and copper-oxide systems is an important one. It has been proposed that, in the bismuth oxide case, pairing is less likely to have a magnetic origin but instead may be one in which phonon [245] or bipolaronic [250] mechanisms are involved.

## VI. CONCLUDING REMARKS

In this article, we have endeavored to present an overview of the progress that has been made during the past two years in characterizing the magnetic, thermodynamic, and transport properties of the new high $T_c$ oxide superconductors as functions of temperature, magnetic field, and external pressure as well as the response of these properties to partial and complete chemical substitution. An important objective of this research is the establishment of systematic relationships between the occurrence of high $T_c$ superconductivity and certain characteristics of the materials, such as the electronic configuration, size, and valence of their atomic consituents, their crystal structures, the charge carrier concentration, and the electronic, lattice, and magnetic properties. These empirical "rules" are useful guides in the search for new high $T_c$ superconducting materials and they provide the "experimental constraints" that a successful theory of high $T_c$ superconductivity is obliged to satisfy.

Based on the rapid developments that have occurred since the discovery of high $T_c$ superconductivity in the La-Ba-Cu-O system, it seems clear that new materials with yet higher $T_c$'s than the present 122 K maximum for compounds in the Tl-Ba-Ca-Cu-O system await to be discovered. Hopefully, we have just seen the "tip of the iceberg." The unexpectedly high values of $T_c$, the proximity of antiferromagnetism, and some unusual superconducting properties suggest that the new high $T_c$ oxide compounds may exhibit a new type of superconductivity involving a novel non-phonon, and possibly magnetic, electron pairing mechanism. Further experimentation will certainly yield new information concerning the nature and origin of the high $T_c$ superconductivity of these remarkable materials, and progress toward developing a viable theory of high temperature superconductivity for the oxides can be anticipated.

## ACKNOWLEDGMENTS

We would like to thank E. A. Early, B. W. Lee, and H. Zhou for assistance in assembling the figures and B. D. Dunlap for useful suggestions. This work was supported by the U. S. Department of Energy under Grant No. DE-FG03-86ER45230 and the U. S. National Science Foundation under Grant. No. DMR-8411839.

# REFERENCES

1. J. G. Bednorz and K. A. Müller, Z. Phys. B **64**, 189 (1986).

2. M. K. Wu, J. R. Ashburn, C. J. Torng, P. H. Hu, R. L. Meng, L. Gao, Z. J. Huang, Y. Q. Wang, and C. W. Chu, Phys. Rev. Lett. **58**, 908 (1987).

3. T. Fujita, Y. Aoki, Y. Maeno, J. Sakurai, H. Fukuba, and H. Fujii, Jpn. J. Appl. Phys. **26**, L202 (1987).

4. H. Takagi, S. Uchida, K. Kitazawa, and S. Tanaka, Jpn. J. Appl. Phys. **26**, L123 (1987).

5. R. J. Cava, R. B. van Dover, B. Batlogg, and E. A. Rietman, Phys. Rev. Lett. **58**, 408 (1987).

6. K. Kishio, K. Kitazawa, S. Kanbe, T. Yasuda, N. Sugii, H. Takagi, S. Uchida, K. Fueki, and S. Tanaka, Chem. Lett. **1987**, 429.

7. R. B. van Dover, R. J. Cava, B. Batlogg, and E. A. Rietman, Phys. Rev. B **35**, 5337 (1987).

8. M. W. Shafer, T. Penney, and B. L. Olson, Phys. Rev. B **36**, 4047 (1987).

9. J. B. Torrance, Y. Tokura, A. I. Nazzal, A. Bezinge, T. C. Huang, and S. S. P. Parkin, Phys. Rev. Lett. **61**, 1127 (1988).

10. J. T. Markert, C. L. Seaman, H. Zhou, and M. B. Maple, Solid State Commun. **66**, 387 (1988); J. T. Markert, S. Ghamaty, B. W. Lee, M. B. Maple, J. J. Neumeier, C. L. Seaman, and H. Zhou, to be published.

11. M. A. Subramanian, J. Gopalakrishnan, C. C. Torardi, T. R. Askew, R. B. Flippen, A. W. Sleight, J. J. Lin, and S. J. Poon, Science **240**, 495 (1988).

12. S. M. Fine, M. Greenblatt, S. Simizu, and S. A. Friedberg, Phys. Rev. B **36**, 5716 (1987).

13. S. A. Shaheen, N. Jisrawi, Y. H. Lee, Y. Z. Zhang, M. Croft, W. L. McLean, H. Zhen, L. Rebelsky, and S. Horn, Phys. Rev. B **36**, 7214 (1987).

14. P. M. Grant, S. S. P. Parkin, V. Y. Lee, E. M. Engler, M. L. Ramirez, J. E. Vazquez, G. Lim, R. D. Jacowitz, and R. L. Greene, Phys. Rev. Lett. **58**, 2482 (1987).

15. K. Sekizawa, Y. Takano, H. Takigami, S. Tasaki, and T. Inaba, Jpn. J. Appl. Phys. **26**, L840 (1987).

16. Y. Nishihara, M. Tokumoto, K. Murata, and H. Unoki, Jpn. J. Appl. Phys. **26**, L1416 (1987).

324

17. J. M. Tarascon, L. H. Greene, B. G. Bagley, W. R. McKinnon, P. Barboux, and G. W. Hull, in *Novel Superconductivity*, S. A. Wolf and V. Z. Kresin, eds. (Plenum Press, New York, 1987), pp. 705–724.

18. J.-M. Tarascon and B. G. Bagley, to appear in MRS Bulletin.

19. R. D. Shannon, Acta. Cryst. **A32**, 751 (1976).

20. K. Kishio, K. Kitazawa, N. Sugii, S. Kanbe, K, Fueki, H. Takagi, and S. Tanaka, Chem. Lett. **1987**, 635.

21. P. Ganguly, R. A. Mohan Ram, K. Sreedhar, and C. N. Rao, Solid State Commun. **62**, 807 (1987).

22. A. P. Malozemoff, Mat. Res. Bull. **22**, 701 (1987).

23. N. P. Ong, Z. Z. Wang, J. Clayhold, J. M. Tarascon, L. H. Greene, and W. R. McKinnon, Phys. Rev. B **35**, 8807 (1987).

24. C. Uher, A. B. Kaiser, E. Gmelin, and L. Walz, Phys. Rev. B **36**, 5676 (1987).

25. R. C. Yu, M. J. Naughton, X. Yan, P. M. Chaikin, F. Holtzberg, R. L. Greene, J. Stuart, and P. Davies, Phys. Rev. B **37**, 7963 (1988).

26. J. M. Ferreira, B. W. Lee, Y. Dalichaouch, M. S. Torikachvili, K. N. Yang, and M. B. Maple, Phys. Rev. B **37**, 1580 (1988).

27. M. Kato, Y. Maeno, and T. Fujita, Physica C **152**, 116 (1988).

28. K. Kumagai, Y. Nakamichi, I. Watanabe, Y. Nakamura, H. Nakajima, N. Wada, and P. Lederer, Phys. Rev. Lett. **60**, 724 (1988).

29. K. Kumagai, Y. Nakamura, I. Watanabe, Y. Nakamichi, and H. Nakajima, preprint.

30. A. R. Moodenbaugh, Y. Xu, and M. Suenaga, Phys. Rev. B **38**, 4596 (1988).

31. T. Hasegawa, K. Kishio, M. Aoki, A. Ooba, K. Kitazawa, K. Fueki, S. Uchida, and S. Tanaka, Jpn. J. Appl. Phys. **26**, L337 (1987).

32. K. Kishio, K. Kitazawa, T. Hasegawa, M. Aoki, K. Fueki, S. Uchida, and S. Tanaka, Jpn. J. Appl. Phys. **26**, L391 (1987).

33. J. M. Tarascon, L. H. Greene, W. R. McKinnon, and G. W. Hull, Solid State Commun. **63**, 499 (1987).

34. G. W. Crabtree, W. K. Kwok, A. Umezawa, L. Soderholm, L. Morss, and E. E. Alp, Phys. Rev B **36**, 5258 (1987).

35. J. M. Tarascon, L. H. Greene, P. Barboux, W. R. McKinnon, G. W. Hull, T. P. Orlando, K. A. Delin, S. Foner, and E. J. McNiff, Jr., Phys. Rev. B **36**, 8393 (1987).

36. W. Kang, H. J. Schulz, D. Jérome, S. S. P. Parkin, J. M. Bassat, and Ph. Odier, Phys. Rev. B **37**, 5132 (1988).

37. K. Muraleedharan and D. Rambabu, Phys. Rev. B **36**, 8918 (1987).

38. Y. Saito, T. Noji, A. Endo, N. Matsuzaki, and S. Katsumata, Jpn. J. Appl. Phys. **26**, L223 (1987).

39. See, for example, J. D. Jorgensen, Jpn. J. Appl. Phys. **26**, 2017 (1987), Suppl. 26-3; and references cited therein.

40. K. N. Yang, Y. Dalichaouch, J. M. Ferreira, B. W. Lee, J. J. Neumeier, M. S. Torikachvili, H. Zhou, M. B. Maple, and R. R. Hake, Solid State Commun. **63**, 515 (1987).

41. K. N. Yang, Y. Dalichaouch, J. M. Ferreira, R. R. Hake, B. W. Lee, M. B. Maple, J. J. Neumeier, M. S. Torikachvili, and H. Zhou, in *Proc. Materials Research Society, Anaheim Meeting*, D. U. Gubser and M. Schlüter, eds. (Materials Research Society, Pittsburgh, PA, 1987), pp. 77–79.

42. E. M. Engler, V. Y. Lee, A. I. Nazzal, R. B. Beyers, G. Lim, P. M. Grant, S. S. P. Parkin, M. L. Ramirez, J. E. Vasquez, and R. J. Savoy, J. Am. Chem. Soc. **109**, 2848 (1987).

43. P. H. Hor, R. L. Meng, Y. Q. Wang, L. Gao, Z. J. Huang, J. Bechtold, K. Forster, and C. W. Chu, Phys. Rev. Lett. **58**, 1891 (1987).

44. M. B. Maple, K. N. Yang, M. S. Torikachvili, J. M. Ferreira, J. J. Neumeier, H. Zhou, Y. Dalichaouch, and B. W. Lee, Solid State Commun. **63**, 635 (1987).

45. Z. Fisk, J. D. Thompson, E. Zirngiebl, J. L. Smith, and S.-W. Cheong, Solid State Commun. **62**, 743 (1987).

46. S. Hosoya, S. Shamoto, M. Onoda, and M. Sato, Jpn. J. Appl. Phys. **26**, L325 (1987).

47. H. Takagi, S. Uchida, H. Sato, H. Ishii, K. Kishio, K. Kitazawa, K. Fueki, and S. Tanaka, Jpn. J. Appl. Phys. **26**, L601 (1987).

48. K. Kitazawa, K. Kishio, H. Takagi, T. Hasegawa, S. Kanbe, S. Uchida, S. Tanaka, and K. Fueki, Jpn. J. Appl. Phys. **26**, L339 (1987).

49. A. R. Moodenbaugh, M. Suenaga, T. Asano, R. N Shelton, H. C. Ku, R. W. McCallum, and P. Klavins, Phys. Rev. Lett. **58**, 1885 (1987).

50. M. B. Maple, Y. Dalichaouch, J. M. Ferreira, R. R. Hake, S. E. Lambert, B. W. Lee, J. J. Neumeier, M. S. Torikachvili, K. N. Yang, H. Zhou, Z. Fisk, M. W. McElfresh and J. L. Smith, In *Novel Superconductivity*, S. A. Wolf and V. Z. Kresin, eds. (Plenum, New York, 1987), pp. 839–853.

326

51. Y. Dalichaouch, M. S. Torikachvili, E. A. Early, B. W. Lee, C. L. Seaman, K. N. Yang, H. Zhou, and M. B. Maple, Solid State Commun. **65**, 1001 (1988).

52. K. N. Yang, B. W. Lee, M. B. Maple, and S. S. Laderman, Appl. Phys. A **46**, 229 (1988).

53. See, for example, M. B. Maple, Y. Dalichaouch, J. M. Ferreira, R. R. Hake, B. W. Lee, J. J. Neumeier, M. S. Torikachvili, K. N. Yang, H. Zhou, R. P. Guertin, and M. V. Kuric, Physica B **148**, 155 (1987).

54. J. J. Neumeier, Y. Dalichaouch, R. R. Hake, B. W. Lee, M. B. Maple, M. S. Torikachvili, K. N. Yang, R. P. Guertin, and M. V. Kuric, Physica C **152**, 293 (1988).

55. T. P. Orlando, K. A. Delin, S. Foner, E. J. McNiff, Jr., J. M. Tarascon, L. H. Greene, W. R. McKinnon, and G. W. Hull, Phys. Rev. B **36**, 2394 (1987).

56. H. Zhou, C. L. Seaman, Y. Dalichaouch, B. W. Lee, K. N. Yang, R. R. Hake, M. B. Maple, R. P. Guertin, and M. V. Kuric, Physica C **152**, 321 (1988).

57. M. B. Maple, in *Magnetism*, Vol. V, G. T. Rado and H. Suhl, eds. (Academic, New York, 1973), Ch. 10.

58. J. J. Neumeier, Y. Dalichaouch, J. M. Ferreira, R. R. Hake, B. W. Lee, M. B. Maple, M. S. Torikachvili, K. N. Yang, and H. Zhou, Appl. Phys. Lett. **51**, 371 (1987).

59. D. O. Welch, M. Suenaga, and T. Asano, Phys. Rev. B **36**, 2390 (1987).

60. A. J. Panson, A. I. Braginski, J. R. Gavaler, J. K. Hulm, M. A. Janocko, H. C. Pol, A. M. Stewart, J. Talvacchio, and G. R. Wagner, Phys. Rev. B **35**, 8774 (1987).

61. K. Takita, T. Ipposhi, and K. Masuda, Jpn. J. Appl. Phys. **26**, L668 (1987).

62. K. Takita, H. Akinaga, H. Katoh, T. Ishigaki, and H. Asano, Jpn. J. Appl. Phys. **26**, 1023 (1987).

63. G. W. Crabtree, private communication, 1988; Wai-Kwong Kwok, Ph. D. Thesis (Purdue University, 1987), unpublished.

64. N. R. Werthamer, E. Helfand, and P. C. Hohenberg, Phys. Rev. **147**, 295 (1966).

65. K. Maki, Phys. Rev. **148**, 362 (1966).

66. S. Foner, E. J. McNiff, Jr., and E. J. Alexander, Phys. Lett. A **49**, 269 (1974).

67. G. W. Crabtree, W. K. Kwok, and A. Umezawa, to appear in *Quantum Field Theory as an Interdisciplinary Basis*, F. C. Khanna, H. Umezawa, G. Kunstatter, and H. C. Lee, eds. (World Scientific, Singapore, 1988).

68. W. C. Lee, R. A. Klemm, and D. C. Johnston, preprint.

69. W. E. Lawrence and S. Doniach, *Proc. Twelfth Int. Conf. Low Temp. Phys.*, Kyoto, Japan, 1970, E. Kanda, ed. (Keigaku, Tokyo, 1971), p. 361.

70. P. Chaudhari, R. H. Koch, R. B. Laibowitz, T. R. McGuire, and R. J. Gambino, Phys. Rev. Lett. **58**, 2684 (1987).

71. C. P. Bean, Rev. Mod. Phys. **36**, 31 (1964).

72. J. M. Ferreira, M. B. Maple, H. Zhou, R. R. Hake, B. W. Lee, C. L. Seaman, M. V. Kuric, and R. P. Guertin, Appl. Phys. A **47**, 105 (1988).

73. J. W. Ekin, A. I. Braginski, A. J. Panson, M. A. Janocko, D. W. Capone II, B. Flandermeyer, O. F. deLima, M. Hong, J. Kwo, and S. H. Liou, J. Appl. Phys. **62**, 4821 (1987).

74. W. Weber, Phys. Rev. Lett. **58**, 1371 (1987).

75. B. Batlogg, G. Kourouklis, W. Weber, R. J. Cava, A. Jayaraman, A. E. White, K. T. Short, L. W. Rupp, and E. A. Rietman, Phys. Rev. Lett. **59**, 912 (1987).

76. T. A. Faltens, W. K. Ham, S. W. Keller, K. J. Leary, J. N. Michaels, A. M. Stacy, H.-C. Loye, D. E. Morris, T. W. Barbee, III, L. C. Bourne, M. L. Cohen, S. Hoen and A. Zettl, Phys. Rev. Lett. **59**, 915 (1987).

77. B. Batlogg, R. J. Cava, A. Jayaraman, R. B. van Dover, G. A. Kourouklis, S. Sunshine, D. W. Murphy, L. W. Rupp, H. S. Chen, A. White, K. T. Short, A. M. Mujsce, and E. A. Rietman, Phys. Rev. Lett. **58**, 2333 (1987).

78. L. C. Bourne, M. F. Crommie, A. Zettl, H. C. Loye, S. W. Keller, K. L. Leary, A. M. Stacy, K. J. Chang, M. L. Cohen, and D. E. Morris, Phys. Rev. Lett. **58**, 2337 (1987).

79. A. V. Inyushkin, N. A. Babushkina, V. V. Florentiev, A. V. Kopylov, V. I. Ozhogin, A. R. Kaul, and I. E. Graboy, in *Proc. Int. Discussion Meeting on High $T_c$ Superconductors*, Schloss Mauterndorf, Austria, Feb. 7–11, 1988, H. Weber, ed. (Plenum, New York, 1988).

80. See, for example, S. V. Vonsovsky, Y. A. Izyumov and E. Z. Kurmaev, *Superconductivity of Transition Metals*, (Springer-Verlag, Berlin, Heidelberg, New York, 1982), p. 180.

81. P. W. Anderson, Science **235**, 1196 (1987); P. W. Anderson, G. Baskaran, Z. Zou, and T. Hsu, Phys. Rev. Lett. **58**, 2790 (1987).

82. V. J. Emery, Phys. Rev. Lett. **58**, 2794 (1987).

83. See, for example, M. B. Maple, to appear in *Proc. Second Annual Conf. on Superconductivity and Applications*, Buffalo, New York, April 18–20, 1988.

84. See, for example, S. K. Sinha, MRS Bulletin **13**, 24 (1988).

85. J.-M. Tarascon, P. Barboux, B. G. Bagley, L. H. Greene, W. R. McKinnon, and G. W. Hull, in *Chemistry of High Temperature Superconductors*, D. L. Nelson, M. S. Wittingham, and T. F. George, eds. (American Chemical Society, Washington, D. C., 1987), p. 198.

86. R. J. Cava, B. Batlogg, C. H. Chen, E. A. Rietman, S. M. Zahurak, and D. Weber, Phys. Rev. B **36**, 5719 (1987).

87. N. Nishida, H. Miyatake, D. Shimada, S. Okuma, M. Ishikawa, T. Takabatake, Y. Nakazawa, Y. Kuno, R. Keitel, J. H. Brewer, T. M. Riseman, D. L. Williams, Y. Watanabe, T. Yamazaki, K. Nishiyama, K. Nagamine, E. J. Ansaldo, and E. Torikai, Jpn. J. Appl. Phys. **26**, L1856 (1987).

88. J. M. Tranquada, D. E. Cox, W. Kunnmann, H. Moudden, G. Shirane, M. Suenaga, P. Zolliker, D. Vaknin, S. K. Sinha, M. S. Alvarez, A. J. Jacobson, and D. C. Johnston, Phys. Rev. Lett. **60**, 156 (1988).

89. P. Burlet, C. Vettier, M. J. G. M. Jurgens, J. Y. Henry, J. Rossat-Mignod, H. Noel, M. Potel, P. Gougeon, and J. C. Levet, Physica C **153–155**, 1115 (1988).

90. J. W. Lynn, W.-H. Li, H. A. Mook, B. C. Sales, and Z. Fisk, Phys. Rev. Lett. **60**, 2781 (1988).

91. R. W. McCallum, D. C. Johnston, C. A. Luengo and M. B. Maple, J. Low Temp. Phys. **25**, 177 (1976).

92. A. Junod, A. Bezinge, D. Cattani, J. Cors, M. Decroux, P. Fischer, P. Genoud, L. Hoffman, J.-L. Jorda, J. Muller and E. Walker, in *Proc. 18th Int. Conf. Low Temp. Physics*, Kyoto, Japan, 1987; Jpn. J. Appl. Phys. **26**, Suppl. 26-3, 1119 (1987).

93. For a review, see R. A. Fischer, J. E. Gordon, and N. E. Phillips, to appear in J. Superconductivity, October, 1988.

94. R. Griessen, Phys. Rev. B **36**, 5284 (1987); and references cited therein.

95. M. W. McElfresh, M. B. Maple, K. N. Yang, and Z. Fisk, Appl. Phys. A **45**, 365 (1988).

96. J. M. Tarascon, W. R. McKinnon, L. H. Greene, G. W. Hull, and E. M. Vogel, Phys. Rev. B **36**, 226 (1987).

97. B. D. Dunlap, J. Magn. Magn. Mat. **37**, 211 (1983).

98. H. Zhou, S. E. Lambert, M. B. Maple, S. K. Malik, and B. D. Dunlap, Phys. Rev. B **36**, 594 (1987).

99. H. Zhou, S. E. Lambert, M. B. Maple, and B. D. Dunlap, to be published.

100. M. B. Maple, H. C. Hamaker, and L. D. Woolf, in *Superconductivity in Ternary Compounds II*, Vol. 34 of *Topics in Current Physics*, M. B. Maple and Ø. Fischer, eds. (Springer-Verlag, New York, 1982), Ch. 4.

101. B. D. Dunlap, M. Slaski, D. G. Hinks, L. Soderholm, M. Beno, K. Zhang, C. Segre, G. W. Crabtree, W. K. Kwok, S. K. Malik, I. K. Schuller, J. D. Jorgensen, and Z. Sungaila, J. Magn. Magn. Mater. **68**, L139 (1987).

102. A. Furrer, P. Brüesh, and P. Unternährer, Phys. Rev. B **38**, 4616 (1988).

103. J. D. Livingston, H. R. Hart, Jr., and W. P. Wolf, to appear in Proc. 4th Joint MMM-Intermag Conf., Vancouver, British Columbia, Canada, July 12–15, 1988.

104. D. E. Farrell, B. S. Chandrasekhar, M. R. DeGuire, M. M. Fang, V. G. Kogan, J. R. Clem, and D. K. Finnemore, Phys Rev. B **36**, 4025 (1987).

105. T. R. Dinger, T. K. Worthington, W. J. Gallagher, and R. L. Sandstrom, Phys. Rev. Lett. **58**, 2687 (1987).

106. G. W. Crabtree, J. Z. Liu, A. Umezawa, W. K. Kwok, C. H. Sowers, S. K. Malik, B. W. Veal, D. J. Lam, M. B. Brodsky, and J. W. Downey, Phys. Rev. B **36**, 4021 (1987).

107. G. Xiao, F. H. Streitz, A. Gavrin, M. Z. Cieplak, J. Childress, M. Lu, A. Zwicker, and C. L. Chien, Phys. Rev. B **36**, 2382 (1987).

108. J. R. Thompson, D. K. Christen, S. T. Sekula, J. Brynestad, and Y. C. Kim, J. Mater. Res. **2**, 779 (1987).

109. B. W. Lee, J. M. Ferreira, Y. Dalichaouch, M. S. Torikachvili, K. N. Yang, and M. B. Maple, Phys. Rev. B **37**, 2368 (1988).

110. A. P. Ramirez, L. F. Schneemeyer, and J. V. Waszczak, Phys. Rev. B **36**, 7145 (1987).

111. S. E. Brown, J. D. Thompson, J. O. Willis, R. M. Aikin, E. Zirngiebl, J. L. Smith, Z. Fisk, and R. B. Schwarz, Phys. Rev. B **36**, 2298 (1987).

112. J. C. Ho, P. H. Hor, R. L. Meng, C. W. Chu, and C. Y Huang, Solid State Commun. **63**, 711 (1987).

113. B. D. Dunlap, M. Slaski, Z. Sungaila, D. G. Hinks, K. Zhang, C. Segre, S. K. Malik, and E. E. Alp, Phys. Rev. B **37**, 592 (1988).

114. M. E. Reeves, D. S. Citrin, B. G. Pazol, T. A. Friedmann, and D. M. Ginsberg, Phys. Rev. B **36**, 6915 (1987).

115. Y. Nakazawa, M. Ishikawa, and T. Takabatake, Physica B **148**, 404 (1987).

116. K. Kadowaki, H. P. Van der Meulen, J. C. P. Klaase, M. van Sprang, J. W. A. Koster, Y. K. Huang, A. A. Menovsky, and J. J. M. Franse, to appear in J. Magn. Magn. Mat.

117. J. A. Hodges, P. Imbert, and G. Jéhanno, Solid State Commun. **64**, 1209 (1987).

118. S. Simizu, S. A. Friedberg, E. A. Hayri, and M. Greenblatt, Jpn. J. Appl. Phys. **26**, 2121 (1987), Supplement 26-3.

119. C.-X. Zhu, G.-M. Zhao, and Q.-Z. Ran, Solid State Commun. **66**, 719 (1988).

120. S. Simizu, S. A. Friedberg, E. A. Hayri, and M. Greenblatt, Phys. Rev. B **36**, 7129 (1987).

121. J. van den Berg, C. J. van der Beek, P. H. Kes, J. A. Mydosh, G. J. Niewenhuys, and L. J. de Jongh, Solid State Commun. **64**, 699 (1987).

122. K. N. Yang, J. M. Ferreira, B. W. Lee, M. B. Maple, W.-H. Lee, J. W. Lynn, and R. W. Erwin, to be published.

123. D. McK. Paul, H. A. Mook, A. W. Hewat, B. C. Sales, L. A. Boatner, J. R. Thompson, and M. Mostoller, Phys. Rev. B **37**, 2341 (1988).

124. A. I. Goldman, B. X. Yang, J. Tranquada, J. E. Crow, and C.-S. Jee, Phys. Rev. B **36**, 7234 (1987).

125. P. Fischer, K. Kakurai, M. Steiner, K. N. Clausen, B. Lebech, R. Hulliger, H. R. Ott, P. Brüesh, and P. Unternährer, Physica C **152**, 145 (1988).

126. J. W. Lynn, W.-H. Li, Q. Li, H. C. Ku, H. D. Yang, and R. N. Shelton, Phys. Rev. B **36**, 2374 (1987).

127. M. B. Maple, J. M. Ferreira, R. R. Hake, B. W. Lee, J. J. Neumeier, C. L. Seaman, K. N. Yang, and H. Zhou, to appear in *Proc. 18th Rare Earth Research Conf.*, held in Lake Geneva, Wisconsin, September 12-16, 1988.

128. B. W. Lee, J. M. Ferreira, S. Ghamaty, K. N. Yang and M. B. Maple, to be published.

129. T. Chattopadhyay, H. Maletta, W. Wirges, K. Fischer, and P. J. Brown, Phys. Rev. B **38**, 838 (1988).

130. D. R. Noakes and G. K. Shenoy, Phys. Lett. **91A**, 35 (1982); B. D. Dunlap and D. Niarchos, Solid State Commun. **44**, 1577 (1982).

131. M. B. Maple, Solid State Commun. **12**, 653 (1973).

132. C. Dong, J. K. Liang, G. C. Che, S. S. Xie, Z. X. Zhao, Q. S. Yang, Y. M. Ni, and G. R. Liu, Phys. Rev. B **37**, 5182 (1988).

133. A. Maeda, T. Yabe, K. Uchinokura, and S. Tanaka, Jpn. J. Appl. Phys. **26**, L1368 (1987).

134. Y. Song, J. P. Golben, S. Chittipeddi, J. R. Gaines, and A. J. Epstein, Phys. Rev B **38**, 4605 (1988).

135. Y. Song, J. P. Golben, X. D. Chen, J. R. Gaines, M.-S. Wong, and E. R. Kreidler, Phys. Rev. B **38**, 2858 (1988).

136. A. Maeda, T. Yabe, K. Uchinokura, M. Izumi, and S. Tonaka, Jpn. J. Appl. Phys. **26**, L1550 (1987).

137. R. Yoshizaki, H. Sawada, T. Iwazumi, Y. Saito, Y. Abe, H. Ikeda, K. Imai, and I. Nakai, Jpn. J. Appl. Phys. **26**, L1703 (1987).

138. I. Nakai, K. Imai, T. Kawashima, and R. Yoshizaki, Jpn. J. Appl. Phys. **26**, L1244 (1987).

139. Y. Song, J. P. Golben, S. Chittipeddi, S. I. Lee, R. D. McMichael, X. D. Chen, J. R. Gaines, D. L. Cox, and A. J. Epstein, Phys. Rev. B **37**, 607 (1988).

140. S. I. Lee, J. P. Golben, S. Y. Lee, X. D. Chen, Y. Song, T. W. Noh, R. D. McMichael, J. R. Gaines, D. L. Cox, and B. R. Patton, Phys. Rev. B **36**, 2417 (1987).

141. L. Soderholm, K. Zhang, D. G. Hinks, M. A. Beno, J. D. Jorgensen, C. U. Segre, and I. K. Shuller, Nature **328**, 604 (1987)

142. A. Matsuda, K. Kinoshita, T. Ishii, H. Shibata, T. Watanabe, and T. Yamada, Phys. Rev. B **38**, 2910 (1988).

143. J. J. Rhyne, J. J. Neumeier, and M. B. Maple, unpublished.

144. B. Okai, M. Kosuge, H. Nozaki, K. Takahashi, and M. Ohta, Jpn. J. Appl. Phys. **27**, L41 (1988).

145. C.-S. Jee, A. Kebede, T. Yuen, S. H. Bloom, M. V. Kuric, J. E. Crow, R. P. Guertin, T. Mihalisin, G. H. Myer, and P. Schlottmann, preprint.

146. E. Moran, U. Amador, M. Barahona, M. A. Alario-Franco, A. Vegas, and J. Rodriguez-Carvajal, Solid State Commun. **67**, 369 (1988).

147. F. Lytle, R. Greegor, E. Marques, E. Larson, J. Wong, and C. Violet, Proposal 1097, Stanford Synchrotron Radiation Laboratory Activity Report for 1987.

148. E. E. Alp, L. Soderholm, G. K. Shenoy, D. G. Hinks, B. W. Veal, and P. A. Montano, Physica B **150**, 74 (1988).

149. J. S. Kang, J. W. Allen, Z-X. Shen, W. P. Ellis, J. J. Yeh, B. W. Lee, M. B. Maple, W. E. Spicer, and I. Lindau, to appear in *Proc. 18th Rare Earth Research Conf.*, held in Lake Geneva, Wisconsin, September 12–16, 1988; Z-X. Shen, P. A. P. Lindberg, C. K. Shih, J. Hwang, H. Guyot, I. Lindau, W. E. Spicer, J. S. Kang, J. W. Allen, B. W. Lee, and M. B. Maple, submitted to Phys. Rev. B.

332

150. J. J. Neumeier, M. B. Maple, and M. S. Torikachvili, to appear in Physica C.

151. M. B. Maple, Appl. Phys. **9**, 179 (1976); and references cited therein.

152. U. Walter, E. Holland-Moritz, A. Severing, A. Erle, H. Schmidt, and E. Zirngiedl, Physica C **153–155**, 170 (1988).

153. C.-S. Jee, A. Kebede, D. Nichols, J. E. Crow, T. Mihalisin, G. H. Myer, I. Perez, R. E. Salomon, and P. Schlottmann, preprint.

154. N. Sankar, V. Sankaranarayanan, L. S. Vaidhyanathan, G. Rangarajan, R. Srinivasan, K. A. Thomas, U. V. Varadaraju, and G. V. Subba Rao, Solid State Commun. **67**, 391 (1988).

155. A. P. Gonzales, I. C. Santos, E. B. Lopes, R. T. Henriques, M. Almeida, and M. O. Figueirado, preprint.

156. A. Kebede, C.-S. Jee, D. Nichols, M. V. Kuric, J. E. Crow, R. P. Guertin, T. Mihalisin, G. H. Myer, I. Perez, R. E. Salomon, and P. Schlottman, preprint.

157. I. Taguchi, Jpn. J. Appl. Phys. **27**, L1058 (1988).

158. T. Ishida, Jpn. J. Appl. Phys. **26**, L1540 (1987).

159. Y. Wadayama, K. Kudo, A. Nagata, K. Ikeda, S. Hanada, O. Izumi, Jpn. J. Appl. Phys. **27**, L561 (1988).

160. T. Ishida, Jpn. J. Appl. Phys. **26**, L1294 (1987).

161. A. Oota, Y. Sasaki, Y. Kiyoshima, M. Ohkubo, and T. Hioki, Jpn. J. Appl. Phys. **26**, L2091 (1987).

162. M. T. Causa, C. Fainstein, G. Nieva, R. Sánchez, L. B. Steren, M. Tovar, R. Zysler, D. C. Vier, S. Schultz, Z. Fisk, and J. L. Smith, Phys. Rev. B **38**, 257 (1988).

163. F. Zuo, X. D. Chen, A. Chakraborty, B. R. Patton, J. R. Gaines, and A. J. Epstein, to appear in Solid State Commun.

164. G. Svensson, Z. Hegedüs, L. Wang, and Ö. Rapp, Physica C **153–155**, 864 (1988).

165. J. P. Franck, J. Jung, and M. A.-K. Mohamed, Phys. Rev. B **36**, 2308 (1987).

166. J. Jung, J. P. Franck, W. A. Miner, and M. A.-K. Mohamed, Phys. Rev. B **37**, 7510 (1988).

167. T. Wada, S. Adachi, T. Mihara, and R. Inaba, Jpn. J. Appl. Phys. **26**, L706 (1987).

168. B. W. Veal, W. K. Kwok, A. Umezawa, G. W. Crabtree, J. D. Jorgensen, J. W. Downey, L. J. Nowicki, A. W. Mitchell, A. P. Paulikas, and C. H. Sowers, Appl. Phys. Lett. **51**, 279 (1987).

169. T. Wada, S. Adachi, O. Inoue, S. Kawashima, and T. Mihara, Jpn. J. Appl. Phys. **26**, L1475 (1987).

170. A. Ono, T. Tanaka, H. Nozaki, and Y. Ishizawa, Jpn. J. Appl. Phys. **26**, L1687 (1987).

171. M. Mehbod, P. Wyder, R. Deltour, Ph. Duvigneaud, and G. Maessens, Phys. Rev. B **36**, 8819 (1987).

172. J. M. Tarascon, P. Barboux, P. F. Miceli, L. H. Greene, G. W. Hull, M. Eibschutz, and S. A. Sunshine, Phys. Rev. B **37**, 7458 (1988).

173. Y. Oda, H. Fujita, H. Toyoda, T. Kaneko, T. Kohara, I. Nalada, and K. Asayama, Jpn. J. Appl. Phys. **26**, L1660 (1987).

174. Y. Maeno, M. Kato, Y. Aoki, and T. Fujita, Jpn. J. Appl. Phys. **26**, L1982 (1987).

175. H. Obara, H. Oyanagi, K. Murata, H. Yamasaki, H. Ihara, M. Tokumoto, Y. Nishihara, and Y. Kimura, Jpn. J. Appl. Phys. **27**, L603 (1988).

176. T. J. Kistenmacher, W. A. Bryden, J. S. Morgan, K. Moorjani, Y-W. Du, Z. Q. Qui, H. Tang, and J. C. Walker, Phys. Rev. B **36**, 8877 (1987).

177. B. R. Zhao, Y. H. Shi, Y. Y. Zhao, and L. Li, Phys. Rev. B **38**, 2486 (1988).

178. H. Tang, Z. Q. Qui, Y.-w. Du, Gang Xiao, C. L. Chien, and J. C. Walker, Phys. Rev. B **36**, 4018 (1987).

179. C. Blue, K. Elgaid, I. Zitkovsky, P. Boolchand, D. McDaniel, W. C. H. Joiner, J. Oostens, and W. Huff, Phys. Rev. B **37**, 5905 (1988).

180. R. Gómez, S. Aburto, M. L. Marquina, M. Jiménez, V. Marquina, C. Quintanar, T. Akachi, R. Escudero, R. A. Barrio, and D. Rios-Jara, Phys. Rev. B **36**, 7226 (1987).

181. T. Tamaki, T. Komai, A. Ito, Y. Maeno, and T. Fujita, Solid State Commun. **65**, 43 (1988).

182. X. Z. Zhou, M. Raudsepp, Q. A. Pankhurst, A. H. Morrish, Y. L. Luo, and I. Maartense, Phys. Rev. B **36**, 7230 (1987).

183. E. Baggio-Saitovitch, I. S. Azevedo, R. B. Scorzelli, H. Saitovitch, S. F. da Cunha, A. P. Guimarães, P. R. Silva, and A. Y. Takeuchi, Phys. Rev. B **37**, 7967 (1988).

184. P. Bordet, J. L. Hodeau, P. Strobel, M. Marezio, and A. Santoro, Solid State Commun. **66**, 435 (1988).

185. Z. Hiroi, M. Takano, Y. Takeda, R. Kanno, and Y. Bando, Jpn. J. Appl. Phys. **27**, L580 (1988).

186. M. Eibschütz, M. E. Lines, J. M. Tarascon, and P. Barboux, Phys. Rev. B **38**, 2896 (1988).

187. G. Xiao, F. H. Streitz, A. Gavrin, Y. W. Du, and C. L. Chien, Phys. Rev. B **35**, 8782 (1987).

188. Y. Maeno, T. Tomita, M. Kyogoku, S. Awaji, Y. Aoki, K. Hoshino, A. Minami, and T. Fujita, Nature **328**, 512 (1987).

189. E. Takayama-Muromachi, Y. Uchida, and K. Kato, Jpn. J. Appl. Phys. **26**, L2087 (1987).

190. B. D. Dunlap, J. D. Jorgensen, W. K. Kwok, C. W. Kimball, J. L. Matykiewicz, H. Lee, and C. U. Segre, Physica C **153–155**, 1100 (1988).

191. Y. Shimakawa, Y. Kubo, K. Utsumi, Y. Takeda, and M. Takano, Jpn. J. Appl. Phys. **27**, L1071 (1988).

192. T. Kajitani, K. Kusaba, M. Kikuchi, Y. Syono, and M. Hirabayashi, Jpn. J. Appl. Phys. **27**, L354 (1988).

193. I. Sankawa, M. Sato, and T. Konaka, Jpn. J. Appl. Phys. **27**, L28 (1988).

194. T. Kajitani, K. Kusaba, M. Kikuchi, Y. Syono, and M. Hirabayashi, Jpn. J. Appl. Phys. **26**, L1727 (1987).

195. D. Shindo, K. Hiraga, M. Hirabayashi, A. Tokiwa, M. Kikuchi, Y. Syono, O. Nakatsu, N. Kobayashi, Y. Muto, and E. Aoyagi, Jpn. J. Appl. Phys. **26**, L1667 (1987).

196. J. F. Bringley, T.-M. Chen, B. A. Averill, K. M. Wong, and S. J. Poon, Phys. Rev. B **38**, 2432 (1988).

197. T. Ichihashi, S. Iijima, Y. Kubo, Y. Shimakawa, and J. Tabuchi, Jpn. J. Appl. Phys. **27**, L594 (1988).

198. Y. Maeno, T. Nojima, Y. Aoki, M. Kato, K. Hoshino, A. Minami, and T. Fujita, Jpn. J. Appl. Phys. **26**, L774 (1987).

199. G. Xiao, M. Z. Cieplak, A. Gavrin, F. H. Streitz, A. Bakhshai, and C. L. Chien, Phys. Rev. Lett. **60**, 1446 (1988).

200. B. Jayaram, S. K. Agarwal, C. V. N. Rao, and A. V. Narlikar, Phys. Rev. B **38**, 2903 (1988).

201. C.-S. Jee, D. Nichols, A. Kebede, S. Rahman, J. E. Crow, A. M. Ponte Gonclaves, T. Mihalisin, G. H. Myer, I. Perez, R. E. Salomon, P. Schlottmann, S. H. Bloom, M. V. Kuric, Y. S. Yao, and R. P. Guertin, J. Superconductivity **1**, 63 (1988).

202. M. Hiratani, Y. Ito, K. Miyauchi, and T. Kudo, Jpn. J. Appl. Phys. **26**, L1997 (1987).

203. P. B. Kirby, M. R. Harrison, W. G. Freeman, I. Samuel, and M. J. Haines, Phys. Rev. B **36**, 8315 (1987).

204. Y. Oda, H. Fujita, T. Ohmichi, T. Kohara, I. Nakada, and K. Asayama, J. Phys. Soc. Jpn. **57**, 1548 (1988).

205. E. Takayama-Muromachi, Y. Uchida, A. Fujimori, and K. Kato, Jpn. J. Appl. Phys. **26**, L1546 (1987).

206. Y. Tokura, J. B. Torrance, T. C. Huang, and A. I. Nazzal, Phys. Rev. B **38**, 7156 (1988).

207. A. Tokiwa, Y. Syono, M. Kikuchi, R. Suzuki, T. Kajitani, N. Kobayashi, T. Sasaki, O. Nakatsu, and Y. Muto, Jpn. J. Appl. Phys. **27**, L1009 (1988).

208. M. Kosuge, B. Okai, K. Takahashi, and M. Ohta, Jpn. J. Appl. Phys. **27**, L1022 (1988).

209. H. Uwe, T. Sakudo, H. Asano, T.-S. Han, K. Yagi, R. Harada, M. Iha, and Y. Yokoyama, Jpn. J. Appl. Phys. **27**, L577 (1988).

210. R. J. Cava, B. Batlogg, R. M. Fleming, S. A. Sunshine, A. Ramirez, E. A. Rietman, S. M. Zahurak, and R. B. van Dover, Phys. Rev. B **37**, 5912 (1988).

211. R. Liang, Y. Inaguma, Y. Takagi, and T. Nakamura, Jpn. J. Appl. Phys. **26**, L1150 (1987).

212. A. Suzuki, E. V. Sampathkumaran, K. Kohn, T. Shibuya, A. Tohdake, and M. Ishikawa, Jpn. J. Appl. Phys. **27**, L792 (1988).

213. E. V. Sampathkumaran, A. Suzuki, K. Kohn, T. Shibuya, A. Tohdake, and M. Ishikawa, Jpn. J. Appl. Phys. **27**, L584 (1988).

214. E. Takayama-Muromachi, Y. Uchida, A. Fujimori, and K. Kato, Jpn. J. Appl. Phys. **27**, L223 (1988).

215. F. Izumi, E. Takayama-Muromachi, M. Kobayashi, Y. Uchida, H. Asano, T. Ishigaki, N. Watanabe, Jpn. J. Appl. Phys. **27**, L824 (1988).

216. K. Zhang, B. Dabrowski, C. U. Segre, D. G. Hinks, I. K. Schuller, J. D. Jorgensen, and M. Slaski, J. Phys. C **20**, L935 (1987).

217. K. Takita, H. Katoh, H. Akinaga, M. Nishino, T. Ishigaki, and H. Asano, Jpn. J. Appl. Phys. **27**, L57 (1988).

218. H. Nozaki, S. Takekawa, and Y. Ishizawa, Jpn. J. Appl. Phys. **27**, L31 (1988).

219. S. Tsurami, T. Iwata, Y. Tajima, and M. Hikita, Jpn. J. Appl. Phys. **27**, L80 (1988).

220. S. Tsurami, T. Iwata, Y. Tajima, and M. Hikita, Jpn. J. Appl. Phys. **27**, L397 (1988).

221. S. Takekawa, H. Nozaki, Y. Ishizawa, N. Iyi, Jpn. J. Appl. Phys. **26**, L2076 (1987).

222. K. Takita, H. Akinaga, H. Katoh, and K. Masuda, Jpn. J. Appl. Phys. **27**, L607 (1988).

223. K. Takita, H. Akinaga, H. Katoh, H. Asano, and K. Masuda, Jpn. J. Appl. Phys. **27**, L67 (1988).

224. S. Li, E. A. Hayri, K. V. Ramanujachary, and M. Greenblatt, Phys. Rev. B **38**, 2450 (1988).

225. H. Akinaga, H. Katoh, K. Takita, H. Asano, and K. Masuda, Jpn. J. Appl. Phys. **27**, L610 (1988).

226. T. Iwata, M. Hikita, Y. Tajima, and S. Tsurumi, Jpn. J. Appl. Phys. **26**, L2049 (1987).

227. Z. Z. Wang, J. Clayhold, N. P. Ong, J. M. Tarascon, L. H. Greene, W. R. McKinnon, and G. W. Hull, Phys. Rev. B **36**, 7222 (1987).

228. C. Michel, M. Hervieu, M. M. Borel, A. Grandin, F. Deslandes, J. Provost, and B. Raveau, Z. Phys. B **68**, 421 (1987).

229. H. Maeda, Y. Tanaka, M. Fukutomi, and T. Asano, Jpn. J. Appl. Phys. **27**, L209 (1988).

230. Z. Z. Sheng and A. M. Hermann, Nature **332**, 138 (1988).

231. S. S. P. Parkin, V. Y. Lee, E. M. Engler, A. I. Nazzal, T. C. Huang, G. Gorman, R. Savoy, and R. Beyers, Phys. Rev. Lett. **60**, 2539 (1988).

232. A. W. Sleight, M. A. Subramanian, and C. C. Torardi, to appear in MRS Bulletin.

233. M. A. Subramanian, C. C. Torardi, J. Gopalakrishnan, P. L. Gai, J. C. Calabrese, T. R. Askew, R. B. Flippen, and A. W. Sleight, to appear in Science.

234. C. C. Torardi, J. B. Parise, M. A. Subramanian, J. Gopalakrishnan, and A. W. Sleight, to appear in Physica C.

235. Y. Gao, P. Lee, P. Coppens, M. A. Subramanian, and A. W. Sleight, Science **241**, 954 (1988).

236. C. C. Torardi, M. A. Subramanian, J. C. Calabrese, J. Gopalakrishnan, E. M. McCarron, K. J. Morrissey, T. R. Askew, R. B. Flippen, U. Chowdhry, and A. W. Sleight, Phys. Rev. B **38**, 225 (1988).

237. P. M. Grant, preprint.

238. C. C. Torardi, M. A. Subramanian, J. C. Calabrese, J. Gopalakrishnan, K. J. Morrissey, T. R. Askew, R. B. Flippen, U. Chowdhry, and A. W. Sleight, Science **240**, 631 (1988).

239. R. A. Fisher, S. Kim, S. E. Lacy, N. E. Phillips, D. E. Morris, A. G. Markelz, J. Y. T. Wei, and D. S. Ginley, to appear in Phys. Rev. B.

240. M. B. Maple, Y. Dalichaouch, E. A. Early, B. W. Lee, J. T. Markert, J. J. Neumeier, C. L. Seaman, K. N. Yang, and H. Zhou, Physica C **153–155**, 858 (1988).

241. L. F. Mattheiss, E. M. Gyorgy, and D. W. Johnson Jr., Phys. Rev. B **37**, 3745 (1988).

242. R. J. Cava, B. Batlogg, J. J. Krajewski, R. Farrow, L. W. Rupp Jr., A. E. White, K. Short, W. F. Peck, and T. Kometani, Nature **332**, 814 (1988).

243 A.W.Sleight, J. L. Gillson, P. E. Bierstedt, Solid State Commun. **17**, 27 (1975).

244. L. F. Mattheiss and D. R. Hamann, Phys. Rev. B **28**, 4227 (1983).

245. L. F. Mattheiss and D. R. Hermann, Phys. Rev. Lett. **60**, 2681 (1988).

246. S. Jin, T. H. Tiefel, R. C. Sherwood, A. P. Ramirez, G. W. Kammlot, and R. A. Fastnatch, submitted to Appl. Phys. Lett.

247. U. Welp, W. K. Kwok, G. W. Crabtree, H. Claus, K. G. Vandervoort, B. Dabrowski, A. W. Mitchell, D. R. Richards, D. T. Marks, and D. G. Hinks, submitted to Physica C.

248. B. Dabrowski, D. G. Hinks, J. D. Jorgensen, R. K. Kalia, P. Vashishta, D. R. Richards, D. T. Marx, and A. W. Mitchell, preprint.

249. D. G. Hinks, D. R. Richards, B. Dabrowski, D. T. Marx, and A. W. Mitchell Nature **335**, 419 (1988).

250. T. M. Rice, Nature **332**, 735 (1988).

# 7

## INFRARED PROPERTIES OF HIGH $T_c$ SUPERCONDUCTORS

Thomas Timusk
*Department of Physics, McMaster University*
*Hamilton, Ontario, Canada L8S 4M1*

and

David B. Tanner
*Department of Physics, University of Florida, Gainesville*
*FL 32611, USA*

340

# I. INTRODUCTION

Soon after the discovery of high temperature superconductivity by Bednorz and Müller[1] attempts were made to find the energy gap and any spectral features responsible for the superconducting pairing by infrared spectroscopy, a technique that had been used with success in the classic superconductors. In addition, several studies of the basic optical properties of these materials were made, with the goal of obtaining information about the electronic band structure, low-lying excitations, phonon frequencies, etc. There has been a large amount of subsequent work, showing that the infrared properties of these materials are extremely complicated, and at the time of the writing of this review our understanding is still incomplete. These experiments are the subject of this review.

According to the BCS theory, a bulk superconductor at low temperature is a perfect reflector of electromagnetic energy at frequencies below its superconducting energy gap. Above this frequency it behaves much like a normal metal. Ordinary superconductors live up to this expectation extremely well. Some of the earliest spectroscopic evidence for an energy gap in ordinary superconductors was obtained by infrared and microwave spectroscopy. In low gap materials such as aluminum Biondi and Garfunkel[2] employed microwave spectroscopy to find an onset of absorption at the gap frequency. Approaching the problem from the infrared side Tinkham's group with Glover[3], Ginsberg[4] and Richards[5] measured the gap in lead and tin. In each case absorption occurred only at or above the superconducting gap frequency. At substantially higher frequencies, the optical properties of a superconductor were found to be independent of the transition to the superconducting state.

A second use of far infrared spectroscopy in superconductivity came from the discovery by Joyce and Richards[6] of phonon structure in the spectrum of the strong-coupling superconductor lead. Predicted by Holstein[7] long before, this process of phonon emission produces a spectrum that is closely related to the Eliashberg function $\alpha^2 F(\Omega)$ seen in tunnelling spectroscopy.[8,9] Here $F(\Omega)$ is the phonon density of states and $\alpha$ is the coupling constant of the electron phonon interaction. Thus, in principle, one can find the spectrum of excitations responsible for superconductivity from a detailed study of the infrared spectrum.

An important final use of infrared and optical spectroscopy in the study of ordinary materials is the determination of electronic and vibrational properties: plasmons, interband transitions, and free-carrier absorption all give a view into the electronic band structure of a material. Lattice vibrations, electron-phonon interactions, etc., also provide fundamental information about the nature of a substance.

However, the new materials present a real challenge to the spectroscopist. They exhibit both metallic and insulating behavior. They show large optical anisotropy. The samples come in forms not ideal for precision spectroscopy: powders, porous ceramics, irregular mosaics of small twinned crystals and thin films on substrates with complicated optical properties of their own. Added to this complexity is the domain structure of $YBa_2Cu_3O_{7-\delta}$ crystals. The presence of a twin pattern in the crystal morphology of this most common high-temperature superconductor means that all measurements (even polarized optical measurements on crystals) will see a complicated effective average of the **a** and **b** dielectric functions.

In this review we survey the application of infrared techniques to the study of high temperature superconductors. Issues include the magnitude of the energy gap, the search for a Holstein spectrum of the superconducting excitations, and the nature of low- lying electronic states. At the time of writing, October 1988, none of these problems has been solved but some progress can be reported. A consensus is emerging that the energy gap is in the neighborhood of $3.5k_BT_c$, the weak coupling BCS value, and that the excitations are at very high frequency and rather weakly coupled to the electrons. The observations are consistent with a weak coupling BCS mechanism involving electronic excitations.

A further consensus exists that the simple Drude model is inadequate to describe the **ab**-plane optical response. The dc and low frequency conductivities are rather different (in terms of temperature dependence, lifetime, and oscillator strength) from the mid- and near-infrared absorption processes.

We start in section II with a review of the elementary theory of the optical constants of solids, particularly metals and superconductors. In section III we deal with the controversial midinfrared absorption, the so-called "0.5 eV" peak. First discovered in the polycrystalline samples of $La_{2-x}Sr_xCuO_4$ this absorption seems to be a common feature of all the high temperature superconductors. We will review the various interpretations of this excess infrared absorption.

Section IV contains a review of some of the work that has been devoted to the search for the energy gap. We cannot in the available space adequately consider the large number of papers that have addressed the topic, particularly on the polycrystalline ceramics; we therefore will focus more on the recent work that is available to us on **ab**-plane oriented samples.

We conclude in section V with an assessment of the current state of the experimental situation and with some comments about work needing to be done.

## II. OPTICAL PROPERTIES OF SUPERCONDUCTORS

### A. Dielectric response function

To begin we consider a medium that is subjected to an electromagnetic field of frequency $\omega$. So long as this material is homogeneous, isotropic, linear, and local in its response, that response may be characterized quite generally by a frequency-dependent complex dielectric function $\epsilon(\omega)$ which we write in terms of its real and imaginary parts as

$$\epsilon(\omega) = \epsilon_1(\omega) + \frac{4\pi i}{\omega}\sigma_1(\omega). \tag{1}$$

Here, the quantity $\epsilon_1(\omega)$ is called the real dielectric function whereas $\sigma_1(\omega)$ is the frequency-dependent conductivity. At zero frequency $\epsilon_1(0)$ becomes the static dielectric constant and $\sigma_1(0)$ is the ordinary dc electrical conductivity, $\sigma_{dc}$.

Other quantities, in addition to $\epsilon_1(\omega)$ and $\sigma_1(\omega)$, are often used to describe the optical properties of a material. These include the complex refractive index, $N = n + i\kappa$, the skin depth, $\delta$, and the complex conductivity, $\sigma = \sigma_1 + i\sigma_2$. These, however, are not independent quantities, being interrelated by

$$\epsilon = N^2 = \left(\frac{c}{\omega\delta}\right)^2 = 1 + \frac{4\pi i}{\omega}\sigma.$$

Thus, for example, the real and imaginary parts of $\epsilon$ and $\sigma$ are related by $\epsilon_2 = 4\pi\sigma_1/\omega$ and $\sigma_2 = \omega(1-\epsilon_1)/4\pi$. (The function $\sigma_2$ is often used in discussions of the low-frequency response of superconductors while $\epsilon_2$ is frequently shown for the optical properties of nonconductors.) Finally, the power absorption coefficient is given by

$$\alpha = 2 \left(\frac{\omega}{c}\right) \kappa = \frac{4\pi}{c} \frac{\sigma_1(\omega)}{n}.$$

The real and imaginary parts of the complex dielectric function are related by the Kramers-Kronig relations and satisfy various sum rules.[10] The Kramers-Kronig relation relates the real dielectric function and the frequency-dependent conductivity as follows:

$$\epsilon_1(\omega) = 1 + 8\mathcal{P} \int_0^\infty \frac{\sigma_1(\omega')}{\omega'^2 - \omega^2} d\omega'. \tag{2}$$

Here, $\mathcal{P}$ means "principal part." There are similar Kramers-Kronig relations between $\sigma_1(\omega)$ and $\epsilon_1(\omega)$, n and $\kappa$, etc.

For the experimental determination of optical properties, the most important of the Kramers-Kronig relations is the one relating the phase shift upon reflection, $\theta$, to the reflectance, R, $viz$[10]

$$\theta(\omega) = \frac{\omega}{\pi} \int_0^\infty \frac{\ln R(\omega') - \ln R(\omega)}{\omega^2 - \omega'^2} d\omega'.$$

The range of the Kramers-Kronig integral extends over all frequencies, requiring extrapolations beyond the measured frequency interval. Typically, power laws are used at high frequency, $R \sim \omega^{-p}$ with $0 \le p \le 4$. At low frequencies the reflectance is assumed to be constant if the sample is an insulator and to follow the Hagen-Rubens relation, $R = 1 - A\sqrt{\omega}$, if the sample is a metal.

A fundamentally important sum rule is the f-sum rule or oscillator-strength sum rule:

$$\int_0^\infty \sigma_1(\omega') d\omega' = \frac{\pi}{2} \frac{ne^2}{m} \tag{3},$$

where $n$ is the total electron density in the material, $e$ the electronic charge, and $m$ the free electron mass. For cases where various types of electronic excitations are well separated in frequency, one may use a partial sum rule to determine the integrated oscillator strength:

$$N_{eff} \frac{m}{m*} = \frac{2m}{\pi n_a e^2} \int_0^\omega \sigma_1(\omega') d\omega'.$$

Here $N_{eff}$ is the number of effective electrons per atom, molecule, or formula unit in the solid participating in optical transitions at frequencies less than $\omega$,

$m^*/m$ is the ratio of their effective mass to the free electron mass, and $n_a$ is the number density of atoms, molecules, or formula units. Because the optical plasma frequency is defined as $\omega_p^2 = 4\pi n e^2/m^*$, the partial sum rule, when applied to the low frequency spectral region dominated by free electrons, is a way to estimate the plasma frequency of the conduction electrons in a metal.

An important consequence of the oscillator-strength sum rule is that the area under the curve of $\sigma_1(\omega)$ vs. $\omega$ is a constant, independent of such factors as temperature, phase transitions, etc. Thus, for example, if $\sigma_{dc}$ increases with decreasing temperature then $\sigma_1(\omega)$ must decrease at higher frequencies so as to conserve the oscillator strength. As another example, a superconductor develops an infinite (delta-function) conductivity at zero frequency; the area under this delta function is removed from $\sigma_1(\omega)$ over $0 < \omega \leq 2\Delta$ (and a little higher); here, $2\Delta$ is the superconducting energy gap.

## B. Anisotropic materials

When the material is anisotropic, as are most high $T_c$ superconductors, the dielectric function becomes a tensor quantity. The response of the medium is well characterized by this tensor dielectric function, although when the directions of the electric field of the electromagnetic wave are not along one of the principal axes of the dielectric tensor, the wave propagation can become rather complicated, with double refraction, rotation of the plane of polarization, etc.[11,12]

Some properties of the dielectric tensor are determined by the symmetry of the crystal. For the orthorhombic crystal class into which some of the high-$T_c$ superconductors fall, the material is biaxial, with different values for the three components of the dielectric tensor; the three principal axes are fixed along the a, b, and c crystallographic directions. For tetragonal crystals, the material is uniaxial, with identical values for the a and b directions and a different value for the c-axis.

Polarized optical measurements made with the electric field vector oriented along one of the principal axis directions will be governed by that particular component of the dielectric tensor. The Kramers-Kronig relations and sum rules discussed above will apply to that component. Clearly, measurements made for all three polarization directions will probe all three components of the dielectric tensor.

## C. Drude dielectric function

The simplest model for the dielectric function of a metal is due to Drude and Sommerfeld (see Kittel[13]). According to this model, when the electrons in a metal are subjected to an external electric field the entire Fermi surface is displaced rigidly in a direction opposite to the field (on account of the negative electronic charge) by an amount proportional to the current density. The amount of the displacement is determined by a balance between two forces: (1) the -eE force exerted by the electric field and (2) the relaxation of electrons towards equilibrium by elastic scattering processes, treated as a viscous damping force. The relaxation rate is $1/\tau = v_F/l$, where $v_F$ is the Fermi velocity and $l$ the electronic mean free path.

The scattering rate is determined by the probability of collisions between electrons and phonons (giving $\tau_p$), impurities ($\tau_i$), the sample surfaces ($\tau_d$), other electrons ($\tau_e$), etc. The scattering processes are generally independent, so that the rates add:

$$1/\tau = 1/\tau_p + 1/\tau_i + 1/\tau_d + 1/\tau_e + \cdots$$

At low frequencies and at low temperatures these scattering processes are quasielastic, with the electrons having both initial and final state energies very close to the Fermi energy. At higher frequencies, even in the limit of zero temperature, emission processes become allowed, wherein an electron absorbs a photon, simultaneously emits one or more phonons or other excitations of the crystal, and scatters, thereby relaxing the current. These Holstein[7] emission processes lead to frequency-dependent scattering rates, as discussed below. For the moment, we take $1/\tau$ to be independent of frequency.

The Drude dielectric function is

$$\epsilon(\omega) = \epsilon_\infty - \frac{\omega_p^2}{\omega^2 + i\omega/\tau}. \tag{4}$$

where $\epsilon_\infty$ is the contribution from high-energy interband processes and the atomic cores (typically $1 \leq \epsilon_\infty \leq 4$) and $\omega_p$ is the plasma frequency, defined above. The real dielectric function, within the Drude model, is

$$\epsilon_1(\omega) = \epsilon_\infty - \frac{\omega_p^2}{\omega^2 + 1/\tau^2}. \tag{5}$$

When $\omega_p \gg 1/\tau$ (the usual case) the real dielectric function is negative for $\omega < \omega_p/\sqrt{\epsilon_\infty}$. It is this negative real dielectric function which causes the high reflectance of a metal below its plasma edge, because a negative $\epsilon_1(\omega)$ (along with $|\epsilon_1| \gg |\epsilon_2|$, i.e., $\omega \gg 1/\tau$) makes the imaginary part of the refractive index much larger than the real part, so that $N \approx i\kappa$. In turn, when $\kappa \gg n$,

$$R = \frac{(n-1)^2 + \kappa^2}{(n+1)^2 + \kappa^2} \approx 1 - \frac{2}{\omega_p\tau} \sim 1$$

At lower frequencies, $\omega \ll 1/\tau$, metals have $\kappa \approx n \gg 1$ and a Hagen-Rubens reflectance,

$$R \approx 1 - \sqrt{\frac{8\omega}{\omega_p^2\tau}}.$$

The Drude-model frequency-dependent conductivity is

$$\sigma_1(\omega) = \frac{\omega_p^2\tau/4\pi}{1 + \omega^2\tau^2} \tag{6}$$

Thus, the dc conductivity is

$$\sigma_{dc} = \frac{\omega_p^2\tau}{4\pi} = \frac{ne^2\tau}{m^*} \tag{7.}$$

The frequency dependent conductivity falls steadily from this value with a characteristic width of $1/\tau$, a frequency typically in the microwave or far infrared region of the spectrum. As a rule of thumb, most metals have plasma frequencies in the 8 eV (60,000 cm$^{-1}$) range and have, at 300 K, $\omega_p\tau \sim 200$. This corresponds to a conductivity of $\sigma_{dc} \sim 2 \times 10^5$ $\Omega^{-1}$cm$^{-1}$ or a resistivity of 5 $\mu\Omega$-cm. (In this chapter equations are written in esu units throughout; the conductivity is thus in sec$^{-1}$. To convert to practical units, $\Omega^{-1}$cm$^{-1}$, divide by $30c = 9 \times 10^{11}$. For quick calculations, note that with $\omega_p$ and $1/\tau$ in cm$^{-1}$, the conductivity in practical units is $\sigma_{dc} = \omega_p^2\tau/60$.)

## D. Frequency-dependent damping

The Drude formula for the frequency dependent conductivity is strictly applicable only at low frequency and low temperature where elastic scattering from impurities that are fixed to the lattice and quasielastic scattering from thermally generated excitations such as phonons dominate. These processes limit the lifetime of an electron and result in a conductivity function with a finite width centered at zero frequency. At higher photon frequency low lying states of the system can be excited and new, inelastic scattering processes become possible. The simple Drude formula with a constant, frequency independent, scattering rate is not valid under these circumstances.

An excitation at frequency $\omega$ can make a contribution to the optical conductivity in two different ways. Free electrons can be scattered inelastically by the excitation or the electromagnetic field can couple directly to the dipole moment of the excitation. An example of the first process is the electron-phonon interaction in ordinary metals where electrons can lose energy by creating phonons.[7,8] Interband transitions provide an example of the second process. Here the energy of the photon is used to make a transition between two eigenstates in the electronic band structure. The first process can be described by adding to the constant damping term in the Drude formula another frequency dependent term. In the second process one adds to the conductivity a direct absorption process between filled and empty states in the band structure.

Inelastic scattering in metals with a frequency dependent damping rate has been treated by a number of authors.[8,14,15,16] It is found that in order to satisfy causality an imaginary part has to be added to the damping rate which can be interpreted as a modification of the mass of the electron that enters the expression for the plasma frequency. Physically, the coupling of the excitation to the electrons changes both the real and imaginary parts of the electron self energy, and therefore both the mass and the lifetime.

One starts by rewriting the Drude dielectric function in terms of a complex damping function, also called a memory function, $G(\omega) = R(\omega) + iI(\omega)$:

$$\epsilon(\omega) = \epsilon_1 + i\epsilon_2 = -\frac{\omega_p^2}{\omega[\omega + iG(\omega)]}, \tag{8}$$

where

$$\omega_p^2 = \frac{e^2}{3\pi^2\hbar} \int v dS_F.$$

Here, $dS_F$ is an element of area on the Fermi surface and $v$ is the velocity of the electron. The simple Drude formula corresponds to $R(\omega) = 1/\tau$ and $I(\omega) = 0$. The quantities $R(\omega)$ and $I(\omega)$ are related by Kramers-Kronig relations, for example if $R(\omega)$ is known:

$$I(\omega) = -\frac{2\omega}{\pi}P \int_0^{\infty} \frac{R(\omega')}{\omega'^2 - \omega^2} d\omega' \tag{9}.$$

The quantities $R(\omega)$ and $I(\omega)$ can be determined experimentally. One can solve for the frequency dependent damping $\tau_{eff}$ and the frequency dependent plasma frequency $\omega_{eff}^2$

$$1/\tau_{eff} = -\frac{\omega\epsilon_2}{(\epsilon_1 - \epsilon_\infty)} \tag{10}$$

and

$$\omega_{eff}^2 = -(\epsilon_1 - \epsilon_\infty)(\omega^2 + 1/\tau_{eff}^2) \tag{11}.$$

The effective damping rate, $1/\tau_{eff}$, and the effective plasma frequency, $\omega_{eff}^2$, can also be expressed in terms of the real and imaginary parts of the memory function:

$$1/\tau_{eff} = \frac{R(\omega)}{1 - I(\omega)/\omega}, \tag{12}$$

$$\omega_{eff}^2 = \frac{\omega_p^2}{1 - I(\omega)/\omega}. \tag{13}$$

In this pair of equations two functions are known, the effective relaxation rate and effective plasma frequency. There are three unknown quantities, R, I and $\omega_p$. Thus to determine I and R some additional information has to be used. Often one can estimate the plasma frequency $\omega_p$ from the sum rule for the conductivity, Eq. (3), if it is exhausted by the available spectrum:

$$\int_0^{\infty} \sigma_1(\omega) d\omega = \frac{\omega_p^2}{8}.$$

From the definition of the plasma frequency in terms of an effective mass $m^*$ one can write

$$\omega_{eff}^2 = \frac{4\pi n e^2}{m^*}.$$

Then, with the definition $m^* = m/(1 + \lambda)$, we find:

$$\lambda(\omega) = -I(\omega)/\omega. \tag{14}$$

The mass renormalization function $\lambda$ can also be calculated from Eq. (9), the Kramers-Kronig relations between $R(\omega)$ and $I(\omega)$, if $1/\tau$ is known over a wide range of frequencies.

In summary, if the plasma frequency is known one can, from the real and imaginary parts of dielectric function, derive the frequency dependent mass and damping rate of the electrons.

There are serious problems with the application of this formalism to real systems. The first requirement is that Kramers-Kronig derived dielectric function be available for the system in the region of interest. The second, more severe, restriction is that the dielectric function should only include intraband, that is free electron processes.

### E. The electron-phonon interaction

An example of frequency dependent damping is the inelastic scattering of the conduction electrons by phonons in ordinary metals, the Holstein process. The photon energy is divided between the change in kinetic energy of the electron and the phonon energy. At zero temperature the process has a definite threshold at the frequency of the phonon—at least this much energy must be delivered to the system by the photon. The expression for the damping rate is[8]

$$1/\tau(\omega) = \frac{2\pi}{\omega} \int_0^\omega \alpha_{tr}^2 F(\Omega)(\omega - \Omega)d\Omega \tag{15}$$

where $\alpha_{tr}^2 F(\Omega)$ is the Eliashberg function, proportional to the phonon density of states $F(\Omega)$ modified by the inclusion of a factor $(1 - cos\theta)$ to weight large scattering angles $\theta$ that make a larger contribution to transport properties such as the frequency dependent conductivity.

In the superconducting state the development of the gap $2\Delta$ at the Fermi level means that the formalism must be modified. Two important changes occur. In the normal state the threshold of the broad structure caused by the phonons occurs at a frequency $\Omega_E$. In the superconducting state an additional energy $2\Delta$ must be available to break the Cooper pairs. As a result this threshold shifts to a higher frequency $\Omega_E + 2\Delta$. In addition the singularities in the superconducting density of states cause the phonon structure to sharpen. As a result an $\alpha^2 F(\Omega)$ function can be extracted from the optical spectrum with a resolution that rivals that of the familiar tunnelling method.[9]

## F. Interband transitions

The general subject of interband transitions is beyond the scope of this review. Ordinary interband transitions in solids are due to excitations of electrons from one set of energy eigenstates to a higher lying set of states; they are thus governed by the details of the electronic band structure.[10] In contrast to the Drude model, which describes relatively accurately the free carrier response, the simple Lorentz oscillator commonly used to model interband transitions does a poor job of representing the lineshape. Nevertheless, this model is widely used to represent the interband and bound-carrier contributions to the dielectric function.

The Lorentzian dielectric function is easily derived in the same spirit as the Drude model by assuming that the electrons are bound by a harmonic potential to a site in the solid, and that motion in this potential is subject to a viscous damping force. The dielectric response function is then given by:

$$\epsilon(\omega) = \epsilon_\infty + \frac{\omega_{pe}^2}{\omega_e^2 - \omega^2 - i\omega\gamma_e}. \tag{16}$$

Here $\epsilon_\infty$ is as usual the high-frequency limiting value of the dielectric function and $\omega_{pe}$ plays the role of an oscillator strength or plasma frequency; it is given by $\omega_{pe} = \sqrt{4\pi n_o e^2/m^*}$, where $n_o$ is the number density of oscillators and $m^*$ their effective mass.

The other terms in Eq. (16) are the center frequency, $\omega_e$, and the linewidth or damping constant, $\gamma_e$. It is tempting to associate $\omega_e$ with the semiconducting energy band gap or interband energy and $\gamma_e$ with the lifetime of excited carriers. This association is valid if the width of the band is caused by lifetime effects, as in atomic absorption. If, however, the width is due to the initial and final states being located in broad bands, so that there is a large range of allowed energies for interband transitions, the association is invalid and will *overestimate* both energy gap and the lifetime. In this case, the gap should be associated with the onset of the absorption rather than with its center frequency.

There are however three features of the Lorentz oscillator which can be interpreted with confidence. First, so long as the damping is not too large there is a range of frequencies over which $\epsilon_1(\omega) < 0$. As in the case of the Drude metal, this negative $\epsilon_1(\omega)$ leads to a high "metallic" reflectance; the frequencies where this occurs are $\omega_e \leq \omega \leq \omega_{pe}\sqrt{1/\epsilon_\infty + \omega_e^2/\omega_{pe}^2}$. Second, the sum rule gives $\omega_{pe}^2/8$ for the oscillator strength of the Lorentzian. Both of these facts justify the use of the Lorentz oscillator model to represent interband transitions.

The third important feature of the Lorentz model is its zero-frequency limit:

$$\epsilon_1(0) = \epsilon_\infty + \frac{\omega_{pe}^2}{\omega_e^2} \tag{17}$$

Thus, interband transitions contribute important terms to the static dielectric constant, which increases as $\omega_{pe}$ increases or $\omega_e$ decreases. This behavior is consistent with observed behavior in semiconductors.[17]

## G. Frequency-dependent conductivity of superconductors

The electromagnetic properties of superconductors are distinguished by perfect dc conductivity, the Meissner exclusion of magnetic fields, and a gap $2\Delta$ in the electronic excitation spectrum. The gap is the minimum energy required to break apart a pair. In weak-coupling BCS theory, this energy at 0 K is given by

$$2\Delta = 3.5 k_B T_c \tag{18}$$

where $k_B$ is Boltzmann's constant and $T_c$ is the superconducting transition temperature. Actual superconductors do not adhere rigidly to this relation, with strong-coupled superconductors (Pb and Hg) showing $2\Delta/k_B T_c \approx 4$.

The infinite dc conductivity is represented in the frequency- dependent conductivity of a superconductor by a delta function at zero frequency. In contrast, at all non-zero frequencies below the gap the real part of the conductivity is zero at zero temperature and exponentially activated at finite temperature. With $\sigma_1(\omega) = 0$ there is no energy absorption for $0 < \omega < 2\Delta$. There is, however, a substantial inductive response of the superconductor in the low frequency range.

To see how this inductive response arises, we consider the London model. According to the London model, Ohm's law is replaced by the London equation

$$\mathbf{j}_s = -\frac{n_s e^2}{m^* c}\mathbf{A}$$

where $\mathbf{j}_s$ is the supercurrent, $\mathbf{A}$ is the magnetic vector potential and $n_s$ is the density of superfluid carriers. (Here we take these to have charge e and mass m; the model can also be written—with identical results—in terms of Cooper pairs, with charge $-2e$, mass $2m^*$, and number density $n/2$.)

If the London equation is combined with one of Maxwell's equations, $\nabla \times \mathbf{H} = (4\pi/c)\mathbf{j}_s$, it is straightforward to show that an external magnetic field decays exponentially inside a superconductor with a characteristic length, the London penetration depth,

$$\lambda_L = \sqrt{\frac{m^*c^2}{4\pi n_s e^2}} \tag{19}$$

Note that if $n_s=n$, then $\omega_L \equiv c/\lambda_L = \omega_p$.

Taking the time derivative of the London equation, one finds

$$\frac{\partial \mathbf{j}_s}{\partial t} = \frac{n_s e^2}{m^*}\mathbf{E}$$

where we have used $\mathbf{E} = -\frac{1}{c}\frac{\partial \mathbf{A}}{\partial t}$. This is just another way of writing $m^*\mathbf{a}_s = -e\mathbf{E}$, where $\mathbf{a}_s$ is the superfluid acceleration and $\mathbf{j}_s = -n_s e\mathbf{v}_s$. We use $e^{-i\omega t}$ time dependences for $\mathbf{j}_s$ and $\mathbf{E}$, and find that

$$\mathbf{j}_s = i\frac{n_s e^2}{m^*\omega}\mathbf{E}$$

or (recalling that the dc conductivity is infinite):

$$\sigma(\omega) = A\delta(\omega) + i\frac{n_s e^2}{m^*\omega} = \frac{\omega[\epsilon(\omega) - 1]}{4\pi} \tag{20}$$

where A is the strength of the zero-frequency delta function in the conductivity. This can be determined by substituting $\sigma_1(\omega)$ in the Kramers-Kronig relation:

$$\epsilon_1(\omega) = 1 + 8A \int_0^\infty \frac{\delta(\omega')}{w'^2 - w^2} dw'$$

to find $A = \pi n_s e^2/2m^* = \omega_{ps}^2/8$. Thus the superfluid in the London model has a real dielectric function given by

$$\epsilon_1(\omega) = 1 - \frac{\omega_{ps}^2}{\omega^2} \tag{21}$$

This dielectric function is that of $n_s$ free carriers, impeded by their inertia but unaffected by scattering. Similarly the frequency dependent conductivity is that of a perfect conductor,

$$\sigma_1(\omega) = \frac{\pi n_s e^2}{2m^*}\delta(\omega) = \frac{1}{8}\omega_{ps}^2\delta(\omega) \tag{22}$$

Note that this conductivity satisfies the f-sum rule if $n_s = n$ so that $\omega_{ps} = \omega_p$. Thus in the London model all of the oscillator strength of the Drude normal-state conductivity ends up under the zero-frequency delta function of the superconductor. This is a consequence of the gap in the London model effectively being infinity.

A more realistic dielectric function for a superconductor is the one developed by Mattis and Bardeen,[18] who calculated the real and imaginary parts of the frequency-dependent conductivity within weak-coupling BCS theory. The calculation is valid either in the extreme clean limit or the extreme dirty limit. Here we define these two limits as

Clean limit : $\qquad \dfrac{1}{\tau} \ll 2\Delta.$

Dirty limit : $\qquad \dfrac{1}{\tau} \gg 2\Delta.$

We set $\hbar = 1.0$. This is the key definition for optics. Because the Pippard coherence length is $\xi_0 = v_F/\pi\Delta$, the above definitions correspond respectively to $l \gg \xi_0$ in the clean limit and $l \ll \xi_0$ in the dirty limit.

Fig. 1 shows the Mattis-Bardeen theory for $\sigma_1(\omega)/\sigma_n$ and $\sigma_2(\omega)/\sigma_n$ as a function of frequency. The real part is zero up to the energy gap and then rises to join the normal-state conductivity at a substantially higher frequency. In actual metallic superconductors, this edge usually is much steeper. The imaginary part displays the $1/\omega$ inductive response discussed above.

It is important to note that Fig. 1 is somewhat misleading because it shows not the conductivity, but the ratio of the conductivity to its normal state value. In the dirty limit, $\sigma_{1n}$ is nearly constant so long as $\omega \ll 1/\tau$, e.g., for $\omega \sim 2\Delta$, and the figure is a reasonable representation of $\sigma_{1s}$ and $\sigma_{2s}$. Much of the Drude oscillator strength exists at frequencies above $2\Delta$, leading to numerical changes from the London model dielectric function discussed above. The area under the

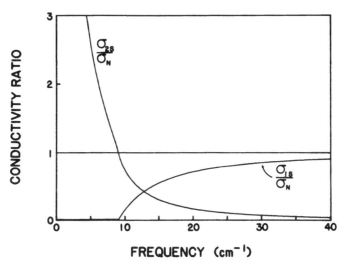

FIG. 1. Real and imaginary parts of the frequency-dependent conductivity of a superconductor, calculated from the theory of Ref. 18.

zero-frequency delta function and the prefactor in the $1/\omega^2$ term in $\epsilon_1(\omega)$ are both substantially reduced. In the dirty limit, these quantities should be multiplied by a factor approximately equal to $2\Delta\tau$, a quantity much smaller than unity.

In the clean limit, most of the area under the Drude $\sigma_1(\omega)$ is contained in the frequency range well below $2\Delta$. Therefore, at frequencies larger than $2\Delta$, $\sigma_{1n}$ is extremely small compared to $\sigma_{dc}$. As a result, the London model accurately describes the infrared response, which is mostly inductive. The zero-frequency delta function contains most of the oscillator strength and the reactive response is essentially that of free carriers at all frequencies above and below $2\Delta$.

One model which attempts to bridge the gap between the extreme clean and extreme dirty limits is that due to Leplae,[19] who put relaxation effects into the Mattis-Bardeen theory in a plausible but not completely rigorous way. The Leplae model has been used by a number of authors, e.g., Ref. 20.

## III. THE MIDINFRARED BAND

We start with a discussion of the normal-state behavior of these materials. As has been generally recognized, the high $T_c$ superconductors have anomalous normal-state electronic properties. Among these we mention the relatively high resistivity of $YBa_2Cu_3O_{7-\delta}$ and $La_{2-x}Sr_xCuO_4$, the linear temperature dependence of the resistivity, the very wide range of temperatures over which this linear temperature dependence is observed, and the insulating behavior of compositions that should be half-filled-band metals, e.g. $La_2CuO_4$ and $YBa_2Cu_3O_6$.

Most measurements of the optical properties of these materials have been reflectance studies. There are not wide differences in the results of these measurements, so it would be possible to construct a generic "high-$T_c$" reflectance curve. Despite this general agreement about the data, a number of controversial issues have arisen, including: (1) the presence (or absence) of a midinfrared absorption band, (2) the use of the Drude model to describe the infrared reflectance, (3) interpretations of doping studies, (4) the frequency dependence of the effective mass and scattering rate, (5) the effects of anisotropy and inhomogeniety, and (6) the importance of electron-vibrational interaction. Each of these will be discussed in this section. Phonon assignments are the subject of the following chapter of this volume and will not be discussed here. We will, however, briefly touch on the high frequency bands seen in the visible and ultraviolet.

### A. Early studies of ceramics

As was the case with all experimental studies of high-temperature superconductors, the first infrared measurements were made on ceramic samples — first $La_{2-x}Sr_xCuO_4$ but were followed shortly by $YBa_2Cu_3O_{7-\delta}$. The first complete infrared spectrum of a high-temperature superconductor that we know of was presented by Tajima et al.[21] It is shown in Fig. 2. (This spectrum was also presented at the "Woodstock" session of the March 1987 APS meeting.) The room-temperature reflectance is shown for ceramic samples of $La_{2-x}Sr_xCuO_4$ with $x = 0.18$, $La_{2-x}Ba_xCuO_4$ with $x = 0.10$, and $La_2CuO_4$. Note that these data are presented on a wavelength scale from 0.5 $\mu$m (20,000 cm$^{-1}$ or 2.5 eV) to 100 $\mu$m (100 cm$^{-1}$ or 12 meV).

There are three characteristic features of these spectra, features which occur in all subsequent spectra of ceramic samples: (1) The doped or substituted

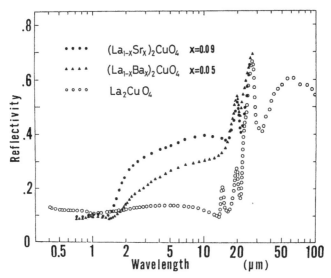

FIG. 2. Infrared reflectance of $La_{1.82}Sr_{0.18}CuO_4$, $La_{1.90}Ba_{0.10}CuO_4$, and $La_2CuO_4$ over 0.5–100 $\mu$. Note that the data are presented as a function of wavelength. (From Ref. 21.)

samples have a reflectance minimum at 1.5 $\mu$ (0.83 eV), followed by a steep rise in reflectance at longer wavelengths. This feature is absent in the non-superconducting $La_2CuO_4$ sample. (2) The intensity of the infrared reflectance rise was strongest in the most heavily-doped sample, but the location of the reflectance minimum did not change with doping level. (3) Strong phonon features are seen in the infrared; one mode which is seen in $La_2CuO_4$ is absent in the doped samples. The spectra were discussed in terms of a plasma edge and metallic reflectance of free carriers by Tajima *et al.*

Shortly afterwards, Orenstein *et al.*[22] presented room-temperature reflectance spectra of $La_{2-x}Sr_xCuO_4$ ($x = 0.175$), $La_2CuO_4$, and $YBa_2Cu_3O_{7-\delta}$ while Herr *et al.*[23] described the reflectance of $La_{2-x}Sr_xCuO_4$ ($x = 0.15$) at 4.2 K, 70 K, and 300 K. Both groups concluded that the optical properties deviated from those of a simple Drude metal, with a strong absorption feature located in the midinfrared and with the free-carrier (Drude) response contributing only a small part of the total infrared oscillator strength. Despite rather different analysis methods, both groups put the midinfrared feature at 0.44–0.45 eV, but with a very broad ($\sim$ 1 eV) linewidth. To reach this conclusion Orenstein *et al.*[22] fit the reflectance of $La_{2-x}Sr_xCuO_4$ using a dielectric function model consisting of a sum of Drude and Lorentzian terms, finding that the Lorentzian part (centered near 0.4 eV) dominated the spectrum. In $YBa_2Cu_3O_{7-\delta}$ the midinfrared band was put at 0.65 eV.

Herr et al.[23] performed a Kramers-Kronig analysis of the reflectance to obtain the frequency-dependent conductivity, shown in Fig. 3. (Note that the 4.2 K conductivity shown in this figure is too large, on account of errors in the low temperature reflectance at high-frequencies. The actual temperature dependence is smaller than indicated here.) In addition to the midinfrared absorption, Herr et al.[23] concluded that the vibrational features (particularly the 240 $cm^{-1}$ mode) were anomalously strong.

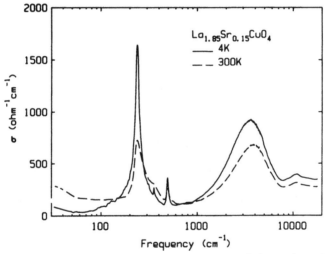

FIG. 3. Frequency-dependent conductivity of $La_{1.85}Sr_{0.15}CuO_4$, as determined by Kramers-Kronig analysis of reflectance. Note the logarithmic frequency scale. (From Ref. 23.)

The far-infrared and midinfrared properties of $YBa_2Cu_3O_{7-\delta}$ ceramic samples were described by Bonn et al.[24] and Kamarás et al.[25]. Kramers-Kronig analysis gave a broad midinfrared band near 0.37 eV, again with a width approaching 1 eV. Comparison of the spectra of superconducting $YBa_2Cu_3O_{6.9}$ and nonsuperconducting $YBa_2Cu_3O_{6.2}$ showed that the midinfrared band was present only in the former sample

The interpretation of these results attributed the midinfrared absorption to a charge-transfer electronic excitation among strongly correlated, perhaps localized electrons, of the sort proposed by Varma *et al.*[26] This "excitonic" absorption is a very controversial issue, as discussed below. It is clear that anisotropy and inhomogeniety distorted these early spectra, making the positions of the bands particularly inaccurate.[27] Nevertheless, as we argue below, it is our belief that the midinfrared oscillator strength is dominated by absorption which is due to some sort of particle-hole excitation and that only a small amount of the absorption arises from the zero-frequency-centered Drude absorption of free carriers.

In contrast, polarized data have made evident, as first pointed out by Schlesinger *et al.*,[28] that the vibrational features are not anomalously strong; instead, the mixing of c-axis phonon structure with ab-plane electronic absorption leads to an erroneous increase in the apparent oscillator strength as determined either by Kramers-Kronig analysis[23,25] or by fits to model dielectric functions.[29]

A number of workers have discussed the effects of anisotropy on the optical properties of ceramic samples, either using effective medium theories,[23,30,31,32] which should apply when the wavelength exceeds the average grain size, or by calculating an average reflectance,[28,32,33] a model which is appropriate when the wavelength is small relative to the grain size. In many ceramic samples the grain size is $\sim 10\mu$, so neither limit is strictly applicable. Qualitative agreement with spectra in the near-infrared and midinfrared regions for ceramic samples of $La_{2-x}Sr_xCuO_4$ can be obtained by taking the reflectance to be a sum of 2/3 ab-plane reflectance and 1/3 c-axis reflectance spectra.

## B. The Drude fits

The reflectance of ordinary metals is dominated by a plasma edge in the visible or ultraviolet, with a rapid rise to high reflectance at longer wavelengths. This spectrum can in general be modelled by a Drude dielectric function, Eq. (4), especially if $\epsilon_\infty$ is made frequency dependent in order to describe the contributions of interband transitions. The value of $\omega_p$ then corresponds closely to the free-carrier or band-theory results for these quantities. Typical metals have $\omega_p\tau \sim 200$, making the reflectance rise at the plasma edge rather steeply. As the scattering rate increases, the width of the edge also increases. It is important to note, however, that semiconductors (e.g., Si and Ge) also display an ultraviolet plasmon edge, with high "metallic" reflectance at frequencies between the plasmon frequency and their semiconducting energy gap.[10] This is due to the negative values of $\epsilon_1(\omega)$ between the gap and the plasmon frequency. (See section II.F.)

The reflectance of the high $T_c$ superconductors resembles that of ordinary materials in many ways, although the "plasma edge" is in the infrared. Many groups have taken the Drude model to describe the infrared properties of the high $T_c$ superconductors, including studies of $La_{2-x}Sr_xCuO_4$ceramics,[21,28,32,34] the ab-plane of $La_{2-x}Sr_xCuO_4$ crystals,[35,36] $YBa_2Cu_3O_{7-\delta}$ceramics,[31,33,37] the ab-plane of $YBa_2Cu_3O_{7-\delta}$crystals,[27,38,39,40] and c-axis-normal $YBa_2Cu_3O_{7-\delta}$ thin films.[41,42]

Leaving aside for a moment the question of the applicability of the Drude model to the high $T_c$ superconductors, we first discuss the results of these fits. Despite the wide range of samples studied, most find rather similar Drude parameters:

(1) In $La_{2-x}Sr_xCuO_4$ $\omega_p \approx$ 1.6–1.7 eV (12,800–13,700 cm$^{-1}$) (independent of $x$!), $\epsilon_\infty \approx$ 4.0–4.5, and $1/\tau \approx$ 0.37–0.60 eV (3000–5000 cm$^{-1}$).

(2) In $YBa_2Cu_3O_{7-\delta}$ these parameters are $\omega_p \approx$ 2.6–3.1 eV (21,000–25,000 cm$^{-1}$), $\epsilon_\infty \approx$ 3.7–4.5, and $1/\tau \approx$ 0.6–1.2 eV (5000–10,000 cm$^{-1}$).

Two examples of Drude model fits are shown in Figs. 4 and 5.

Fig. 4 (from Ref. 38) shows the room temperature reflectance and Kramers-Kronig-derived conductivity for a crystal of $YBa_2Cu_3O_{7-\delta}$. The data are shown as points, $\sigma_1(\omega)$ as a dashed curve, and the Drude model reflectance and conductivity as solid lines. The Drude parameters are $\omega_p =$ 3.1 eV (25,000 cm$^{-1}$), and $1/\tau =$ 0.93 eV (7500 cm$^{-1}$). ($\epsilon_\infty$ was not specified; from the location of the minimum in the reflectance fit, it must have been $\epsilon_\infty \sim$ 4.)

Fig. 5 (from Tajima *et al.*[36]) presents the reflectance, R, the real part of the dielectric function, $\epsilon$, the loss function, -Im(1/$\epsilon$), and the frequency-dependent

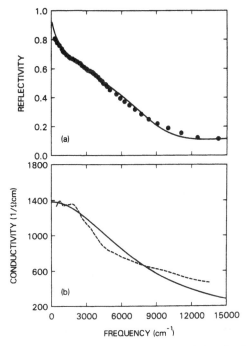

FIG. 4. The upper panel shows the room-temperature infrared reflectance of a crystal of $YBa_2Cu_3O_{7-\delta}$; the lower curve the Kramers-Kronig-derived conductivity. The solid lines are fits assuming a simple Drude dielectric function. (From Ref. 38.)

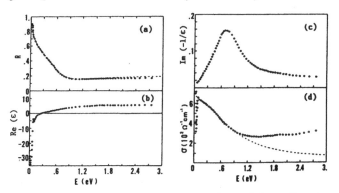

FIG. 5. The infrared room-temperature reflectance (upper left panel), real part of the dielectric function (lower left), loss function (upper right), and frequency-dependent conductivity (lower right) of of a crystal of $La_{1.8}Sr_{0.2}CuO_4$. (From Ref. 36.)

conductivity, $\sigma_1(\omega)$, of a crystal of $La_{1.8}Sr_{0.2}CuO_4$. The data, measured at 300 K, were taken with the electric field parallel to the **ab**-plane. It should be noted, however, that the superconducting transition of this sample began at $\sim$ 30 K and was not complete until $\sim$ 14 K, even though the Sr concentration of 0.2, in this ceramic, would give $T_c \approx$ 38 K. Drude parameters were $\omega_p$ = 2.4 eV (19,000 cm$^{-1}$), $\epsilon_\infty$ = 7, and $1/\tau$ = 0.93 eV (7500 cm$^{-1}$). Here, $\omega_p$ and $\epsilon_\infty$ are both somewhat larger than the typical results for $La_{2-x}Sr_xCuO_4$ quoted above; however, the screened plasmon frequency, which is $\omega_p/\sqrt{\epsilon_\infty}$ = 0.89 eV in the Drude model, is very close in all samples.

The good fit of the reflectance with a Drude dielectric function was used to argue against the midinfrared absorption reported in $YBa_2Cu_3O_{7-\delta}$ ceramics by Kamarás et al.[25] and in $La_{2-x}Sr_xCuO_4$ ceramics by Orenstein et al.[22] and Herr et al.[23] Based on the fit shown above in Fig. 4, Schlesinger et al.[38] claimed that there was no evidence for a mode at 0.5 eV in their crystals. Bozovic et al.[42] measured the reflectance and transmittance of an oriented $YBa_2Cu_3O_{7-\delta}$ film (**c**-axis normal to the film surface), evidently in a search for narrow, semiconductor-like, excitonic lines. Not finding any, they concluded that the reflectance spectrum could be well fit with a simple Drude model.

The Drude dielectric function is characterized by a mode at zero frequency, of oscillator strength $\omega_p^2/8$, proportional to carrier concentration, broadened to $1/\tau$ by scattering processes. Considerable evidence exists against the applicability of this model to the high $T_c$ superconductors and in favor of a strong *non-Drude* absorption in the mid infrared.[43] In the Drude model, the same scattering rate applies to both dc transport and high-frequency absorption. This is inconsistent with the known temperature dependence of the dc resistivity and midinfrared reflectance. The dc resistivity of many samples follows[44,45] $\rho = \rho_0 + aT$. This linear temperature dependence has been shown[45] to extend from $T_c$ to 1000 K in $La_{2-x}Sr_xCuO_4$ and up to 600 K in $YBa_2Cu_3O_{7-\delta}$. In the best samples, $\rho_0 \approx 0$. In contrast, the midinfrared reflectance is essentially temperature independent. This is illustrated in Fig. 6, which shows the reflectance of a ceramic sample of $YBa_2Cu_3O_{7-\delta}$ at 55 K, 105 K, and 300 K. The lack of temperature dependence is evident.

The magnitude of both the plasma frequency and the scattering rate ob-

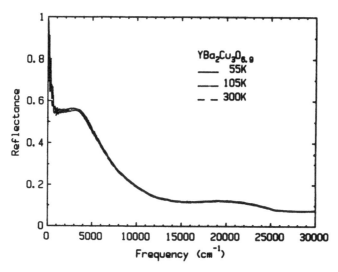

FIG. 6. The infrared reflectance of a $YBa_2Cu_3O_{7-\delta}$ ceramic sample at 55 K, 105 K, and 300 K. (From Ref. 46.)

tained by fits to the Drude model are unreasonably large. The London penetration depth, measured by muon spin resonance[47,48] or other techniques,[49] gives $\omega_p \approx 1.1$ eV in $YBa_2Cu_3O_{7-\delta}$ and $\omega_p \approx 0.6$ eV in $La_{2-x}Sr_xCuO_4$, i.e., approximately 40 % of the Drude-model result. The relaxation rate of 0.4–1.2 eV found by *all* Drude-model fits corresponds to a relaxation time of $\tau \sim 0.5$–$1.5 \times 10^{-15}$ sec. In turn, taking $v_F = 2 \times 10^7$ cm/sec, the mean free path would be $l = v_F\tau = 1$–3 Å. This unphysically short value for $l$ is by itself inconsistent with the strong temperature dependence of the conductivity.[45]

Other evidence against the simple Drude model comes from studies of the optical properties as a function of Sr or Ba doping levels in $La_{2-x}Sr_xCuO_4$ and $La_{2-x}Ba_xCuO_4$, and from detailed fits to the reflectance or conductivity at a variety of temperatures. These experiments are discussed in the following subsections.

It should be pointed out that the shortcomings of the simple Drude model are only clear when low-frequency and low-temperature measurements are made; at 300 K over the range 0.1–2 eV the Drude dielectric function gives a reasonably good reproduction of the measured dielectric function. Furthermore, the Drude model dc conductivity is $\sim 1200$ $\Omega^{-1}cm^{-1}$ for

$YBa_2Cu_3O_{7-\delta}$, not far from the actual room-temperature dc conductivity. Finally, the free carrier concentration estimated from the plasma frequency agrees rather well with simple estimates for the hole concentration, taking the carrier mass equal to the free electron mass:

(1) In the case of $La_{2-x}Sr_xCuO_4$, a plasma frequency of approximately 13,000 $cm^{-1}$ gives $n = 1.6 \times 10^{21}$ $cm^{-3} \approx 0.15n_{CuO_2}$.

(2) For $YBa_2Cu_3O_{7-\delta}$, with $\delta \sim 0$, a plasma frequency of 24,000 $cm^{-1}$ gives $n = 5.6 \times 10^{21}$ $cm^{-3} \approx 0.5n_{CuO_2}$.

Here, $n_{CuO_2}$ is the number density of two-dimensional (plane) $CuO_2$ units in the material. As discussed below, the simple Drude-model plasma frequency is related to the integrated oscillator strength of both the far-infrared Drude and midinfrared Lorentzian contributions to the dielectric function.

### C. Doping studies

Tajima et al.[21] observed that the location of the 0.83 eV (6600 $cm^{-1}$) plasmon minimum of $La_{2-x}Ba_xCuO_4$ was independent of the Ba concentration $x$ over a range $0.06 \leq x \leq 0.18$. Similar observations have been made by several other groups.[34,39,50,51,52] The most complete study, by Etemad and coworkers,[50,51,52] covered the concentration range $0 < x \leq 0.30$ in $La_{2-x}Sr_xCuO_4$. The results are shown in Fig. 7. Even though the low frequency reflectance changes with doping level, the location of the reflectance minimum is "pinned" at 0.9 eV.

This invariance of the plasmon frequency is inconsistent with the Drude model, for which $\omega_p \sim \sqrt{n}$. It is generally assumed that the carrier concentration in $La_{2-x}Sr_xCuO_4$ equals the Sr concentration $x$ for $x < 0.18$, and increases more slowly with $x$ at higher doping levels on account of partial compensation by oxygen vacancies.[53] (The early argument put forth by Tajima et al.,[21] that the system is a tight-binding metal near the half-filled limit, making the plasma frequency nearly independent of $x$, has been ruled out by the observation that $La_2CuO_4$ is an antiferromagnetic insulator.) Even if the change above $x = 0.18$ were minimal, a change by a factor greater than 2 should have been observed if the material were described by a Drude model. It seems to us that this conclusion is valid even if the relaxation rate and effective mass are frequency dependent.

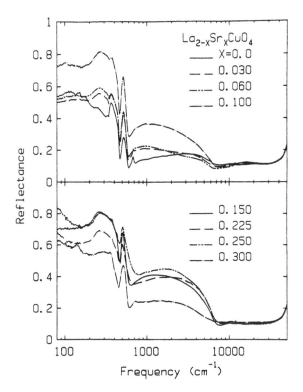

FIG. 7. Reflectance of $La_2CuO_4$ and $La_{2-x}Sr_xCuO_4$ at Sr concentrations between 0.02 and 0.30. The reflectance minimum at 0.9 eV is independent of Sr concentration. (From Ref. 52.)

More interesting than the pinned plasma frequency is the correlation between the intensity of the infrared absorption and the occurrence of superconductivity. This correlation, observed in $YBa_2Cu_3O_{7-\delta}$ by Kamarás et al.[25,54] and in $La_{2-x}Sr_xCuO_4$ by Orenstein et al.[39] and by Etemad and coworkers,[50,51,52] is illustrated by Fig. 8, which shows the concentration dependence of the transition temperature and the Meissner susceptibility in its upper panel and the infrared oscillator strength in its lower panel. Two measures of the oscillator strength are shown: the open circles are from the oscillator strength sum rule evaluated over the entire range below the plasmon minimum. The second (asterisks) show the apparent intensity of the c-axis polarized $240$ cm$^{-1}$ phonon of $La_2CuO_4$. In ceramics this intensity appears enhanced

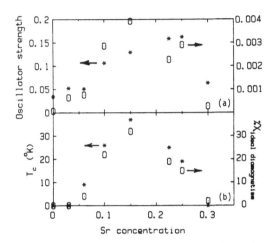

FIG. 8. Concentration dependence of superconductivity and infrared oscillator strength in La$_{2-x}$Sr$_x$CuO$_4$. The upper panel shows the infrared oscillator strength (see text); the lower the transition temperature (open triangles) and Meissner susceptibility (closed triangles). (From Ref. 52.)

because of the growth of the Drude and midinfrared **ab**-plane electronic absorption. All of the quantities plotted in Fig. 8 are seen initially to increase as the dopant concentration increases, to reach a maximum at $x \approx 0.18$ and to decrease at higher concentrations. Thus, a clear correlation exists between superconductivity and the infrared absorption.

Note that the normal-state dc conductivity is *not* correlated with the superconductivity. At high Sr concentrations, $\sigma_{dc}$ continues to increase with $x$,[55] whereas the transition temperature, Meissner susceptibility, and the infrared oscillator strength all decrease.

## D. The midinfrared absorption—textured ceramics

Several recent studies have shown that the infrared dielectric function of the high $T_c$ superconductors may be modelled with a sum of a Drude part (governing the far-infrared properties) and a broad electronic absorption band (dominating the midinfrared properties). These studies are discussed in this and the following subsection.

The properties of "textured" ceramic samples have been discussed in detail by the McMaster–Florida group. Bonn *et al.*[56] showed that the surfaces of as-prepared ceramic samples can have a high degree of orientation, with the crystallographic c-axis normal to the surface of the ceramic sample. In order to achieve this texture, the ceramic pellets have to be of high density, must be annealed at high temperatures for sufficient time, and must *not* be polished or mechanically worked in any way. If these conditions are met, then x-ray and electron microscope analysis shows that approximately 80 % of the surface is covered with small crystals having their c-axes oriented normal to the surface. Although the surfaces are rather rough and therefore appear to be of lower optical quality than well-polished ceramic samples, It was shown[56] that the textured samples have much larger far-infrared reflectance (corrected for scattering) than do polished samples.

Optical studies on these textured samples have been made[56,57,58,59,60,61] over a wide range of frequencies and at temperatures between 2 K and 300 K. Fig. 9 shows the conductivity (determined from a Kramers-Kronig analysis of reflectance) at 100 K. Note that the conductivity has a narrow peak at low frequencies, six sharp lines (attributed to phonons), and a broad wing to high frequencies. The normal-state data cannot be fit with a simple Drude model alone; instead, a dielectric function which is a sum of Drude, (Eq. 4), Lorentzian, (Eq. 16), phonon, and high-frequency terms must be used:

$$\epsilon(\omega) = -\frac{\omega_{pD}^2}{\omega^2 + i\omega/\tau} + \frac{\omega_{pe}^2}{\omega_e^2 - \omega^2 - i\omega\gamma_e} + \sum_{j=1,N} \frac{S_j\omega_j^2}{\omega_j^2 - \omega^2 - i\omega\gamma_j} + \epsilon_\infty \qquad (23)$$

There are four contributions to this dielectric function:

(1) Drude, characterized by a plasma frequency $\omega_{pD} = 0.74$ eV (5900 cm$^{-1}$) and a relaxation rate $1/\tau = 0.037$ eV (300 cm$^{-1}$);

(2) midinfrared, having a strength $\omega_{pe} = 2.6$ eV (21,000 cm$^{-1}$), a center frequency $\omega_e = 0.26$ eV (2100 cm$^{-1}$), and a width $\gamma_e = 1$ eV (8400 cm$^{-1}$);

(3) a high frequency part, with a constant $\epsilon_\infty = 3.8$; and

(4) the phonons, represented by $N = 6$ sharp oscillators. (For the parameters of these phonon modes, see Ref. 56.)

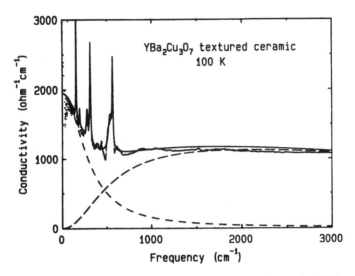

FIG. 9. Optical conductivity at 100 K of a textured ceramic sample of $YBa_2Cu_3O_{7-\delta}$. Data are shown as points while the fit to the model dielectric function of Eq. (23), which includes a Drude low-frequency part, a midinfrared band, and sharp phonons, is shown as the smooth curve. The individual contributions to $\sigma_1(\omega)$ of the Drude and midinfrared terms are shown as the dashed lines. (From Ref. 59.)

The fit to Eq. (23) is shown as the smooth curve in Fig. 9. The contribution of the Drude and the midinfrared parts are shown as the dashed lines. Although the Lorentzian representation of the midinfrared part is centered at 0.26 eV, the band has a relatively rapid onset, becoming dominant at $\sim$ 500 cm$^{-1}$ (0.06 eV). Note that the total infrared oscillator strength is given by $\omega_p^2 = \omega_{pD}^2 + \omega_{pe}^2$, with $\omega_p$ the plasma frequency found by fits of the Drude model alone to infrared data; also, $\gamma_e$ equals the relaxation rates found in such fits.

The zero-frequency intercept of the conductivity is $\sim 2000$ $\Omega^{-1}$cm$^{-1}$, somewhat below the measured 100 K conductivity, $\sigma_{dc} \approx 3000$ $\Omega^{-1}$cm$^{-1}$. Despite this discrepancy, the temperature dependence of the infrared conductivity is in *excellent* agreement with the temperature dependence of the dc resistivity. This is illustrated in Fig. 10, where the main part shows the infrared $\sigma_1(\omega)$ at three temperatures. The strongest temperature dependence is in the far infrared, with a substantially weaker high frequency behavior. The inset shows the resistivity measured in the far infrared; the resistivity is the reciprocal of the 100 cm$^{-1}$ conductivity. The resistivity follows the well-known[44,45] linear behavior of high-quality samples: $\rho = \rho_0 + aT$, with a relatively low value of $\rho_0$.

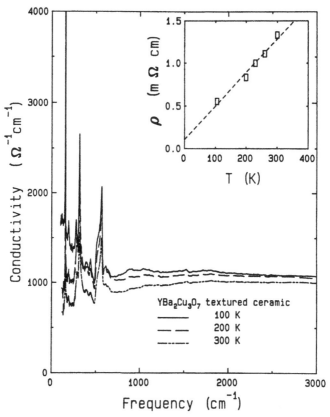

FIG. 10. Optical conductivity at 100, 200, and 300 K of a textured ceramic sample of YBa$_2$Cu$_3$O$_{7-\delta}$. The inset shows the temperature dependence of the resistivity measured by far-infrared means: the inverse of the 100 cm$^{-1}$ conductivity. (From Ref. 59.)

Fig. 11 shows $\epsilon_1(\omega)$ at 100 and 300 K. Note that $\epsilon_1(\omega)$ is negative at low frequencies (as appropriate for a conductor) but then rises and is nearly zero above about 800 cm$^{-1}$ (0.1 eV). (At 300 K, $\epsilon_1(\omega)$ becomes positive for a range of frequencies 800 cm$^{-1} < \omega < 2000$ cm$^{-1}$ whereas at 100 K $\epsilon_1(\omega)$ is negative but small in magnitude over this frequency region. It is this drop to a near zero value in $\epsilon_1(\omega)$ which causes the reflectance to display a plasmon-like drop in the

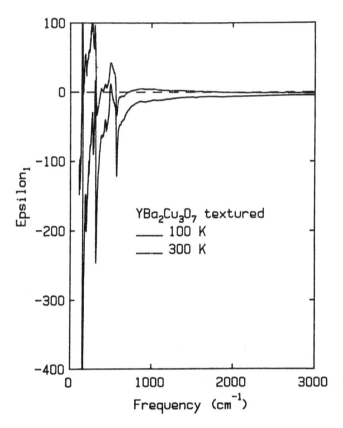

FIG. 11. Real part of the frequency-dependent dielectric function at 100 and 300 K of a textured ceramic sample of $YBa_2Cu_3O_{7-\delta}$. (From Ref. 60.)

far infrared, as discussed in section IV.

In the superconducting state, the optical properties are consistent with the Drude contribution collapsing to a delta-function at the origin, but with only minimal changes in the other parts of $\epsilon(\omega)$. As discussed by Timusk *et al.*[59] and Tanner *et al.*[60] the dielectric function is then well described by:

$$\epsilon(\omega) = -\frac{\omega_{ps}^2}{\omega^2} + i\frac{\pi\omega_{ps}^2}{2\omega}\delta(\omega) + \frac{\omega_{pe}^2}{\omega_e^2 - \omega^2 - i\omega\gamma_e} + \sum_{j=1,N}\frac{S_j\omega_j^2}{\omega_j^2 - \omega^2 - i\omega\gamma_j} + \epsilon_\infty \quad (24)$$

Here, $\omega_{ps}$ is the plasma frequency or oscillator strength of the superfluid electrons, and is proportional to $\sqrt{n_s}$, where $n_s$ is the superfluid electron density. In the clean limit, $n_s \approx n$. (See section IV of this review) The Drude contribution appropriate to the normal state has been replaced in Eq. (24) by the London-model contribution from the superconducting electrons, Eqs. (20)–(22). The first represents the inertial response at finite frequency while the second is the delta function conductivity of the superconductor. The remaining terms in Eq. (24) are the same as in Eq. (23). Most importantly, the onset of the midinfrared absorption does *not* shift to higher frequencies on entering the superconducting state. See Fig. 19 for the conductivity of the textured ceramic sample below $T_c$.

### E. The midinfrared absorption—crystals and films

The midinfrared absorption band is also evident in the **ab**-plane infrared properties of $YBa_2Cu_3O_{7-\delta}$ crystals[57,58,59,62,63] and thin films.[60,64]

For $YBa_2Cu_3O_{7-\delta}$ the reflectance of crystal samples can be well described by Eq. (23) above $T_c$. The fitting parameters differ somewhat for crystal samples, but in an understandable way. At 100 K, the McMaster-Florida experiments[57,58,59] find:

(1) a Drude contribution characterized by a plasma frequency $\omega_{pD} = 1.2$ eV (9,700 cm$^{-1}$) and relaxation rate $1/\tau = 0.03$ eV (240 cm$^{-1}$);

(2) a midinfrared part having a strength $\omega_{pe} = 2.5$ eV (20,000 cm$^{-1}$), a center frequency $\omega_e = 0.25$ eV (2000 cm$^{-1}$), and a width $\gamma_e = 0.6$ eV (5000 cm$^{-1}$); and

(3) a high frequency part modelled by a constant $\epsilon_\infty = 3.8$.

Phonon modes can be discerned in the reflectance but, on account of the substantially higher conductivity of the crystals (and somewhat lower signal-to-noise ratio of the data), phonon parameters have not been extracted.

The dc conductivity calculated from the Drude contribution to the dielectric function at 100 K gives a dc conductivity $\sigma_{dc} = 6500$ $\Omega^{-1}$cm$^{-1}$, and agrees with the measured dc conductivity. In comparison to the textured ceramics, the crystals have a considerably higher Drude plasma frequency $\omega_{pD}$ and a slightly lower scattering rate $1/\tau$. The higher plasma frequency can be qualitatively understood in terms of a greater degree of orientation of the surface of the crystal samples.

The crystals used in the AT&T Bell Labs experiments,[62,63] had a depressed transition temperature of $T_c = 50$ K. The data were fit using Eq. (23) for the dielectric function; the results were very similar to the McMaster-Florida parameters, finding at 100 K,

(1) a Drude contribution characterized by a plasma frequency $\omega_{pD} = 1.0$ eV (8200 cm$^{-1}$) and relaxation rate $1/\tau = 0.015$ eV (120 cm$^{-1}$); and

(2) the midinfrared part having a strength $\omega_{pe} = 3.0$ eV (24,000 cm$^{-1}$), center frequency $\omega_e = 0.2$ eV (1700 cm$^{-1}$), and width $\gamma_e = 0.9$ eV (7500 cm$^{-1}$).

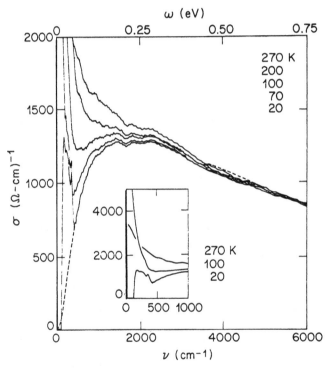

FIG. 12. Optical conductivity at five temperatures of a crystal of YBa$_2$Cu$_3$O$_{7-\delta}$ having a transition temperature of 50 K. Data are shown as lines while the midinfrared contribution to the fit is shown as the dashed line. (From Ref. 63.)

While some phonon modes can be seen, they were not included in the fit. The conductivity spectra are shown in Fig. 12 at five temperatures; the midinfrared part of the conductivity is the dashed line. The inset shows the low frequency conductivity on a compressed scale. The data were interpreted in terms of a frequency-dependent relaxation rate and effective mass, as discussed below.

Van der Marel et al.[64] studied the room temperature reflectance of a thin film of $YBa_2Cu_3O_{7-\delta}$ in combination with XPS and BIS (inverse photoemission) to determine the densities of occupied and unoccupied states. Their results are shown in Fig. 13. The conductivity shows a clear peak at 0.4 eV, with additional bands at 1.4 and 2.3 eV. The convolution of the XPS and BIS spectra, which to first order corresponds to an experimentally determined joint density of states, is also shown. This function has an onset between 2 and 3 eV with a rapid rise above 4 eV. Thus the low-energy bands cannot be due to ordinary interband transitions. Van der Marel et al.[64] speculate that they could indicate either charge transfer ($O \rightarrow Cu$) or a forbidden transition of Cu.

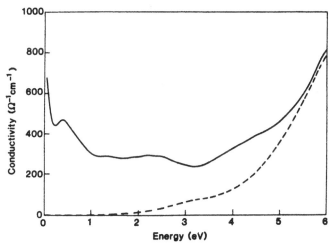

FIG. 13. Optical conductivity at room temperature of a $YBa_2Cu_3O_{7-\delta}$ thin film. The conductivity is shown as the solid line, while the joint density of states calculated from a convolution of XPS and BIS spectra is the dashed line. (From Ref. 64.)

Further independent evidence for a midinfrared absorption band comes from the *single-domain* reflectance of a crystal of $EuBa_2Cu_3O_{7-\delta}$ presented by Tanaka et al.[65,66], shown in Figs. 14–16. All other measurements in the 1–2–3 system are only averages over the **ab**-plane, on account of the well-known twinning of crystals of these high $T_c$ superconductors. (On account of this twinning, the phrase "single crystal" should in general be avoided in discussing experiments on the orthorhombic compounds, as twinned crystals are not single crystals.)

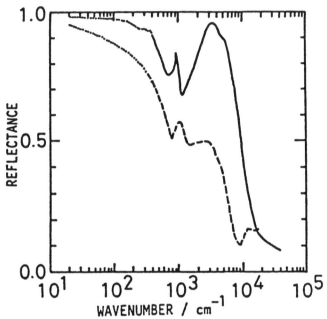

FIG. 14. Room-temperature reflectance of a *single domain* of $EuBa_2Cu_3O_{7-\delta}$. Polarization parallel to the b axis (i.e., the chains) is shown as the solid line while the a-axis polarization is shown as the dashed line. (From Ref. 65.)

Tanaka *et al.*[65,66] used a microscope to isolate a single domain on their $EuBa_2Cu_3O_{7-\delta}$ crystal, and then made polarized measurements parallel to both the **a** and **b** crystal axes. These are thus the first measurements of a single component of the dielectric tensor in the **a**- or **b**-direction. The reflectance is shown in Fig. 14. It is very interesting that these measurements show rather different spectra for the a-axis and b-axis polarizations. With the electric field along **b**, i.e., the direction of the one-dimensional Cu-O chains, the reflectance is considerably higher and the plasmon edge is both higher in frequency and sharper than for the a-axis polarization. Considerable structure is seen in the

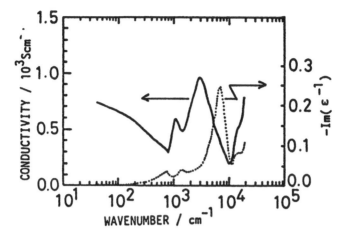

FIG. 15. Frequency-dependent conductivity (solid line) and loss function (dashed line) for the **a** axis of a *single domain* of $EuBa_2Cu_3O_{7-\delta}$. (From Ref. 65.)

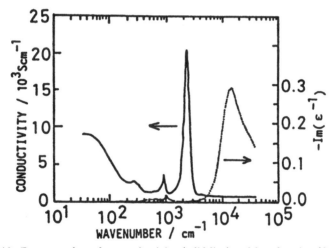

FIG. 16. Frequency-dependent conductivity (solid line) and loss function (dashed line) for the **b** axis of a *single domain* of $EuBa_2Cu_3O_{7-\delta}$. (From Ref. 65.)

$400\text{–}2000$ $cm^{-1}$ region.

Kramers-Kronig analysis reveals that there are strong midinfrared bands in both polarizations, particularly for $\mathbf{E} \parallel \mathbf{b}$. These are shown in Figs. 15 and 16. These figures give the conductivity, $\sigma_1(\omega)$ and loss function $-Im(1/\epsilon)$ for both polarizations. The peak in the **b** axis polarization is especially noteworthy.

This very intense band is located at 0.3 eV (2200 cm$^{-1}$) and has a width smaller than 0.1 eV (800 cm$^{-1}$).

Evidence of a strong midinfrared absorption has also been seen in studies of La$_2$CuO$_4$ crystals[67] and La$_{2-x}$Sr$_x$CuO$_4$ films.[68] Gervais et al.[67] expanded on the beautiful polarized measurements of La$_2$NiO$_4$ single crystals presented by Bassat et al.,[69] which demonstrated the strong anisotropy in both the electronic and vibrational contributions to the optical properties of that material. From their analysis they concluded that an electronic mode, polarized in the **ab** plane occurs both in La$_2$NiO$_4$ and La$_{2-x}$Sr$_x$CuO$_4$.

## F. Frequency-dependent scattering and effective mass

Thomas et al.[63] and Collins et al.[70] have recently presented an alternative to the decomposition of the dielectric function into a sum of Drude and midinfrared bands: use of a frequency-dependent scattering rate, $\Gamma(\omega)$, and effective mass, $m^*(\omega)$. (For a discussion of the use of $\Gamma(\omega)$ and $m^*(\omega)$ in the Drude formula, see subsection II.D of this review.) The application of this formalism gives an effective mass which is close to the free electron mass at high frequencies and which becomes an order of magnitude larger at low frequencies. Note that the high-frequency value for the effective mass is based on the assumption that the carrier concentration is $\sim 6 \times 10^{21}$ cm$^{-3}$. (See the discussion in section III.B above.)

The scattering rate is large and nearly temperature independent at high frequencies: $\Gamma(\omega) \approx 0.9$ eV (7000 cm$^{-1}$) for $\omega > 0.25$ eV (2000 cm$^{-1}$). At low frequencies, the rate becomes smaller and strongly temperature dependent; Thomas et al.[63] find $\Gamma(0)$ varying from 0.4 eV (3200 cm$^{-1}$) at room temperature to 0.12 eV (1000 cm$^{-1}$) at 100 K. Not surprisingly, these high- and low-frequency limiting values are comparable to the corresponding midinfrared band and far-infrared Drude contributions to the dielectric function, as described in subsection III.E of this review. The scattering rate values differ because Thomas et al.[63] have defined the scattering rate so that $\Gamma(\omega) = [m^*(\omega)/m][1/\tau(\omega)]$.

The changeover from low to high-frequency behavior occurs between 0.03–0.1 eV (250–800 cm$^{-1}$). This is taken[63,70] to suggest that the electrons are strongly coupled ($\lambda \approx 9$) to a set of excitations which occur in this approximate energy range. The frequency-dependent scattering rate and mass are interpreted as due to the Holstein effect,[7] in which a carrier can absorb a photon of energy $\omega$, emit an excitation of energy $\mathcal{E}$ and scatter, so long as $\omega > \mathcal{E}$. This phenomenon leads to an enhanced scattering rate at high frequencies, as discussed in subsection II.D of this review.

There are two observations in the high $T_c$ superconductors which work against this explanation. First, if the increase in scattering rate is due to the Holstein effect, and if the high $T_c$ superconductors have a superconducting energy gap, then the midinfrared band should shift upwards by $2\Delta$ in the superconducting state. The reason for the shift is straightforward: in the superconducting state the absorbed photon must have sufficient energy to break a pair as well as create the excitation. In the high $T_c$ superconductors no upshift is seen. Fig. 12 shows that the edge of the mid infrared band is at the same location in the superconducting and normal state.

Second, any excitation which is strongly coupled to the conduction carriers will affect the dc conductivity once temperature is high enough to thermally populate the excitation spectrum, e.g., once $k_B T \sim \mathcal{E}$. One therefore expects to see a significant modification of the temperature dependence of the conductivity at $T \geq 350$ K. In the high $T_c$ superconductors no change in the temperature coefficient occurs at these temperatures,[45] suggesting that excitations in this energy range do not contribute to the scattering of conduction carriers.

### G. High frequency bands

A number of authors have studied the high-frequency ($> 1$ eV) properties of $La_{2-x}Sr_xCuO_4$ ceramics[23,50], $La_{2-x}Sr_xCuO_4$ crystals[35], $YBa_2Cu_3O_{7-\delta}$ ceramics,[25,46,71,72] $YBa_2Cu_3O_{7-\delta}$ films,[64,73] and $YBa_2Cu_3O_{7-\delta}$ crystals.[72] A number of electronic bands are seen, some of which have interesting dependences on $x$ or $\delta$, whereas others are present at all concentrations of Sr or O.

The concentration-independent bands in $La_{2-x}Sr_xCuO_4$ occur at 1.4, 2.7, and 5 eV, with the 5 eV feature being the onset of strong interband absorption. $YBa_2Cu_3O_{7-\delta}$ has absorptions at 2.5–2.7 eV and above 6 eV, with the latter energy corresponding to the onset of strong interband transitions.

Etemad $et$ $al.$[50] studied the Sr concentration dependence of high-frequency absorption in $La_{2-x}Sr_xCuO_4$ using ellipsometry. They found that a band at

2.0 eV, which is quite strong in undoped $La_2CuO_4$, becomes weaker at higher doping concentrations and is almost absent for $x > 0.10$. Similarly, Kelly *et al.*[72] and Garriga *et al.*[71] find that the spectrum of $YBa_2Cu_3O_6$ has two rather narrow absorption bands at 1.7 eV and 4.1 eV, which decrease in strength as oxygen is added, until they disappear at $\delta \approx 0.5$. The band at 4.1 eV is particularly narrow, being only 0.2 eV wide in a $\delta \approx 0.9$ sample. This band is interpreted as an excitation of $Cu^+(1)$ atoms of the tetragonal-phase, large $\delta$ material. The oscillator strength which is removed from these high-frequency features on doping presumably appears as dc and midinfrared oscillator strength.

Consistent with these studies is a photoconductivity study,[74] which determined the onset of direct transitions in $La_{2-x}Sr_xCuO_4$ to be at 2.6 eV, while the onset of direct transitions in $La_2CuO_4$ was at 2.2 eV. The observation of photoconductivity in conducting $La_{2-x}Sr_xCuO_4$ samples is interesting, suggesting that the carriers which participate in interband transitions have little interaction with the carriers responsible for the conductivity.

### H. Photoinduced absorption

We end this section with a brief discussion of photoinduced absorption experiments.[75,76,77,78] In these studies, a sample of the nonsuperconducting, insulating phase, $La_2CuO_4$ or $YBa_2Cu_3O_{6.2}$ are illuminated with a high-frequency light source, typically a laser, and the absorption ratio between light on/light off is recorded. This ratio represents the excess absorption (or—if less than unity—the bleaching) due to the photoexcited carriers.

A photoinduced peak at 0.5 eV ($4000\ cm^{-1}$) in $La_2CuO_4$ has been observed by two groups.[75,76] Additional features include photoinduced infrared vibrational modes[75] and at higher frequencies, photoinduced absorption at 1.4 eV and photoinduced bleaching at 2eV and above.[76] The former effect suggests strong electron-vibration coupling while the bleaching is consistent with a 2 eV gap in $La_2CuO_4$.

Similar effects are observed[77,78] in $YBa_2Cu_3O_{6.2}$. Here the photoinduced midinfrared band has a maximum at 0.12 eV (1000 $cm^{-1}$) but extends to quite high frequencies. Several phonon features are also seen in the photoinduced spectra. Thus the photoinduced spectra of the insulating phases of high $T_c$ superconductors resemble in many ways the ordinary absorption spectra of the conducting phases.

## IV. THE ENERGY GAP

### A. Search for the gap in the ceramics

Guided by analogy to ordinary superconductors, which, at low temperature, have a region of essentially perfect reflectance at frequencies below the energy gap, much of the early infrared work on the new high temperature superconductors was aimed at finding the energy gap by reflectance techniques on pressed ceramic pellets. Within months of the announcement of the discovery of high $T_c$ superconductivity a number of groups submitted reports on the low temperature far-infrared spectra.

In the normal state, the $La_{2-x}Sr_xCuO_4$ pellets showed the high far infrared reflectivity characteristic of free carriers, as expected from the metallic dc resistance behavior. In the superconducting state a prominent edge appeared in the 50 $cm^{-1}$ region.[79-85] The position of the edge had a temperature dependence that was very suggestive of the BCS energy gap in the dirty limit. Fig. 17 shows the reflectance spectrum in $La_{2-x}Sr_xCuO_4$ ceramic as measured by Sherwin *et al.*[86] as a function of temperature. The edge at 60 $cm^{-1}$ dominates the spectrum. If we were to identify it as superconducting gap, $2\Delta/k_BT_c \approx 1.6$–2.5, a very low value.

This interpretation of the 60 $cm^{-1}$ edge as the onset of absorption at the energy gap of a dirty BCS superconductor leads to several problems. The first is the unexpectedly low energy of the gap found this way. The ratio $2\Delta/k_BT_c$ ranging from $\approx 1.6$ to $\approx 2.5$ is surprising since one had anticipated, in view of the very large transition temperature, a value suggestive of a strong coupling superconductivity, that is a value much larger than 3.5.

The second problem as shown by Bonn *et al.*[87] and by Sherwin *et al.*[86], is that a close examination of the optical constants of the effective medium of the

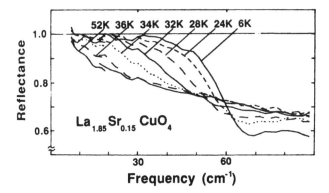

FIG. 17. Reflectance of a $La_{2-x}Sr_xCuO_4$ ceramic at various temperatures. The prominent temperature dependent edge has been interpreted as the superconducting energy gap and as a plasma edge. (After Sherwin et al.[86])

ceramic revealed that the edge at 60 cm$^{-1}$ did not correspond to an onset of electronic absorption. Instead it arose from a zero-crossing of $\epsilon_1$ of the composite material, with its mixture of grains with various orientations of the low conductivity c axis . Such a reflectance edge, when caused by a sign change in the real part of the dielectric function, is called a plasma edge in ordinary conductors, and it occurs at the screened plasma frequency. In a normal metal this frequency is rather high, occurring in the ultraviolet.

In the oxide ceramics the conduction electrons have a low density and their negative contribution to $\epsilon_1$ at low frequency is relatively weak. A positive contribution from *any* absorption band can cause $\epsilon_1$ to become positive at low frequency resulting in a premature zero crossing of this quantity. It is at such a zero crossing that the reflectivity develops a characteristic edge. In the case of the $La_{2-x}Sr_xCuO_4$ ceramics there are two sources of absorption that make a positive contribution to $\epsilon_1$: the strong midinfrared band and a strong phonon absorption at 250 cm$^{-1}$, polarized in the c direction.[67]

Sherwin et al.[86] have analysed in detail the temperature dependence of such a plasma edge in a hypothetical BCS superconductor but with normal state optical constants appropriate to the $La_{2-x}Sr_xCuO_4$ ceramics. The position of the edge depends on an interplay between the positive contribution of the finite frequency oscillators and the negative contribution from the pure inductive response of the superconducting condensate at zero frequency. As the temperature is increased from absolute zero the condensate fraction decreases and with it the negative contribution to $\epsilon_1$. As a result the edge moves to lower frequencies and the position of the edge mimics the temperature dependence of the BCS energy gap.

This can be seen from the London model. From Eq. (21) in Section II, the superconductor has a dielectric function given by

$$\epsilon_1 = \epsilon_\infty - \frac{\omega_{ps}^2}{\omega^2}$$

where for this discussion, the first term, $\epsilon_\infty$ is dominated by the contribution from the oscillators, assumed to be constant at low frequency, and the second is the contribution from the superfluid condensate, i.e., the inertial response of the superconducting electrons. This expression is zero at

$$\omega = \sqrt{\frac{4\pi n_s e^2}{m^* \epsilon_\infty}}$$

where $n_s$ is the density of superconducting electrons.

In the absence of strong absorption from the lattice or from interband effects, the reflectance will have a deep minimum at this frequency. Taking $\epsilon_\infty$ to be temperature independent, this expression can be used to determine the temperature dependence of $n_s$. Fig. 18 shows as a function of reduced temperature both the BCS gap and the reflectance edge based on this model. The temperature dependences of the two functions are very similar. There is some ambiguity in the density of the condensate determined by this simple argument since in the presence of damping it is not always easy to determine the zero crossing point of $\epsilon_1$ from an examination of various features of the reflectance edge.

Sherwin et al.[86] conclude that it is not possible, on the basis of far infrared measurements of the reflectance edge, to determine whether the low frequency feature in the $La_{2-x}Sr_xCuO_4$ is the BCS energy gap or a plasma edge caused by the zero crossing of $\epsilon_1$.

A similar edge appears in the $YBa_2Cu_3O_{7-\delta}$ ceramics at 220 cm$^{-1}$ giving $2\Delta/k_BT_c \approx 3.5$ (there is another, sharper one associated with the 150 cm$^{-1}$ phonon). A large number of investigators[29,88-98] have interpreted this as evi-

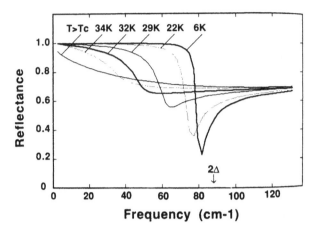

FIG. 18. Calculated reflectance at various temperatures from a model for a BCS super-conductor that has a strong phonon in addition to the free carriers. The arrow denotes the position of the assumed superconducting gap. There is no obvious feature in the calculated reflectance spectrum at the gap frequency. (From Sherwin et al.[86]).

dence of a BCS energy gap at this frequency. The position of this edge depends on the method of sample preparation[56] but in ceramics with well annealed and undamaged surfaces the reflectance does drop rapidly in the region of 220 cm$^{-1}$ as shown in Fig. 19.

Kramers-Kronig analysis[87,56,99] shows that this shoulder, observed in reflectance in $YBa_2Cu_3O_{7-\delta}$ceramics at low temperature, is not due to a well defined gap in the conductivity as expected for a BCS superconductor.[18] Instead it is partly caused by the strong dispersion contributed by the phonon pair at 277 cm$^{-1}$ and 311 cm$^{-1}$ that affect the effective-medium optical properties. The strong midinfrared absorption adds another important positive term to $\epsilon_1$. The net effect is the pronounced reflectance decrease seen at 220 cm$^{-1}$. The underlying continuous electronic conductivity of the ceramic does not exhibit a gap in this region. The conductivity of this sample is shown in Fig. 20. The conductivity has no BCS-like gap at the frequency where the reflectance edge is observed. The sharp features seen in the reflectance spectrum transform to sharp phonon peaks superimposed on a smoothly rising continuous background; thus, they cannot be associated with the gap either.

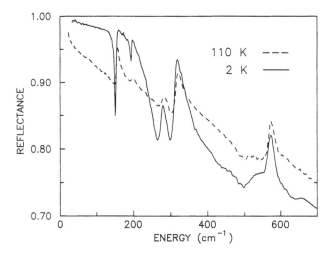

FIG. 19. Reflectance of a ceramic pellet of $YBa_2Cu_3O_{7-\delta}$ at two temperatures, one above the superconducting transition and the other well below it. The prominent edge at 220 cm$^{-1}$ has been interpreted as a signature of the superconducting energy gap. Kramers Kronig analysis shows however that this feature is associated with the strong dispersion associated with the phonons and the midinfrared band.

Nevertheless it is important to note that the frequency dependent conductivity is reduced in the far infrared region in the superconducting state. The characteristic frequency range of this reduction is of the order of 3.5 kT$_c$ in both $La_{2-x}Sr_xCuO_4$ and $YBa_2Cu_3O_{7-\delta}$.[82,56,99] Unlike a BCS superconductor where a region of true zero conductivity is observed,[100] the conductivity in the ceramics of the high temperature superconductors is not zero at far-infrared frequencies.

Although the phonon peaks confuse the reflectance spectrum, they also provide an indirect measurement of the energy gap. Phonon absorption features have finite widths caused by impurities, anharmonicity, and electron phonon interaction. In the superconducting state, phonons with frequency smaller than $2\Delta$ lose the width caused by the electron phonon interaction because they do not have enough energy to break Cooper pairs. Hence phonons with $\omega < 2\Delta$ narrow upon entering the superconducing state. This has been observed[56] in $YBa_2Cu_3O_{7-\delta}$ where the phonons at 155 cm$^{-1}$ and 195 cm$^{-1}$ narrow upon cooling, but the phonons above 279 cm$^{-1}$ do not. This suggests that the gap

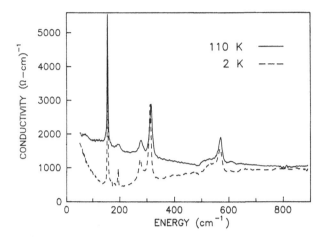

FIG. 20. The optical conductivity of a YBa$_2$Cu$_3$O$_{7-\delta}$ ceramic pellet. In the normal state at 110 K the conductivity is characterized by a rising free carrier part superimposed on a uniform background associated with the midinfrared band. Superimposed are several sharp phonon bands. On entering the superconducting state the free carrier conductivity is markedly reduced in the 100 to 300 cm$^{-1}$ region but there is no evidence of a sharp gap. At the lowest frequencies the conductivity rises, as expected from the low-frequency behavior of the reflectance, shown in Fig. 19.

lies between 195 and 279 cm$^{-1}$ $(3.0 < 2\Delta/k_B T_c < 4.3)$.

In summary, the far infrared reflectance spectra of ceramic samples of the high temperature superconductors show shoulders that superficially resemble the energy gap of a BCS superconductor. However, these shoulders can be attributed to changes in the real part of the dielectric function associated with phonons and with the gradual onset of strong midinfrared absorption. Kramers-Kronig analysis gives a more detailed picture showing that while the far infrared conductivity is depressed in the superconducting state no true gap in the conductivity can be discerned.

The failure to observe a true energy gap by far-infrared spectroscopy in the ceramics can be traced to several causes. Firstly, non-superconducting material present on the surfaces of the specimens. Surface analysis techniques show small concentrations of minority phases in some samples, materials that originate from the crucible such as Al or Pt, as well as components of unreacted material such as CuO.[59] Aluminum is known to suppress the transition temperature[101] and presumably, the energy gap.

Secondly there is the problem of c axis anisotropy. Measurements of dc electrical conductivity show that at low temperature the electron transport is very anisotropic: metallic if the currents flow in the **ab** plane and almost insulating in the **c** direction.

Finally the $YBa_2Cu_3O_{7-\delta}$ material is often heavily twinned with very small regions of alternating **a** and **b** direction domains. It is possible that the domain wall region is one with reduced oxygen content which could depress the gap locally.

## B. The gap in ab-plane oriented crystals

To eliminate the uncertainties arising from anisotropy in polycrystalline, unoriented samples, single crystal studies are desirable. Radiation polarized along the various crystallographic direction can be used to obtain the corresponding optical conductivities. For simple reflectance spectroscopy in the energy gap region, 100 - 600 $cm^{-1}$ , measurements on flat surfaces several millimeters in size, are needed. Crystals large enough to study both the two **ab**-plane directions and the **c** axis in the far infrared are not available at the time of writing of this review. The typical crystal that does exist is a very thin plate up to 3 by 3 mm in size in the **ab** plane and less than 0.1 mm thick in the **c** direction. In the **ab** plane there is considerable twinning: the largest untwinned regions are less than 100 $\mu$m in size.[102,65] It is for this reason that all the far-infrared measurements on crystals to date are unpolarized averages over the **a** and **b** directions of twinned specimens. No far-infrared spectra have been reported for light polarized in the **c** direction for superconducting samples.

The first absolute reflectance curves for $YBa_2Cu_3O_{7-\delta}$ in the long wavelength region where the energy gap is expected to appear are the measurements of Bonn *et al.*[57] on a mosaic of **ab** plane twinned crystals. Fig. 21 shows the reflectance at two temperatures, in the normal and superconducting states. There is no evidence of unit reflectance in the superconducting state that could be used to identify an energy gap. The reflectance is high at low frequency but there is residual absorption of the order of a few percent in the 200 – 500 $cm^{-1}$

region. The material was grown in an alumina crucible, and the unintentional Al doping suppresses $T_c$[101] somewhat, but this kind of residual absorption is also seen in most crystals grown in zirconia crucibles.[103] There is no sign of the characteristic signature of a BCS type of energy gap in this material: a region of unit reflectance followed by a rapid decrease to the normal state reflectance.

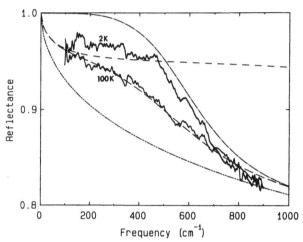

FIG 21. Reflectance of a mosaic of crystals oriented with the **ab** plane normal to the plane of incidence. The normal state reflectance is dominated by a region of negative curvature in the 400 cm$^{-1}$ region, which becomes more prominent in the superconducting state. There is no evidence of a superconducting gap in this spectrum. The two dashed lines with positive curvature are fits to Drude models with different damping constants. The two curves with negative curvature include a midinfrared oscillator; the lower one is a least squares fit to the 100 K reflectance, the upper one uses the same parameters, but the Drude damping has been reduced to zero.

A detailed analysis of the optical properties of this sample[58,59] shows that in the normal state the conductivity can be described by a Drude term with a damping of 250 cm$^{-1}$ at 100 K plus an oscillator to account for the midinfrared band. The onset of the midinfrared band is responsible for the region of negative curvature in the 500 cm$^{-1}$ region. This hump is characteristic of many crystal samples of $YBa_2Cu_3O_{7-\delta}$.[103]

Calculations using the Mattis Bardeen formula show that any superconducting gap of the order of 200 cm$^{-1}$ or larger does not produce an observable feature in the reflectance spectrum.[59] This is because already at 100 K the material is close to the clean limit, where $2\Delta > 1/\tau$. On account of the temperature dependence of the conductivity, one expects an even smaller damping rate at lower temperatures. In the dirty limit a strong shoulder develops in the reflectance at the gap frequency whereas in the clean limit there is no discernible feature at $2\Delta$. The physical reason for the disappearance of the gap in the case of very pure materials is the lack of collision processes that conserve momentum during the absorption of a photon. In contrast to the case of a semiconductor, where the gap is tied to the lattice potential, the superconducting gap is of electronic origin and cannot absorb the momentum of an excited electron[104].

Perhaps the strongest case for the observation of a superconducting gap by infrared techniques is the work of Thomas et al.[63] who report on the ab-plane properties of crystals of $YBa_2Cu_3O_{7-\delta}$ deliberately treated to yield a sample with reduced oxygen content and depressed transition temperature. Fig. 22 shows the absolute reflectance in the low frequency region of one of the samples. The authors argue that the region of high reflectance with the characteristic break at 150 cm$^{-1}$ for the 68 K material represents the superconducting gap.

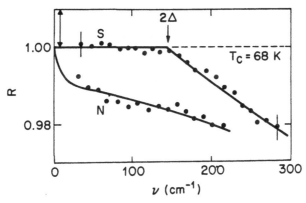

FIG. 22. Reflectivity of an oxygen depleted superconductor $YBa_2Cu_3O_{7-\delta}$. The authors interpret the region of unit reflectance below 150 cm$^{-1}$ as the superconducting gap. (After Thomas et al.[63])

There are several conditions that must be met before a reflectance shoulder can legitimately be identified as a superconducting gap. First the absolute reflectance should be unity up to the gap. Within the quoted 2 % error limits, the curve of Thomas *et al.*[63] meets this condition . Secondly the feature should not be present in the normal state. There seems to be no region of positive curvature in the $100 - 200$ cm$^{-1}$ region that sharpens up to become the break at 150 cm$^{-1}$ in the superconducting state. As a final check the material should not be in the clean limit. This last condition is difficult to test since we do not know what the Drude relaxation rate is in the superconducting state at low temperature. Thomas *et al.*[63] claim that in the normal state $1/\tau \sim 2k_BT$. At 100 K this gives a value of $1/\tau = 140$ cm$^{-1}$. If the normal-state resistivity levelled off below 100 K, so that the value $1/\tau = 140$ cm$^{-1}$ persisted to low temperatures, then the sample would not be in the clean limit and the feature observed at 150 cm$^{-1}$ by Thomas *et al.*[63] has a good chance of being the superconducting gap. A value of $2\Delta/k_BT_c \approx 3.5$ is obtained from this gap value and the 68 K transition temperature of the sample.

A much higher value of $2\Delta/k_BT_c$ is reported by Schlesinger *et al.*[38,105] Only reflectance ratios between the superconducting and normal states, $R_s/R_n$ of a mosaic of aligned twinned crystals are given. The authors identify a shoulder that occurs at 480 cm$^{-1}$ as the superconducting energy gap, so $2\Delta/k_BT_c \approx 7$ to 8. They do not observe unit reflectance at this frequency as one would expect for a bulk BCS superconductor. From their description it appears that the sample still has about 10 % absorption at 480 cm$^{-1}$. They suggest that presence of normal material causes this residual absorption.

The 480 cm$^{-1}$ shoulder can be seen in the absolute reflectance curves of other investigators. Timusk *et al.*[58] associate it with the onset of the midinfrared absorption, whereas Thomas *et al.*[63] interpret it in terms of changes in the frequency dependent scattering rate.

Against the interpretation of shoulders such as the one at 480 cm$^{-1}$ as superconducting gaps is an argument raised by Thomas *et al.*:[63] the reflectance does not rise to unity below this frequency as expected for a superconductor with a full BCS gap. Unit reflectance by itself does not guarantee that a shoulder is due to a superconducting gap. As we have seen in the discussion of the ceramics, the plasmon model of Sherwin *et al.*[86] yields a plateau of high reflectance with a gap-like shoulder, without the necessity of a gap. The same mechanism appears to be operating at 480 cm$^{-1}$: the onset of the strong mid-infrared absorption pulls down the reflectivity and gives rise to an apparent shoulder.

The influence of the midinfrared absorption on reflectance in the 500 cm$^{-1}$ region can be seen clearly in Fig. 23, from Thomas *et al.*[63] The reflectance shoulder that signals the onset of non-Drude behavior in the 400 – 500 cm$^{-1}$ range can be seen at 200 K in the normal state and it seems to sharpen continuously until at 12 K, in the superconducting state, it becomes the sharp break at 480 cm$^{-1}$. In other words, if the 480 cm$^{-1}$ feature is a gap, it is a gap that is already present in the normal state, only becoming sharper and better defined at low temperature.

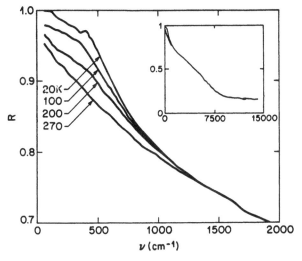

FIG. 23. Reflectivity of a crystal of YBa$_2$Cu$_3$O$_{7-\delta}$ with a transition temperature of 50 K. Note the characteristic region of positive curvature seen at low temperature in the 500 cm$^{-1}$ region of the spectrum. It persists up to 200 K while at room temperature the spectrum takes on the characteristic Drude shape with a continuous positive curvature. (After Thomas *et al.*[63])

Recent work on Bi$_2$Sr$_2$CaCu$_2$O$_8$ by Reedyk *et al.*[106] gives further support to this view. Fig. 24 shows the reflectance of this material in the normal and superconducting states in the far infrared. In the normal state the conductivity can be fit with a model that has two broad oscillators at 400 and 1000 cm$^{-1}$ in addition to a Drude term centered at zero frequency.

In the superconducting state, within the experimental error of 0.5 %, the sample has unit reflectance below 300 cm$^{-1}$. This behavior is suggestive of a superconducting gap. However the oscillator needed to fit the normal state conductivity is close to this frequency, and quite possibly the sharp threshold is just a result of the sharpening of the low frequency edge of this oscillator, not the superconducting gap. Thus the 300 cm$^{-1}$ shoulder in the Bi$_2$Sr$_2$CaCu$_2$O$_8$ has several of the necessary hallmarks of a superconducting gap, unit reflectance, a ratio $R_S/R_n$ that superficially resembles the Mattis-Bardeen shape, and a marked sharpening in the superconducting state. Nevertheless it is very difficult to argue that this compound has a *superconducting* gap at this frequency since the structure is already present in the normal state.

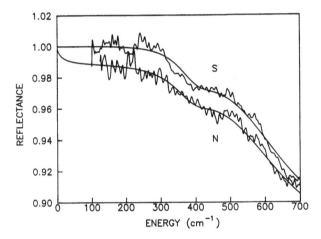

FIG. 24. Reflectance of Bi$_2$Sr$_2$CaCu$_2$O$_8$ in the far infrared. The normal state reflectance, marked N is very high at low frequency. The solid curve is a fit with with two oscillators at high frequency and a Drude term. The superconducting state, marked S, can be fit with the same oscillators but a Drude damping that is zero (solid smooth curve). In the 300 cm$^{-1}$ region the experimental onset of absorption appears steeper than predicted by the model.

The difficulties of locating a superconducting gap in the crystals available at the time of writing this review are best illustrated in the recent work of the

AT&T group.[103] Reflectance curves for a large variety of crystal samples are given. There is a substantial amount of sample-to-sample variation but certain common features can be identified in the spectra:

(1) Good samples, defined as those having greater than 95 % absolute reflectance below 500 cm$^{-1}$, tend to show the 480 cm$^{-1}$ shoulder in both the normal and the superconducting state.

(2) Poorer samples, those with lower reflectance, are more Drude-like: the reflectance curve has the uniform positive curvature expected for a Drude metal in the low frequency region. This is not too surprising: in the simple Drude picture, a high absorption at low frequency corresponds to a high scattering rate, which would tend to wash out any low-lying spectral features such as the shoulder at 480 cm$^{-1}$. In agreement the better samples tend to have a more washed out shoulder at higher temperature.

(3) The energy gap cannot be seen clearly in any of the samples. Having ruled out the 480 cm$^{-1}$ shoulder as the gap, there is no evidence of a lower shoulder in any of the curves.

(4) There is some evidence for additional non-Drude behavior at the very lowest frequencies in the 50 to 100 cm$^{-1}$ range in many of the samples. The reflectance seems to rise to unity at a finite, nonzero frequency. This could be consistent with a very low gap in the surface regions; it might be depressed by impurities or other defects.

At the time of writing of this review the gap question has not been settled: there is evidence in the work of Thomas et al.[63] for a rather low gap, of the order of 100 to 150 cm$^{-1}$ magnitude in some oxygen deficient samples. These samples are characterized by very high normal state reflectances $R_n > 0.98$ at 200 cm$^{-1}$ and perhaps have surface layers that are closer to the pure bulk superconductor than other samples. In other, perhaps less ideal, samples no gap can be identified in the 200 - 600 cm$^{-1}$ region.

### C. Other techniques

All the data presented so far was obtained by single-bounce reflectance, the most popular technique used in the study of the infrared properties of the high temperature superconductors. In this section we will discuss other infrared techniques for finding the energy gap.

An alternative method of great potential sensitivity is the direct measurement of absorbed power by recording the temperature rise of a sample with a sensitive bolometer. The information provided by this technique is in principle not different from what is gained by reflectance, but the method is capable of great precision with small samples and with small absorption levels, as shown in studies of lead by Joyce and Richards[6] and in the organic conductor TTF-TCNQ by Eldridge and Bates[107]. The method has the disadvantage that the measurements have to be carried out in a narrow temperatures range matched to the region of sensitivity of the bolometer.

So far only ceramic samples have been measured. Gershenzon $et$ $al.$[92] find in $YBa_2Cu_3O_{7-\delta}$ a threshold of absorption above 40 cm$^{-1}$ that they identify with the energy gap. The material is not fully reflecting below this frequency but appears to have a residual absorption that can be estimated to be of the order of 1 % if we assume that the 200 cm$^{-1}$ absorption is 30 %, typical of $YBa_2Cu_3O_{7-\delta}$ ceramics. Similarly Lee $et$ $al.$[108] find that their sample of $YBa_2Cu_3O_{7-\delta}$ ceramic has an absorption on the order of 20 % below 200 cm$^{-1}$, rising strongly above this frequency. The characteristic phonon lines at 150 cm$^{-1}$ and 195 cm$^{-1}$ can be discerned.[108]

The bolometric method, when further refined, will be particularly useful with small crystals of very high reflectivity since the measurement of absolute reflectance to an accuracy better than 1 % is difficult by conventional single-bounce reflectance techniques.

Whereas bolometric measurements are made at constant temperature, microwave loss measurements are generally performed at constant frequency. In most cases the temperature dependence is studied. To extract an energy gap from surface impedance, a detailed theory is needed.[109] At low temperatures one can write the ratio of surface resistance in the superconducting state $R_s$ to the normal state $R_n$ as

$$\frac{R_s}{R_n} \sim e^{-\Delta/kT}.$$

This formula implies that the surface resistance would drop to zero at low temperature; in principle, one could find the energy gap from the temperature dependence. In practice, even conventional superconductors fail to follow this formula to the lowest temperature, always with some residual sample-dependent surface resistance. This residual absorption is much higher in the oxide superconductors.[110–113] Beyermann *et al.*[110] find, for example, a residual $R_s/R_n \approx 0.01$ at low temperature. They model this with a Lorentzian distribution of gap centered around $2\Delta/k_B T_c = 3.5$.

Porch *et al.*[113] studied $YBa_2Cu_3O_{7-\delta}$ powders diluted with $Al_2O_3$ at 5 GHz. Because of the difficulties with shape corrections, the absolute values of the conductivities $\sigma_1$ and $\sigma_2$ are subject to considerable error. Good results are obtained for the temperature dependence of $\sigma_2$. For example they find that $\sigma_2$, closely follows the clean limit BCS curve below $T_c$, (Fig. 25.) this behavior suggests that $2\Delta/k_B T_c$ is within 30% of the BCS weak coupling behavior. The real part of the conductivity $\sigma_1$ is more difficult to determine, especially as large additional losses above the expected BCS surface resistance were found. This behavior is consistent with other studies, both infrared and microwave.

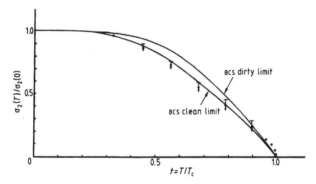

FIG. 25. The temperature dependence of the imaginary part of the conductivity ($\sigma_2$) of $YBa_2Cu_3O_{7-\delta}$ powders obtained by microwave measurements. The authors interpret these data with the clean limit BCS model, finding a gap within 30 % of the BCS value. (From Ref. 113.)

## V. SUMMARY

### A. The midinfrared absorption

Most recent infrared studies of the high $T_c$ superconductors show that the far-infrared and the mid infrared properties are governed by different phenomena, with the far-infrared conductivity being closely related to the dc transport properties whereas the midinfrared conductivity is dominated by some sort of low-lying excitation process. The crossover between the two regions occurs in the 0.05–0.13 eV (400–1000 cm$^{-1}$) region.

In discussing the midinfrared absorption, it is important to note that it occurs not only in the copper-oxide based superconductors, but also in many other systems. Barker,[114] in a study of the infrared properties of SrTiO$_3$ which had been made superconducting by reduction in hydrogen, observed that there was a strong midinfrared absorption in the superconducting materials. T$_c$ in this material is low, but in other ways it resembles the copper-oxide materials. The broad infrared absorption, located at 0.25 eV (2000 cm$^{-1}$) has a peak conductivity of 100 $\Omega^{-1}$cm$^{-1}$, about a factor of ten smaller than in the high T$_c$ superconductors.

A second material in which a midinfrared absorption occurs is the "old" high-T$_c$ material, BaPb$_{1-x}$Bi$_x$O$_3$, which was studied in detail by Tajima $et$ $al$.[115] It is important to note that unlike the new high T$_c$ superconductors, BaPb$_{1-x}$Bi$_x$O$_3$ is a cubic material; hence, its dielectric function should be a scalar quantity. None of the issues of anisotropy or inhomogeniety arise in the discussion of this material.

BaPb$_{1-x}$Bi$_x$O$_3$ is a semiconductor for $x > 0.35$, a superconducting metal with T$_c \approx 10$ K for $0.20 \leq x \leq 0.35$, and a semimetal for small $x$. Samples in the superconducting range have a significant peak in the conductivity at approximately 0.5 eV (4000 cm$^{-1}$).[115] Fits to a model dielectric function similar to Eq. (23), including Drude, midinfrared, and $\epsilon_\infty$ terms, shows that the midinfrared absorption has about 75 % of the oscillator strength; the Drude 25 %. The midinfrared absorption is absent or much reduced in the low-T$_c$, $x \approx 0$ regime, although the $\sigma_{dc}$ is still rather large. A final point of similarity is that when $x$ is well into the semiconducting regime, the midinfrared band has been replaced by a 1.5–2 eV peak. In contrast to the copper-oxide high T$_c$ superconductors, however, the location of the plasmon minimum in the reflectivity, which occurs near 1.5 eV (12,000 cm$^{-1}$) for $x \approx 0.27$, is strongly concentration dependent. Fits to Drude (or Drude plus Lorentzian) dielectric function models show that the square of the plasma frequency is a linear function of Bi concentration, *x*. However, much of the increase at large x is clearly related to the growth and blue shift of the midinfrared absorption, and does not represent a linear increase in carrier concentration with *x*.

Finally, Kaplan *et al.*[116] have studied the infrared and optical properties of $La_4BaCu_5O_{13}$ and $La_2SrCu_2O_{6+\delta}$. As discussed by Torrance *et al.*[53] these materials have square-planar copper-oxide structures, are metallic (when $\delta > 0$), but are not superconducting. Kaplan *et al.* observed that $La_4BaCu_5O_{13}$ has a broad maximum at 1 eV (12,000 cm$^{-1}$) with a quasi-Drude, zero-frequency-centered peak dominant below 0.3 eV (3000 cm$^{-1}$). Unannealed samples of $La_2SrCu_2O_{6+\delta}$ (that is, samples with $\delta \approx 0$) have insulating behavior and no midinfrared absorption; those with increasing $\delta$ have a progressively stronger midinfrared absorption at 0.6 eV (5000 cm$^{-1}$). The center frequency of the infrared absorption is higher in these non-superconducting oxides than the 0.2–0.4 eV value found in $YBa_2Cu_3O_{7-\delta}$.

It seems that the common feature of the materials that display a strong midinfrared band is their mixed-valence nature. In $La_2CuO_4$, $YBa_2Cu_3O_6$, $SrTiO_3$, $BaBiO_3$, and $La_2SrCu_2O_6$ the valence of all ions is satisfied at their nominal values; in particular all of the oxygens are $O^{2-}$. When Sr or Pb are substituted for other ions or when oxygen is added or removed, the situation becomes complicated. The consensus at the present time seems to be that in the high $T_c$ superconductors the holes reside on the planar oxygen atoms; if this is the case also in the other compounds, then the midinfrared absorption can be regarded as a signature of a significant concentration of $O^-$ centers in the solid.

There are at least three ways of interpreting the midinfrared absorption:

(1) The effect is due to an emission of the excitation by the conduction carriers, i.e., a Holstein[7] process. In this indirect process, everything takes place in the conduction band and the *same* carriers are involved in both regimes, albeit with differing effective masses and lifetimes. This is the perspective taken by Thomas *et al.*[63] and Collins *et al.*[70]

(2) The midinfrared absorption is due to a direct excitation of some sort, which we call a "particle-hole" excitation. This excitation involves either a different set of carriers, a different initial state, or a different final state than the

dc transport. Herr *et al.*,[23] Kamarás *et al.*,[25] and, more recently, Timusk *et al.*[59] have taken this point of view.

(3) The absorption is due to a parallel channel of charge carriers with a very short, temperature independent relaxation time.

Two difficulties faced by the first interpretation were discussed in some detail in section IV. Briefly, the onset $\mathcal{E}$ of the Holstein processes should shift by $2\Delta$ to higher frequencies when the material becomes superconducting; the midinfrared absorption, if anything, shifts to lower frequencies.[63,117] In addition, at temperatures where $k_B T \sim \mathcal{E}$ the dc resistivity should be strongly affected due to scattering of charge carriers by a thermal population of the excitations; the resistivity, however, remains linear over this temperature range.[45]

The interpretation in terms of a particle-hole excitation does not suffer from these problems. The superconducting gap would appear in the spectrum of the Drude carriers but would not necessarily affect the onset energies of particle-hole excitations. Similarly, if these excitations are weakly coupled to the conduction electrons, they would not affect the dc transport processes in any drastic way.

The third interpretation would lead to Drude response, governed at low frequencies by the long relaxation time but at high frequencies by the short relaxation time (remembering that for $\omega \gg 1/\tau$ the $\tau$-dependence of the conductivity is $\sigma_1 \sim 1/\tau$). However, this notion is ruled out[59] by the lack of saturation of the dc resistance at high temperature. The high- frequency Drude parameters imply the temperature-independent part of the conductivity would be $\sigma_o \approx 1200$ $\Omega^{-1}\mathrm{cm}^{-1}$. This is substantially larger than permitted by the measured high temperature conductivity,[45] which would give $\sigma_o \ll 300 \ \Omega^{-1}\mathrm{cm}^{-1}$.

There are several theories for the high $T_c$ superconductors which give low-lying electronic transitions. It is beyond the scope of this review to give even an overview of the theory, but we mention here, in no particular order, a few papers which have addressed directly the optical properties.

Electronic band structure calculations can be used in principle to find the optical properties. This has been done for the $YBa_2Cu_3O_{7-\delta}$ system by Chui *et al.*[118] with quite interesting results. For $\delta \approx 0$ a strong midinfrared absorption is found, while for $\delta \approx 1$ this band is substantially reduced in intensity. This result agrees qualitatively with the measurements of Kamarás et al.[25] on ceramic $YBa_2Cu_3O_{7-\delta}$ samples. Moreover, Chui et al. find that the midinfrared absorption, which is rather broad in unpolarized **ab**-plane spectra actually occurs at different frequencies for the two polarizations: 0.5 eV (4000 cm$^{-1}$) for **E** ∥ **a** and 0.35 eV (2800 cm$^{-1}$) for **E** ∥ **b**. These values agree well with the results of Tanaka et al.[65]

The higher energy bands seem to be able to be well accounted for within ordinary electronic band structure pictures, with most calculations giving approximately 4–5 eV for the onset of many strong transitions and with the minimum direct interband gap being in the 1.7–2.2 eV range.[119,120]

The other approaches are generally based on the idea that the charge carriers suffer from strong Coulomb repulsion, making double occupancy of sites (especially Cu) unlikely. This approach leads to a correlated ground state, with low-lying electronic excitations of the charge transfer type, consisting of electrons being promoted from their ground-state orbital either to a neighboring site or to a different orbital on the same site. A wide variety of such transitions have been considered.

Shortly after the initial discovery of high $T_c$ superconductors Varma, et al.[26] proposed that the superconductivity is mediated by charge transfer excitons from oxygen to copper. Taking an extended Hubbard picture, they proposed that the excitation consist of $Cu^{2+}O^{2-} \rightarrow Cu^+O^-$, etc., and predicted a band at $\approx 0.5$ eV, governed by V, the near-neighbor term in the extended Hubbard model for two sites per unit cell.

A different charge-transfer picture has been presented by Weber[121] and by Jarrell et al.,[122] who consider charge transfer transitions between two $d$ levels on Cu. They also find a band in the 0.4 eV region.

Interactions with the lattice have been considered by a number of workers. Here the picture is that a hole introduced by doping interacts strongly with the nearby Cu–O bonds, leading to the formation of a self-trapped polaron. The optical absorption associated with polarons has been discussed by Laughlin and Hanna[123] and by Rice and Wang.[124] Rice and Wang consider hole polarons, with the holes located mostly on the square-planar oxygen atoms surrounding a particular Cu site. The lowest optical transition is then an excitation from a non-bonding O level to the half-filled Cu-O antibonding level. With the parameters chosen, this occurs at 0.3–0.5 eV (2000–4000 $cm^{-1}$). In addition, they showed that electron-phonon coupling can lead to enhancements of the phonon absorption oscillator strength.

Magnetic ground states have been investigated widely, on account of the antiferromagnetic ground states of $La_2CuO_4$ and $YBa_2Cu_3O_6$. Schrieffer *et al.*[125] have constructed a "spin-bag" model by starting with a spin-density-wave (SDW) ground state for the insulating phases. This SDW state has a relatively large energy gap. On the introduction of holes through doping or increased oxygen content, the SDW gap is reduced. This reduction is not uniform; instead, it occurs only in the vicinity of the hole, forming a local distortion and trapping the hole in a manner analogous to a polaron. It is relatively straightforward to speculate about the optical absorption within this picture. The undoped state would be an insulator, and transitions across its SDW gap would be seen. Upon doping, a new absorption would appear at low frequencies, at an energy equal to the value of the gap within the bag. The oscillator strength of this absorption would grow with doping until the doping level becomes large enough to destroy the SDW order, when the strength should begin to decrease. Accompanying the initial increase in the low-energy band would be a corresponding decrease in the intensity of transitions across the original SDW gap. It is tempting to interpret the changes seen in $La_{2-x}Sr_xCuO_4$ upon doping[50], where a decrease in absorption in the 2.0 eV (16,000 $cm^{-1}$) range goes along with the increase in midinfrared absorption, in terms of this model, but no detailed calculations have as yet been carried out.

In summary, the midinfrared absorption in the high $T_c$ superconductors, first seen in ceramic samples,[22,23,25] is also present in oriented **ab**-plane[56,59,63] samples. The nature of this absorption at the time of writing this review is undecided. We have discussed it in terms of a direct particle-hole excitation, but this still leaves many possibilities open. The working out of this fascinating puzzle, which itself is only part of the overall, multidimensional puzzle presented by the high $T_c$ superconductors, will require further and better optical measurements, improved sample quality, and, no doubt, a number of new ideas.

## B. The energy gap

At the time of the completion of this review there is little conclusive that can be said about the energy gap in the high $T_c$ superconductors. Like the search for the Loch Ness monster, there have been many sightings but little tangible evidence. In this section we will briefly summarize our view of reasons for the lack of progress so far and suggest a possible route to the solution of the problem.

The main problem, from the beginning of the far infrared work on these fascinating materials, has been the complexity of the optical response. The copious structures produced by phonons in the polycrystalline materials and the strong midinfrared absorption in the **ab** plane oriented samples have hidden the subtle effects of the superconducting gap.

Even when these perturbations are taken into account, the gap is difficult to discern. The very pure materials approach the clean limit, and the threshold of conductivity that is predicted to appear at $2\Delta$ by Mattis and Bardeen is very weak and overwhelmed by the inevitable noise. In the very dirty materials on the other hand, both $T_c$ and the gap appear to be depressed. As a result, in the region of the expected gap $\approx 250$ cm$^{-1}$, there is a residual absorption of a few percent, resulting possibly from surface regions where the average gap is well below this frequency.

From these considerations it appears that a compromise has to be struck. For the gap to be visible in the conductivity, the sample cannot be too pure and it cannot be too dirty. Of the data so far, the first polycrystalline ceramic samples of Bonn et al.[82,87] and the oxygen depleted samples of Thomas et al.[63] seem to fall into the middle ground, and show some evidence of a gap in the region expected for a BCS superconductor.

Other samples appear to be either in the clean limit (examples are the **ab** oriented ceramics of Bonn et al.[56] and of Ose et al.[99]) or else have contaminated surface layers where the gap, if it present at all, is depressed below the far infrared spectral region. This contamination is seen in most of the single crystal work to date.

It is puzzling that all the samples nevertheless show high $T_c$ superconductivity according to four-probe resistance measurements. It appears that current-carrying channels of undisturbed good superconductor exist between domains of low-gap material.

The problem of the observation of the gap is related to the very short coherence length, a few lattice spacings at most. To increase the scattering rate to the point where the dirty limit is approached, the mean free path has to become of the order of the coherence length, in this case only a few lattice spacings. In classic superconductors the defects that scatter the electrons have no effect on the gap and the true gap can be observed in the dirty limit. In high $T_c$ superconductors, however, many defects have a very detrimental effect on $T_c$. Thus it appears that one cannot make the scattering length too short and still expect to see the intrinsic gap of the pure material.

## Acknowledgements

"May you live in interesting times!" is supposed to be an ancient Chinese curse, but for those of us fortunate enough to have worked in the area of high $T_c$ superconductors during the past two years, these have been interesting times indeed. The research from the McMaster-Florida collaboration that we have described here is the result of the skill and hard work of our students and postdoctoral associates: Doug Bonn, Mona Doss, Steve Herr, Kati Kamarás, Charles Porter, and Maureen Reedyk. John Greedan with Jim Garrett and Carl Stager have been an invaluable source of many well characterized high-quality samples. We are grateful to Siu-Wai Chan, Shahab Etemad, Laura Greene, and J.-M. Tarascon for their collaboration in this research. We have had stimulating discussions with G.L. Carr, J.A. Berlinsky, P. Hirshfeld, S. Jeyadev, C. Kallin, P. Kumar, A.J. Millis, J. Orenstein, M.J. Rice, J.R. Schrieffer, G.A. Thomas, C. Varma, and P. Wölfle.

Research at Florida was supported initially by the National Science Foundation—Solid State Chemistry—DMR-8416511 and is currently supported by DARPA grant MDA972-88-J-1006. At McMaster the support has come from the National Science and Engineering Research Council (NSERC) and The Canadian Institute for Advanced Research (CIAR).

# REFERENCES

1. J.G. Bednorz and K.A. Müller, *Z. Phys. B* **64,** 189 (1986)

2. M.A. Biondi and M.P. Garfunkel, *Phys. Rev. Lett* **2,** 143, (1959); see also *Phys. Rev.* **116,** 853, (1959); *Phys. Rev.* **116,** 862, (1959).

3. R.E. Glover, III, and M. Tinkham, *Phys. Rev.* **108,** 243 (1957).

4. D.M. Ginsberg and M. Tinkham, *Phys. Rev.* **118,** 990 (1960).

5. P.L. Richards and M. Tinkham, *Phys. Rev.* **119,** 575 (1960).

6. R.R. Joyce and P.L. Richards, *Phys. Rev. Lett.* **24,** 1007 (1970).

7. T. Holstein, *Phys. Rev* **96,** 539, (1954); *Ann. Phys. (N.Y.)* **29,** 410, (1964).

8. P.B. Allen, *Phys. Rev. B* **3,** 305 (1971).

9. B. Farnworth and T. Timusk, *Phys. Rev. B* **14,** 5119 (1976).

10. Frederick Wooten, *Optical Properties of Solids* (Academic Press, New York, 1972).

11. Max Born and Emil Wolf, *Principles of Optics* (Pergamon Press, Oxford, 1980).

12. Miles V. Klein and Thomas E. Furtak, *Optics*, Second Ed. (Wiley, New York, 1986).

13. Charles Kittel, *Introduction to Solid State Physics*, Fifth Ed. (Wiley, New York, 1976).

14. H. Mori, *Prog. Theor. Phys.* **34,** 399, (1965).

15. W. Götze and P. Wölfle, *Phys. Rev. B* **6,** 1226 (1972).

16. J.W. Allen and J.C. Mikkelsen, *Phys. Rev. B* **15,** 2952 (1977).

17. D. Penn, *Phys. Rev.* **128,** 2093 (1962).

18. D.C. Mattis and J. Bardeen, *Phys. Rev.* **111,** 412 (1958).

19. L. Leplae, Ph.D. Thesis, University of Maryland, (1962); *Phys. Rev. B* **27,** 1911 (1983).

20. D.R.Karecki, G.L. Carr, S. Perkowitz, D.U. Gubser, and S.A. Wolf, *Phys. Rev. B* **27,** 5460, (1983), D.M. Ginsberg, *Phys. Rev.* **151,** 241 (1966).

21. S. Tajima, S. Uchida, S. Tanaka, S. Kanbe, K. Kitazawa, and K. Fueki, *Jpn. J. Appl. Phys.* **26,** L432 (1987).

22. J. Orenstein, G.A. Thomas, D.H. Rapkine, C.B. Bethea, B.F. Levine, R.J. Cava, E.A. Reitman, and D.W. Johnson, Jr., *Phys. Rev. B* **36,** 729 (1987).

23. S.L. Herr, K. Kamarás, C.D. Porter, M.G. Doss, D.B. Tanner, D.A. Bonn, J.E. Greedan, C.V. Stager, and T. Timusk, *Phys. Rev. B* **36,** 733 (1987).

24. D.A. Bonn, J.E. Greedan, C.V. Stager, T. Timusk, M.G. Doss, S.L. Herr, K. Kamarás, and D.B. Tanner, *Phys. Rev. Lett.* **58**, 2249 (1987).

25. K. Kamarás, C.D. Porter, M.G. Doss, S.L. Herr, D.B. Tanner, D.A. Bonn, J.E. Greedan, A.H. O'Reilly, C.V. Stager, and T. Timusk, *Phys. Rev. Lett.* **59**, 919 (1987).

26. C.M. Varma, S. Schmitt-Rink, and E. Abrahams *Solid State Comm.* **62**, 681 (1987).

27. J. Orenstein and D.H. Rapkine, *Phys. Rev. Lett.* **60**, 968 (1988).

28. Z. Schlesinger, R.T. Collins, M.W. Schafer, and E.M. Engler, *Phys. Rev. B* **36**, 5275 (1987).

29. G.A. Thomas, A.J. Millis, R.N. Bhatt, R.J. Cava, and E.A. Rietman, *Phys. Rev. B* **36**, 736 (1987).

30. P.E. Sulewski, T.W. Noh, J.T. McWhirter, and A.J. Sievers, *Phys. Rev. B* **36**, 5735 (1987).

31. T.W. Noh, P.E. Sulewski, and A.J. Sievers, *Phys. Rev. B* **36**, 8866 (1987).

32. G.L. Doll, J. Steinbeck, G. Dresselhaus, M.S. Dresselhaus, A.J. Strauss, H.J. Steiger, *Phys. Rev. B* **36**, 8884 (1987).

33. X. Wang, T. Nanba, M. Ikezawa, Y. Isikawa, K. Mori, K. Kobayashi, K. Kasai, K. Sato, and T. Fukase, *Jpn. J. Appl. Phys.* **26**, L1391 (1987).

34. G.L. Doll, J.T. Nicholls, M.S. Dresselhaus, A.M. Rao, J.M. Zhang, G.W. Lehmann, P.C. Eklund, G. Dresselhaus, and A.J. Strauss, *Phys. Rev. B* submitted.

35. T. Koide, H. Fukutani, A. Fujimori, R. Suzuki, T. Shidara, T. Takahshi, S. Hosoya, and M. Sato, *Jpn. J. Appl. Phys.* **26-3**, 915 (1987).

36. S. Tajima, S. Uchida, H. Ishii, H. Takagi, S. Tanaka, U. Kawabe, H. Hasegawa, T. Aita, and T. Ishiba, *Mod. Phys. Lett. B* **1**, 353 (1988).

37. P.E. Sulewski, T.W. Noh, J.T. McWhirter, A.J. Sievers, S.E. Russek, R.A. Buhrman, C.S. Jee, J.E. Crow, R.E. Salomon, amd G. Myer, *Phys. Rev. B* **36**, 2357 (1987).

38. Z. Schlesinger, R.T. Collins, D.L. Kaiser, and F. Holtzberg, *Phys. Rev. Lett.* **59**, 1958 (1987).

39. J. Orenstein, G.A. Thomas, D.H. Rapkine, C.B. Bethea, B.F. Levine, B. Batlogg, R.J. Cava, D.W. Johnson, Jr., and E.A. Rietman, *Phys. Rev. B* **36**, 8892 (1987).

40. I. Bozovic, K. Char, S.J.B. Yoo, A Kapitulnik, M.R. Beasley, T.H. Geballe, Z.Z. Wang, S. Hagen, N.P. Ong, D.E. Aspnes, and M.K. Kelly, *Phys. Rev. B* **38**, 5077 (1988).

41. R.T. Collins, Z. Schlesinger, R.H. Koch, R.B. Laibowitz, T.S. Plaskett, P. Freitas, W.J. Gallagher, R.L. Sandstrom, and T.R. Dinger, *Phys. Rev. Lett.* **59**, 704 (1987).

42. I. Bozovic, D. Kirillov, A. Kapitulnik, K. Char, M.R. Hahn, M.R. Beasley, T.H. Geballe, Y.H. Kim, and A.J. Heeger, *Phys. Rev. Lett.* **59**, 2219 (1987).

43. K. Kamarás, C.D. Porter, M.G. Doss, S.L. Herr, D.B. Tanner, D.A. Bonn, J.E. Greedan, A.H. O'Reilly, C.V. Stager, and T. Timusk, *Phys. Rev. Lett.* **60**, 969 (1988).

44. J.M. Tarascon, W.R. McKinnon, L.H. Greene, G.W. Hull, and E.M. Vogel, *Phys. Rev. B* **36**, 226 (1987).

45. M. Gurvitch and A.T. Fiory, *Phys. Rev. Lett.* **59**, 1337 (1987).

46. D.A. Bonn, J.E. Greedan, C.V. Stager, T. Timusk, K. Kamarás, C.D. Porter, M.G. Doss, S.L. Herr, and D.B. Tanner, *Rev. Solid State Sci.* **1**, 349 (1987).

47. D.R. Harshman, G. Aeppli, E.J. Ansoldo, B. Batlogg, J.H. Brewer, J.F. Carolan, R.J. Cava, M. Celio, A.C.D. Chaklader, W.N. Hardy, S.R. Kreitzman, G.M. Luke, D.R. Noakes, and M. Sheba, *Phys. Rev. B* **36**, 2368 (1987).

48. Y.J. Uemura, V.J. Emery, A.R. Moodenbaugh, M. Suenaga, D.C. Johnston, A.J. Jacobsen, J.T. Lewandowski, J.H. Brewer, R.F. Kiefl, S.R. Kreitzman, G.M. Luke, T. Riseman, C.E. Stronach, W.J. Kossler, J.R. Kempton, X.H. Yu, D.Opie, and H.E. Schone, *Phys. Rev. B* **38**, 909 (1988).

49. A.T. Fiory, A.F. Hebard, P.M. Mankiewich, and R.E. Howard, *Phys. Rev. Lett.* **61**, 1419 (1988).

50. S. Etemad, D.E. Aspnes, M.K. Kelly, R. Thomson, J.-M. Tarascon, and G.W. Hull, *Phys. Rev. B* **37**, 3396 (1988).

51. S. Etemad, D.E. Aspnes, P. Barboux, G.W. Hull, M.K. Kelly, J.M. Tarascon, R. Thompson, S.L. Herr, K. Kamarás, C.D. Porter, and D.B. Tanner, *Mat. Res. Soc. Symp. Proc.* **99**, 135 (1988).

52. S.L. Herr et al. in *High Temperature Superconducting Materials: Preparations, Properties and Processing,* edited by William Hatfield and J.J. Miller (Marcel Dekker, Inc., New York, 1988), p. 275.

53. J.B. Torrance, Y. Tokura, A.I. Nazzal, A. Bezinge, T.C. Huang, and S.S. Parkin, *Phys. Rev. Lett.* **61**, 1127 (1988).

54. K. Kamarás, M.G. Doss, S.L. Herr, J.S. Kim, C.D. Porter, G.R. Stewart, D.B. Tanner, D.A. Bonn, J.E. Greedan, A.H. O'Reilly, C.V. Stager, T. Timusk, B. Keszei, S. Pekker, Gy. Hutiray, and L. Mihály, *Mat. Res. Soc. Symp. Proc* **99**, 777 (1988).

55. J.M. Tarascon, L.H. Greene, W.R. McKinnon, G.W. Hull, and T.H. Geballe, *Science* **235**, 1373 (1987).

56. D.A. Bonn, A.H. O'Reilly, J.E. Greedan, C.V. Stager, T. Timusk, K. Kamarás, and D.B. Tanner, *Phys. Rev. B* **37**, 1547 (1988).

57. D.A. Bonn, A.H. O'Reilly, J.E. Greedan, C.V. Stager, T. Timusk, K. Kamarás, and D.B. Tanner, *Mat. Res. Soc. Symp. Proc.* **99**, 227, (1988).

58. T. Timusk, D.A. Bonn, J.E. Greedan, C.V. Stager, J.D. Garrett, A.H. O'Reilly, M. Reedyk, K. Kamarás, C.D. Porter, S.L. Herr, and D.B. Tanner, *Physica C* **153–155**, 1744 (1988).

59. T. Timusk, S.L. Herr, K. Kamarás, C.D. Porter, D.B. Tanner, D.A. Bonn, J.D. Garrett, C.V. Stager, J.E. Greedan, and M. Reedyk, *Phys. Rev. B* **38**, 6683 (1988).

60. D.B. Tanner, S.L. Herr, K. Kamarás, C.D. Porter, T. Timusk, D.A. Bonn, J.D. Garrett, J.E. Greedan, C.V. Stager, M. Reedyk, S. Etemad, and S.-W. Chan, *Synth. Met.* in press.

61. D.B. Tanner, T. Timusk, S.L. Herr, K. Kamarás, C.D. Porter, D.A. Bonn, J.D. Garrett, C.V. Stager, J.E. Greedan, and M. Reedyk, *Physica A* in press.

62. J. Orenstein, G.A. Thomas, D.H. Rapkine, A.J. Millis, L.F. Schneemeyer, and J.V. Waszczak, *Physica C* **153–155**, 1740 (1988).

63. G.A. Thomas, J. Orenstein, D.H. Rapkine, M. Capizzi, A.J. Millis, L.F. Schneemeyer, and J.V. Waszczak, *Phys. Rev. Lett.* **61**, 1313 (1988).

64. D. van der Marel, J. van Elp, G.A. Sawatzky, and D. Heitmann, *Phys. Rev. B* **37**, 5136 (1988).

65. J. Tanaka, K. Kamya, M. Shimizu, M. Simada, C. Tanaka, H. Ozeki, K. Adachi, K. Iwahashi, F. Sato, A. Sawada, S. Iwata, H. Sakuma, and S Uchiyama, *Physica C* **153–155**, 1752 (1988).

66. J. Tanaka, K. Kamiya, and S. Tsurumi, *Physica C* **153–155**, 653 (1988).

67. F. Gervais, P. Echegut, J.M. Bassat, and P. Odier, *Phys. Rev. B* **37**, 9364 (1988).

68. J. Tanaka, M. Shimada, U. Mizutani, and M. Hasegawa, *Physica C* **153–155**, 651 (1988).

69. J. Bassat, P. Odier, and F. Gervais, *Phys. Rev. B* **35**, 7126 (1987).

70. R.T. Collins, Z. Schlesinger, F. Holtzberg, P. Chaudhari, and C. Feild, preprint.

71. M. Garriga, U. Venkateswaren, K. Syassen, J. Humlicek, M. Cardona, Hj. Mattausch, and E. Schönherr, *Physica C* **153–155**, 643 (1988).

72. M.K. Kelly, P. Barboux, J.-M. Tarascon, D.E. Aspnes, W.A. Bonner, and P.A. Morris, *Phys. Rev. B* **38**, 870 (1988).

73. H.P. Geserich, B. Koch, G. Scheiber, J. Geerk, H.C. Li, G. Linker, W. Weber, and W. Assmus, *Physica C* **153–155**, 661 (1988).

74. F.A. Benko, G. Abdussalam, F.S. Razavi, and F.P. Koffyberg, *Phys. Rev. B* **38**, 2820 (1988).

75. Y.H. Kim, A.J. Heeger, L. Acedo, G. Stuckey, and F. Wudl, *Phys. Rev. B* **36**, 7252 (1987).

76. J.M. Ginder, M.G. Roe, Y. Song, R.P. McCall, J.R. Gaines, E. Ehrenfreund, and A.J. Epstein, *Phys. Rev. B* **37**, 7506 (1988).

77. Y.H. Kim, C.M. Foster, A.J. Heeger, S. Cox, and G. Stucky, *Phys. Rev. B* **38**, 6478 (1988).

78. C. Taliani, R. Zamboni, G. Ruani, F.C Matacotta, and K.I. Pokhodnya, *Solid State Comm.* **66**, in press (1988).

79. P.E. Sulewski, A.J. Sievers, S.E. Russek, H.D. Hallen, D.K. Lathrop, and R.A. Buhrman, *Phys. Rev. B* **35**, 5330 (1987).

80. U. Walter, M.S. Sherwin, A. Stacy, P.L. Richards, and A. Zettl, *Phys. Rev. B* **35**, 5327 (1987).

81. Z. Schlesinger, R.L. Greene, J.G. Bednorz, and K.A. Müller, *Phys. Rev. B* **35**, 5334 (1987).

82. D.A. Bonn, J.E. Greedan, C.V. Stager, and T. Timusk, *Solid State Comm.* **62**, 838 (1987).

83. G.A. Thomas, M. Capizzi, J. Orenstein, D.H. Rapkine, L.F. Schneemeyer, J.V. Waszczak, A.J. Millis, and R.N. Bhatt, *Jpn. J. Appl. Phys.* **26-3**, 1001 (1987).

84. Z. Schlesinger, R.T. Collins, and M.W. Schafer, *Phys. Rev. B* **35**, 7232 (1987).

85. L. Degiorgi, E. Kaldis, and P. Wachter, *Solid State Comm.* **64**, 873 (1987).

86. M.S. Sherwin, P.L. Richards, and A. Zettl, *Phys. Rev. B* **37**, 1587 (1988).

87. D.A. Bonn, J.E. Greedan, C.V. Stager, T. Timusk, M.G. Doss, S.L. Herr, K. Kamarás, C.D. Porter, D.B. Tanner, J.M. Tarascon, W.R. McKinnon, and L.H. Greene, *Phys. Rev. B* **35**, 8843 (1987).

88. T.H.H. Vuong, D.C. Tsui, V.J. Goldman, P.H. Hor, R.L. Meng, and C.W. Chu, *Solid State Comm.* **63**, 525 (1987).

89. L. Genzel, A. Wittlin, J. Kuhl, Hj. Mattausch, W. Bauhofer, and A. Simon, *Solid State Comm.* , **63**, 843 (1987).

90. J.M. Wrobel, S. Wang, S. Gygax, B.P. Clayman, and L.K. Peterson, *Phys. Rev. B* **36**, 2368 (1987).

91. A.V. Bazhenov, A.V. Gorbunov, N.V. Klassen, S.F. Kondakov, I.V. Kukuskin, V.D. Kulakovskii, O.V. Misochlo, V.B. Timofeev, L.I. Chernyshova, and B.N. Shepel', *Jpn. J. Appl. Phys.* **26-3**, 893 (1987).

92. E.M. Gershenzon, G.N. Gol'tsman, B.S. Karasik, and A.D. Semenov, *Pis'ma Zh. Eksp. Teor. Fiz.*, **46**, 122 (1987); *JETP Lett.*, *48*, 151 (1987).

93. S. Perkowitz, G.L. Carr, B. Lou, S.S. Yom, R. Sudharsanan, and D.S. Ginley, *Solid State Comm.* **64**, 721 (1987).

94. M. Cardona, L. Genzel, R. Liu, A. Wittlin, Hj. Mattausch, F. García-Alvarado, and E. García-González, *Solid State Comm.* **64**, 727 (1987).

95. Y. Saito, H. Sawada, T. Iwazumi, Y. Abe, H. Ikeda, and R. Yoshizaki, *Solid State Comm.* **64**, 1047 (1987).

96. H. Ye, W. Lu, Z. Lu, X. Shen, B. Miao, Y. Cai, and Y. Qian, *Phys. Rev. B* **36**, 8802 (1987).

97. A. Wittlin, L. Genzel, M. Cardona, M. Bauer, W. König, E. Garcia, M. Barahona, and M.V. Cabañias, *Phys. Rev. B* **37**, 652 (1988).

98. P.J.M van Bentum, L.E.C. van de Leemput, L.W.M. Schreurs, P.A.A. Teunissen, and H. van Kempen, *Phys. Rev. B* **36**, 843 (1987).

99. W. Ose, P.E. Obermayer, H.H. Otto, T. Zetterer, H. Lengfellner, N. Tasler, J. Keller, and K.F. Renk, *Physica C* **153–155**, 639 (1988).

100. S.L. Norman and D.H. Douglass, *Phys. Rev. Lett.* **10**, 339 (1967).

101. J.P. Franck, J. Jung, and M.A.-K. Mohamed, *Phys. Rev. B* **36**, 2308 (1987).

102. N.P. Ong, Z.Z. Wang, S. Hagen, T.W. Jing, J. Clayhold, and J. Horvath, *Physica C* **153–155**, 1072 (1988).

103. G.A. Thomas, M. Capizzi, T. Timusk, S.L. Cooper, J. Orenstein, D. Rapkine, S. Martin, L.F. Schneemeyer, and J.V. Waszczak, *J. Opt. Soc. Am., Special Issue on Superconductivity* (submitted) 1989.

104. J. Hopfield, *Phys. Rev.* **139**, A419 (1964).

105. Z. Schlesinger, R.T. Collins, D.L. Kaiser, and F. Holtzberg, G.V. Chandrashekhar, M.W. Scafer, and T.M. Plaskett, *Physica C* **153–155**, 1734 (1988).

106. M. Reedyk, D.A. Bonn, J.D. Garrett, J.E. Greedan, C.V. Stager, T. Timusk, K. Kamarás, and D.B. Tanner, *Phys. Rev. B* in press.

107. J.E. Eldridge and F.E. Bates, *Phys. Rev. B* **28**, 6972 (1983).

108. M.W. Lee, T. Pham, H.D. Drew, S.M Bhagat, R.E. Glover III, S.H. Mosley, K.P. Stewart, and C. Lisse, *Solid State Comm.* **65**, 1135 (1988).

109. J.R. Waldram, *Advances in Physics* **13**, 1, (1964).

110. W.P. Beyermann, B.Alavi, and G. Grüner, *Phys. Rev. B* **35**, 8826 (1987).

111. S. Sridhar, C.A. Shiffman, and H. Hamdeh, *Phys. Rev. B* **36**, 2301 (1987).

112. M. Poirier, G. Quiron, K.R. Poppelmeier,and J.P. Thiel, *Phys. Rev. B* **36**, 3906 (1987).

113. A. Porch, J.R. Waldram, and L. Cohen, *J. Phys. F: Met. Phys.* **18**, 1547, (1988)

114. A.S. Barker, in *Optical Properties and Electronic Structure of Metals and Alloys,* edited by F. Abelés (North Holland, Amsterdam, 1965)

115. S.Tajima, S. Uchida, A. Masaki, H. Tagaki, K. Kitazawa, and S. Tanaka, *Phys. Rev. B* **32**, 6302 (1985).

116. S.G. Kaplan, T.W. Noh, P.E. Sulewski, H. Xia, A.J. Sievers, J. Wang, and R. Raj, *Phys. Rev. B* **38**, 5006 (1988).

117. C.D. Porter, S.L. Herr, K. Kamarás, D.B. Tanner, D.A. Bonn, J.E. Greedan, A.H. O'Reilly, and T. Timusk, unpublished.

118. S.T. Chui, R.V. Kasowski, and W.Y. Hsu, *Phys. Rev. Lett.* **61**, 885 (1988).

119. L.F. Mattheiss, *Phys. Rev. Lett.* **58**, 1028 (1987).

120. S. Massida, J. Yu, A.J. Freeman, and D.D. Koelling, *Phys. Lett.* **A122**, 198 (1987).

121. Werner Weber, *Z. Phys. B* **70**, 323 (1987).

122. Mark Jarrell, H.R. Krishnamurthy, and D.L. Cox, *Phys. Rev. B* **38**, 4584 (1988).

123. R.B. Laughlin, C.B. Hanna, in *Novel Superconductivity*, edited by S.A. Wolf and V. Z. Kreskin (Plenum, New York, 1987).

124. M.J. Rice and Y.R. Wang, *Phys. Rev. B* **36**, 8794 (1987).

125. J.R. Schrieffer, X.G. Wen, and S.C. Zhang, *Phys. Rev. Lett.* **60**, 944 (1988).

# 8
# RAMAN SCATTERING IN HIGH-$T_c$ SUPERCONDUCTORS

Christian Thomsen and Manuel Cardona
*Max-Planck Institut für Festkörperforschung*
*Heisenbergstrasse 1, D-7000 Stuttgart 80*
*Federal Republic of Germany*

410

# I. INTRODUCTION

## A. Theory

### 1. Light scattering by phonons

Infrared absorption, light scattering, and thermal neutron scattering are the most commonly used techniques for investigating low-energy elementary excitations in solids, in particular phonons. The optical spectroscopies, however, are severely limited by the selection rule which arises from conservation of wavevector $\vec{k}$ (also called crystal momentum) in crystalline solids: the magnitude of the wavevector of light is very small compared with that of typical elementary excitations in crystals with a small number of atoms per primitive cell. The latter is of the order of $2\pi/a_o$ with $a_o$ a typical lattice constant (of the order of $\approx 10$ A in high-$T_c$ superconductors) while the former is $4\pi/\lambda$ or smaller, where $\lambda$ is the wavelength of light in the medium. Hence, of the many excitations which correspond to all points in the first Brillouin zone (BZ) first-order optical spectroscopies only reveal those with $\vec{k} \cong 0$. The typical wavelengths of thermal neutrons, however, are of the order of 5 A and hence the corresponding scattering wavevectors can be made to cover the whole BZ of a typical crystal. Thus, neutron spectroscopy has been profusely used to investigate dispersion relations of phonons[1] and also of magnetic excitations (magnons)[2] in many crystals.

In spite of this obvious advantage, optical spectroscopies, in particular Raman scattering, are very useful for investigating low-frequency elementary excitations. The accuracy and resolution that can be reached in light-scattering spectroscopy ($<0.1$ cm$^{-1}$, reminder: 1 cm$^{-1} = 1.2 \times 10^{-4}$ eV = 1.5 K) is one to two orders of magnitude better than that obtained in neutron spectroscopy. It is, for instance, easy to obtain information about the natural linewidths of phonons and their temperature dependence[3] with light scattering while the corresponding linewidths observed in neutron spectroscopy are usually determined by instrumental resolution. Also, the volume sampled in light scattering can be made very small: it is easy to focus the laser to a spot of less than 10 μm diameter while the penetration depth for metallic samples measured in a backscattering configuration is about 1000 A. Because of the small cross section for interaction of neutrons with atoms the typical volumes needed for inelastic scattering work are larger than 1 $cm^3$, a fact which renders the technique inapplicable to presently available single crystals of most high-$T_c$ superconductors. A Raman spectrometer can be used in conjunction with a microscope illuminator and scattered light collector to scan the Raman spectra on the surface of a sample.[4] In this manner, a great wealth of information can be obtained for inhomogeneous samples, including ceramic polycrystalline high-$T_c$ superconductors. Single crystallites can even be selected out and, with some care, the need for isolated single crystals may be circumvented.[5] Such Raman microprobes are nowadays commercially available.

Light-scattering and ir-absorption or reflection spectrometers offer the advantage of being considerably more economical (and ecological) than neutron setups. They are faster and can sometimes be used to follow reactions and transformations in real time. The equipment is totally under the control of a single operator. In the case of light scattering, the laser frequency can be varied from the near ir to the near uv, thus changing the penetration depth in the sample and, when it concides with the frequency of electronic interband transitions, showing a rich phenomenology of resonance effects. Recent work on light scattering in solids is covered in the five volumes of Ref. 6.

The $\vec{k}$-conservation selection rule is, strictly speaking, only valid for perfect, infinite crystals since it is a consequence of the existence of a network of translations which bring the crystal onto itself. Deviations from the infinite solid, such as point defects (e.g. vacancies) and finite crystal size[7] may make modes with larger $|\vec{k}|$ = k optically active. In particular, for amorphous solids all 'phonons' are observed.[8,9] Disorder-induced large-k phonon modes should be present in the optical spectra of high-$T_c$ superconductors but have yet to be identified as such. Size effects become important for grain dimensions $\leq$ 100 A, much smaller than the grains of typical ceramic superconductors (a few μm).[7] Size-effect-induced Raman modes can be excited in a controllable manner in artificially made superlattices.[10]

Modes with large $|\vec{k}|$ can also be excited in higher-order optical spectroscopies such as second- (or higher) order Raman scattering. In the latter, two phonons participate in the scattering process, usually both being emitted (Stokes spectrum). Phonon emission is the only type of process allowed at low temperatures. At higher temperatures phonons can also be destroyed, and anti-Stokes spectra, in which one or two phonons are destroyed, can be obtained. Stokes spectra, in which one phonon is emitted and one of lower frequency destroyed (difference processes) are also possible. These processes can, in principle, be distinguished from each other by their dependences on temperature T. For each phonon-creation process a Bose-Einstein statistical factor:

$$1 + n(\omega) = 1 + (e^{\frac{\hbar\omega}{\beta T}} - 1)^{-1} \qquad (1)$$

appears in the scattered intensity (ω is the angular frequency of the phonon, $\hbar$ and β the Planck and Boltzmann constants, respectively). For each phonon-annihilation process the factor is simply n(ω), i.e., it vanishes for $\hbar\omega \gg kT$. Note that the ratio of an anti-Stokes to the corresponding Stokes intensity may be used to determine the temperature of the scattering volume of the sample, which may differ from the nominal one because of heating by the laser. In the case of first-order processes this ratio equals the Boltzmann factor $\exp(-\hbar\omega/kT)$. This type of manipulation implies, of course, that the other factors which appear in the scattering efficiency[9] remain the same for Stokes and anti-Stokes, a condition that may not hold when the laser is near an electronic resonance of the solid.[11]

As mentioned above, all phonons at general points can participate in second-order scattering. The corresponding spectra represent the combined two-phonon density of states, with participation of overtones (two phonons from the same branch with opposite $\vec{k}$'s) and combinations of different branches. By judiciously choosing the polarization of the laser and scattered fields, it is sometimes possible to select scat-

414

tering by overtone modes.[9] Such modes have been observed in $La_2CuO_4$.[12]
We should add that phonon densities-of-states of high-$T_c$ superconductors are obtainable with neutron-scattering techniques (as opposed to
the full dispersion curves) since one can use large randomly oriented
polycrystalline samples for the neutron measurements. An example obtained for $YBa_2Cu_3O_7$ above and below $T_c$ is shown in Fig. 1.[13]

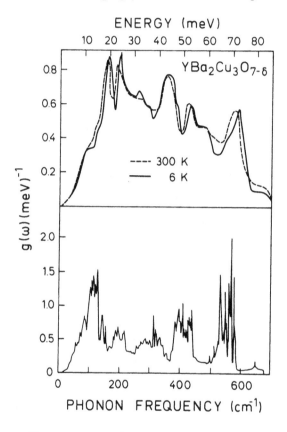

**Fig. 1.** Comparison of the total phonon density-of-states (PDOS) of $YBa_2Cu_3O_{7-\delta}$ ($\delta \approx 0$) (a) experimental, amplitude and cross-section weighted PDOS (b) PDOS obtained from a lattice dynamical calculation presented in Sect. II.A.2. The data (taken from Ref. 13) have, for comparison to the theoretical curve (Ref. 53), been normalized to a total of 39 modes in the interval 0 to 80 meV.

The response function for light scattering relates the incident
(laser) field to the scattered field. It can thus be represented by a
second-rank tensor (the Raman tensor, $\underset{\approx}{R}$) which connects both fields as
a 'transfer matrix'. There is one such tensor associated with each
particular excitation; each of them contains the corresponding polarization selection rules for light scattering. Away from electronic re-

sonances and for scattering by phonons, the Raman tensor is proportional to the derivative of the electric susceptibility with respect to the phonon coordinate: the phonon modulates the susceptibility at its frequency $\omega$. This susceptibility, when multiplied by the laser field produces scattered fields at the frequencies $\omega_L - \omega$ (Stokes) and $\omega_L + \omega$ (anti-Stokes). Since the susceptibility $\underset{\approx}{\chi}$ is a <u>symmetric</u> second-rank tensor, so will be $\underset{\approx}{R}$. The symmetrical nature of $\underset{\approx}{R}$ may be lost near an electronic resonance.[14] Effects of asymmetry in $\underset{\approx}{R}$ have not been observed in high-$T_c$ superconductors although electronic resonances are known to exist in the visible and uv regions (2.8, 4.1, and 4.7 eV in $YBa_2Cu_3O_7$[15]). These resonances are, however, broader than the phonon frequencies and thus should not lead to asymmetries in $\underset{\approx}{R}$.

Each $\vec{k} \approx 0$ phonon transforms under operations of the point group of the crystal like one of the irreducible representations of this group.[16] For the phonon (or any other $\vec{k} \cong 0$ excitation) to be Raman active it must transform like one of the irreducible representations which correspond to the components of a second-rank tensor (usually symmetric) and thus have the symmetry of $x^2$, $y^2$, $z^2$, xy, yz, and zx. If the point group of the crystal has a center of inversion, parity is a good quantum number and the irreducible representations can be classified into odd and even ones. In order for the phonons to be Raman allowed they must have <u>even parity</u>, a necessary but not sufficient condition. Conversely, ir-allowed phonons must have odd parity since in an ir-absorption process a photon (odd parity) is converted into the phonon. Most of the high-$T_c$ superconductors known so far have a center of inversion and thus these rules apply in principle. (A possible exception regarding the Bi-based superconductors is discussed in Sect. II.B) The parity selection rules may, however, be violated because of disorder, so that ir-active modes (odd) may appear in the Raman spectra. Such modes, induced by oxygen defects or disorder, seem to have been observed in $YBa_2Cu_3O_7$ single crystals.[17] From observations in semiconductors (one should keep in mind that the electron density of high-$T_c$ superconductors is similar to that of heavily doped semiconductors) it is possible to suggest one mechanism for the observation of forbidden, ir-allowed modes in Raman spectra. This is the Fröhlich interaction of the phonons with the electrons through the electrostatic potential which accompanies ir-active LO phonons.[9] Two types of such processes are observed <u>near the resonances</u>. One of them, of the electric-quadrupole type, leads to cross sections proportional to the square of q (the magnitude of the scattering wavevector). The other is a fourth-order process induced by Fröhlich interaction and interaction of the electrons with impurities.[18] The latter can be described in terms of the former (q-induced) except that instead of the small q provided by the scattering wavevector, a much larger q is provided by the scattering by impurities. Although these mechanisms should be screened in conducting materials, they may contribute to the appearance of forbidden phonons in high-$T_c$ superconductors. No positive evidence is, however, available at this point. We note that these processes usually lead to diagonal Raman tensors, i.e., $A_g$ symmetry scattering in the orthorhombic high-$T_c$ superconductors.

## 2. Electronic scattering

We have, so far, discussed only scattering by phonons. Raman scattering can also reveal low-frequency electronic excitations, usually of the intraband type, in heavily doped semiconductors. In materials with simple, free-electron-like band structures (e.g. n-type GaAs) the Hamiltonian for the coupling of electrons with the incident and scattered field is:

$$H_{er} = \frac{e^2}{c^2} \vec{A}_L \cdot \frac{1}{\overset{\star}{\underset{\approx}{m}}} \cdot \vec{A}_S \qquad (2)$$

where $\vec{A}_{L,S}$ are the vector potentials of the incident and scattered fields and $\overset{\star}{\underset{\approx}{m}}$ the effective-mass tensor of the quasi-free electrons. Equation (2) can be viewed as a scalar perturbation equivalent to an electrostatic potential. The response function for such 'longitudinal' perturbation must therefore take into account screening by the free electron gas in a self-consistent manner: the response is thus proportional to the imaginary part of $\varepsilon^{-1}$, where $\varepsilon$ is the dielectric constant of the system including free carriers. As such it exhibits a peak at the plasma frequency $\omega_p$ and nearly no low-frequency (i.e. single particle) scattering: the latter is screened by the collective response of the electron gas. Hence no scattering would be expected in high-$T_c$ superconductors in the region well below the plasma frequency in the normal phase, although such scattering may become observable in the superconducting phase at and slightly above the gap.[19] YBa$_2$Cu$_3$O$_7$, however, has been shown to display considerable electronic scattering below the gap in the normal phase.[20-24] Two possible mechanisms for such scattering exist.[25] One of them is related to electronic resonances and the corresponding excitations involve spin flips. It should only be observed for crossed incident and scattered polarization. It can thus be excluded for YBa$_2$Cu$_3$O$_7$ since in this case the scattering appears for parallel polarizations. The other involves 'intervalley' excitations in many-valley systems. Note that the planes and chains in YBa$_2$Cu$_3$O$_7$ may constitute such a system.

Raman scattering also reveals magnetic excitations (spin waves, magnons) the corresponding modulation of $\underset{\approx}{\chi}$ usually taking place through spin-orbit or exchange mechanisms.[26] The latter, which leads to second-order scattering by two magnons, is often stronger than the former. It seems to have been observed[27] in the insulating modifications of high-$T_c$ superconductors and reveals the antiferromagnetic order of these materials seen also in elastic neutron scattering.[28]

## B. Instrumentation

The penetration depth of light in high-$T_c$ superconductors is about 2000 A,[15] hence the volume of interaction of light with radiation is very small. For a typical resolution of 2 cm$^{-1}$ count rates of 0.1 to 1 counts/sec are obtained for phonon peaks. This corresponds to a scattering efficiency of $10^{-10}$/sterad cm with a typical phonon linewidth of 10 cm$^{-1}$. Because of these small scattering efficiencies, and the large amount of Rayleigh scattering present, in particular in ceramic samples, high quality Raman instrumentation is needed. This means double or triple monochromators and cooled, low dark-current detectors. If single-channel photomultipliers are used, long measuring times become necessary. Recently, multichannel detectors have become commercially available.[29] They shorten the measuring time typically by two orders of magnitude. Three types of these detectors exist: silicon diode arrays, charge coupled device systems (CCD) and MEPSICRONS. A common disadvantage to all of them is poor resolution, a fact which does not seem to be important for high-$T_c$ superconductors. For a comparative analysis see Ref. 29.

Multichannel detectors are used in place of the exit slit of the monochromator. If a double monochromator is employed it must thus be of the additive dispersion type (the dispersions of the two gratings add) with the center slit wide open so as to project a broad spectrum onto the detector. Another popular possibility is to use a triple monochromator system consisting of a double monochromator with subtractive dispersion (used only as a filter to set the spectral range) followed by a single monochromator with the detector in the exit focal plane. The range of the collected spectrum can be simply changed by changing the grating of the single monochromator.

Argon or krypton-ion lasers are the most commonly used light sources. They have a number of discrete spectral lines which permit a coarse exploration of possible resonance effects. A more detailed investigation of such effects would require continuously-tunable dye or color-center lasers. When using the 5145 A line of the argon-ion laser elastically scattered light can be significantly reduced with the help of an iodine-vapor filter.[30]

## II. VIBRATIONAL ANALYSIS AND LATTICE DYNAMICS

In order to apply the above concepts to a real material we will first take a group-theoretical look at the new class of high-$T_c$ super-conductors. In this section we analyze the symmetries of the different optical modes that correspond to the structure of various superconductors. Due to the related structure, it is natural to expect the eigenmodes of $RBa_2Cu_3O_{7-\delta}$ (Sect. II.A.1) to resemble those of cubic perovskites. There are, however, important differences, e.g. in the optical selection rules. We will see that most interesting in connection with the superconductivity mechanism is the 'silent mode' of oxygen in the perovskites. It becomes Raman active through the doubling of the $CuO_2$ planes in the unit cell of $RBa_2Cu_3O_{7-\delta}$, compared to perovskites, and shows a number of interesting properties.

In addition to a symmetry analysis, it is desirable to have a theoretical estimate of the energy of the expected modes. The experimentally observed symmetries and frequencies combined with the theoretical results will generally make a reliable assignment to specific eigenmodes possible. In Sect. II.A.2 we present such a lattice dynamical calculation for the $RBa_2Cu_3O_{7-\delta}$ structure. In Sect. II.B we discuss the optically active modes of the Bi and Tℓ-based high-$T_c$ superconductors and present results of recent lattice dynamical calculations while in Sect. II.C we discuss the equivalent features of the Zurich superconductors.

## A. Optical Modes of $RBa_2Cu_3O_{7-\delta}$

Most of the scientific interest has focussed so far on orthorhombic $RBa_2Cu_3O_{7-\delta}$ ($T_c \approx 90$ K). We present here a detailed symmetry and lattice dynamical analysis of these compounds. Since they are related to the corresponding tetragonal semiconductors with $\delta \approx 1$, and important conclusions can be drawn from the near tetragonality of the superconductor, both symmetries will be considered.

### 1. Symmetry analysis of the $\vec{k}$ = 0 modes

It is well known that the $RBa_2Cu_3O_{7-\delta}$ compounds have an orthorhombic unit cell with lattice constants (a = 3.822 A, b = 3.891 A, and c = 11.677 A for R = Y) which increase slightly with increasing ionic radius of $R^{3+}$.[31] They belong to the symmorphic space group Pmmm ($D_{2h}^1$ in Schoenfliess notation) and point group mmm ($D_{2h}$).[32] These groups have an inversion center, so that optical modes near $\vec{k}$ = 0 cannot be both Raman and ir active. From the symmorphic character we may conclude that there is at least one site which has the full symmetry of the point

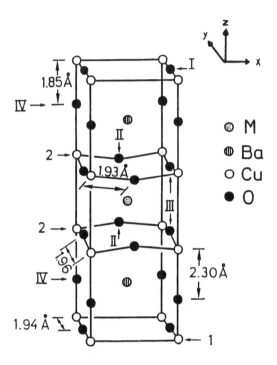

Fig. 2. Structure of $YBa_2Cu_3O_7$.

group. It may be seen from Fig. 2 that this is indeed the case for the
sites of the OI, Cu1, and Y atoms. The pairs of atoms OII, OIII, OIV,
Ba, and Cu2 have lower, $C_{2v}$ symmetry.[33] Permutation of the equivalent
atoms in the $C_{2v}$ pairs leads to the irreducible representations $A_g$ and
$B_{1u}$ of $D_{2h}$. The corresponding modes are obtained by multiplying these
representations by those of the displacement vectors in x, y, and z
directions (representations $B_{3u}$, $B_{2u}$, and $B_{1u}$). From standard tables[16]
one finds three odd (ungerade) modes $B_{3u}$, $B_{2u}$, $B_{1u}$, and three even
(gerade) modes $A_g$, $B_{2g}$, and $B_{3g}$ for each of the members of a pair. For
the vibrations involving the atomic positions with full symmetry we
find only three odd modes for each atom. After subtracting the three
acoustic modes we arrive at 36 optically active modes to be considered
in $RBa_2Cu_3O_{7-\delta}$: $5A_g + 5B_{2g} + 5B_{3g} + 7B_{1u} + 7B_{2u} + 7B_{3u}$. Of these, 15 are

Raman active according to symmetry and 21 ir active. We shall see that in many cases these activities are negligibly small. In Table I we give the character table for these modes and their optical activities. Many workers have used a group-theoretical analysis of the $RBa_2Cu_3O_7$ structure as a starting point in the analysis of their data.[23,33-44] Due to the simple structure of the k = 0 modes in these materials, this was a straightforward task.

Table I: Character table and optical activities of representations of the point group $D_{2h}$ of the orthorhombic $RBa_2Cu_3O_7$. Under activities we list the non-zero tensor components $\alpha_{ij}$ and the non-zero dipole vector components ir (i) where i and j refer to directions relative to the crystal axes. The group-theoretical symbols have their usual meanings.[16]

| Representation | E | $C_2^z$ | $C_2^y$ | $C_2^x$ | i | $\sigma^{xy}$ | $\sigma^{xz}$ | $\sigma^{yz}$ | activity |
|---|---|---|---|---|---|---|---|---|---|
| $A_g$ | 1 | 1 | 1 | 1 | 1 | 1 | 1 | 1 | $\alpha_{xx}$, $\alpha_{yy}$, $\alpha_{zz}$ |
| $A_u$ | 1 | 1 | 1 | 1 | -1 | -1 | -1 | -1 | silent |
| $B_{1g}$ | 1 | 1 | -1 | -1 | 1 | 1 | -1 | -1 | $\alpha_{xy}$ |
| $B_{1u}$ | 1 | 1 | -1 | -1 | -1 | -1 | 1 | 1 | ir(z) |
| $B_{2g}$ | 1 | -1 | 1 | -1 | 1 | -1 | 1 | -1 | $\alpha_{xz}$ |
| $B_{2u}$ | 1 | -1 | 1 | -1 | -1 | 1 | -1 | 1 | ir(y) |
| $B_{3g}$ | 1 | -1 | -1 | 1 | 1 | -1 | -1 | 1 | $\alpha_{yz}$ |
| $B_{3u}$ | 1 | -1 | -1 | 1 | -1 | 1 | 1 | -1 | ir(z) |

Which of the expected modes will actually be observed in an experiment depends on a number of additional physical aspects of these modes. To be observed in ir spectra, e.g., a mode must show a significant charge displacement (dipole moment), a requirement for which group-theoretical ir activity is necessary but not sufficient. It can sometimes be concluded from specific models for the ionic charges and the expected displacement of those ions that a mode is actually nearly silent i.e., it shows no appreciable activity although it is ir allowed. An important example is a mode related to the silent (ir and

Raman forbidden) mode of the perovskites. The physical reason for the latter being ir forbidden is that the oxygen ions situated on contiguous faces of the perovskite unit cell vibrate out of phase in the z-direction thus yielding no net dipole moment. A mode similar to this one exists in $RBa_2Cu_3O_{7-\delta}$, but with the amplitudes for the oxygen displacements somewhat different for the two contiguous oxygen ions (a result of the lower, orthorhombic symmetry, see discussion below). The resulting dipole moment remains small, yielding a nearly ir-forbidden mode in the superconductor.

Sometimes it is useful to analyze a system with higher symmetry if the system under consideration possesses approximately that symmetry, broken by a weak perturbation of lower symmetry. Transitions or modes forbidden for higher symmetry may be expected to appear only weakly, even if they are allowed, in the actual system. The higher-symmetry modes may have different selection rules which will be fulfilled approximately in the lower-symmetry system, a fact which should help to identify some modes. The approximate higher symmetry for $RBa_2Cu_3O_{7-\delta}$ is tetragonal since the lattice constants a and b are nearly equal in the superconductor (the approximation, of course, breaks down for the ir-active modes of OI and may not be very good for the modes of OIV). Furthermore, if the equivalent of one oxygen is removed from $RBa_2Cu_3O_7$, the system becomes exactly tetragonal (and semiconducting), so the Raman and ir phonons can be experimentally studied separately for both symmetries[37,45] (except those of the removed OI).

Following the same procedure as indicated above for the orthorhombic system, we arrive at the modes for tetragonal $RBa_2Cu_3O_6$ [a = b = 3.853 A and c = 11.780 A for R = Y;[46] space group P4/mmm $(D_{4h}^1)$, and point group 4/mmm $(D_{4h})$]: $A_{1g} + 2E_g + A_{2u} + 2E_u$ for $C_{4v}$ site symmetry (Cu2, OIV, Ba atoms) and $A_{1g} + 4E_g + B_{1g} + A_{2u} + 4E_u + B_{2u}$ for the lower $C_{2v}$ site occupied by O in the copper-oxygen planes. The fully symmetric (in $D_{4h}$) sites (Cu1 and Y) contribute the usual 3 ir-active modes, in this case $A_{2u} + 2E_u$. Adding these modes with the multiplicity of the sites and subtracting the three acoustic modes yields the 33 optical modes of the tetragonal system. We have also included a character table of the $D_{4h}$ group and the compatibility of its representa-

423

tions with those of $D_{2h}$ in Table II.

Since the notational differences in the two point groups for equivalent modes may slightly obscure the similarities between the mode sym-

Table II: Character table and activities of the representations (Reps.) of the point group $D_{4h}$. For an explanation of the symbols see Table I. The compatibility relations (Comp.) for the lower-symmetry $D_{2h}$ group have also been added.

| Comp. $D_{2h}$ | Reps. | E | $2C_4$ | $C_2^z$ | $2C_2$ | $2C_2'$ | i | $2S_4$ | $\sigma_h$ | $2\sigma_v$ | $2\sigma_d$ | activity |
|---|---|---|---|---|---|---|---|---|---|---|---|---|
| $A_g$ | $A_{1g}$ | 1 | 1 | 1 | 1 | 1 | 1 | 1 | 1 | 1 | 1 | $\alpha_{xx}+\alpha_{yy},\alpha_{zz}$ |
| $A_u$ | $A_{1u}$ | 1 | 1 | 1 | 1 | 1 | -1 | -1 | -1 | -1 | -1 | |
| $B_{1g}$ | $A_{2g}$ | 1 | 1 | 1 | -1 | -1 | 1 | 1 | 1 | -1 | -1 | |
| $B_{1u}$ | $A_{2u}$ | 1 | 1 | 1 | -1 | -1 | -1 | -1 | -1 | 1 | 1 | ir(z) |
| $A_g$ | $B_{1g}$ | 1 | -1 | 1 | 1 | -1 | 1 | -1 | 1 | 1 | -1 | $\alpha_{xx}-\alpha_{yy}$ |
| $A_u$ | $B_{1u}$ | 1 | -1 | 1 | 1 | -1 | -1 | 1 | -1 | -1 | 1 | |
| $B_{1g}$ | $B_{2g}$ | 1 | -1 | 1 | -1 | 1 | 1 | -1 | 1 | -1 | 1 | $\alpha_{xy}$ |
| $B_{1u}$ | $B_{2u}$ | 1 | -1 | 1 | -1 | 1 | -1 | 1 | -1 | 1 | -1 | |
| $B_{2g}+B_{3g}$ | $E_g$ | 2 | 0 | -2 | 0 | 0 | 2 | 0 | -2 | 0 | 0 | $\alpha_{xz},\alpha_{yz}$ |
| $B_{2u}+B_{3u}$ | $E_u$ | 2 | 0 | -2 | 0 | 0 | -2 | 0 | 2 | 0 | 0 | ir(x),ir(y) |

metries we discuss now in some more detail two modes that are fundamentally different in the two systems. They concern vertical oxygen displacements in the copper-oxygen planes. In Fig. 3a we show two eigenmodes of a rectangular plane lattice, considered to be equivalent to the OII-CuII-OIII planes of $RBa_2Cu_3O_7$, with the four corners fixed and four atoms placed in the middle of the edges. Opposite atoms are to be regarded as the same. The two sites on adjacent edges, though, have different displacement amplitudes and energies for $D_{2h}$ symmetry (e.g. they could be atoms of different mass) and hence there are, in general, two distinct eigenmodes of $A_g$ (even symmetry). When we include the

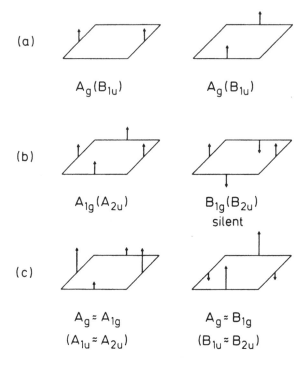

(a)

$A_g(B_{1u})$        $A_g(B_{1u})$

(b)

$A_{1g}(A_{2u})$      $B_{1g}(B_{2u})$
silent

(c)

$A_g \approx A_{1g}$       $A_g \approx B_{1g}$
$(A_{1u} \approx A_{2u})$     $(B_{1u} \approx B_{2u})$

**Fig. 3.** Schematic illustration of the near tetragonality of the two oxygen-related $A_g$ modes in the $CuO_2$ planes: (a) two modes with $A_g$ symmetry in an orthorhombic system. (b) Only the atomic displacements shown here have the correct symmetry in a tetragonal system. (c) In nearly tetragonal systems like $YBa_2Cu_3O_{7-\delta}$ the modes are as shown. The approximate-tetragonal symmetry allows the use of more stringent optical selection rules, greatly simplifying the analysis of the Raman spectra. The corresponding ir symmetries are given in parentheses.

second such plane of the $RBa_2Cu_3O_7$ unit cell each of the modes generates an $A_{1g}$ and a $B_{1u}$ mode. In fourfold symmetry (Fig. 3b) the four edge sites have the same environment (identical atoms, equal bond distances etc.) and their displacement amplitudes must be the same since the OII and OIII atoms are related by a fourfold rotation $C_4$, i.e., they are equivalent. The in-phase $A_{1g}$ ($A_{2u}$) and out-of-phase $B_{1g}$ ($B_{2u}$) modes shown in Fig. 3b fulfill this requirement. For an approximately tetragonal system the eigenmodes will be similar but now the displacement of OII and OIII need not to be exactly equal. In Fig. 3c we depict such modes. This is, schematically, what the OII, OIII vibrational modes along z of the nearly tetragonal, orthorhombic superconductor are expected to look like.

The <u>approximate</u> $B_{1g}$ ($D_{4h}$) symmetry of one of the modes in the orthorhombic ($D_{2h}$) $CuO_2$ planes is of great importance because of different selection rules for Raman scattering in the $A_{1g}$ and $B_{1g}$ symmetries. The approximate $B_{1g}$ in the orthorhombic system is hence expected to stand out among the other $A_g$ modes and can serve as a reference in identifying those. It is interesting to note that when considering <u>one</u> $CuO_2$ plane the displacement in this mode (for $D_{4h}$ symmetry) is the same as that of the silent perovskite mode already discussed. In fact, the corresponding $B_{2u}$ mode also has similar displacements of OII and OIII within each plane. It is the second $CuO_2$ plane in the superconductor – separated by the R-atom – that splits the perovskite mode into odd and even components: in one of them upper and lower $CuO_2$ planes are out-of-phase ($B_{1g}$) and in the other they are in-phase ($B_{2u}$). The ir activity is not affected by adding another plane with no dipole moment, and $B_{2u}$ remains silent. The $B_{1g}$ mode, however, becomes symmetry-allowed for Raman activity and is indeed observed. The corresponding Raman tensors have the form:

$$D_{4h}: \quad A_{1g}: \begin{pmatrix} a & & \\ & a & \\ & & b \end{pmatrix} \quad B_{1g}: \begin{pmatrix} d & & \\ & -d & \\ & & 0 \end{pmatrix}$$

$$D_{2h}: \quad A_g: \begin{pmatrix} a+\alpha & & \\ & a-\alpha & \\ & & c \end{pmatrix} \quad A_g: \begin{pmatrix} d-\beta & & \\ & -d-\beta & \\ & & \gamma \end{pmatrix} \tag{3}$$

where $\alpha$, $\beta$, and $\gamma$ should be small.

As we shall see below, only modes of $A_g$ symmetry have been <u>clearly</u> identified (i.e. appear strongly) in the Raman spectra of $RBa_2Cu_3O_7$. It is easy to justify heuristically why modes of other symmetries are weak. Let us consider the modes of $B_{2g}$ ($\equiv xz$) symmetry which correspond to motions of either OII or OIII along x. If we assume that the Raman polarizability is of the nearest-neighbor bond-stretching nature, the 'allowed' xz component of the Raman tensor vanishes since the oxygen atom is at a center of inversion for the planes: The polarizability of the Cu-O bond to the right of OII exactly cancels that of the bond to the left. For the $A_{1g}$ mode of Fig. 3, however, $A_{1g}$ contributions of the Raman tensor due to stretching of O-Ba, O-Y, and O-O bonds in a set of adjacent 'planes' do not cancel. We should point out that the number of fully symmetric phonons ($A_{1g}$, $A_g$) corresponds to the number of parameters required to give the positions of the atoms within the unit cell.[47]

Another case of interest is that of $RBa_2Cu_3O_{6.5}$, with the approximate oxygen concentration at which superconductivity disappears. This material corresponds to the exact formal valence of $Cu^{+2}$ and $O^{-2}$ and seems to have ordered oxygen vacancies, so that one out of two alternating chains of Cu1–OI (Fig. 2) is missing.[48] Its space group is obviously the same orthorhombic $D_{2h}^1$, with the OI and Cu1 atoms still in full $D_{2h}$ symmetry sites but R having lowered, $C_{2v}$ symmetry. The symmetry of the Ba and OII atoms becomes $C_s$, all other atoms retaining $C_{2v}$ symmetry and thus the same vibrational optical selection rules (as do the OI and Cu1 atoms of $RBa_2Cu_3O_7$). The R atoms should thus generate <u>Raman-active</u> modes of xy ($B_{1g}$, forbidden in the $O_7$ materials) $A_g$, and zx ($B_{2g}$) symmetry. The vibrational analysis of the atoms of $C_s$ symmetry is straightforward (see Table 8B of Ref. 16).

## 2. Lattice dynamical calculation for $RBa_2Cu_3O_7$ and $RBa_2Cu_3O_6$

Having identified the symmetries of the optical phonons in the orthorhombic and tetragonal $YBa_2Cu_3O_{7-\delta}$ systems it is now important to obtain a physical estimate of the frequencies (i.e. energies) of the expected modes. Simple considerations of the mass of the vibrating atoms or inspection of similar vibrations in other, well understood compounds may provide a first feeling for the eigenvectors at hand. In order, however, to obtain a good basis for the assignment of the observed phonon frequencies it is necessary to approximate the interaction between the different ions in the unit cell, possibly taking into account force constants beyond nearest neighbors (e.g. Coulomb interaction). A number of such calculations for $RBa_2Cu_3O_7$ have been published[34,43,49-53] including simple spring-and-masses approaches[34,43,49,50] and shell-model calculations.[52,53] The former suffer from the difficulty of determining the appropriate force constants; they are often adjusted to obtain experimentally observed frequencies. This procedure has the disadvantage that it implies assumptions about the assignments of peaks which, ideally, should be made as a result of the calculation. In this section we present shell-model calculations where the interaction parameters are taken from other, known oxides. In this way, a fundamentally empirical but conceptually independent procedure of obtaining lattice vibrational energies is followed. As we shall show below, this approach yields reasonably good frequency estimates for $YBa_2Cu_3O_7$, $YBa_2Cu_3O_6$, $Bi_2Sr_2CaCu_2O_8$, $Tl_2Ba_2CaCu_2O_8$, and $La_2CuO_4$.

The basis for the shell-model calculation presented here consists of the short-range repulsive interaction potentials between two ions i and j. They are of the so-called Born-Mayer type and defined as:

$$V_{ij} = a_{ij} \exp ( -b_{ij}r) \qquad (4)$$

Radial and tangential force constants may be obtained from these potentials by taking derivatives.

$$A_{ij} = \frac{\partial^2}{\partial r^2} V_{ij}(r) \Big|_{r=r_0}$$

$$\qquad (5)$$

$$B_{ij} = \frac{1}{r} \frac{\partial}{\partial r} V_{ij}(r) \Big|_{r=r_0}$$

It remains, therefore, to choose the appropriate constants $a_{ij}$ and $b_{ij}$. In the calculation presented here, the O-O and Ba-O potential parameters were taken from fits to the perovskite $BaTiO_3$[54,55] and those of Cu-O were taken from known potentials[56] of Ni-O assuming that the short-range forces are similar. The Y-O parameters were chosen from SrO[57] because Sr and Y have a similar ionic radius. The ionic charges in the shell model, accounting for long-range Coulomb interactions and initially chosen to be +3, +2, +2 for Y, Ba, and Cu, respectively, were refined to +2.85, +1.90, and +2.00 to obtain a stable lattice (phonon frequencies positive) over the entire Brillouin zone.[53] The charges on the oxygen ions were then determined by the charge-neutrality condition leading to a charge of -1.81. Other schemes involving for instance different valence of Cu and/or O in the chains and formal charges in the planes were tried. They did not lead to significant changes or improvement of the model.

The shell model allows the inclusion of a strong dynamic polarizability of the ions which is represented by the shell charge and the shell-ion force constant $\kappa$. In the case of oxygen the polarizability can be very strong because of the instability of the $O^{2-}$ ions. It is customary, in fact, to introduce for the Cu-O bonds a transverse polarizability $k_\perp$ much larger (6-7 times) than the longitudinal one which is reduced by the ionic confinement. In the initial calculations this was done for all Cu-O bonds and the agreement with experiments probing $\vec{k} \cong 0$ phonons was good, with the exception of the in-phase displacement (z-direction) of oxygen in the Cu-O plane. Due to strong Coulombic attraction between the copper and oxygen planes this mode had a high vibrational energy (above 600 $cm^{-1}$). The effect of dynamical carrier screening in the copper-oxygen planes was simulated in an 'ad hoc' fashion, by introducing a large isotropic polarizability for the Cu-O bonds in the Cu-O planes. Without any further adjustment this model gives quite close agreement with the experimentally observed ir and Raman frequencies. We should note that the use of an improper (strongly anisotropic) polarizability also led to too high oxygen vertical stretching modes in $La_2CuO_4$ (see also Sect. II.C).[58]

In Fig. 4 we show the eigenmodes of $RBa_2Cu_3O_7$ in the upper half of the unit cell. The motion of the atoms in the lower half may be easily obtained by extending the displacements of atoms symmetrically through the inversion center (R-atom), with even parity for the Raman modes and odd parity for ir modes. We note the occurrence of the near-$B_{1g}$-tetragonal $A_g$ mode already discussed, with a frequency of 355 $cm^{-1}$. As we shall show in Sect. III.A this mode is observed at an experimental frequency of 335 $cm^{-1}$, a remarkably good agreement (<6% error) considering that, apart from adjustments to obtain lattice stability and the modification of the bond polarizability to simulate carrier screening, the model uses no fits to experimental data. We note that the group-theoretical classification of modes already given in Sect. II.A.1 is obtained again as a by-product of this calculation. Frequencies which are slightly different from those in Fig. 4 (<20% variation), because of the non-uniqueness of the lattice-stability requirement were also reported as a result of a similar calculation with somewhat different

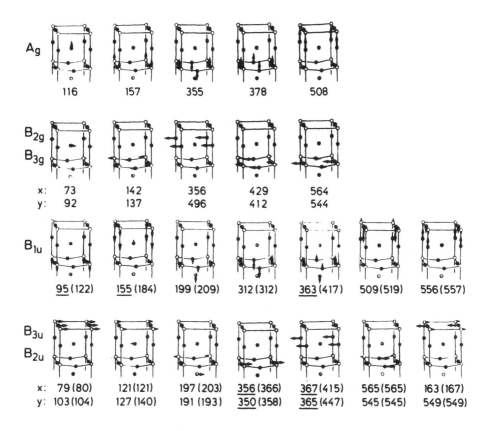

$A_g$    116    157    355    378    508

$B_{2g}$
$B_{3g}$

| x: | 73 | 142 | 356 | 429 | 564 |
| y: | 92 | 137 | 496 | 412 | 544 |

$B_{1u}$

95(122)   155(184)   199(209)   312(312)   363(417)   509(519)   556(557)

$B_{3u}$
$B_{2u}$

| x: | 79(80) | 121(121) | 197(203) | 356(366) | 367(415) | 565(565) | 163(167) |
| y: | 103(104) | 127(140) | 191(193) | 350(358) | 365(447) | 545(545) | 549(549) |

Fig. 4. Calculated atomic displacements of $RBa_2Cu_3O_{7-\delta}$ and the corresponding theoretical vibrational frequencies. The atoms shown form the upper half of the unit cell in Fig. 2. The displacements can be extended easily to the lower half according to the symmetry (even or odd with respect to the center of inversion). The displacements are taken from Ref. 59 but the frequencies, calculated with slightly different parameters, are those of Ref. 53. The differences in frequencies are small (<20%) and should be taken as a measure of the relative insensitivity of the calculation to the detailed parameters used.

parameters.[59] The similarity of these results shows that the eigenmodes and frequencies obtained in the calculation may be used as a basis set for interpreting experimental findings with a fair degree of confidence.

A variety of properties of the structure may be inferred from this calculation, apart from the $\vec{k} \cong 0$ modes. By evaluating the frequencies at many points in the Brillouin zone the dispersion curves and hence the one-phonon density-of-states were obtained.[52,53] In Fig. 1 the latter is compared to results from neutron scattering[13] on polycrystalline samples, yielding the amplitude- and cross-section-weighted one-phonon density-of-states. Single crystals large enough for neutron scattering experiments have recently become available; they have yielded preliminary results for the low-frequency branches of the dispersion relations.[92] The lattice specific heat of $RBa_2Cu_3O_7$ calculated from this model[53] is compared to experiment[60] in Fig. 5.

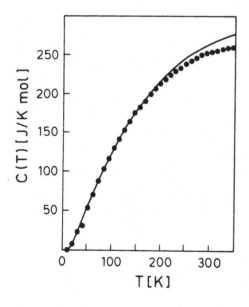

Fig. 5. Comparison of experimental and calculated specific heat of $YBa_2Cu_3O_{7-\delta}$. The figure is taken from Ref. 53.

For the identification of optical modes it is furthermore important to note that effects on the lattice frequencies caused by substitution of one or several atoms may easily be calculated. Of particular interest in this respect is the decrease in vibrational energy when standard oxygen ($^{16}O$) is replaced by the isotope $^{18}O$. Frequency shifts may be calculated quite accurately since no first-order changes in the long-range Coulomb interactions take place.[61,62] For comparison of the

calculated isotopic energy changes of the phonons with the measured
ones see the discussion in Sect. III.A.4. Second-order corrections to
the Coulomb interactions may arise from changes in the zero-point motion of some ions, as was pointed out by Fisher et al.,[63] and will be
discussed more in Sect. III.A.4. Using the lattice dynamical calculation presented here, it is also possible to obtain Grüneisen parameters
which are in good agreement with observed frequency changes under hy-
drostatic pressure.[64] The frequencies for $YBa_2Cu_3O_6$ may be found simply
by removing the chain oxygen (OI) and adjusting the lattice parameters
appropriately. The calculated changes in the Raman frequencies agree
qualitatively with those observed. See Ref. 45 for the detailed results
of that calculation.

   We note that the lattice dynamical calculations also yield, in principle, the TO-LO splittings of the ir-active modes. These splittings
are related to the 'dynamical' charges which, in the model described,
are determined by the ionic and shell charges, the latter being much
less important than the former. Because of the absence of dynamical
charge transfer between ions in the model, it is not expected to yield
very reliable data for the oscillator strengths. We calculate from the
model an oscillator strength S defined as (in atomic units):

$$S = \frac{4\pi}{1836 \ V_c} \ (\sum_i \frac{e_i^* q_i}{\sqrt{M_i}})^2 \tag{6}$$

where $V_c$ is the volume of the primitive cell, $e_i^*$ the charge of the i-th
ion, $M_i$ its atomic mass, and $q_i$ its displacement vector. The sum runs
over all ions in the unit cell. For the $B_{1u}$ mode at 155 $cm^{-1}$, for instance, we find from Eq. (6) S = 4, while the experimental value obtained from reflectivity data for the ceramic case is S ≅ 30.[65] It is
hard to attribute this large discrepancy to dynamic charge transfer. An
interesting question is whether it may be related in any way to the
mechanism of high-$T_c$ superconductivity.

### B. The Bismuth and Thallium Compounds

These materials have formula units:

$$Bi_2Sr_2Ca_{n-1}Cu_nO_{4+2n+\delta}$$

$$T\ell_2Ba_2Ca_{n-1}Cu_nO_{4+2n+\delta} \qquad (n = 1, 2, \ldots)$$

$$(7)$$

whereby $\delta$ is close to zero and some Ca can be replaced by Sr (Bi compound) or Ba (T$\ell$ compound) and vice versa. Values of n as high as 3 have been reported for both the T$\ell$ and the Bi compounds.[66,67] Note that these structures contain n $CuO_2$ planes perpendicular to the c-axes per formula unit, similar to those of $RBa_2Cu_3O_7$ (Fig. 2). $T_c$ is known to increase with increasing n for a given type of material. Compounds with only one T$\ell$ atom per formula unit ($T\ell Ba_2Ca_{n-1}Cu_nO_{2n+3+\delta}$) and $T_c$ as high as 90 K have also been reported.[68,69] One of them, with n = 4, may exhibit a $T_c$ as high as 160 K.[70]

The structure of the compounds of Eq. (7) is related to that shown in Fig. 6, which possesses the space group I4/mmm ($D_{4h}^{17}$) and a body centered tetragonal Bravais lattice (we disregard here possible Sr-Ca or Ba-Ca disorder and consider only ordered structures). Figure 6 describes rather accurately $T\ell_2Ba_2CaCu_2O_8$, the full tetragonal prism being the primitive cell with the upper, incompletely shown prism completing the crystallographic unit cell. Other values of n can be obtained by adding or removing $CuO_2$ layers and Ca planes. The tetragonal structure applies rather accurately to the T$\ell$ compounds (see, however, small deviations in Ref. 71).

The primitive cell of the Bi compounds of Eq. (7) is similar to that in Fig. 6 except for an orthorhombic distortion along the x-y diagonal: the new lattice constants a' and b' thus approximate that of the tetragonal structure a = b times $\sqrt{2}$ [a' = 5.41 A, b' = 5.42 A, a(T$\ell$) = b = 3.87 A]. Also, a significant periodic, but incommensurate distortion along the b-axis, coupled with a commensurate distortion along c, seems to occur.[72] We shall neglect this distortion in the discussion of Raman

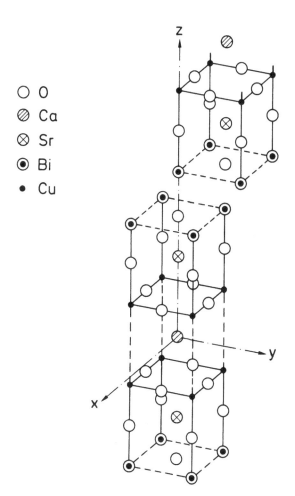

O  O
⊘  Ca
⊗  Sr
◉  Bi
•  Cu

Fig. 6. Approximate structure of the Bi and Tℓ
superconductors. The upper
portion is shifted with
respect to the lower one
by $(\frac{1}{2},\frac{1}{2},\frac{1}{2})$. The crystallographic unit cell extends
in z-direction twice the
z-separation between the
two Ca atoms shown, the a'
and b' lattice parameters
lie along x+y, x−y direc-
tions in the figure and
amount to the shortest Cu
distances in those direc-
tions. The detailed loca-
tion of oxygens in the BiO
planes is subject to dis-
cussion in the literature
(see also text and Fig.
7). For the Tℓ supercon-
ductor the structure shown
appears to be an excellent
approximation.

spectra.

There is still a major controversy about the structure of the
$BiO_{1+\delta/2}$ planes in the Bi-superconductors. Some authors thought that
these planes are NaCl-like $(\delta \cong 0)$[73] (see Fig. 7a) while others proposed a defect $CuO_2$-like structure $(\delta \cong 0)$[74] (Fig. 7b). More recent
work[75] has suggested a modification of the oxygen sites of Fig. 7b

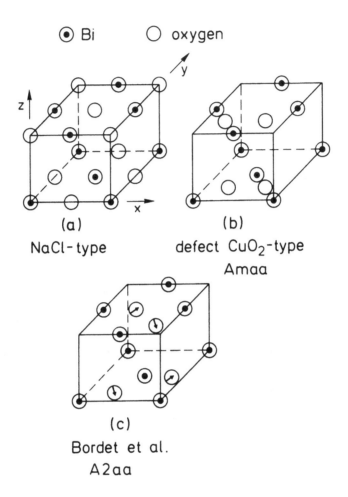

⊙ Bi    ◯ oxygen

(a)
NaCl-type

(b)
defect $CuO_2$-type
Amaa

(c)
Bordet et al.
A2aa

<u>Fig. 7.</u> Different models for the placement of oxygen in the BiO planes
(a) the NaCl-type has been suggested by Ref. 73 (b) the defect $CuO_2$-
type by Ref. 74 and (c) by Ref. 75.


leading to the structure of Fig. 7c in which the oxygens move to be
off-center of the Bi-Bi lines, placing themselves almost (but not
quite) under or over Bi sites of the adjacent planes. The resulting
non-centrosymmetric space group is A2aa ($C_{2v}^{13}$), with simultaneously
Raman and ir-allowed vibrational modes. The mixture of these two types

of modes (i.e. the lifting of the inversion symmetry) should be felt mostly by the vibrations of the BiO planes, while the modes involving other atoms should remain either ir or Raman allowed. This should apply particularly to the central Ca and $CuO_2$ planes which, except for the BiO planes, see an environment of $D_{2h}$ symmetry but with twice as many atoms per primitive cell as in the case of $RBa_2Cu_3O_7$ (Ca playing the role of R). Thus, as far as Ca and $CuO_2$ are concerned, the optical activity of the vibrations should be similar to that of R and $CuO_2$ in $RBa_2Cu_3O_7$, the additional modes resulting from the doubling of the unit cell remaining nearly forbidden (NF). The x and y axes should be rotated by 45° with respect to $RBa_2Cu_3O_7$ and consequently the tetragonal $B_{1g}$-like modes (Fig. 3) should now appear for crossed polarization with incident and scattered fields parallel to the a' and b'-axes, respectively ([110], [1$\bar{1}$0] of the tetragonal phase).

The point group of the A2aa Bi-superconductor is thus $C_{2v}$. We present in Table III its character table, ir and Raman activity, and the compatibility to the $D_{2h}$ representations. Note the effect of the 45° axis rotation between the $D_{4h}$ and the lower-symmetry $D_{2h}$ groups which results in compatibility relations different from those of Table II. The $(D_{4h})$ $B_{1g}$ mode corresponding to that of Fig. 3 has now $(D_{2h})$ $B_{1g}$ or $(C_{2v})$ $B_1$ symmetry and, correspondingly, $\alpha_{xy}$ polarization with respect to the orthorhombic axes (but still $\alpha_{xx} = -\alpha_{yy}$ polarization with respect to the $D_{4h}$ axes). Contrary to the case of the $D_{2h}$ selection rule of $RBa_2Cu_3O_7$, the $\alpha_{xy}$ polarization is now <u>exact</u> (for $RBa_2Cu_3O_7$ $\alpha_{xx}$ was only <u>approximately</u> equal to $-\alpha_{yy}$).

The central Ca atoms of the A2aa structure of Eq. (7) with <u>n even</u> have $C_2^x$ point symmetry and two inequivalent sites per primitive cell. A schematic diagram of its possible normal modes and their symmetries is given in Fig. 8. For odd values of n the role of the central Ca is played by one of the Cu atoms of Eq. (7). The two oxygen atoms in the central $CuO_2$ plane have the same $C_2^x$ symmetry and their vibrations should be handled as such. According to Fig. 8 no Raman-active modes can exist for <u>central</u> atoms other than those forbidden in the tetragonal case (polarized along x and y). Hence the important Raman-active

Table III: Character table and activities of the representations (Reps.) of the $C_{2v}^x$ point group. The compatibility relations (Comp.) to the $D_{2h}$ group of the Bi compounds and the corresponding $D_{4h}$ group of the Thallium compounds are given with the inactive representations of $D_{4h}$ and $D_{2h}$ in brackets. Note that the $D_{4h} - D_{2h}$ compatibilities differ from those in Table II because of the 45° axes rotation discussed in the text.

| Comp. $D_{4h}$ | Comp. $D_{2h}$ | Reps. | $E^x$ | $C_2^x$ | $\sigma^{yx}$ | $\sigma^{zx}$ | activity |
|---|---|---|---|---|---|---|---|
| $E_u$ | $B_{3u}$ | $A_1$ | 1 | 1 | 1 | 1 | ir(x) |
| $A_{1g}$, $B_{2g}$ | $A_g$ | | | | | | $\alpha_{xx}, \alpha_{yy}, \alpha_{zz}$ |
| $(A_{1u}, B_{2u})$ | $(A_u)$ | $A_2$ | 1 | 1 | -1 | -1 | |
| $E_g$ | $B_{3g}$ | | | | | | $\alpha_{yz}$ |
| $E_u$ | $B_{2u}$ | $B_1$ | 1 | -1 | 1 | -1 | ir(y) |
| $(A_{2g}), B_{1g}$ | $B_{1g}$ | | | | | | $\alpha_{yx}$ |
| $A_{2u}, (B_{1u})$ | $B_{1u}$ | $B_2$ | 1 | -1 | -1 | 1 | ir(z) |
| $E_g$ | $B_{2g}$ | | | | | | $\alpha_{zx}$ |

mode of Fig. 3 should not exist for the central oxygens found when n is odd. All other atoms (Bi, Sr, O, non-central Ca, Cu, and O) have 'general' symmetry and four sites each per primitive cell. The four equivalent atoms of a given kind can be placed as pairs in two parallel planes, perpendicular to c, see similar arrangements in Figs. 2 and 6. Hence for each vibrational direction we find four modes which are

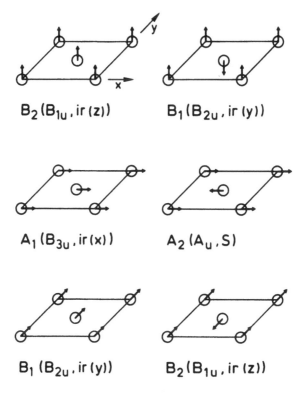

Fig. 8. Normal modes for the central atoms (Ca for n even, Cu and O for n odd) in the A2aa structure of Eq. (7) together with their irreducible representations and, in brackets, those of the corresponding Amaa structure and the activity for the latter (S means silent). The modes to the right are edge-of-the-zone modes for the $D_{4h}$ group of the Tℓ compounds.

sketched in Fig. 9 together with their $C_{2v}$ and $D_{2h}$ symmetries. The more stringent selection rules of the latter symmetry (or the even more stringent ones of $D_{4h}$) should yield approximate selection rules for all modes except those associated with the strongly distorted BiO planes (Fig. 7). Thus for the O of these planes one would expect to see one $A_1$ mode corresponding to bond stretching along z (in the 500 - 600 cm$^{-1}$ region) and two which are tetragonal-forbidden and polarized along x and y, corresponding to bond bending (around 300 cm$^{-1}$). These modes

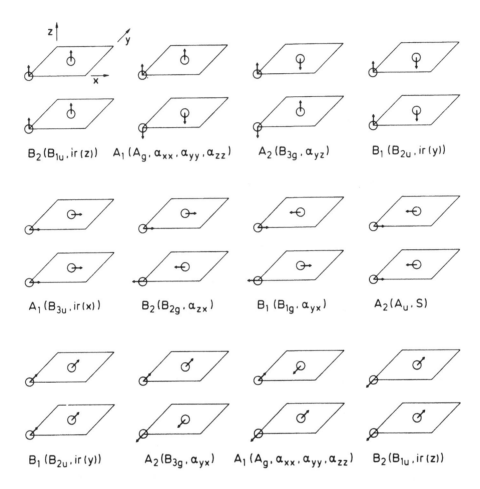

$B_2(B_{1u}, ir(z))$  $A_1(A_g, \alpha_{xx}, \alpha_{yy}, \alpha_{zz})$  $A_2(B_{3g}, \alpha_{yz})$  $B_1(B_{2u}, ir(y))$

$A_1(B_{3u}, ir(x))$  $B_2(B_{2g}, \alpha_{zx})$  $B_1(B_{1g}, \alpha_{yx})$  $A_2(A_u, S)$

$B_1(B_{2u}, ir(y))$  $A_2(B_{3g}, \alpha_{yx})$  $A_1(A_g, \alpha_{xx}, \alpha_{yy}, \alpha_{zz})$  $B_2(B_{1u}, ir(z))$

Fig. 9. Normal modes of vibrations of atoms in general sites of the A2aa structure of Eq. (7) together with their irreducible representations and, in brackets, those of the corresponding Amaa structure and the activity in the latter case (S means silent). The two columns to the right represent edge of the zone (forbidden) modes of the $D_{4h}$ group of Tℓ compounds.

have indeed been observed and identified.[76] Similar modes related to the motion of Bi should be found at very low frequencies ($<100$ cm$^{-1}$).[77]

A final complication may arise from the existence of the incommensurate periodic distortion along y. This distortion, which is expected to affect mainly the BiO planes, should partially break down the $\vec{k}$ conservation for the corresponding BiO modes and lead to broad bands with two peaks at their ends, corresponding to one-dimensional densities-of-states.

By appropriately modifying the analysis just performed, that of the Tℓ-based tetragonal phase [$D_{4h}$, Eq. (7)] becomes straightforward. The normal modes are those of the first column of Fig. 8 for central atoms and the two first columns of Fig. 9 for all others. The corresponding representations and activities can be inferred from the compatibility relations of Table III.

The structure of the Tℓ compounds with only one Tℓ-O plane per primitive cell[68,69,70] is , to this point, not certain. In Ref. 77 a simple P4/mmm structure is attributed to the $TℓBa_2CaCu_2O_7$ phase. It corresponds to the primitive cell of Fig. 6 (only lower part) and thus the same space group as $RBa_2Cu_3O_6$ ($D_{4h}^1$). The vibrational analysis given in Sect. II.A.1 for $RBa_2Cu_3O_6$ can be applied to this Tℓ compound provided one replaces CuI by Tℓ, Y by Ca, and one adds one oxygen at the center of the diagonals in the Tℓ squares. Thallium thus forms octahedral coordination polyhedra, rather than Tℓ-O chains; the additional oxygen has the same site and vibrational symmetry as Y (Ca) and Cu1 (Tℓ). The space group of the $Tℓ_1$ compounds with n = 3[69] seems to be body centered tetragonal (I4/mcm),[69] that for n = 4 primitive tetragonal (P/mcc).[70]

In order to illustrate the vibrational eigenvectors of these compounds we present the results of calculations[78], similar to those shown in Fig. 4, performed for $Bi_2Sr_2CaCu_2O_8$ under the assumption of NaCl-type BiO layers (Fig. 7a) and $D_{4h}$ symmetry in Fig. 10.

## C. The Zurich Superconductors

The 'Zurich Oxides' have typical formula units $La_{1.85}Sr_{0.15}CuO_4$ and $La_{1.85}Ba_{0.15}CuO_4$. They are disordered crystals derived from the ordered $La_2CuO_4$ which has a tetragonal perovskite-related structure above 432 K (space group I4/mmm or $D_{4h}^{17}$).[79] Below this temperature it undergoes a small orthorhombic distortion leading to the Abma ($D_{2h}^{18}$) space group. A

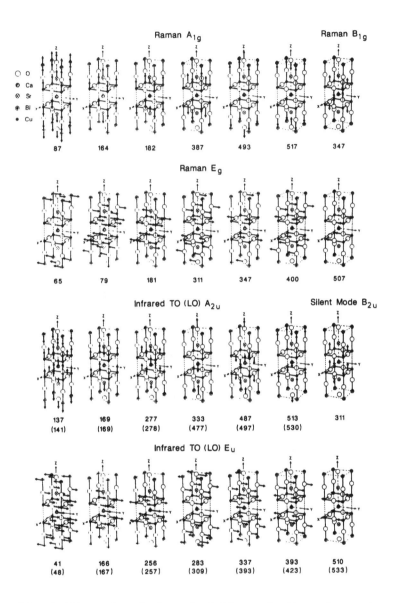

Fig. 10. Lattice dynamical calculation of the eigenvectors of $Bi_2Sr_2CaCu_2O_8$. The BiO planes are assumed to be NaCl-like (see Fig. 7a), a structure probably more appropriate to the compound in which Bi is replaced by Tℓ. Modes not related to the BiO planes are expected to be in agreement with the experiment. Note the occurrence of the $B_{1g}$ mode in the upper right-hand corner. Frequencies are given in $cm^{-1}$. From Ref. 78.

schematic diagram of these structures is shown in Fig. 11. It represents, in the I4/mmm case, the projection of the unit cell on a plane

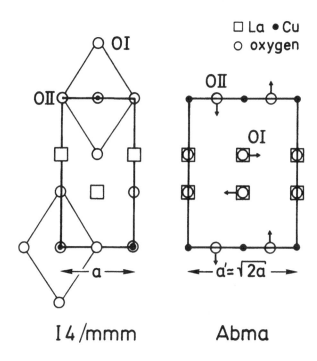

Fig. 11. Schematic representations of the primitive cells of the tetragonal and orthorhombic phases of $La_2CuO_4$. The tetragonal phase is shown
projected on the (010) plane while the orthorhombic one is projected on
the (010) plane of the orthorhombic structure, i.e., the tetragonal
(110) plane. Arrows indicate the orthorhombic distortions of the La,
OI, and OII atoms.

perpendicular to y (y-plane), the carets corresponding to $CuO_6$ octahedra, the upper one being shifted by a/2 perpendicular to the paper
with respect to the lower one. Note the presence of <u>one</u> $CuO_2$ plane per
unit cell.

The lowering of the I4/mmm symmetry into Abma produces a doubling of the volume of the primitive cell: the basal primitive translations of the new structure (a', b') are rotated by 45° with respect to those of I4/mmm; hence a' $\cong$ b' $\cong$ 2√a. Simultaneously, rotations of the octahedra around the b' axis take place. These rotations are equivalent to a frozen phonon with $\vec{k}$ at the $\frac{\pi}{a}$(110) point of the <u>tetragonal</u> Brillouin zone.[79] The pure $La_2CuO_4$ compound is a ferromagnetic insulator. The Sr-doped material has an 'average' tetragonal → orthorhombic transition at around 200 K[80], thus the presence of Sr stabilizes the cubic phase. The Ba-doped compound, however, seems to make a transition to monoclinic symmetry, whose details are not well established, at about 150 K.[81] We discuss below the symmetries of the $\vec{k}$ = 0 normal modes of the tetragonal and orthorhombic phases.

The normal-mode analysis of the tetragonal phase of $La_2CuO_4$ ($K_2NiF_4$) structure (centrosymmetric) have been discussed by several authors.[82] There is only one copper atom per <u>primitive</u> cell which gives rise to three ir-active (Raman forbidden) modes of $D_{4h}$ symmetries $A_{2u}$ and $E_u$ (see Table II). The two La atoms are transformed into each other by the inversion, hence they give rise to three Raman active ($A_{1g}$, $E_g$) and three ir-active modes ($A_{2u}$, $E_u$). The four oxygen atoms break up into two pairs of equivalent ones: OI and OII. The pair of La atoms has the same properties as OI; these atoms generate each one $A_{1g}$, two degenerate $E_g$ modes (Raman active), an $A_{2u}$ and two $E_u$ modes (ir active). The OII atoms generate five ir-active modes ($A_{2u}$ and $E_u$) and a silent mode ($B_{2u}$). Note that no mode of $B_{1g}$ symmetry, similar to that of Fig. 3, exists in this case since there is only one $CuO_2$ plane per primitive cell. It should be easy for the reader to draw the corresponding normal modes.

We discuss next the normal modes for the Abma phase. The compatibility relations between the representations of its $D_{2h}$ point group and those of the tetragonal $D_{4h}$ phase are those of Table III. The latter has one Cu atom per primitive cell while the orthorhombic one has two. We can place the three displacement vectors in both atoms with the same sign and we obtain ir-active modes equivalent to those of the tetragonal phase ($\vec{k}$ = 0 of tetragonal Brillouin zone). New optically active modes arise when we place the displacement vectors with opposite signs: they correspond to tetragonal $\vec{k} = \frac{\pi}{a}(110)$ (edge of the zone, forbidden) modes which now move to $\vec{k} = 0$ and become allowed as a result of the doubling of the primitive cell. We show in Fig. 12 these modes and their representations. Their optical activities are expected to be small, corresponding to the smallness of the orthorhombic distortions. Note that there is no new Cu mode of $A_g$ symmetry, hence the distortion <u>does not affect the position of the Cu atoms</u> which remains fixed by symmetry.

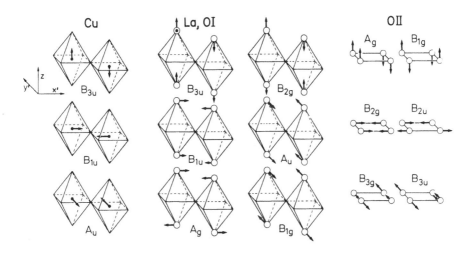

**Fig. 12.** Schematic illustration of the additional $k \approx 0$ modes of orthorhombic $LaCuO_4$. These modes are zone-edge modes in the tetragonal, undoubled unit cell.

The La and OI atoms have the same site symmetry, and their orthorhombic normal modes can be discussed together. The <u>new</u> modes which correspond to $\vec{k} = \frac{\pi}{a}(110)$ in the tetragonal structure are shown in Fig. 12. Note that one of these modes, corresponding to staggered rotations about the tetragonal ($1\bar{1}0$) axis, has $A_g$ symmetry: it corresponds to the frozen-in orthorhombic distortion.

Finally we discuss the normal modes of the four OII atoms. The additional modes introduced by the doubling of the cell and their activities are shown in Fig. 12. Note that there is also an $A_g$ mode which corresponds to the rotations just discussed and thus represents the frozen-in distortion of these atoms.

Lattice dynamical calculations, similar to those presented for the yttrium and Bi-based superconductors, are also available[58] for $La_2CuO_4$. The frequencies obtained agree with the experimentally observed ones with the exception of modes where O and Cu vibrate out of phase in z-direction: the calculated frequency is too high because an unscreened Coulomb force was used. When dynamical carrier screening is introduced in form of a large effective isotropic polarizability in $La_2CuO_4$ those calculations, like in $RBa_2Cu_3O_7$, presumably will yield better frequencies for these modes. Recently we became aware of a fully microscopic calculation (based on the total energy obtained from the electronic band structure vs. phonon displacement) which predicts the frequencies of some of the modes quite accurately.[83]

## III. EXPERIMENTAL RESULTS ON $RBa_2Cu_3O_{7-\delta}$

We arrange experimental results obtained by light scattering and their consequences for the superconductivity mechanism into topics according to the physical origin of the scattering process. Within each topic we present results on $RBa_2Cu_3O_{7-\delta}$-type crystals, ceramics, and thin films as available and appropriate. First we discuss scattering processes of phononic origin. The discussion includes an assignment of eigenmodes to peaks observed in the spectra of crystalline and ceramic samples, and the isotope effect (Sect. III.A.4) which has been taken by many workers to reveal important information about the superconductivity mechanism. Light scattering here plays the important role of verifying and quantifying atomic substitutions. Then we survey results on electronic scattering, a phenomenon which is currently receiving much attention since information about the superconducting energy gap may be obtained from it (Sect. III.A.5). The anomalous softening of one of the phonon modes at $T_c$, direct evidence for the transition to superconductivity in Raman spectra, is discussed in Sect. III.A.6. Finally, in Sect. III.A.7 we show results on scattering by magnons.

## A. Raman Scattering by Phonons

Most experimental Raman work on high-$T_c$ superconductors has revolved around the identification and assignment of intrinsic peaks in the Raman spectra to vibrational eigenmodes of the lattice. As a consequence of the low scattering intensities and of semiconducting impurity phases present in ceramic materials, reliable experimental results and interpretations were scarce at first. In particular the assignments of observed peaks, a prerequisite for insight into most physical aspects of the lattice vibrations, could be made reliable only with the availability of single crystals of sufficient size to analyze the polarization selection rules. Contrary to the chronological development, we first present results on single crystals which led to the nearly unambiguous assignment of eigenmodes to Raman peaks, and later discuss important results obtained on ceramics.

### 1. Single crystals of $YBa_2Cu_3O_{7-\delta}$

Two types of crystals of this superconductor have been referred to in the literature as 'single': twinned and untwinned ones. Twinning in the a-b plane along [110] and [1$\bar{1}$0] directions is a common occurrence in the crystal growth of these materials today; this explains the somewhat sloppy usuage of the term 'single'. We will, in this discussion, refer to twinned and untwinned crystals whenever the distinction is needed; otherwise 'crystal' or 'single crystal' mean 'twinned crystal'. Most results on crystals were reported on such twinned crystals due to the scarcity of untwinned specimens. Because of the near tetragonality, results on twinned and untwinned crystals are expected to be similar, differences arising primarily from influences of the OI (the chain oxygen) on the amplitudes of the Raman-active modes and from defect-induced OI-related modes (as shown in Sect. II.A.1 OI itself is Raman inactive in a perfect crystal with the chain sites occupied). Raman scattering data on twinned crystals will be shown and discussed first and the specific differences to results on untwinned crystals presented later.

To relate the experiment to the discussion of Sect. II.A.1 it is necessary, as a first step, to determine the symmetries of the observed phonon peaks. Then, together with the results of the lattice dynamical calculation presented in Sect. II.A.2, an assignment may be made. By 'symmetries of phonon peaks' is meant the form of the Raman tensor $\underset{\approx}{R}$ for each phonon, which is related to an irreducible representation of the point group of the crystal. The specific form of a tensor for a particular point group with the number of independent non-zero components can be found in Table 2.1 of Ref. 9. The number of components is further reduced over that given in the table, assuming that the exciting laser is away from an electronic resonance, i.e., that the Raman tensors are symmetric ($\alpha_{ij} = \alpha_{ji}$). Experimentally, the relative magnitude of each tensor component $\alpha_{ij}$ may be found by adjusting the incident, polarized electric field ($\vec{E}_L$) along the j-direction of the coordinate system defined by the crystal axes and, using a polarization analyzer, measuring the i-th component of the scattered light ($\vec{E}_S$) with a monochromator. The scattering intensity I is

$$I \sim |\vec{E}_S^* \cdot \underset{\approx}{R} \cdot \vec{E}_L|^2 \ . \tag{8}$$

The absolute magnitude of the Raman-tensor components may be found by comparing the observed intensities with known Raman scattering cross sections in other materials (e.g. diamond).[9]

The relative sign of the various non-zero tensor components is also of importance. For the identification of the $B_{1g}$ ($D_{4h}$) perovskite-like mode (or its analogue in $D_{2h}$ symmetry) discussed in Sect. II.A.1 with the Raman tensors of Eq. (3) and the displacements of oxygen depicted in Fig. 3, the relative sign of the components is needed. If s = sgn ($\alpha_{xx} \cdot \alpha_{yy}$) is found to be negative <u>and</u> $|\alpha_{xx}| \approx |\alpha_{yy}|$ (exactly equal in $D_{4h}$) this symmetry can be positively identified. Experimentally, s may be obtained by 'mixing' $\alpha_{xx}$ and $\alpha_{yy}$ under polarizations parallel to the diagonals of the a-b plane of the superconductor unit cell ([110] and [1$\bar{1}$0]). For both incident and scattered polarizations along [110] Eq. (8) is easily seen to yield a scattering intensity proportional to $|\alpha_{xx} + \alpha_{yy}|^2$ while for crossed polarization ($\vec{E}_L \parallel [110]$ and $\vec{E}_S \parallel [1\bar{1}0]$), $|\alpha_{xx} - \alpha_{yy}|^2$ results. Since we already know from Eq. (3) that $|\alpha_{xx}| \approx |\alpha_{yy}|$ (exactly equal in $D_{4h}$) the scattering intensity relative to that for polarizations $\vec{E}_L \parallel \vec{E}_S$ along [100] or [010] is

$$I_{dia} = \frac{1}{2} [1 \pm s + 0(\frac{\alpha}{a}, \frac{\beta}{d})] \tag{9}$$

where the + sign (-) describes parallel (perpendicular) polarizations along [110] and [1$\bar{1}$0]. Equation (9) holds for both the $A_g$ and $B_{1g}$ modes and allows them to be distinguished from each other because s is different. (The terms of order $\frac{\alpha}{a}$, $\frac{\beta}{d}$ vanish in $D_{4h}$ symmetry.)

In Fig. 13 we show the Raman spectra of twinned crystals of $YBa_2Cu_3O_{7-\delta}$.[59] Indicated in the figure are the polarizations $\vec{e}_L$ and $\vec{e}_S$ of the incident and scattered light with respect to the crystal axes. Curves I through VI are taken with the incident and scattered light vectors parallel to the c-axis (electric fields in the $CuO_2$ plane). Special care had to be exercised to obtain curves VII and VIII where, without the use of a microscope, the laser beam was focussed onto a narrow side of a crystal. Comparison of curves I and II shows that all features seen in curve I in polarizations parallel to [100] disappear when viewed under crossed polarizations along the crystal axes. Consequently, all modes observed in this frequency range are of the $A_g$ type (in $D_{2h}$), in agreement with most reports in the literature.[5,40,84-90] While only $\vec{e}_L \parallel \vec{e}_S \parallel [100]$ is shown in the figure, it is clear that for twinned crystals the same spectrum must be obtained for $\vec{e}_L \parallel \vec{e}_S \parallel [010]$. There have been reports of phonons with xz symmetry,[86,91] but amplitudes significantly lower than those of $A_g$ peaks cast some doubt on the reliability of those results. Recently, however, two low lying $B_{2g}$, $B_{3g}$ pairs of modes ($\sim$ at 60 and 150 cm$^{-1}$) have been found with neutron scattering on relatively large $YBa_2Cu_3O_{7-\delta}$ single crystals.[92]

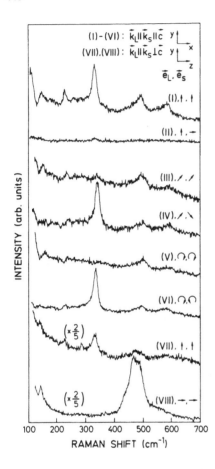

Fig. 13. Polarized Raman spectra of a single crystal of $YBa_2Cu_3O_{7-\delta}$. The various polarization directions of incident ($\vec{e}_L$) and scattered ($\vec{e}_S$) light with respect to the crystal axes are indicated. The changes in the spectra for various configurations allow the determination of the symmetries of the vibrations, i.e. the elements of the Raman tensor.

Curve III corresponds to $\vec{e}_S \| \vec{e}_L \| [110]$ and we find that the strongest peak at 330 cm$^{-1}$ has almost disappeared with respect to curve I but that otherwise the spectrum is nearly unchanged. From Eq. (9) for scattering intensity along the diagonals of the crystal axes we find — neglecting terms of order $\alpha/a$ and $\beta/d$ — that s must equal −1. For perpendicular polarizations along the diagonals [minus sign in Eq. (9)] the peak should then be observable and is indeed seen in curve IV of Fig. 13. From the fact that s ≃ −1 we thus <u>prove</u> that this mode indeed has $B_{1g}$-like symmetry. The mode at 145 cm$^{-1}$, on the other hand, is easily seen to have s ≃ +1, and it follows that it is of the $A_g$ type. Further

confirmation of this analysis comes from the spectra taken with circularly polarized light (curves V and VI). The peaks at 330 and
145 cm$^{-1}$ are seen again to have opposite selection rules which can be
traced back to s. Representing right-hand circularly-polarized light as
$\vec{e}_{rh} = 2^{-\frac{1}{2}}$ (1, i, 0), left-hand circularly-polarized light as $\vec{e}_{\ell h} =$
$2^{-\frac{1}{2}}$ (1, -i, 0), and contracting the $B_{1g}$ Raman tensor as in Eq. (8), we
find that Eq. (9) holds for equal (plus sign) and opposite (minus sign)
helicities (helicities are given in an absolute, laboratory frame, not
with respect to the propagation vector of light). The presence of the
330 cm$^{-1}$ peak in curve VI, its absence in curve V, and the opposite
behavior of the 145 cm$^{-1}$ peak are consistent only with s $\simeq$ +1 for the
latter mode and s $\simeq$ -1 for the former.

It is important to have proven the near-$B_{1g}$ symmetry of the peak at
330 cm$^{-1}$ for a number of reasons: Only one eigenmode can have this
symmetry in the $RBa_2Cu_3O_{7-\delta}$ superconductor. Thus even without a lattice
dynamical calculation we already know the displacements involved; the
assignment as out-of-phase, z-direction motion of the two oxygen in the
$CuO_2$ plane follows directly from group theory and the experimental
polarization selection rules, independent of any further approximation.
Having identified one of the five z-direction modes the task of assigning the remaining 'true' $A_g$ modes is now greatly simplified. As has
been already pointed out, a number of properties link the $B_{1g}$-like mode
to the phenomenon of superconductivity and it is necessary to have a
clear identification for it. Lastly, the mode originates from the $CuO_2$
planes which are common to most high-$T_c$ superconductors. Where they are
Raman active (in materials with pairs of $CuO_2$ planes, e.g.
$Bi_2(Sr_{1-x}Ca_x)_3Cu_2O_{8+\delta}$ or $T\ell_2Ba_2Ca_3Cu_4O_x$ with two pairs and hence two
$B_{1g}$ modes) they are always easily identified, may have common properties, and should lead towards an understanding of the role these planes
play for superconductivity.

Of the modes we have mentioned so far we have discussed the xx and
yy components of the Raman tensor. We shall now look at $\alpha_{zz}$ by considering curves VII and VIII, taken with the light incident perpendicular
to the c-axis on the thin face of a single-crystal platelet. Curve VII

should, of course, be identical to curve I and thus serves as an amplitude reference for curve VIII. (Curves I and VIII are taken on a dif-
ferent crystal face.) For $\vec{e}_L \| \vec{e}_S \| [001]$ (curve VIII) we find that $\alpha_{zz} = 0$
for the 330 cm$^{-1}$ mode; it should be exactly zero in tetragonal symmetry
[see Eq. (3)]. The mode at 145 cm$^{-1}$ is still present and we find for it
$|\alpha_{zz}| \approx 1.2\ |\alpha_{xx}|$. The higher frequency modes at 430 cm$^{-1}$ and 500 cm$^{-1}$
in good superconductors have high amplitudes only in the zz configura-
tion and thus, for them, $|\alpha_{zz}| \gg |\alpha_{xx}|,\ |\alpha_{yy}|$. Furthermore, it can be
seen that the 500 cm$^{-1}$ mode in the spectra of Fig. 13 is rather broad
and centered around 475 cm$^{-1}$. This is characteristic of a distribution
of oxygen content $1 \le \delta \le 0.4$ in the sample and is typical for many
crystals that were not oxygen treated. The dependence of the Raman frequencies on $\delta$ is strongest for this line and has been calibrated in
ceramic samples with carefully determined  content.[93-95]

Two different situations must be considered when discussing inhomogeneities in $\delta$. The oxygen deficiency $\delta$ may vary due to conditions
related to the preparation; its variation may then persist throughout
the specimen. For $\delta \to 1$, such a sample as a whole, e.g. in a magnetic
measurement, will behave like a semiconductor. Secondly, $\delta$ may vary
near the surface due to effusion of oxygen and surface sensitive
methods will integrate over a range of $\delta$'s even though a bulk measurement may yield a well-defined $\delta$. How much is averaged over depends on
the characteristic depth $\zeta$ of the probing method, but generally it can
be said, that the smaller $\zeta$, the less consistent bulk and surface measurements will be. For Raman measurements with visible light as an excitation source a depth of 500 Å $< \zeta <$ 1000 Å is probed, a region over
which variations in $\delta$ are likely to occur and have been reported.[96]
Hence it cannot be said what the intrinsic (i.e. one-$\delta$ value) width of
the 500 cm$^{-1}$ mode is.

In order to estimate the approximate ratio of magnitudes of the
Raman-tensor elements for, say, the 330 and 500 cm$^{-1}$ mode, we compare
the integrated intensity under the two peaks. The linewidth of the
500 cm$^{-1}$ mode in $\alpha_{zz}$ configuration is about three times larger than
that of the 330 cm$^{-1}$ mode, its amplitude three times larger as well,
thus $\alpha_{zz}(500$ cm$^{-1}) \approx 3\ \alpha_{xx}(330$ cm$^{-1})$ follows. Similarly, $\alpha_{zz}(430$ cm$^{-1})$

$\approx \alpha_{xx}(330\ cm^{-1})$. Due to the approximations made, this should however only be considered to be a rough estimate. Note that the considerations above do not take into account the fact that the absorption coefficient $\zeta^{-1}$ and index of refraction n should be anisotropic. From ellipsometric measurements it is known that the $\zeta_x^{-1}$ is about twice as large as $\zeta_z^{-1}$ (subscript denotes the component of the light vector in the directions determined by crystal-axes).[97,98] Since the scattering intensity is proportional to the scattering volume,[99] each Raman-tensor component may be corrected by the appropriate $\zeta^{-1}$ for incident and scattered polarizations according to[99]

$$\alpha'_{ij} = \alpha_{ij}\left[\frac{\zeta_i^{-1}+\zeta_j^{-1}}{(1-R_i)(1-R_j)}\right]^{1/2} \tag{10}$$

where $R_i = |(n_i-1)/(n_i+1)|^2$ in the usual fashion. Equation (10) assumes that all light is absorbed in the sample; in transparent thin films corrections for interference from multiple reflections in the film must be made. Equation (10) assumes that the depth of focus of the collecting lens and spectrometer is large compared to $\zeta_i$ and $\zeta_j$, a condition generally fulfilled for the highly-absorbing superconducting oxides. $\alpha'_{ij}$ then represents the Raman polarizability per unit length in the sample. We shall present the relative magnitude of the xx and yy components in connection with the study of an untwinned single crystal.

The assignment of the highest-frequency Raman mode to vertical vibrations of the bridging oxygen (OIV) (see Fig. 4) is generally agreed upon in the literature. For the 430 $cm^{-1}$ mode we see from the lattice dynamical calculation of Fig. 4 that the in-phase vibration of oxygen in the $CuO_2$ planes is the only possible assignment. The two $A_g$ peaks on the low-energy side of the $B_{1g}$-like mode are therefore the Ba and Cu vibrations. (The lowest-energy $A_g$ peak may be better observed in Fig. 14.) From the lattice dynamics we conclude that the heavy Ba has the lowest $A_g$ frequency, and that Cu in the $CuO_2$ planes vibrates vertically with an eigenenergy of 145 $cm^{-1}$ in Fig. 13. The reverse assignment of the two low-energy modes has been made in Ref. 23, based on the

452

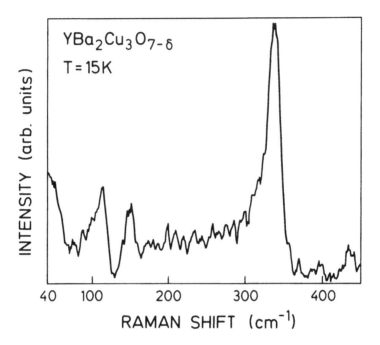

<u>Fig. 14.</u> Raman spectra of a single crystal of $YBa_2Cu_3O_{7-\delta}$. The Fano asymmetry discussed in the text is seen for the phonon peaks at 120 cm$^{-1}$ and 335 cm$^{-1}$. Spectra similar to this one have been reported by Ref. 23.


lineshapes of the observed phonons. This possibility is discussed later.

Of further interest in Fig. 13 are two modes at 220 cm$^{-1}$ and 580 cm$^{-1}$. The only configuration (apart from curve II) where neither of them appears to be present is perhaps curve VIII (this is definitely true for the lower-frequency mode but not quite so clear for the other mode because of the background present in that region). No single Raman tensor with $\alpha_{xx} \approx \alpha_{yy}$ can explain the absence of a clear selection rule. Since $\alpha_{xx} = \alpha_{yy}$ is required by tetragonal symmetry, we conclude that both modes stem from a strongly orthorhombic element in the unit cell or that they may be defect induced. Since all Raman $A_g$ modes have been identified in the spectrum, these two modes are likely to be associated with defect-induced ir-active vibrations. As we will show below for untwinned crystals, the modes have indeed a strong a-b anisotropy and can be satisfactorily explained by vibrations of the chain oxygen, defect-induced for δ not equal to 1 or 0. This mode shows up fairly strongly in the spectra of Fig. 13. It is important to note that the assignment of both modes to an OI-related vibration implies that they should be present or absent together. This implication allows the defect-induced modes to be separated from modes of the impurity[101] $BaCuO_2$ which, among other peaks, has one also at 580 $cm^{-1}$.

Having established the general form of the $RBa_2Cu_3O_{7-\delta}$ Raman tensors and their assignments to lattice vibrations, we discuss now the fine differences in xx and yy components which should be imposed by the deviation of the crystal from tetragonality. To do a reliable analysis it is necessary to obtain an untwinned crystal with a surface equally large or larger than the focussed spot of the investigating laser. By using the slow-cooling method described in more detail in Refs. 22 and 100 it was found that (Zn, Ti)-doped $SnO_2$ crucibles produced the best results for crystal growth. With x-ray precession analysis a largely untwinned crystal was found with unit cell parameters a = 3.81 A and b = 3.88 A. Without further annealing after growth the crystal investigated here had a sharp superconducting transition temperature at $T_c$ = 89 K ($\Delta T_c$ ≈ 3 K as measured by the Meissner effect for a field of 16 Gauss).

In Fig. 15 we show the Raman spectrum of this untwinned single crystal for xx, yy, and xy polarization geometries (light vector $\vec{k}$∥z-axis). Two significant differences between the xx and yy spectra are noticeable: 1) The variation in amplitudes of the various peaks and 2) the presence of two peaks at 220 and 580 $cm^{-1}$, exclusively in yy geometry. The variation in amplitude, in principle, has at least two possible origins. The variation of scattering volume with absorption depth in the presence of anisotropic absorption in the x or y directions is the more trivial one since it affects all Raman peaks equally. It is,

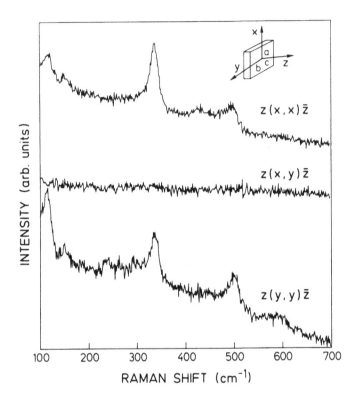

<u>Fig. 15.</u> Polarized Raman spectra of an untwinned single crystal of $YBa_2Cu_3O_{7-\delta}$. The differences in peak amplitudes result from the orthorhombic nature of the crystal.

however, at present not known how large the x-y absorption anisotropy is in the visible region and thus we do not attempt to correct for it. It is clear from Fig. 15, however, that the amplitude variation is different for the different phonon peaks. For example, the lowest energy peak, at 120 $cm^{-1}$, increases by a factor of 3 - 4, whereas the $B_{1g}$-like mode decreases by about 0.7 in switching from xx to yy polarization. The peak at 154 $cm^{-1}$ appears to remain approximately constant. These differences may be represented by the elements $\alpha$ and $\beta$ in the Raman tensors of Eq. (3).

The physical origin of these terms, which are probably small compared to a and d, is difficult to ascertain without a microscopic calculation of the differential susceptibilities involved. It is most
likely related to the different displacements of the OII and OIII atoms
(Fig. 2) which is induced by long-range Coulomb interaction with the
symmetry-breaking OI: In Ref. 53 one finds that OII moves nearly twice
as much as OIII for the 335 $cm^{-1}$ mode. In the case of the 500 $cm^{-1}$ mode
(and the 120 $cm^{-1}$ Ba mode) the xx-yy anisotropy must be due to a direct
effect of OI on the polarizability. The Cu2 mode (150 $cm^{-1}$) is nearly
isotropic as expected from the large distance to OI.

Of some controversy has been the origin of a Raman feature at
580 $cm^{-1}$ reported in many publications, particularly for ceramic material. Frequently the mode at 580 $cm^{-1}$ occurred together with a
roughly twice as strong mode at ~640 $cm^{-1}$. It was realized that $BaCuO_2$,
a phase which is formed during the preparation and frequently remains
in the sample in amounts too small to be detected by x-ray scattering,
was likely to be the origin of these peaks.[101] There is, however, a
peak, broader and much weaker than that of $BaCuO_2$ near 580 $cm^{-1}$ which
is intrinsic to $RBa_2Cu_3O_{7-\delta}$. As is seen in Fig. 15, the peak occurs
together with a peak at 220 $cm^{-1}$ for yy polarizations only. The absence
of the peak in xx polarizations identifies the mode as stemming from an
element with the same crystallographic symmetry (with respect to x,y
orientations) as the $YBa_2Cu_3O_{7-\delta}$ crystal and not from a randomly
ordered impurity phase. Since all $A_g$ modes have been assigned already,
we may look among the ir-active modes of Fig. 4 for those with strong
orthorhombic character which, furthermore, are likely to be defect-
induced. Within the bond-polarizability model all the above conditions
are fulfilled for the modes involving in-phase and out-of-phase vibrations of the chain oxygen against the chain copper. The out-of-phase
vibration in y direction has been calculated to be at ~550 $cm^{-1}$ while
the in-phase mode is expected to be at ~160 $cm^{-1}$. Both modes, strictly
speaking ir active only, should become Raman active in the presence of
oxygen vacancies in the chains, i.e. if $\delta > 0$. Due to the bond direction we expect $\alpha_{xx} \ll \alpha_{yy}$, in accordance with experiment (Fig. 15). In
summary, the modes at 220 $cm^{-1}$ and 580 $cm^{-1}$, when occurring together,

presumably stem from an oxygen-vacancy-induced vibration of the chain oxygen OI. In the context of the issue of vacancy ordering[48] for $\delta$ = 0.5, the amplitude of these two peaks versus a well-characterized oxygen content may provide additional insight into the occurrence of ordering: Should only every second chain be completely occupied, as suggested by Ref. 48, the unit cell doubles and the vibrations of OI should remain Raman inactive as discussed in Sect. II.A.1. Studies on ceramic samples with gravimetrically determined oxygen content did not show this effect of ordering but variations in $\delta$ near the surface, as discussed above, may be responsible for that.[93] Note that the disorder-activated peaks may not correspond to k = 0 phonons but rather to some appropriately convoluted density-of-states.[9]

This concludes the discussion about the identification and assignment of the Raman-active and two disorder-activated ir-active phonons in $RBa_2Cu_3O_{7-\delta}$. With the availability of single crystals, an initially confused picture of phonons in the literature has become clear. These crystals now provide the basis for any discussion tying observations made on these phonons to the superconducting transition.

## 2. Ceramic materials, impurity phases, and thin films

Ceramic samples of the high-$T_c$ superconductors have a number of drawbacks for spectroscopic studies that exploit symmetry properties of the chemical unit cell. As long as the diameter of the interrogating beam averages over different orientations of crystallites, selection rules cannot be used to identify modes with certainty. This is true for Raman and neutron scattering, as well as for fir spectroscopy. Since the exciting lasers in Raman measurements can - with careful consideration of the power density - be focussed down to a few microns, and single crystals of this size are available, interest in ceramics has diminished quickly in this field. This is not so for ir and neutron spectroscopy, where mosaics consisting of several crystalline specimen have just started to approximate larger crystals. Ceramics, however, played an important role in the early understanding of the Raman spectra as well. Their advantages are many: For example, samples can be easily prepared, the oxygen content can be well controlled, $\delta$ can reversibly be driven from 1 to 0 many times if so desired, substitutions may be obtained readily. There is, however, one disadvantage which, in the beginning, severely impeded the identification of the true $RBa_2Cu_3O_{7-\delta}$ modes, and that is the presence of impurity phases. These are usually semiconducting and thus have Raman-scattering cross-sections 1 to 2 orders of magnitude larger than the metallic superconductor. In particular, mistakes were made when samples appeared to be single phase from x-ray analysis, but still contained small amounts of strongly scattering impurities. Now that the Raman spectra of the phases most commonly occurring in the preparation of superconductors are known, light scattering can be used as a test for the impurities with a sensitivity <u>at the surface</u> equal to or better than x-ray scattering. $BaCuO_2$, $Y_2BaCuO_5$ ('green' phase), $R_2Cu_2O_5$, $R_2O_3$, BaO, CuO are probably the most common ones.

We refer the reader to the literature (Refs. 101-105) for examples of the impurity spectra and discuss here only a few selected phases which have attracted attention in connection with the study of $RBa_2Cu_3O_{7-\delta}$. The case of $BaCuO_{2-y}$ (y $\cong$ 0) has already been mentioned in Sect. III.A.1. It has a peak at 580 cm$^{-1}$ concurring with a <u>defect</u>-induced intrinsic mode of $YBa_2Cu_3O_{7-\delta}$(0 < $\delta$ < 1), which makes the latter hard to identify. Furthermore, $BaCuO_2$ has been shown to have a somewhat variable oxygen content (y $\neq$ 0) and its strongest mode at 640 cm$^{-1}$ (probably O-vibrations) may therefore shift by a small amount; this has caused more confusion in the literature than any other impurity phase.

Also interesting is the case of cuprous oxide, $Cu_2O$. It has a strongly resonant Raman spectrum, and its spectra for different laser excitations are well known[106] from single-crystal studies on $Cu_2O$. Studies at low temperatures of samples of $RBa_2Cu_3O_6$ annealed from the superconducting state with various excitation energies from the Ar-ion laser, found[107] that the Raman spectra of $RBa_2Cu_3O_6$ were swamped by the blue-exciton resonance of $Cu_2O$. Upon annealing back to the superconducting state these resonant spectra disappeared. It is reasonable to explain these findings in terms of minute amounts of $Cu_2O$ formed under

Ar annealing. The strong resonant enhancement makes Raman scattering a most sensitive probe of this phase in the superconductor.

In spite of the pitfalls in the use of the ceramic superconductors, there are a number of interesting properties that can be studied. We mention here the dependence of the Raman frequencies on the ionic radius of the rare earth[108,109] atom of R. This can, in principle, also be studied in crystalline samples, but the preparation is cheaper and faster for ceramic materials. From the discussion in Sect. II.A.1 it should be clear that the R atom is not Raman active. Contrary to the fir, where the vibrations involving predominantly R are active and are seen[108] to shift like $\omega_i/\omega_j \cong (m_j/m_i)^{1/2}$, the Raman spectra are expected to change only slightly. The effect on Raman-active vibrations of the change from yttrium to various lanthanides lies predominantly (in the case of isovalent replacements) in the change of the lattice constants and other unit cell parameters. It has been shown in a systematic study of most rare earths forming the $RBa_2Cu_3O_{7-\delta}$ compound (not Ce and Tb) and having valence 3+ only (not Pr) that the frequency of the Cu1-OIV bond is most strongly affected. This bond gets shortened the most for large-ionic-radius lanthanides[110] and thus the highest frequency is observed for those elements.[108,109]

In this context it is interesting to look for the corresponding Raman frequencies of $PrBa_2Cu_3O_{7-\delta}$, the only non-superconducting $RBa_2Cu_3O_{7-\delta}$. In single crystals of this material the Raman vibrations of the Cu1-OIV bond were found at a frequency (554 cm$^{-1}$) much higher than the ionic radius dependence for 3+ coordinated ions would suggest.[111] It is therefore likely from these studies that the formal charges are distributed differently from those in R$^{3+}$ compounds. This may either indicate a 4+ coordination and/or a differently occupied oxygen-Cu1 plane. Thermogravimetric experiments on ceramic samples have indeed shown[112] that annealing may vary the range of $\delta$ outside of the usual interval [0,1] and recent photoemission experiments[113] suggest that the additional oxygen sits in the R-planes. On the other hand, Raman spectra of ceramic samples of R = Pr have been reported where the peak frequencies do fall correctly (516 cm$^{-1}$) on the ionic-radius dependence for a 3+ coordinated Pr ion.[109,114] None of the authors have reported superconductivity in the R = Pr samples. It remains to be seen whether different preparation conditions may affect the charge and oxygen distribution inside the superconducting unit cell for R = Pr in a way that we do not understand at this point.

Substitution of Cu by other metals ($Cu_{1-x}M_x$) has also been investigated by Raman scattering. An observation of frequency shifts could, in principle, reveal whether plane sites or chain sites or both are being replaced. Indeed, it has been found that Co, Fe, and Al affect the Cu1-OIV bond frequency but Ni in place of Cu (up to $x \leq 0.2$) leaves that frequency constant.[115-118] While it is tempting to attribute frequency changes in the 500 $cm^{-1}$ mode to replacement of chain-copper atoms, a number of simultaneously occurring effects have prevented a quantitative understanding of the shifts. In particular, $\delta$ may become negative because additional oxygen atoms fulfill a preferred octahedral or square-pyramidal coordination of, for example, $Co^{3+}$ and it is not known what sign the resulting frequency shift of the Cu1-OIV bond may have. Using neutron scattering Tarascon et al.[119] have indeed shown that (for $x \geq 0.07$) each additional Co atom is associated with 0.5 additional oxygen atoms on the a-axis. They also find that $Co^{3+}$ does substitute in the chains, while for Ni they suggest that the planes constitute a more adequate replacement site.[119] Thus neutron measurements agree with what can be seen qualitatively using Raman scattering, demonstrating the ability to act as a local probe in the unit cell.

Thin films of superconducting material are useful for technological applications, and they come in single crystalline, partially oriented, or random polycrystalline form. Raman scattering may perform a dual task here: First, it is capable of identifying Raman-active impurity phases just like in ceramics; secondly, it may determine the possible orientation of the film. Using the $B_{1g}$-selection rule discussed in Sect. II.A.1 the authors of Ref. 120 report the identification of oriented growth of $YBa_2Cu_3O_{7-\delta}$ films on $SrTiO_3$ substrates. In Fig. 16 we show the Raman spectrum (under crossed polarizations in backscattering geometry) of an $YBa_2Cu_3O_{7-\delta}$ film ($d \cong 0.45$ μm) grown on a Si

460

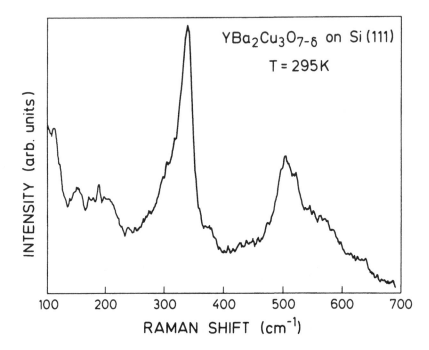

**Fig. 16.** Raman spectrum of a thin superconducting film ($T_c$ = 75 K) deposited on Si(111). The Raman spectra show a high degree of orientation of the c-axis perpendicular to the substrate.

(111) surface.[121] Although not epitaxial, because of a completely mismatching substrate, the film shows clearly a strong peak at 335 $cm^{-1}$ and only a small one at 500 $cm^{-1}$. From the evaluation and discussion of the Raman-tensor elements in Sect. III.A.1 it should be clear to the reader that this film is strongly c-axis oriented, i.e. the c-axis is perpendicular to the substrate in the region of the probing depth of the laser. (The spectrum of Fig. 16 may also be compared to that of a single crystal in Fig. 13, curve I.) From Fig. 16 we may conclude that there is no preferential orientation in the x-y plane.

In the ways just exemplified, Raman spectroscopy provides a useful tool for the characterization of superconductors even in a production environment. The examples given in this section illustrate some of the problems which arise, and properties of superconductors that may be investigated.

### 3. Oxygen deficiency

The oxygen content of $RBa_2Cu_3O_{7-\delta}$ has been of great importance for almost all studies of these superconductors, since many properties vary appreciably for $\delta$ in the interval [0,1]. By increasing $\delta$ above ~0.5 the substance becomes semiconducting and thus allows one through a relatively small variation of the chemical composition (one oxygen may be simply removed from the primitive cell by annealing the superconductor in argon above 700 K) to compare superconducting and normal state. Magnetic ordering effects, for example, may best be studied for $\delta < 0.5$, permitting inferences to be made about the superconductor (see Sect. III.A.7). Why the seventh oxygen atom is so important for superconductivity has not been completely resolved, but is is generally agreed that it predominantly acts as a dopant introducing free holes in the valence band. Associated with the removal of the oxygen from the superconductor are, however, structural changes as well as changes in the carrier concentration, and Raman scattering is sensitive to both of those. A number of groups have studied the dependence of the Raman frequencies on oxygen content, and we briefly review the most important conclusions here.

First, the oxygen which is removed in the transition $\delta = 0$ to $\delta = 1$ (OI in Fig. 2) occurs only once per unit cell and therefore is not Raman active. Secondly, the structural change from orthorhombic to tetragonal is so small that no _appreciable_ changes of most phonon eigenvectors are expected. (See also discussion in Sect. II.A.1.) The influence on the phonon spectra is therefore indirect but nevertheless quite noticeable.

The mode expected to be most strongly affected by the removal of the OI is, due to its proximity, the Cu1-OIV vertical vibration at ~500 cm$^{-1}$. This has been confirmed by most authors.[41,44,93-95,122-124]

Since the lattice parameter c increases upon removal of oxygen, one might expect the frequency of this mode to decrease correspondingly. This is indeed observed experimentally, and even quantitatively agrees with a reasonable Grüneisen parameter of ~1.9.[93] This simple picture is, however, not correct. When the internal coordinates of all the atoms in the unit cell became known it was found that the Cu1-OIV distance actually decreases[123,125] with increasing δ while the frequency is lowered. Kourouklis et al.[123] have suggested that the change in valence may be the dominanting effect causing the frequency shift. More generally, Macfarlane et al.[94] remark that it is possible to think of covalent $(Cu-O)^+$ complexes sharing the oxygen deficiency. It is worth mentioning an experiment that, similar to the one described above, changes the Cu1-OIV bond length. Upon replacement of Y by rare earths with increasing ionic radii the c-axis increases, but the Cu2-OIV bond length increases even more. The bond shortening of Cu1-OIV is – as one would expect intuitively – reflected in a frequency increase with ionic radius.[108,109] (See also discussion in Sect. III.A.2.) The dependence of the Cu1-OIV frequency on oxygen content is therefore anomalous, as Macfarlane et al.[94] pointed out. A further anomaly of this bond frequency has been discussed in connection with R = Pr (see Sect. III.A.2). In Figs. 17 and 18 we display the results of Refs. 93 and 94, partly to show the qualitatively similar but in detail different behavior of two Raman modes and partly because the authors made attempts to determine the actual oxygen content independently. Both groups of authors have made thermogravimetric δ determinations; the authors of Ref. 94 have checked their δ values against those determined from iodometric titration. Good agreement exists for the modes at 500 $cm^{-1}$ (δ ≈ 0) (see Figs. 17 and 18) and at 145 $cm^{-1}$. The dependences of the Raman modes at 335 and 430 $cm^{-1}$ are quite different, though. The 340 $cm^{-1}$ mode shows no variation with δ in in the work of Fig. 18 while it hardens by ~10 $cm^{-1}$ for δ from 0 to 1 in Fig. 17. Even more curious is the dependence of the other $CuO_2$-plane mode which, in Fig. 18, appears to harden rapidly between δ = 0 and δ ≈ 0.5 and then remains constant up to δ = 1. This behavior would nicely reflect a strong change in carrier screening from δ = 0 to δ ≈ 0.5, the transition of

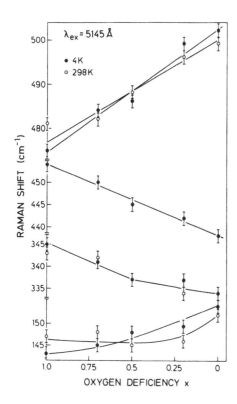

**Fig. 17.** Dependence of four Raman-active mode frequencies on gravimetrically-determined oxygen content $\delta$ (from Ref. 93). The 500 cm$^{-1}$ mode shows the strongest, but anomalous, variation. See Fig. 18 for comparison with the results of a second group.[94]

$RBa_2Cu_3O_{7-\delta}$ to a semiconductor. Reference 122 shows abrupt changes as a function of quenching temperature of the 500, 430, and 150 cm$^{-1}$ modes. Unfortunately, these results have not been directly correlated to oxygen content. The question remaining is why the results of different groups differ so much. Inhomogeneities in $\delta$, especially near the surface, are likely to be at the origin of this problem, as has been discussed earlier.

Regardless of the more subtle and not yet understood differences in the $\delta$-dependence of the $CuO_2$-plane modes, the frequency of the Cu1-OIV mode, once calibrated, serves as a characterization tool for the $RBa_2Cu_3O_{7-\delta}$ superconductors. From a single room-temperature Raman spectrum a number of impurity phases may be identified and the oxygen con-

464

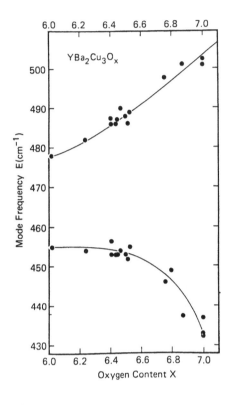

Fig. 18. Same as Fig. 17 but data are from Ref. 94. Note the difference in the functional form of the dependence on oxygen content. The inconsistencies of these results on ceramic samples presumably stem from variations in sample preparation and near-surface oxygen concentration.

tent determined.

The effect of defect-induced ir-active modes on Raman spectra, vacancy ordering, and new Raman-active modes has already been discussed; for details the reader is referred to the respective sections (II.A.1 and III.A.1).

### 4. Isotope effect

A great deal of attention has been focussed on the possibility of an isotope effect in the high-$T_c$ superconductors. Historically this has been an important indicator of the phononic origin of the conventional pairing mechanism. Experiments performed on $RBa_2Cu_3O_{7-\delta}$ with oxygen isotopically substituted yielded first none,[126] then a small isotopic shift[127] of $\Delta T_c \sim 0.4$ K $\pm$ 0.1 K. Before it could be concluded with certainty that the non-BCS value of $\alpha = 0.04$ (instead of $\alpha = 0.5$, $T_c \sim M^{-\alpha}$, M - ionic mass) was correct, complete substitution of $^{18}O$ on all lattice sites had to be ascertained. Raman scattering may do just that. The authors of Ref. 126 have monitored the shift of the Raman vibrations under exchange of isotopes and found that most of the oxygen is exchanged. However, they were not certain about the sites that were exchanged for lack of reliable Raman-peak assignments. The expected frequency change upon substitution for modes involving mainly the motion of one atom may be approximated by $\omega_1/\omega_2 = (m_2/m_1)^{1/2}$, an approximation which holds relatively well for many Raman modes and not so well for fir modes. In the latter, at least two different atoms in the unit cell are involved in each vibration and the detailed knowledge of the eigenvectors is required to make an accurate estimate. The study of the fir modes in connection with the isotope shift is important because the OI atom is not Raman active, and a lack of substitution on that site would not be noticed in Raman spectra (the defect-induced modes will have small intensities because $\delta \to 0$). While it is unlikely that sub-stitution does not occur at all on the OI site - the exchange occurs through that site - it cannot be entirely excluded that it occurs in-completely. Also, it is interesting to find out how the generally ac-cepted small isotope shift of $T_c$ is affected if only selected sites are exchanged.

In a detailed study Cardona et al.[61] have found 86% substitution homogeneously on all oxygen sites, including the OI site which was analyzed by fir-reflectivity measurements. They have presented a table with the shifts for all modes as calculated using the lattice dynamical model of Sect. II.A.2. More interesting, they have found that when only the equivalent of <u>one</u> oxygen atom is exchanged in the samples, this does not happen exclusively on the OI site. Rather, other oxygen sites begin to be replaced as well, in particular the OIV sites. The $CuO_2$ planes remain largely unchanged, a fact for which the detailed under-standing of the assignment of the lattice vibrations was required. It is significant to add that the isotopic shift the authors of Ref. 61 found for the fully (86%) exchanged sample was $\Delta T_c = 0.3$ K $\pm$ 0.1 K while there was no shift observable when only non-plane sites were exchanged. Infrared studies for $YBa_2Cu_3{}^{16}O_7$ and $YBa_2Cu_3{}^{18}O_7$ have been also recently reported by Crawford et al.[128]

In an experiment complementary to this, $YBa_2Cu_3{}^{18}O_{7-\delta}$ was prepared by first burning metals to oxides in $^{18}O$ atmosphere and then preparing the superconductor.[129] The isotope shift of this pristine sample was nevertheless found to be consistent with all previous results of a small or non-existent shift. The experimental phonon frequencies compared favorably to those calculated from the lattice dynamical model.

Measurements of the isotope shift continue to raise the issue of the actual significance of the absence or presence of a shift. As Ref. 126 points out, the transition metals Ru and Zr have $\alpha = 0$ but explanations for that behavior related to the band structure cannot readily be carried over to high-$T_c$ superconductors because of a low density-of-states at the Fermi level. On the other hand, a number of non-BCS mechanisms may also result in an isotope effect. Fisher et al.[63] have suggested several such mechanisms, one of which they estimate from existing data: The change in the zero-point motion of $^{18}O$, due to the heavier mass compared to $^{16}O$, combined with lattice anharmonicities may have an effect on the Cu-O overlap integrals along the bond direction. Estimating the change in $T_c$ with the overlap integrals from the pressure on $T_c$, Ref. 63 finds an $\alpha$ six times smaller than the experimental value. While this $\alpha$ still has the correct order of magnitude, it remains significantly smaller than the observed value and it is thus doubtful that lattice anharmonicities could explain the observed $\alpha$. For a discussion of the reduction in $\alpha$ due to strong coupling see Ref. 130. Model calculations based on McMillan's equation indicate that it is possible, in principle, to explain the isotope effect observed for the Zurich oxides but not that of $YBa_2Cu_3O_7$.[130a]

Copper and barium have also been isotopically replaced and no appreciable isotope shift has been reported.[130,131,132] Partly this is due to the small relative difference in mass between the isotopes for the heavier metal atoms ($\Delta m/m = 0.12$ for $^{16}O \rightarrow {}^{18}O$ whereas $\Delta m/m = 0.03$ for $^{63}Cu \rightarrow {}^{65}Cu$ and $\Delta m/m = 0.02$ for $^{135}Ba \rightarrow {}^{138}Ba$). Isotopic replacement for the rare earths is approximated by placing various lanthanides in the R-position. To a first approximation the f-electrons do not affect the outer electronic configuration (except for the _average_ ionic radius), so lanthanide replacement simulates isotopic replacement. No systematic variation in $T_c$ with the mass of the R-atom has been reported.

The isotopic shift $\Delta T_c$ upon replacement of $^{16}O$ by $^{18}O$ has also been measured in the high-$T_c$ Bi compounds and was also reported to be $\Delta T_c \sim$ 0.3 K for both the 110 K phase and the 80 K phase.[133] The similarity of $\Delta T_c$ for the two materials with quite different $T_c$ suggests that there is a common (small) phononic contribution to the high transition temperatures. In fact, also the Zurich oxides[134] and $RBa_2Cu_3O_{7-\delta}$ have a $\Delta T_c$ of the same size. A noticeable exception may be the superconductor $Ba_{1-x}K_xBiO_3$ ($T_c$ = 30 K).[135] Batlogg et al.[136] report $\alpha = 0.2 - 0.25$ for this compound as well as for $Ba_{1-x}Pb_xBiO_3$, a circumstance which may suggest a conventional pairing mechanism. The authors claim, however, that viewed in terms of the ratio of $T_c$ to the density-of-states at the Fermi level, $Ba_{1-x}K_xBiO_3$ is truely a high-$T_c$ superconductor. They speculate that electron pairing is mediated by electronic excitations and that lattice deformations caused by charge redistributions are the reason for phonon-related effects in these materials.

## 5. Electronic scattering and the gap problem

Ever since laser Raman spectrometers have become available to solid state physicists, attempts have been made to observe the opening of a gap for electronic excitations below $T_c$ in superconductors (for a review see Ref. 137). These attempts were motivated in part by theoretical predictions of a singular behavior of the scattering at $\omega = 2\Delta$[137,138] ($2\Delta$ = gap)). They were hampered by the strong reflectivity and the small gap of conventional superconductors, which brought the phenomenon into the experimental region of strong Rayleigh scattering.[139,140]

The first convincing observations of across-the-gap excitations in superconductors were performed for $Nb_3Sn$, $V_3Si$[141,142] materials with $T_c$ around 18 K. Comparable evidence has been presented recently for

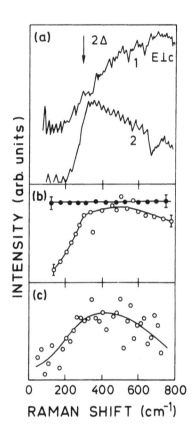

**Fig. 19.** Electronic scattering in the high-$T_c$ superconductor $YBa_2Cu_3O_{7-\delta}$. The data are taken by different groups (curve a: Ref. 21, curve b: Ref. 143, curve c: Ref. 20) in order to search for a superconducting gap.

$YBa_2Cu_3O_7$ samples[20-24,143], including thin films[20], ceramic material[21], twinned single crystals[21,23,24] and untwinned ones.[22,143] We display the data of Refs. 20,21,143 in Fig. 19. All data show peaks in the scattering of the superconducting phase at around 400 cm$^{-1}$ ($\approx$50 meV $\approx$600 K) which, if taken to represent the superconducting gap $\Delta$, would correspond to $2\Delta/kT_c \cong 7$, twice as large as the value predicted by the BCS theory (see below). However, we should keep in mind that below this gap, excitations still seem to be observable in most of the published data, in spite of the large experimental scatter. This is contrary to the few observations in conventional superconductors.[137,138,141,142] It suggests the existence of a distribution of gaps, either in $\vec{k}$-space around the Fermi sphere or in real space in the thin surface layer where the scattering takes place. The latter could be due to oxygen deficiency. The 500 cm$^{-1}$ Raman phonon of these samples, however, does not show sufficient signs of the required oxygen deficiency, a fact which has led some authors[20,23] to conclude that the scattering below the gap, which seems to extend to zero frequency, is an intrinsic property of high-T$_c$ superconductors. Cooper et al.[23] go even one step beyond and conclude that this phenomenon cannot be due to gap anisotropy around the Fermi surface with points or lines of zero gap: a fraction of the superconducting material must have zero gap. We believe that the available data are not sufficiently accurate to clarify this matter, although there is no doubt that gap-like behavior is seen and that the gap is not a sharp one. In Ref. 138, 142, for instance, the spectra measured on the a-b face of single crystals for several polarizations are fitted to predictions for the differential scattering efficiency based on the BCS theory:

$$\frac{d^2 s}{d\omega d\Omega} \sim \frac{\Delta^2}{\omega} (\omega^2 - 4\Delta^2)^{-1/2} \tag{11}$$

convoluted with a Gaussian distribution of gaps $\Delta$. The curves fitting the experimental data obtained in this manner are shown in Fig. 20. The average gaps obtained for the four polarization configurations measured yield $2\Delta/kT$ varying between 3.0 and 5.5. We should point out, however, that it is not possible to explain the anisotropies of Fig. 20 on the basis of standard Raman tensors. Moreover, in Refs. 20,23 no scattering was found for crossed polarizations parallel to x,y while strong scattering is seen in Fig. 20. In Ref. 23b measurements on single domain crystals show stronger scattering for parallel polarizations along the chains than for those perpendicular to them. We believe that painstaking work is still needed to disentangle the symmetry properties of this scattering.

An interesting problem is posed by the scattering observed in the normal phase (Fig. 19). It is well known that electrons in simple parabolic bands (not necessary isotropic) lead to scattering via the "longitudinal" Hamiltonian of Eq. (2).[144] We assume that $\omega_L$ and $\omega_S$ are

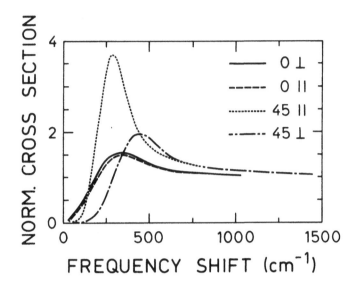

<u>Fig. 20.</u> Polarization-selective spectra of electronic scattering in YBa$_2$Cu$_3$O$_{7-\delta}$. The fits are to Eq. (11) and discussed in the text.

away from strong interband resonances. H$_R$ acts as a standard scalar potential and produces a charge density wave which will be screened by the dielectric function $\epsilon$ of the free electrons. The result is a Stokes-scattering efficiency:

$$\frac{d^2S}{d\omega d\Omega} \propto -(\hat{e}_L \cdot \frac{1}{\underset{\approx}{m}} \cdot \hat{e}_S) \; q^2 \; [n(\omega)+1] \; Im \; [\epsilon^{-1}(\omega)]. \qquad (12)$$

The function $-Im \; [\epsilon^{-1}(\omega)]$ has a strong peak at the screened plasma frequency of the free electrons which, in typical high-T$_c$ superconductors amounts to several thousand cm$^{-1}$.[145] This plasma frequency is, however, overdamped and should not be observable in Raman spectra. The small value of the scattering vector q [Eq. (12)] reduces it beyond observability, even for backscattering.

Another possibility of overcoming the screening which brings the scattering peak to $\omega_p$ and eliminates the low-frequency scattering arises if the conduction band is not parabolic and thus fluctuations in

$\underset{\approx}{m}$ occur over the Fermi surface. In this case carrier-density fluctuations can take place between parts of the Fermi surface with different masses, without producing any charge density fluctuations. Such excitations are not screened and low-frequency scattering results. This scattering should, in the limit of mean free path $\ell$ larger than the Fermi wavelength $\lambda_F$, also be proportional to $q^2$ and thus remain weak. Scattering efficiencies two orders of magnitude larger than those so predicted have been observed in heavily doped Ge and Si, with free-electron concentrations close to those for high $T_c$ superconductors.[146] The unexpectedly strong scattering is attributed to the fact that $\ell < \lambda_F$ and thus in Eq. (12) an effective $q_{eff}$-vector much larger than that supplied by the backscattering of a phonon must be used. It is reasonable to assume that a similar phenomenon is responsible for the nearly $\omega$-independent free-carrier scattering observed in the normal phase of $YBa_2Cu_3O_7$, the mass fluctuations being given either by the strong x-y anisotropy of the $CuO_2$ planes or by differences in the masses between planes and chains. Measurements on the Bi compounds, which do not have chains, should help to clarify this point.

We note that two types of calculations, in two extreme limits, have been performed for the scattering by mass fluctuations in metals and heavily doped semiconductors. In the limit of infinite mean free path the efficiency is proportional to $q^2$ (q = scattering vector) while in the collision-limited regime this $q^2$ dependence disappears.[147] The available experiments correspond to the latter case[146] and it is reasonable to assume that this will also be true for high-$T_c$ superconductors in the normal state because of their short mean free paths, a fact which leads to stronger scattering. Available calculations for the superconducting phase include Coulomb-screening corrections, a fact which is of importance only for scattering configurations which correspond to $A_g$ symmetry. Such corrections may lead to different lineshapes for different scattering configurations such as those in the preliminary results of Fig. 20. For Raman shifts near $2\Delta$, pairing leads to a scattering peak independent of q in the collisionless case. Unfortunately, calculations in the collision-limited regime for $\omega > 2\Delta$ have not

yet been reported. The appearance of scattering for q = 0 can be qualitatively understood as due to the formation of pairs with a coherence
length $\xi$, which is equivalent to a q-uncertainty of $\xi^{-1}$. It is of interest to compare the scattering <u>rate</u> for q = 0 and $\omega \gg 2\Delta$ induced by
the pairing in the collisionless case:[137,138]

$$d\sigma = r_0^2 (\frac{2\Delta}{\omega})^2 F\delta N_d d\omega, \tag{13}$$

where $r_0$ is the Thomson radius, $\delta$ the laser-penetration depth, $N_d$ the
density-of-states at the Fermi surface, and F the dimensionless mass
and gap fluctuation, with that obtained in the normal state in the
collision regime

$$d\sigma = \frac{r_0^2}{3\pi}(\frac{\omega_c}{\omega})F\delta N_d d\omega \;, \tag{14}$$

where $\omega_c$ is a collision frequency and F represents the mass fluctuation
without gap contribution. Note the striking similarity between Eqs.
(13) and (14). Nevertheless the scattering in Eq. (13) decays like $\omega^{-2}$
while Eq. (14) does it like $\omega^{-1}$, the former decay being faster as it is
induced by the presence of the gap and should disappear away from it,
at a point where the collision-limited effects must take over. Hence
the near equality of normal and superconducting scattering for $\omega \gg \Delta$
proves that we are indeed in the collision-limited regime, which has
yet to be treated for superconductors.

We have pointed out in Sect. III.A.1 the importance of the unique
selection rules of the $B_{1g}$ ($D_{4h}$)-like 340 cm$^{-1}$ phonon. This phonon has
other interesting features related to that near-symmetry and to the
fact that its frequency is close to the gap or to the centroid of the
distribution of gaps. Its line shape is not a symmetric Lorentzian but
rather a Fano profile (Fig. 21)

$$\sigma(\omega) \sim \frac{(\varepsilon+q)^2}{1+\varepsilon^2} \tag{15}$$

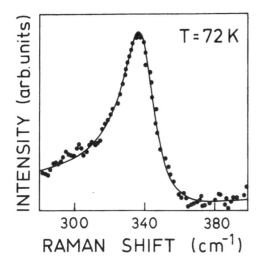

Fig. 21. Asymmetric line shape of the $B_{1g}$-like phonon at 335 cm$^{-1}$ in YBa$_2$Cu$_3$O$_{7-\delta}$. The fit is to Eq. (15) with q = -4.3 and $\Gamma$ = 9.5 cm$^{-1}$ and represents the interaction of the phonon with the electronic continuum. Note that the phonon at 120 cm$^{-1}$ also has such an asymmetric line shape. From Ref. 22.

where $\varepsilon = (\omega-\omega_0)/\Gamma$, $\Gamma$ is the line width and q the Fano parameter. Equation (15) is the signature of a discrete excitation (phonon) interacting strongly with a continuous background, such as that discussed above. The parameters q are determined by $\Gamma$, the scattering <u>amplitude</u> for phonons $T_p$, and that for electrons $T_e$:

$$q \approx \frac{T_p}{\Gamma T_e} \qquad (16)$$

since $\Gamma$ is proportional to the density of continuum excitations, it should change around $T_c$. From the fit of Fig. 21 we obtain for the $B_{1g}$ ($D_{4h}$) phonon q = -4.3 and $\Gamma$ = 9.5 cm$^{-1}$. The temperature dependences of q and $\Gamma$ have been reported in Ref. 148; they agree roughly with values determined from Fig. 21 near $T_c$.

Figure 14 also displays our measurements of the Fano line shape of the $A_{1g}$ phonon at 110 cm$^{-1}$ which was recently reported by Cooper et al.[23] These authors concluded from that line shape, observed even <u>below</u> $T_c$, that electronic excitations must occur at 110 cm$^{-1}$ even below $T_c$

i.e., that gapless or very-low-gap regions exist. They also suggested that the 110 cm$^{-1}$ peak is the $A_{1g}$ modes of Cu2, assuming that only $CuO_2$-plane vibrations will affect carriers believed to be located in these planes. As mentioned in Sect. III.A.1, however, lattice dynamical calculations and other evidence (Ba → Sr substitution) weigh heavily in favor of the Ba ($A_{1g}$) assignment: the corresponding Cu modes are seen at 150 cm$^{-1}$.

Raman scattering thus suggests the existence of several gaps or, more likely, a continuous distribution of them with values of $\Upsilon = 2\Delta/kT_c$ as high as 5.5. Tunneling measurements, a classical technique for the determination of gaps, give even a wider range including $\Upsilon$'s from 0.7 to 13.[149] Nuclear-spin relaxation yields two gaps, with $\Upsilon = 8.3$ and 2.4, respectively.[150,151] IR measurements by several groups give evidence[145] of $\Upsilon$'s around 3 while some data[152] seem to suggest the existence of gaps with $\Upsilon$ as high as 8. While all experiments concur in the existence of a gap, the values vary widely around the BCS value of 3.5. There seems to be regions either in real or in $\vec{k}$-space where the gap vanishes even at T ≈ 0 K. Recently, Cooper et al.[23b] have reported the measurement of two gap values for $A_g$ and $B_{2g}$ symmetries on un-twinned single crystal. They report an anisotropy of 35% and speculate that the higher value (~530 cm$^{-1}$, $B_{1g}$) is related to the Cu-O chains and the lower one (340 cm$^{-1}$, $A_g$) to the $CuO_2$ planes.

6. Softening of the 335 cm$^{-1}$ phonon below $T_c$ or $H_{c2}$: the gap problem

The effect of a conducting electronic system on phonons is directly apparent in the anomalous frequency softening of one of the Raman pho-nons in $RBa_2Cu_3O_{7-\delta}$. First reported in Ref. 35 for ceramic samples of $YBa_2Cu_3O_{7-\delta}$, this result has been followed up by a number of in-depth experimental and theoretical investigations aimed at clarifying whether it is related to strong electron-phonon interactions in the supercon-ductors. While the experimental evidence for the existence of the sof-tening is clear (in both Raman and ir spectroscopy) it cannot be said a priori what the causal relationship between it and the condensation of electrons into pairs, if any, might be, i.e. the experiments do not imply that the condensation mechanism is of the conventional BCS-type. On the other hand, if we assume that it is, strong-coupling theory may be applied and coupling parameters estimated:[153] the near-absence of isotope effect remains then unexplained.

It should be noted that a softening of modes in that energy range is not observed in neutron scattering.[13] This is, however, not in contradiction to the Raman results since neutron measurements in that energy range have so far been possible only for ceramic samples. Those measurements probe the (amplitude and cross-section weighted) total phonon density-of-states (i.e. all of the BZ) and are not sensitive enough to detect small changes in specific modes. It would be interesting to see - on crystalline samples - how far into the BZ the softening extends, but it remains questionable whether sufficient experimental resolution will ever be attainable. (For similar effects in heavily doped silicon see Ref. 154.)

In this section we present the experimental evidence for the softening from light scattering, and its dependence on a magnetic field, discuss its origin, and compare its behavior in $YBa_2Cu_3O_{7-\delta}$ with that in two non-superconducting systems ($\delta = 1$ and $R = Pr$). Upon cooling of a material, in the absence of phase transitions phonon frequencies usually increase. This behavior is indeed observed in $RBa_2Cu_3O_{7-\delta}$ for the OI-Cu1-OI vibration at 500 $cm^{-1}$.[155] In striking contrast to that is, however, the temperature dependence of the Raman-active phonon at 340 $cm^{-1}$. In Fig. 22 we show how this phonon frequency remains roughly constant between room temperature and $T_c$ where, within ~10 K, it drops by ~1% (softens) to remain at the lower frequency down to the lowest temperature measured.[22] This is the sharpest such effect reported so far and was obtained on an untwinned single crystal with a narrow transition to superconductivity ($\Delta T_c \approx 3$ K as measured by the Meissner effect). Other reports generally show softening transitions that extend down to between 50 K and all the way to 5 K.[23,35,108,120,155,156]

Softening measurements have also been made on (twinned) crystals as well as on well-characterized ceramic samples; the sharpness observed

476

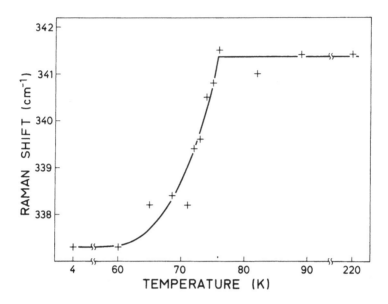

<u>Fig. 22.</u> Anomalous softening transition in $YBa_2Cu_3O_{7-\delta}$ near $T_c$ of the $B_{1g}$-like phonon as evidence for the coupling between the $CuO_2$ planes and the electron gas. The frequencies $\omega_0$ in the figure were taken from fits to a Fano lineshape [Eq. (15)]. From Ref. 22.

in Fig. 22 does not necessarily stem from the untwinned nature of that particular crystal. Rather, from Ref. 156 it may be concluded that the anomaly depends sensitively on the oxygen content of a particular sample (in $RBa_2Cu_3O_{7-\delta}$; the situation may be different in Bi and Tℓ-related superconductors). Contrary to what they had expected, the authors of Ref. 156 found the anomaly to disappear for $\delta > 0.2$, whereas $RBa_2Cu_3O_{7-\delta}$ remains superconducting until $\delta \approx 0.5$. This concurs with the observation that crystals when illuminated at too high laser power densities tend to lose an initially present anomaly irreversibly[157], presumably due to oxygen effusion in the proximity of the surface. As discussed above in connection with $\delta$-dependent mode frequencies, this poses the question as to what the true, maximum softening might be, and for which $\delta$-value it occurs. Variations of $\delta$ near the surface, and the shallow penetration depth of the exciting laser, cause averaging over various δ-values in a particular sample. Presumably a high degree of homogeneity in δ, also reflected in the sharp transition to supercon-ductivity of that crystal without post annealing in oxygen, is the cause for the well defined softening behavior seen in Fig. 22.

There is, however, a second reason for the observation of broad softening transitions. It is a <u>reversible</u>, power-dependent broadening as shown in Fig. 23, where we plot the softening for various incident laser-power levels. It is evident that the range over which the sof-tening occurs increases noticeably for higher laser powers. In addi-tion, it can be seen that the onset of the softening moves to lower temperatures as well, making it likely that laser heating of the sample

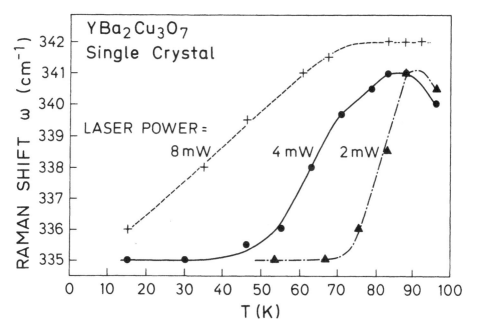

<u>Fig. 23.</u> Frequency softening for different incident laser powers. The broadening changes reversibly and is due to laser heating effects. From Ref. 157.

478

is responsible for onsets apparently occurring below $T_c$. In Fig. 24 the onset of the softening is plotted versus the incident laser power. A straight line is seen to extrapolate to $T_c$ for zero incident laser power, establishing that the softening occurs indeed <u>at</u> $T_c$. (A lower temperature for the softening transition could not have been <u>a priori</u> excluded.)

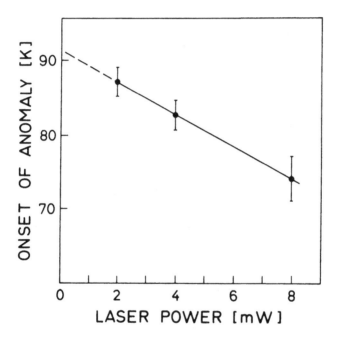

<u>Fig. 24.</u> Onset of the softening transition of the $B_{1g}$-like phonon in a single crystal as a function of the incident laser power. The onset extrapolates to $T_c$ for zero laser power. From Ref. 157.

While the occurrence of the anomalous softening at $T_c$ might be coincidental, the experiments described in the following show that it is not. It was possible to show, in particular through the dependence on an applied magnetic field, that a superconducting transition is indeed a prerequisite for the frequency softening in high-$T_c$ superconductors. When a magnetic field of 10 T is applied[158] the softening transition of this phonon is lowered by 5 - 6 K (Fig. 25). The lowering of the super-conducting transition is the dominant effect of the application of the magnetic field and structural transition e.g., a lattice distortion or a Peierls transition, can be excluded as the origin of the phonon softening. The dependence of the softening transition on magnetic field was shown to be compatible with the relationship between $H_{c2}$ and $T_c$.[158] In an experiment complementary to the one described above, the magnetic field was varied at a fixed temperature.[158] As the temperature is lowered below $T_c$ the magnetic field H necessary to quench the softening increases just as much as it does to quench superconductivity in the crystal. For $|T-T_c| \geq 8$ K $(T < T_c)$ it was not possible to bring the sample to the normal-conducting state with $H \leq 12.7$ T, and no frequency

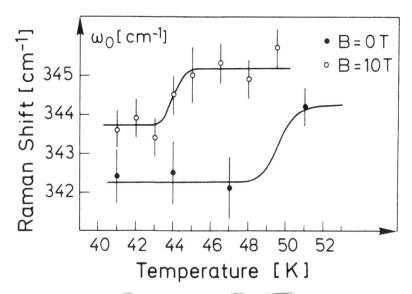

Fig. 25. Lowering of the phonon-softening transition under the application of a magnetic field in a (twinned) single crystal. The frequencies in this figure have been extracted from the experiment by fits to the lineshape of the Fano form. Plotted is $\omega_0$ of Eq. (15). From Ref. 158.

change was observed for the phonon at 340 cm$^{-1}$. It might be argued that the phonon softening should broaden considerably for magnetic fields $H_{c1} \leq H \leq H_{c2}$ due to the penetration of flux in the sample. However, because of the short coherence length of the extreme type-II supercon- ductor the resulting normal regions should contribute only negligibly to the total scattering cross-section. In these experiments the field of a split-coil superconducting magnet was kept perpendicular to the $CuO_2$ planes of the crystal, the direction in which the anisotropic $dH_{c2}/dT$ is small. It would be interesting to investigate the situation $\vec{B}\|x,y$ even though higher fields and well-controlled temperatures are necessary to quench superconductivity in this direction.[159]

The experiments in magnetic fields are particularly important be- cause they allow comparison of superconducting and normal state at the same temperature in one specimen. It has also been reported[45,156] that there is no phonon softening in semiconducting $YBa_2Cu_3O_{7-\delta}$ ($\delta \geq 0.5$). By itself this does not, however, say very much, mainly because Krantz et al. (Ref. 156) have shown that already for $\delta > 0.15$ the softening disappears. If this were not a surface effect (see discussion above), it would be a puzzle why the softening already disappears for such small $\delta$ whereas the sample remains strongly superconducting.

An interesting variation of this experiment is the study of $RBa_2Cu_3O_{7-\delta}$ ($\delta \approx 0$) with R = Pr. Upon replacement of Y by this rare earth, superconductivity does not occur down to liquid-He tempera- ture.[160,161] This material may be tested with $\delta \approx 0$ for a possible pho- non anomaly. As was shown in Ref. 162 there is no phonon anomaly (in either Raman or ir measurements) and, also, no line shape asymmetry. The structure $RBa_2Cu_3O_7$ therefore does not alone cause the phonon sof- tening; all experiments conducted so far indicate that superconducti- vity is a necessary requirement.

The phonon softening is not unique to Raman-active phonons. In far- infrared-reflection (fir) spectroscopy two phonons have been reported to lower their frequency significantly between RT and 10 K;[155,163] in the case of a thin film on a sapphire substrate the softening was par- ticularly strong (13 cm$^{-1}$).[163] Most affected by the transition to superconductivity are two phonons around 300 $cm^{-1}$, although small frequency decreases (~0.3%) have also been reported[65] from a Kramers-Kronig analysis of the ir phonon at 570 $cm^{-1}$. No statements can be made for the $B_{2u}$ ($D_{4h}$) mode, corresponding to the $B_{1g}$ mode, from fir measurements because it is silent.

What does the observation of an anomalous phonon-frequency softening imply? First of all, it provides direct evidence that the $CuO_2$ planes are the unit-cell elements most connected to the superconductivity mechanism. As we know from the discussion of the phonon symmetries and their assignments, the related phonon has near-$B_{1g}$ ($D_{4h}$) symmetry. In crystals this symmetry is easily identified and, as shown before, there is only one phonon and one possible eigenmode in $RBa_2Cu_3O_{7-\delta}$ (out of 15 Raman-active modes) with that symmetry. (In Fig. 3c the out-of-phase, vertical displacements of the two oxygen atoms in each $CuO_2$ plane were shown.) The anomaly indicated even before the discovery of the Bi-related superconductor without 'chains' that $CuO_2$ planes may play a central role in the superconductivity mechanism. It is of course interesting to ask whether a similar phonon and the corresponding softening is observed in the Bi compound. The phonon has indeed been identified[76] at 282 $cm^{-1}$ (see also below, Sect. III.B) but because of its larger width, it has not been possible to detect an anomalous shift of 1% in frequency as of yet.[164] There has been a report[165] claiming an anomaly of an $A_g$ mode [incorrectly assigned as $B_{1g}$ ($D_{4h}$) in that reference] at 464 $cm^{-1}$, but we have not been able to confirm this observation.

A second implication derived from the anomalous softening comes from an analysis of its physical origin. It is possible that the softening is caused by a (possibly anisotropic) gap opening at or above the phonon frequency: a value of $\Upsilon = 2\Delta/kT_c \geq 5$ would then follow. From a theoretical point of view, calculations have been reported[153] showing that the softening rules out weak-coupling BCS theory as a good description of the superconductivity phenomenon since it predicts hardening of the phonons when cooling below $T_c$. The well-established softening of some phonons is rather indicative of strong interactions between phonons and electrons, and Ref. 153 has shown that it may be described by strong-coupling BCS theory and a coupling parameter $\lambda^* \geq$ 1.7 (but, we repeat, the small isotope effect remains then unexplained).

## 7. Scattering by magnons

It is well established that in the semiconducting modification of $RBa_2Cu_3O_{7-\delta}$ ($\delta \leq 0.5$) the $Cu^{+y}$ ($y \cong 2$) spins are antiferromagnetically ordered, with Neel temperatures $T_N \approx 400$ K for $\delta \cong 0.9$.[28] Nearest Cu-ions in the $CuO_2$ planes have opposite spins. The ions that occupy equivalent positions in adjacent planes have also opposite spins (antiferromagnetic stacking along z). While long-range order disappears at $T_N$, shorter range antiferromagnetic correlations exist up to much higher temperatures: they are particularly strong within the $CuO_2$ planes. Such phenomena seem to be common to all semiconducting phases of $CuO_2$-based high-$T_c$ superconductors. They have been profusely studied in $La_2CuO_4$ ($T_N \approx 200$ K with rather long-range correlations in the $CuO_2$ planes above $T_N$).[166] $T_N$ decreases strongly with increasing 'doping', i.e. increasing $\delta$ in $RBa_2Cu_3O_{7-\delta}$, and x in $La_{2-x}Sr_xCuO_4$.[167,168] In the case of $RBa_2Cu_3O_6$ a second magnetically ordered phase also involving the CuO chains has been found at low temperatures for very low values of $\delta$ and R = Y, Nd.[169] For R = Gd, and possibly for other R's, a very low temperature (< 2K) phase involving the magnetic ordering of the R atoms has been reported.[170,171] Very recently, inelastic neutron data have been obtained for a large single crystal (7x7x3 mm) of $YBa_2Cu_3O_{6.2}$ showing clearly the two-dimensional spin excitation corresponding to a spin-wave velocity in agreement with that determined from the light scattering data reported here.[172] We discuss here light scattering by spin-wave excitations (magnons) of the Cu spins of the $CuO_2$ planes.

Light scattering by single magnons is expected to be very weak in antiferromagnets while scattering by two magnons can be rather strong.[26] The latter involves the flipping of two neighboring spins which can be accomplished by means of an optical <u>dipole</u> transition at one of the ions, followed by _exchange_ interaction between the excited electron and the corresponding electron at the other ion induced by the Hamiltonian

$$H_{ex} = J \, \vec{S}_i \cdot \vec{S}_j \tag{17}$$

followed by another dipole transition.[173] In Eq. (17) J is the exchange constant and $S_i$, $S_j$ are spin operators on neighboring (opposite spin if antiferromagnetic) lattice sites. The scattering process involves _approximately_ the simultaneous flipping of both (opposite) spins at a pair of adjacent sites, which for $CuO_2$ planes should require an energy of 3 J (more precise calculations yield 2.7 $J$[174]). Considering a nearest neighbor Cu pair ($Cu_1 - Cu_2$) and the bridging O atom we find the following steps for the process, involving dipole transitions from p (of O) to 3d states of Cu and a d↑-hole in $Cu_1$ together with a d↓-hole in $Cu_2$.

$$
\begin{array}{lll}
p\!\uparrow \rightarrow d_{1h}\!\uparrow & \text{electric dipole, laser field} & \\
d_2\!\uparrow, p\!\downarrow \rightarrow p_h\!\uparrow, d_{2h}\!\downarrow & \text{exchange J} & (18) \\
d_1\!\downarrow \rightarrow p_h\!\downarrow & \text{electric dipole, scattered field} &
\end{array}
$$

where the subindex h respresents holes. The total effect of the sequential process of Eq. (18) is:

$$d_{1h}\!\uparrow, d_{2h}\!\downarrow \rightarrow d_{1h}\!\downarrow, d_{2h}\!\uparrow,$$

i.e., a simultaneous flip of the spin of the hole pair. An evaluation of the scattering matrix elements for in-plane order only leads to a Raman tensor for the predominant two-magnon (two spin flips) process:[175,177]

$$
\begin{pmatrix}
1 & 0 & 0 \\
0 & -1 & 0 \\
0 & 0 & 0
\end{pmatrix}
\tag{19}
$$

referred to the tetragonal x,y axes. Hence the two-magnon scattering should have the same polarization selection rules as the $B_{1g}$ $(D_{4h})$ phonon found at 340 cm$^{-1}$. Order along the z-direction should result in z-components of the Raman tensor.

We note that, strictly speaking, the two-magnon excitation is not simply a flip of neighboring spins but the superposition of two spin waves with opposite $\vec{k}$. The magnon dispersion relation $\omega_M(\vec{k})$ is expected to be fairly isotropic if the main terms in the Hamiltonian are those of Eq. (17). This dispersion relation is given (in the absence of an external magnetic field) by[174]

$$E_M^2(\vec{k}) = (\tfrac{1}{2}JZ)^2[1- \tfrac{1}{4}(\cos k_x a + \cos k_y a)^2]\qquad(20)$$

where Z is the number of nearest neighbors involved (z = 4 for $CuO_2$ planes). Thus at the edge of the magnetic Brillouin zone we find for the energy of a _pair_ of magnons $E_M$ = 4J. Since this value is constant all around the Brillouin zone it should lead to a one-dimensional density-of-states with a singularity of the type $(\omega_M - \omega)^{-\frac{1}{2}}$ at the edge of that zone. Magnon-magnon interaction reduces the value from 4J to 2.8J and broadens the singularity into a somewhat asymmetric peak. The interaction parameter is determined solely by J and Z; in the case under consideration the line width (FWHM) should amount to ~10% of the peak frequency.[175]

We display in Fig. 26 the scattering which has been assigned to two magnons by Lyons and Fleury[175] in a $YBa_2Cu_3O_{7-\delta}$ ($\delta \cong 0.3$) together with the polarization dependence, which clearly identifies its symmetry as $B_{1g}$ $(D_{4h})$-like. This material is superconducting ($T_c$ ~ 60 K), a fact which suggests that some form of short-range antiferromagnetic order exists even in the superconducting state. The peak frequency in Fig. 26 is ~1800 cm$^{-1}$. It is known to decrease rapidly with decreasing $\delta$ and the peak is not observable for $\delta \cong 0.1$, except as a very broad background. For $\delta \cong 1$ the peak occurs at 3600 cm$^{-1}$, which corresponds to J = 1300 cm$^{-1}$, a value much higher than that found for other similar non-superconducting compounds (J = 77 cm$^{-1}$ for $K_2NiF_4$, with antiferromagnetism in $NiF_2$ planes[176]). However, these large values of J seem to

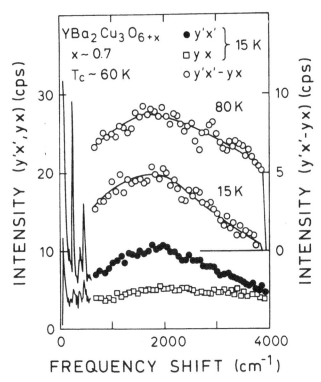

<u>Fig. 26</u>. Scattering by two magnons in $YBa_2Cu_3O_{6.7}$ showing its $B_{1g}$ ($D_{4h}$) symmetry. Note also the decrease in background below $T_c$ = 60 K. From Ref. 175.

be characteristic of $CuO_2$ planes: J = 1100 $cm^{-1}$ is found from the 3000 $cm^{-1}$ peak observed in $La_2CuO_4$.[177,178]

An interesting, unexplained observation has been made in connection with Fig. 26. A broad scattering background exists above 4000 $cm^{-1}$ for $T > T_c$. However, this background nearly disappears below $T_c$.

From the anomalies just reported, specific to the high-$T_c$ supercon-ductors, it is tempting to infer that some connection must exist bet-ween the ferromagnetic order and the mechanism of high-$T_c$ superconduc-tivity.

Note that a peak is usually also observed in these materials at about 1200 cm$^{-1}$ in the A$_{1g}$ configuration. It has been attributed in Ref. 177 to impurity-induced scattering by one magnon, while in Refs. 178,179 it is assigned to scattering by two phonons. The results just described may be particularly relevant to the mechanism of superconductivity if the magnon pairing interaction recently proposed by W.A. Goddard III et al. (to be published) is confirmed.

## B. Bismuth and CuO$_2$-Based Materials: Single-Crystal Spectra and Phonon Assignments

Soon after their discovery, the Bi superconductors were available in crystalline form. The Raman spectra could therefore be measured with their polarization dependence, thus largely avoiding the difficulties associated with ceramic samples. Due to the similarity of the structure with that of RBa$_2$Cu$_3$O$_7$ (see Figs. 2 and 6) tentative assignments were based on analogies to the 90 K superconductor. Lattice dynamical calculations of the Bi compound (Fig. 10) are largely consistent with those assignments, in particular for those vibrations not involving the BiO planes. Because of uncertainties in the structure in the BiO planes (see Fig. 7 and discussion in text), and possibly statistically varying oxygen occupation, these calculations are not expected to be very accurate. Since the Bi (and also Tℓ) superconductor exists with different numbers of CuO$_2$-planes [n in Eq. (7)], conclusions have been drawn by comparing the spectra of different n's. Raman spectra on ceramic[180-182] as well as on (nearly) single-crystal[76,166,183,184] samples have been reported. Popovic et al.[181] show the Raman and fir-reflectivity spectra of the n = 1 phase where some of the Sr was replaced by Ca. The material was superconducting with T$_c$ = 50 K (resistivity ρ = 0) and showed two steps in the Meissner and resistivity curve, at 80 K and 60 K. Even though nearly single phase according to x-ray diffraction, the sample evidently had two transition temperatures, perhaps related to different Sr-Ca distributions within the n = 1 unit cell. The Raman spectrum of this superconductor has peaks at 130, 205 (weak, broad), 310 (weak, broad), 462, 492, 587, and 650 cm$^{-1}$, where the lowest and highest frequencies reported have the highest intensity. Noticeable in the fir spectrum is the softening of a phonon at 485 cm$^{-1}$ (300 K) to ~470 cm$^{-1}$ (10 K). The authors also report Raman and fir-reflectivity spectra of the major impurity phases $SrCO_3$, $CaCO_3$, $Bi_2O_3$, and CuO. Reference 182 reports a similar spectrum of the superconductor which covers a slightly different frequency range. Two strong peaks at 98 and 64 cm$^{-1}$ are reported while frequencies above 550 cm$^{-1}$ were not recorded.

When single crystals became available it was possible to obtain information about the symmetry of the phonons. Most important was the identification of a $B_{1g}$ phonon (one for n = 2, 3). Just as in $RBa_2Cu_3O_{7-\delta}$ its eigenvector is known entirely from symmetry considerations and it serves as a reference in the assignment of the other phonons. (The eigenvector is shown schematically in Fig. 3b and involves predominantly oxygen displacements in the $CuO_2$ planes.) The phonon was correctly identified in Ref. 76 and the spectra proving its $B_{1g}$ symmetry are shown in Fig. 27. As has been pointed out in Sect. II.B the $CuO_2$ planes are rotated with respect to the crystal axes by 45° compared to $RBa_2Cu_3O_{7-\delta}$ so this mode should appear in the $z(xy)\bar{z}$ configuration. Figure 27 shows just that for a frequency of 282 cm$^{-1}$, together with the absence of that peak for $z(x+y,x-y)\bar{z}$ configuration, as required. All other modes are of the $A_{1g}$ type. Furthermore, Cardona et al.[76] report the absence of such a $B_{1g}$ mode for the corresponding spectra of crystals with n = 1; for n = 1 the $CuO_2$ plane is Raman inactive. The remaining peaks may now be divided into two groups: those with frequencies $\omega < \omega(B_{1g}$-mode), mostly, Bi, Sr, and Cu vibrations and those with higher frequencies than the $B_{1g}$ mode, predominantly oxygen vibrations. Of the latter, the authors of Ref. 76 assign the strongest mode (480 cm$^{-1}$) to vertical $O_3$ oxygen vibrations ($O_3$ bridges the $CuO_2$ planes with the BiO planes) similar to those of OIV (500 cm$^{-1}$) observed in $RBa_2Cu_3O_{7-\delta}$. The two modes resolved at the highest frequencies vary in intensity and frequency somewhat from sample to sample and for different publications. They probably stem from BiO vibrations and cannot yet be calculated accurately because of the structural uncertainties in the BiO planes. The peaks observed at low frequencies probably belong

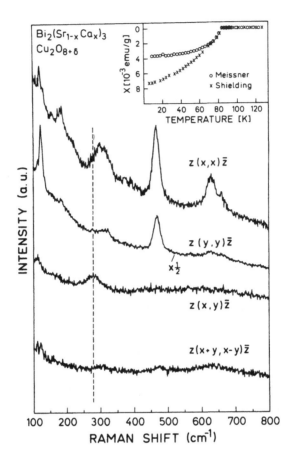

**Fig. 27.** Polarized Raman spectra of the Bi superconductor with three $CuO_2$ layers (n = 3). The top two spectra show the x-y anisotropy of the $A_{1g}$ phonons excited for $\vec{k} \parallel$ c-axis. The vertical dashed line indicates the position of a $B_{1g}$ mode at 282 $cm^{-1}$ with its unique polarization-selection rule. Its eigenvector corresponds to a vibration of oxygen in the $CuO_2$ planes. From Ref. 76.

to Bi (122 $cm^{-1}$), Sr and Cu (156 and 183 $cm^{-1}$) $A_{1g}$ modes. Broadening of peaks (183 $cm^{-1}$) may be caused by statistical replacement of Sr by Ca or mixture of those two $A_g$ modes as indicated e.g. in the calculation of Ref. 77 (Fig. 10, 1st row).

In Ref. 183 a spectrum of a n = 2 material is reported with a strong peak at ~630 cm$^{-1}$, assigned to BiO vibrations as well as a peak at 460 cm$^{-1}$. Both peaks are observed only for parallel polarizations. Burns et al.[165] find four peaks in their n = 2 crystals with frequencies and intensities rougly coinciding with those of Fig. 27 (164, 212, 464, and 625 cm$^{-1}$) but they seem to have measured with lower resolution than in Fig. 27. The $B_{1g}$ mode assignment of these authors to the peak 464 cm$^{-1}$ is, however, incorrect for reasons discussed above. The spectra of Ref. 184 are in good agreement with those of Fig. 27. In addition to the phonon part of the spectrum, the authors of Ref. 184 report the observation of overdamped two-magnon scattering, which they infer from a broad scattering range extending to over 4000 cm$^{-1}$. Kirillov et al.[185] have also reported spectra similar to those of Fig. 27 and, in addition, have found that the peak at 480 cm$^{-1}$ (463 cm$^{-1}$ in Ref. 185) is enhanced when the excitation source is changed from the green to the blue line of the argon laser. These authors have also attempted to derive a superconducting gap of $2\Delta \approx 32$ meV from the electronic scattering component of their spectra.

In summary, Raman spectra of the Bi superconductor are in general agreement in the literature. This is due to the early availability of large single crystals. Because of the morphology of those crystals it has, however, not been possible so far, to obtain reliable spectra for zz polarization. [The spectrum of Ref. 185 reported for that configuration appears to be very similar to the of xx polarization (Ref. 185) and is likely to be indeed that.] The $B_{1g}$ mode expected for the Bi superconductor with n ≥ 2 has been identified; an anomalous temperature dependence has not been reported for it so far (see also Sect. III.A.6). This mode is currently the only one with a definite assignment in the Bi superconductor with n = 2.

## C. Thallium and CuO$_2$ Based Materials

Because the structure of the Tℓ superconductors closely resembles that of the Bi-compound it is expected to lead to similar Raman spectra. The compound with one Tℓ-O layer (see Sect. II.B) is almost iso-

morphic to $RBa_2Cu_3O_{7-\delta}$ and Raman modes should therefore be easily identified. On the other hand, due to the health hazards of Tℓ compounds
only few spectra have been reported as of yet. In Ref. 186 spectra of
microcrystalline $Tℓ_2Ba_2Ca_{n-1}Cu_nO_x$ with n = 2 ($T_c$ = 107 K) and n = 3
($T_c$ = 114 K) are shown. For n = 2 the authors report peaks at 102, 132,
148, 230, 410, 495, 519, and 595 $cm^{-1}$, however, in a different specimen
of the same composition they find only peaks at 102, 148, and 519 $cm^{-1}$.
The strongest features in those spectra are the peaks near 500 $cm^{-1}$,
whereas the region between 150 and 400 $cm^{-1}$ contains only very weak
peaks. Similarly for the n = 3 compound, strong peaks exist at 132,
495, and 595 $cm^{-1}$, weak ones at 96, 230, and 410 $cm^{-1}$. Our measurement
of the relatively uncomplicated, polarized spectrum of a small n = 3
crystal is shown in Fig. 28a and it agrees with that of Ref. 186. We
compare it to the spectrum of a ceramic n = 3 specimen (Fig. 28b). Well
reproduced are the strong peaks of the crystal at 497 and 600 $cm^{-1}$
while additional strong peaks appear at 552 and 612 $cm^{-1}$. Without
knowing the Raman spectra of the relevant impurity phases one cannot
exclude the possibility that the extra peaks, in particular weak ones,
originate from them. It may, however, be hypothesized that the peaks at
552 and 612 $cm^{-1}$ are related to Tℓ-O vibrations. Fluctuations in the
oxygen content, most likely different for the samples in Fig. 28a and
28b, will make some forbidden modes appear. Some evidence for this
hypothesis comes from the study of a Bi (n = 2) ceramic. There, an
additional mode ($\sim$ 560 $cm^{-1}$) appeared at nearly the same frequency as
in the Tℓ ceramic (552 $cm^{-1}$) when the sample was annealed in argon.
Interestingly, $T_c$ was lowered but did not disappear as it does in
$RBa_2Cu_3O_{7-\delta}$. Upon annealing in oxygen (returning $T_c$ back to the original value) the additional peak was reduced in intensity, even though
it did not disappear.[187] This would suggest that the additional peak is
indeed related to varying oxygen occupation in the Bi(Tℓ)-O planes.
Further extended studies of the dependence of the Raman spectra on
oxygen content are required to clarify this issue. In Fig. 28c we also
display our spectrum of a Tℓ ceramic with n = 2 $CuO_2$ planes; it is
similar to that of the n = 3 ceramic. It is instructive to compare

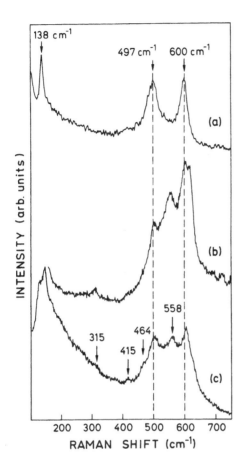

**Fig. 28.** Polarized Raman spectra of $Tl_2Ba_2Ca_2Cu_3O_x$ of (a) a small crystal and (b) a ceramic sample. The ceramic has $T_c$ = 122 K and its spectrum displays a number of additional peaks compared to that of the crystal. Because of strong scattering of the laser light the spectral region below 150 cm$^{-1}$ was not be recorded for the ceramic. The spectrum of ceramic $Tl_2Ba_2CaCu_2O_x$ is shown in c) for comparison. (Liu et al., unpublished data).

these spectra to those of the Bi-superconductors (Fig. 27). We tentatively assign the peak at ~500 cm$^{-1}$ (Tl) to the same, i.e. vertical, $O_3$ vibration as in the Bi-compound. Higher-lying modes are thus likely to be Tl-O vibrations, lower ones those of Tl, Ba, Cu, and oxygen in the $CuO_2$ planes. In particular the mode around 140 cm$^{-1}$ is likely to originate from vertical vibrations of Ba or Tl. The identity of the $B_{1g}$ mode which exists for both the n = 2 and n = 3 compound has not yet been established positively. While from depolarization studies on ceramic samples it appeared that the mode at 315 cm$^{-1}$ had indeed that symmetry, single crystal measurement have not yet been able to confirm this assignment.[188] No lattice dynamical calculations have been published so far to support these tentative assignments.

McCarty et al.[68] reported Raman spectra of the compound with <u>one</u> Tℓ-O layer and n = 2: TℓBa$_2$CaCu$_2$O$_7$, a structure the authors find is stabilized by the addition of lead at the cost of a lower T$_c$. By studying various polarization geometries the authors find that these spectra are quite similar to that of YBa$_2$Cu$_3$O$_7$, as expected. The authors readily assign peaks at 120, 148, and 525 cm$^{-1}$ to A$_{1g}$ modes similar to those of Ba, Cu, and O in YBa$_2$Cu$_3$O$_7$, being aware of a possible interchange of the assignment of the two lowest modes. They also find a weak band of B$_{1g}$ symmetry at 278 cm$^{-1}$, which they identify with the CuO$_2$-plane vibration of oxygen. On one side of their sample the authors observed Raman peaks, which they tentatively suggest to originate from magnetic scattering.

It is not surprising that the spectra of the superconductor with one Tℓ layer are readily understood. They resemble quite closely the thoroughly investigated RBa$_2$Cu$_3$O$_{7-\delta}$, the displaced oxygen (compared to RBa$_2$Cu$_3$O$_{7-\delta}$) not being Raman active. The two-Tℓ-layer compounds await further analysis and the understanding of the Bi superconductor to which it is structurally more closely related.

## D. The Zurich Oxides

A large number of the spectra reported in the literature contain peaks originating from impurities like La$_2$O$_3$ and Cu$_2$O[107] (105, 190, and 390 cm$^{-1}$); these peaks can be stronger than those of the nominal material.[189] Extensive reviews of results of vibrational spectroscopies of these materials have been prepared by Weber et al.[189] and by Maroni and Ferraro.[190] In Table II of the latter, frequencies and their assignments observed by most workers are summarized. In the tetragonal phase, as discussed in Sect. II.C, only two Raman-active modes for each La and OI are expected. They have A$_{1g}$ and E$_g$ symmetry and correspond to vibrations of those atoms along the c and x,y-axes, respectively. It is generally agreed upon[179,189-192] that the peak seen in crystals at ~425 cm$^{-1}$ corresponds to the O ($A_{1g}$) mode, and the one at 230 cm$^{-1}$ to the La ($A_{1g}$) mode. This is true for the tetragonal phase and also holds for the orthorhombic one in which these peaks continue to exist. The assignment of the O ($A_{1g}$) vibration is also supported by the observed shift in peak frequency when $^{16}$O is replaced by $^{18}$O.[193] The $E_g$ (tetragonal) modes are expected to split into $B_{2g}$ and $B_{3g}$ in the orthorhombic phase and Weber et al.[191] report such a splitting, or at least a broadening, for a peak at ~230 cm$^{-1}$ (see Table IV). The location of the La ($E_g$) peak has been reported[192] at 156 cm$^{-1}$, but other authors[189] believe it to be below 100 cm$^{-1}$ where it has not been observed yet. Additional $A_g$ peaks exist in the orthorhombic phase (see Sect. II.C, Fig. 12) which disappear at the phase transition to tetragonal. Weber et al.[191] report two of those at 104 and 271 cm$^{-1}$ whereas a third, at 522 cm$^{-1}$, continues to exist into the tetragonal phase and is hence not

Table IV: Experimental and theoretical phonon frequencies for $La_2CuO_4$. Cohen et al.[83] have performed a total-energy calculation while Prade et al.[58] obtained the theoretical values from a lattice dynamial calculation described in more detail in the text. The experimental frequencies were taken from Ref. 189.

| Experimental Weber et al. (Ref. 189) | Theoretical Cohen et al. (Ref. 83) | Prade et al. (Ref. 58) | Assignment tetragonal | orthorhombic |
|---|---|---|---|---|
| 104 | | | soft mode | OI,II, $A_g$ |
| 220 | 233 | | | OI, $B_{1g}$ |
| 280 | 220 | 218 | La, $A_{1g}$ | $A_g$ |
| 230 (250) | 233 | 333 | OII, $E_g$ | $B_{2g}$ ($B_{3g}$) |
| 271 | | | - | OI,II, $A_g$ |
| 320 | | | - | OII, $B_{1g}$ |
| 429 | 417 | 553 | OI, $A_{1g}$ | OI, $A_g$ |

assigned. Kourouklis et al.[193] have observed this behavior as well, and attribute it to local symmetry breaking of the mode at 522 cm$^{-1}$. Sugai[192], on the other hand, assigns peaks at 126, 156, 229, 273, and 426 cm$^{-1}$ to A$_g$ symmetry, thus avoiding having to invoke symmetry breaking. An assignment of the experimental frequencies to the eigenmodes displayed in Fig. 12 and a comparison with theoretically obtained frequencies are shown in Table IV. Of particular interest is the observation of a soft mode which corresponds to the orthorhombic distortion at the structural phase transition. As discussed in Sect. II.C the distortion comes from a tilting of the oxygen octahedra and thus corresponds to a combination of the OI and OII A$_g$ modes displayed in Fig. 12. Figure 29, taken from Ref. 189, shows how the soft-mode frequency at ~100 cm$^{-1}$ goes to zero at the phase-transition temperature of ~500 K (in undoped material). Curious is, however, the fact that in Sr-doped

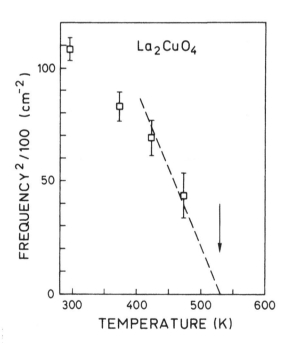

Fig. 29. Behavior of the frequency of the soft mode of La$_2$CuO$_4$ versus temperature. From Ref. 189.

$La_2CuO_4$ this peak is not observed, even well below the phase-transition temperature in that material;[189] it remains to be seen whether a connection to the superconductivity mechanism exists. For detailed spectra of $(La_{1-x}Sr_x)_2CuO_4$ as the phase transition orthorhombic → tetragonal is crossed for both $x = 0$ and $x \neq 0$ see e.g. Refs. 191 and 192.

## IV. BRIEF COMPARISON WITH IR DATA

We do not intend to review the existing literature on fir-spectra of high-$T_c$ superconductors. Rather, we just point out how information from Raman spectra is complemented by that from fir spectra. The bearing of fir spectra on the issue of the superconducting gap is not touched upon here.

The center of inversion of $RBa_2Cu_3O_{7-\delta}$ (and other materials) makes all phonons either Raman active or ir active (or silent). Hence it is interesting to look at the phonons in the fir spectra and, perhaps, find parallels in their behavior to the Raman-active phonons. Most important, we may consider the peaks related to atoms that do not participate in Raman vibrations, i.e. OI (chain oxygen), Cu1, and R. It is obvious that in order to draw reliable conclusions, an assignment of peaks to particular vibrations must be made, a formidable task because there are 21 modes of odd symmetry and crystals of the $\delta \approx 0$ compounds have yet not been available in sufficient size to distinguish those symmetries. Moreover, because of the large in-plane conductivity in-plane polarized modes should be very weak in reflection spectra. In one attempt, several small crystals with the c-axis perpendicular to the surface were arranged in a mosaic.[152] Fir reflection spectra so obtained for the x-y plane indeed contained no obvious features related to phonons.[152,194] Thus all phonons observed in the fir for ceramic samples may have vibrations in z-direction. However, effective-medium average of grains oriented at random could also make x-y polarized modes observable.

Spectra taken on semiconducting single crystals ($\delta \approx 1$, c-axis perpendicular to surface) show numerous phonon peaks.[65] The frequencies of these phonons and their oscillator strengths [Eq. (6)] have been calculated using the lattice dynamical model of Sect. II.A.2. Good agreement in both quantities are found for nearly all phonons, most being vibrations in x and y directions.[65] In ceramic samples with $\delta \approx 1$ similar spectra have been reported[37,45] with comparable assignments. It has already been pointed out that the experimental oscillator strength increases dramatically (7 to 8 times) in superconducting ceramics.[65] The origin of the phenomenon is not fully understood, nor where it stands in connection with the superconductivity mechanism.

As for vibrations of the R atom, they are easily identified by comparing spectra of compounds with lanthanides in place of R = Y. By calculating the change in the reduced mass of the vibration, frequency changes due to a different mass of the R-atom were predicted.[45,195] The identity of the R-related vibration is thus well established; it lies at ~190 $cm^{-1}$ for R = Y and $\delta \approx 0$.

The OI atom has a high frequency ir-active vibration near 560 $cm^{-1}$ (also a conclusion from the calculation in Sect. II.A.2, see Fig. 4). Experimentally this peak is found at ~570 $cm^{-1}$. It was particularly important to identify this vibration in connection with studies of the isotope shift. The fir spectra alone could show that upon gas exchange of $^{16}O \rightarrow {}^{18}O$, this site participates in the substitution (see Sect. III.A.4 and Refs. 61 and 129 for more details).

Temperature anomalies in the frequency of two phonons near 300 $cm^{-1}$ have been seen in the fir spectra. The anomaly of a Raman-active phonon has been discussed in detail in Sect. III.A.6; we just note here that while only one particular Raman phonon shows this behavior, at least two ir-phonons (at 275 and 315 $cm^{-1}$), curiously enough with nearly the same frequency as the Raman phonon, soften below $T_c$. Unfortunately the ir eigenvectors are not known with certainty, but it is likely that they involve the $CuO_2$ planes as well. The absolute softening is larger by a factor of 2-3 than that of the Raman-active phonon. Similarly to the latter, the anomaly is absent in semiconducting $RBa_2Cu_3O_{7-\delta}$ as well as for R = Pr with $\delta \approx 0$.

Finally, we mention the possible observation of two Raman-active phonons in photoinduced ir absorption where small changes in the ir transmission ($\Delta T/T \sim 10^{-3}$) are detected.[196]

## V. CONSEQUENCES CONCERNING THE MECHANISM OF HIGH-$T_c$ SUPERCONDUCTIVITY
## AND CONCLUSION

Raman spectroscopy has yielded a reliable picture of some of the zone-center phonon frequencies and eigenvectors, especially those of $A_g$ symmetry, in high-$T_c$ superconductors. This picture will be completed as better and larger single crystals of the various species become available. Knowledge of these phonon eigenstates, and those obtained by extrapolation and interpolation with lattice dynamical models, is important to ascertain the role of phonons in the pairing mechanism. They have also been utilized as an experimental check of 'ab initio' lattice dynamical estimates based on electronic total-energy calculations.

The phonons so identified can be used for the characterization of the superconducting materials, e.g., to determine in a non-destructive way the oxygen content of $YBa_2Cu_3O_{7-\delta}$ and related compounds. They also play an important role in measurements of the isotope effect in crystals in which, for instance, $^{16}O$ is replaced by $^{18}O$. It is even possible to substitute specific oxygen sites and to check which have been substituted by means of Raman spectroscopy. Total substitution of $^{16}O$ by $^{18}O$ leads to a decrease of about 0.3 K in the $T_c$ of most high-$T_c$ superconductors, including the perovskites which do not contain $CuO_2$ planes ($Ba_{0.6}K_{0.4}BiO_3$, $T_c \simeq 30$ K and $BaPb_{0.75}Bi_{0.25}O_3$, $T_c \simeq 10$ K). In all cases the isotope shift in $T_c$ is thus much smaller than that predicted by the weak-coupling BCS theory, the difference being larger the higher $T_c$ is. From this one can conclude that the mechanism, if BSC-like, must be in the strong-coupling limit. Other inconsistencies, however, speak against this. In view of the finite, though small, isotope effect we must nevertheless conclude that the ionic motion must play some role, possibly indirect, possibly additive, in the pairing mechanism.

Raman spectroscopy has been also successful in detecting soft phonon modes, such as those responsible for the orthorhombic-tetragonal transition in the Zurich oxides. This softening is most likely related to anharmonic terms in the phonon Hamiltonian. Another type of softening, induced by electron-phonon interaction, has been observed in

$YBa_2Cu_3O_{7-\delta}$ for $\delta < 0.2$ when crossing $T_c$. From this softening details of the electron-phonon coupling constant can be obtained.

Besides phonons, Raman spectroscopy yields information about electronic excitations, both in the normal and the superconducting state. In the latter, evidence for an anisotropic gap, with excitations which extend in part to very low frequencies, has been obtained. If it turns out that these effects are intrinsic of the perfect material and not due to compositional fluctuations, the Raman results will pose rather stringent predictive requirements on any theoretical mechanism.

Another type of 'electronic' excitations seen in the Raman spectra involve scattering by two magnons, i.e., the flip of two antiferromagnetically ordered spins in neighboring $Cu^{2+}$ ions in the $CuO_2$ planes. These measurements yield precise values for the antiferromagnetic coupling constant which plays an important role in mechanisms of superconductivity involving pairing through exchange of magnetic excitations. Such mechanisms enjoy increasing favor at the time of closing this article.

## ACKNOWLEDGEMENTS

We owe special thanks to S. Birtel for her expert typing of this manuscript and to innumerable colleagues at the Max-Planck-Institute für Festkörperforschung who have collaborated with us in the original research reported here.

1. Bilz, H., and Kress, W., <u>Phonon Dispersion Relations in Insulators</u> (Springer, Heidelberg, 1979).

2. Shirane, G., et al., Phys. Rev. Lett. <u>59</u>, 1613 (1987).

3. Menéndez, J. and Cardona, M., Phys. Rev. B <u>29</u>, 2051 (1984).

4. Fauchet, P.M., IEEE Circuits and Devices Mag. <u>2</u>, 37 (1986).

5. Note, however, that it is easy to make mistakes in this manner. See for instance Hemley, R.J. and Mao, H.K., Phys. Rev. Lett. <u>58</u>, 2340 (1987), and compare with Sect. 3.1.

6. <u>Light Scattering in Solids</u>, Vols. I to V, ed. by Cardona, M. and Güntherodt, G. (Springer, Heidelberg, 1975-1988).

7. Richter, H., Wang, Z.P., and Ley, L., Solid State Commun. <u>39</u>, 625 (1981); Campbell, I.M., and Fauchet, P.M., Solid State Commun. <u>58</u>, 739 (1986).

8. Brodsky, M.H., in Vol. I of Ref. 6, p. 205.

9. Cardona, M., in Vol. II of Ref. 6, p. 19.

10. see Vol. V of Ref. 6.

11. Compaan, A. and Trodahl, J., Phys. Rev. B <u>29</u>, 793 (1984).

12. Weber, W.H., Peters, C.R., Wanklyn, B.M., Chen, C., and Watts, B.E., Phys. Rev. B <u>38</u>, 917 (1988).

13. Renker, B., Gompf, F., Gering, E., Roth, G., Reichardt, W., Ewert, D., Rietschel, H., and Mutka, H., Z. Phys. B <u>71</u>, 437 (1988).

14. Cerdeira, F., Anastassakis, E., Kauschke, W., and Cardona, M., Phys. Rev. Lett. <u>57</u>, 3209 (1986).

15. Humlíček, J., et al., Solid State Commun. <u>66</u>, 1071 (1988).

16. See, for instance, Rousseau, D.L., Baumann, R.P., and Porto, S.P.S., J. Raman Spectr. <u>10</u>, 253 (1981).

17. Liu, R., et al., Phys. Rev. B <u>37</u>, 7971 (1988).

18. See, for instance, Menéndez, J. and Cardona, M., Phys. Rev. B <u>31</u>, 3696 (1985), and references therein.

19. Abrikosov, A.A. and Fal'kovskii, L.A., Pis'ma ZhETP. <u>46</u>, 236 (1987); (ZhETP letters <u>46</u>, 298 (1988); Physica C <u>156</u>, 1 (1988)).

20. Lyons, K.B., et al., Phys. Rev. B <u>36</u>, 5592 (1987).

21. Ossipyan, Y.A., Timofeev, V.B., and Schegolev, I.F., Physica C <u>153-155</u>, 1133 (1988).

500

22. Thomsen, C., et al., Phys. Rev. B $\underline{37}$, 9860 (1988).

23. Cooper, S.L., et al. Phys. Rev. B $\underline{37}$, 5920 (1988). b) Cooper, S.L., Slakey, F., Klein, M.V., et al., Phys. Rev. B, in press.

24. Hackl, R., Gläser, W., Müller, P., et al., to be published.

25. Abstreiter, G., Cardona, M., and Pinzcuk, A., in Ref. 10.

26. Cottam, M.G. and Lockwood, D.J., Light Scattering in Magnetic Solids (Wiley, New York, 1986).

27. Lyons, K.B., Fleury, P.A., Schneemeyer, L.F., and Waszczak, J.V., Phys. Rev. Lett. $\underline{60}$, 732 (1988); Sugai, S., Shamoto, S.-I., and Sato, M., Phys. Rev. B, to be published.

28. Tranquada, J.M., et al., Phys. Rev. Lett. $\underline{60}$, 156 (1988).

29. Chang, R.K., and Long, M.B., in Vol. II of Ref. 6, p. 179. Tsang, J., in Vol. V of Ref. 6.

30. Devlin, G.E., Davis, J.L., Chase, L., and Geschwind, S., Appl. Phys. Lett. $\underline{19}$, 138 (1971).

31. Cava, R.J., et al., Phys. Rev. Lett. $\underline{58}$, 1676 (1987).

32. LePage, Y., et al., Phys. Rev. B $\underline{35}$, 7245 (1987).

33. Liu, R., et al., Solid State Commun. $\underline{63}$, 839 (1987).

34. Stavola, M., et al., Phys. Rev. B $\underline{36}$, 850 (1987).

35. Macfarlane, R.M., Rosen, H., and Seki, H., Solid State Commun. $\underline{63}$, 831 (1987).

36. Iqbal, Z., et al., Phys. Rev. B $\underline{36}$, 2283 (1987).

37. Burns, G., et al., Solid State Commun. $\underline{64}$, 471 (1987).

38. Morioka, Y., Kikuchi, M., Syono, Y., Jpn. J. of Appl. Physics $\underline{26}$, L1499 (1987).

39. Sanjurjo, J.A., et al., Solid State Commun. $\underline{64}$, 505 (1987).

40. Kulakovskii, V.D., et al., Pis'ma ZhETF $\underline{46}$, 460 (1987).

41. Sugai, S., Phys. Rev. B $\underline{36}$, 7133 (1987).

42. Collman, J.P., et al., Phys. Rev. B $\underline{37}$, 3660 (1988).

43. McMullan, W.G., Gygax, S., and Irwin, J.C., Solid State Commun. $\underline{66}$, 165 (1988).

44. Hangyo, M., et al., Solid State Commun. $\underline{65}$, 835 (1988).

45. Thomsen, C., et al., Solid State Commun. $\underline{65}$, 1139 (1988).

46. Gallagher, P.K., O'Bryan, H.M., Sunshine, S.A., and Murphy, D.W., Mat. Res. Bull. $\underline{22}$, 995 (1987).

47. Anastassakis, E., and Cardona, M., phys. stat. sol. (b) 104, 589 (1981).

48. Chaillout, C., et al., Solid State Commun. 65, 283 (1988).

49. Feile, R., Schmitt, U., and Leiderer, P., Physica C 153-155, 292 (1988).

50. Bates, F.E., and Eldridge, J.E., Solid State Commun. 64, 1435 (1987).

51. Chaplot, S.L., Phys. Rev. B 37, 7435 (1988).

52. Kress, W., et al., Physica C 153-155, 221 (1988).

53. Kress, W., Schröder, U., Prade, J., Kulkarni, A.D., and de Wette, F.W., submitted to Phys. Rev.

54. Jannot, B., Escribe-Filippini, C., and Bouillot, J., J. Phys. C. 17, 1329 (1984).

55. Bouillot, J., Escribe, C., Fitzgerald, W.J., and Gnininvi, L., Solid State Commun. 30, 521 (1979).

56. Reichardt, W., Wagner, V., and Kress, W., J. Phys. C 8, 3955 (1975).

57. Rieder, K.H., Migoni, R., and Renker, B., Phys. Rev. B 12, 3374 (1975).

58. Prade, J., et al., Solid State Commun. 64, 1267 (1987).

59. Liu, R., et al., Phys. Rev. B 37, 7971 (1988).

60. Lang, M., et al., Phys. B 69, 459 (1988).

61. Cardona, M., et al., Solid State Commun. 67, 789 (1988).

62. Thomsen, C., et al., Solid State Commun. 67, 1069 (1988).

63. Fisher, D.S., Millis, A.J., Shrainman, B., and Bhatt, R.N., Phys. Rev. Lett. 61, 482 (1988).

64. Syassen, K., et al., Physica C 153-155, 264 (1988).

65. Genzel, L., Bauer, M., Wittlin, A., Cardona, M., Schönherr, E., and Simon, A., to be published.

66. Torardi, C.C., and Subramanian, M.A., Science 240, 631 (1988); Parkin, S.S.P., Phys. Rev. Lett. 60, 2539 (1988).

67. Zandbergen, H.W., Huang, Y.K., Menken, M.J., Li, J.N., Kadowaki, K., Menovsky, A.A., Van Tendeloo, G., and Amelinckx, S., Nature 332, 620 (1988).

68. McCarty, et al., Physica C 156, 119 (1988); Ganguli, A.K.,

Subbanna, G.N., and Rao, C.N.R., Physica C 156, 116 (1988).

69. Wu, D.T., et al., Physica C 156, 109 (1988).

70. Liu, R.S., Wu, P.P., Liang, J.M., and Chen, L.J., to be published.

71. Fitz Gerald, J.D., and Withers, R.L., Phys. Rev. Lett. 60, 2797 (1988).

72. Chen, C.H., et al., Phys. Rev. B 37, 9834 (1988).

73. Sunshine, S.A., Siegrist, T., Schneemeyer, L.F., et al., to be published.

74. von Schnering, H.G., et al., Angew. Chem. Int. Ed. Engl. 27, 574 (1988).

75. Bordet, P., et al., Physica C 156, 189 (1988).

76. Cardona, M., et al., Solid State Commun. 66, 1225 (1988).

77. Morosin, B., et al., Physica C 152, 413 (1988).

78. Prade, J., Kulkarni, A.D., de Wette, F.W., Schröder, U., and Kress, W., Phys. Rev., to be published.

79. Birgeneau, R.J., et al., Phys. Rev. Lett. 59, 1329 (1987)

80. Cava, R.J., Santoro, A., Johnson, D.W., Jr., and Rhodes, W.W., Phys. Rev. B 35, 6716 (1987).

81. Moss, S.C., et al., Phys. Rev. B 35, 7195 (1987).

82. see for instance Burns, G., Dacol, F.H., and Shafer, M.W., Solid State Commun. 62, 687 (1987).

83. Cohen, R.E., Pickett, W.E., and Krakauer, H., Phys. Rev. Lett., submitted.

84. Krol, D.M., et al., Phys. Rev. B 36, 8325 (1987).

85. Yamanaka, A., et al., Jpn. J. Appl. Phys. 26, L1404 (1987).

86. Burns, G., Dacol, F.H., Holtzberg, F., and Kaiser, D.L., Solid State Commun. 66, 217 (1988).

87. Bhadra, R., et al., Phys. Rev. B 37, 5142 (1988).

88. Hadjiev, V.G., Iliev, M.N., and Vassilev, P.G., Physica C 153-155, 290 (1988).

89. Mihailović, D., and Brničević, Physica C 153-155, 147 (1988).

90. Knoll, P., and Kiefer, W., Proceedings of the Meeting on High-$T_c$ Superconductors, Mauterndorf, 7-11.2.1988, Ed. Weber, H.W., Plenum

Press, London, 1988, to be published.

91. Hadjiev, V.G. and Iliev, M.N., Solid State Commun. <u>66</u>, 451 (1988) and Gasparov, L.V., Emel'chenko, G.A., Kulakovskii, V.D., Misochko, O.V., Rashba, E.I., and Timofeev, V.B., submitted.

92. Reichardt, W., Pintschovius, L., Hennion, B., and Collin, F., to be published.

93. Thomsen, C., et al., Solid State Commun. <u>65</u>, 55 (1988).

94. Macfarlane, R.M., et al., Phys. Rev. B <u>38</u>, 284 (1988).

95. Kuzmany, H., et al., Solid State Commun. <u>65</u>, 1343 (1988).

96. Barros Leite, C.V., Patnaik, B.K., Freire Jr., F.L., et al., <u>Lectures on Surface Science</u>, Proc. Fourth Latin-American Symposium, Bogotá, Colombia, 1988, to be published.

97. Humlíček, J., and Garriga, M., private communication.

98. Lu, F., Perry, C.H., Chen, K., et al., JOSA, submitted.

99. Ref. 6, vol. II, p. 40.

100. Schneemeyer, L.F., et al., Nature <u>328</u>, 601 (1987).

101. Rosen, H., et al., Phys. Rev. B <u>36</u>, 726 (1988).

102. Popović, Z.V., et al. Solid State Commun. <u>66</u>, 43 (1988).

103. Popović, Z.V., et al., Solid State Commun. <u>66</u>, 965 (1988).

104. Udagawa, M., et al., Jpn. J. Appl. Phys. <u>26</u>, L858 (1987).

105. Baran, E.J., Cicileo, G.P., Punte, G., Lavat, A.E., and Trezza, M., J. Mater. Sci. Letters, to be published.

106. Compaan, A., and Cummins, H.Z., Phys. Rev. B <u>6</u>, 4753 (1972); Yu, P.Y., Shen, Y.R., and Petroff, Y., Solid State Commun. <u>12</u>, 973 (1973); Compaan, A., Solid State Commun. <u>16</u>, 293 (1975).

107. Liu, R., Thomsen, C., Cardona, M., and Mattausch, Hj., Solid State Commun. <u>65</u>, 67 (1988).

108. Cardona, M. et al., Solid State Commun. <u>65</u>, 71 (1988).

109. Rosen, H.J., et al., Phys. Rev. <u>38</u>, 2460 (1988).

110. Fig. 6 in Ref. 109 and references therein.

111. Thomsen, C., et al., Solid State Commun. <u>67</u>, 271 (1988).

112. Schönherr, E., private communication.

113. Wu, N.J., Xie, K., Zhao, L.H., et al., Solid State Commun., in press.

114. Liu, R., private communication.

504

115. Hillebrecht, F.U., et al., Solid State Commun. 67, 379 (1988).

116. Hangyo, M., et al., Solid State Commun. 67, 1171 (1988).

117. Kirby, P.B., et al, Phys. Rev. B 36, 8315 (1987).

118. Morioka, Y., Tokiwa, A., Kilzuchi, M., and Syonon, Y., Solid State Commun. 67, 267 (1988).

119. Tarascon, J.M., et al., Phys. Rev. B 37, 7458 (1988).

120. Feile, R., et al., Z. Phys. B, Cond. Matter 72, 161 (1988).

121. Berberich, P., Tate, J., Dietsche, W., and Kinder, H., Appl. Phys. Lett. 53, 925 (1988); Thomsen, C., and Tate, J., private communication.

122. Nishitani, R., Yoshida, N., Sasaki, Y., and Nishina, Y., Jpn. J. Appl. Phys. 27, L1284 (1988).

123. Kourouklis, G.A., et al., Phys. Rev. B 36, 8320 (1988).

124. Kirillov, D., et al., Phys. Rev. B 37, 3660 (1988).

125. Henry, J.Y., et al., Solid State Commun. 64, 1037 (1987).

126. Batlogg, B., et al., Phys. Rev. Lett. 58, 2333 (1987).

127. Leary, K.J., Phys. Rev. Lett. 59, 1236 (1987).

128. Crawford, M.K., et al., Phys. Rev. B, in press.

129. Thomsen, C., et al., Solid State Commun. 67, 1069 (1988).

130. Bourne, L.C., et al., Phys. Rev. B 36, 3990 (1987).

130a de Wette, F.W., et al., to be published.

131. Quan, L., et al., Solid State Commun. 65, 869 (1988).

132. Katayama-Yoshida, H., Jpn. J. Appl. Phys. L2085 (1987).

133. Katayama-Yoshida, H., Hirooka, T., Oyamada, A., et al., to be published.

134. Batlogg, B., et al., Phys. Rev. Lett. 59, 912 (1988); Faltens, T.A., et al., Phys. Rev. Lett. 59, 915 (1988).

135. Mattheiss, L.F., Gyorgy, E.M., and Johnson, D.W., Jr., Phys. Rev. B 37, 3745 (1988); and Cava, R.J., et al., Nature 332, 814 (1988).

136. Batlogg, B., et al., Phys. Rev. Lett. 61, 1670 (1988).

137. Klein, M.V., and Dierker, S.B., Phys. Rev. B 29, 4976 (1984).

138. Abrikosov, A.A., and Falkovsky, L.A., Sov. Phys. JEPT 13, 79 (1961).

139. Tong, S.Y, and Maradudin, A.A., Mat. Res. Bull, 4, 563 (1969).

140. Tilley, D.R., Z. Phys. 254, 71 (1972).

141. Dierker, S.B., et al., Phys. Rev. Lett. 50, 853 (1983).

142. Hackle, R., Kaiser, R., and Schickentanz, S., J. Phys. C. 16, 1729 (1983).

143. Thomsen, C., et al., Physica C 153-155, 1756 (1988).

144. Abstreiter, G., et al., in Vol. IV of Ref. 6, p. 5ff.

145. Wittlin, A., et al., Phys. Rev. B 37, 652 (1988); Takayi, H., et al., Nature 332, 236 (1988).

146. Contreras, G., Sood, A.K., and Cardona, M., Phys. Rev. B 32, 924 (1985) and Phys. Rev. B 32, 930 (1985).

147. Gantsevitch, S.V., Katityus, R., and Ustinov, N.G., Soviet Phys. Solid State 16, 711 (1974); Ipatova, I.P., Subashiev, A.V., and Voitenko, V.A., Solid State Commun. 37, 893 (1981); see also Ref. 144.

148. Feile, R., Leiderer, P., Kowalewski, J., et al., Z. Phys. B, to be published.

149. Gallager, M.C., Phys. Rev. B 37, 7847 (1988); Bulaevskii, L.N., Dolgov, O.V., Kazakov, I.P., Supercond. Science and Techn., to be published.

150. Warren, W.W., Jr., et al., Phys. Rev. Lett. 59, 1860 (1987)

151. Pennington, C.H., et al., Phys. Rev. B 37, 7944 (1988).

152. Schlesinger, Z., Collins, R.T., Kaiser, D., and Holtzberg, F., Phys. Rev. Lett. 59, 1958 (1987).

153. Zeyher, R., and Zwicknagel, G., Solid State Commun. 66, 617 (1988).

154. Pintschovius, L., Vergés, J.A., and Cardona, M., Phys. Rev. B 26, 5658 (1982).

155. Wittlin, A., et al., Solid State Commun. 64, 477 (1987).

156. Krantz, M., Rosen, H.J., Macfarlane, R.M., and Lee, V.Y., Phys. Rev. B 38, 4992 (1988).

157. Ruf, T., private communication.

158. Ruf, T., Thomsen, C., Liu, R., and Cardona, M., Phys. Rev. B, Dec. 1 (1988).

159. Worthington, T.K., Gallagher, W.J., and Dinger, T.R., Phys. Rev. Lett. 59, 1160 (1987).

160. Tarascon, J.M., et al., Phys. Rev. B 36, 226 (1987).

161. Morán, E., et al., Solid State Commun. 67, 369 (1988).

162. Thomsen, C., Liu, R., Cardona, M., Amador, U., and Morán, E., Solid State Commun. 67, 271 (1988).

163. Thomsen, C., Dietsche, W., and König, W., unpublished.

164. Liu, R., private communication

165. Burns, G., et al., Solid State Commun. 67, 603 (1988) and private communication.

166. Endols, Y., et al., Phys. Rev. B 37, 7443 (1988).

167. Budnik, J.I., et al., Europhys. Lett. 5, 651 (1988).

168. Kitazawa, H., Katsumata, K., Tozikai, E., and Nagamine, K., Solid State Commun. 67, 1191 (1988).

169. Lynn, J.W., et al. Phys. Rev. Lett. 60, 2781 (1988).

170. Chattopadhyay, T., et al., Phys. Rev. B, in press.

171. Dunlap, B.D., et al., J. Mag. Mag. Mat. 68, L139 (1987).

172. Sato, M., et al., Phys. Rev. Lett. 61, 1317 (1988).

173. Fleury, P.A., and Loudon, R., Phys. Rev. 166, 514 (1976).

174. Parkinson, J.B., J. Phys. C 2, 2012 (1969).

175. Lyons, R.B., and Fleury, P.B., to be published.

176. Fleury, P.A., and Guggenheim, H.J., Phys. Rev. Lett. 24, 1346 (1987).

177. Lyons, K.B., et al., Phys. Rev. Lett. 37, 2353 (1988).

178. Sugai, S., et al., submitted to Phys. Rev. B.

179. Weber, W.H., et al., Phys. Rev. B 38, 917 (1988).

180. Leising, G., et al., Physica C 153-155, 886 (1988).

181. Popović, Z.V., et al., Solid State Commun. 66, 965 (1988).

182. Farrow, L.A., et al., Phys. Rev. B 28, 752 (1988).

183. Kostadinov, I.Z., et al., Physica C 153-155, 627 (1988).

184. Sugai, S., Takagi, H., Uchida, S., and Tanaka, S., Jpn. J. Appl. Phys., to be published.

185. Kirillov, D., Bozovic, I., Geballe, T.H., et al., Phys. Rev. B, submitted.

186. Kostadinov, I.Z., Hadjiev, V.G., Tihov, J., et al., Physica C, to be published.

187. Mattausch, Hj., and Thomsen, C., to be published.

188. Thomsen, C., and Liu, R., private communication.

189. Weber, W.H., Peters, C.R. and Logothetis, E.M., JOSA, to be published.

190. Maroni, V.A., and Ferraro, J.R., in FT—IR Spectroscopy in Industrial and Laboratory Analyses, Ferraro, J.R., and Krishman, K., Eds., to be published.

191. Weber, W.H., Peters, C.R., Wanklyn, B.M., et al., Solid State Commun., to be published.

192. Sugai, S., Phys. Rev. B, submitted.

193. Kourouklis, G.A., et al., Phys. Rev. B $\underline{36}$, 7218 (1987).

194. Thomas, G.A., et al., Phys. Rev. Lett. $\underline{61}$, 1313 (1988).

195. Cardona, M., et al., Solid State Commun. $\underline{64}$, 727 (1987)

196. Taliani, C., et al., Solid State Commun. $\underline{66}$, 487 (1988).

# SUBJECT INDEX